Evolution "On Purpose"

Vienna Series in Theoretical Biology
Gerd B. Müller, editor-in-chief
Thomas Pradeu and Katrin Schäfer, associate editors

Transformations of Lamarckism, edited by Snait B. Gissis and Eva Jablonka, 2011

Convergent Evolution: Limited Forms Most Beautiful, by George McGhee, 2011

From Groups to Individuals, edited by Frédéric Bouchard and Philippe Huneman, 2013

Developing Scaffolds in Evolution, Culture, and Cognition, edited by Linnda R. Caporael, James Griesemer, and William C. Wimsatt, 2013

Multicellularity: Origins and Evolution, edited by Karl J. Niklas and Stuart A. Newman, 2016

Vivarium: Experimental, Quantitative, and Theoretical Biology at Vienna's Biologische Versuchsanstalt, edited by Gerd B. Müller, 2017

Landscapes of Collectivity in the Life Sciences, edited by Snait B. Gissis, Ehud Lamm, and Ayelet Shavit, 2017

Rethinking Human Evolution, edited by Jeffrey H. Schwartz, 2018

Convergent Evolution in Stone-Tool Technology, edited by Michael J. O'Brien, Briggs Buchanan, and Metin I. Erin, 2018

Evolutionary Causation: Biological and Philosophical Reflections, edited by Tobias Uller and Kevin N. Lala, 2019

Convergent Evolution on Earth: Lessons for the Search for Extraterrestrial Life, by George McGhee, 2019

Contingency and Convergence: Toward a Cosmic Biology of Body and Mind, by Russell Powell, 2020

How Molecular Forces and Rotating Planets Create Life, Jan Spitzer, 2021

Rethinking Cancer: A New Understanding for the Post-Genomics Era, edited by Bernhard Strauss, Marta Bertolaso, Ingemar Ernberg, and Mina J. Bissell, 2021

Levels of Organization in the Biological Sciences, edited by Daniel S. Brooks, James DiFrisco, and William C. Wimsatt, 2021

The Convergent Evolution of Agriculture in Humans and Insects, edited by Ted R. Schultz, Richard Gawne, and Peter N. Peregrine, 2022

Evolvability: A Unifying Concept in Evolutionary Biology?, edited by Thomas F. Hansen, David Houle, Mihaela Pavlicev, and Christophe Pélabon, 2023

Evolution "On Purpose": Teleonomy in Living Systems, edited by Peter A. Corning, Stuart A. Kauffman, Denis Noble, James A. Shapiro, Richard I. Vane-Wright, and Addy Pross, 2023

Properties of Life: Toward a Theory of Organismic Biology, Bernd Rosslenbroich, 2023

Evolution "On Purpose"

Teleonomy in Living Systems

edited by Peter A. Corning, Stuart A. Kauffman, Denis Noble,
James A. Shapiro, Richard I. Vane-Wright, and Addy Pross

The MIT Press
Cambridge, Massachusetts
London, England

The MIT Press would like to thank the anonymous peer reviewers who provided comments on drafts of this book. The generous work of academic experts is essential for establishing the authority and quality of our publications. We acknowledge with gratitude the contributions of these otherwise uncredited readers.

This book was set in Times New Roman by Westchester Publishing Services. Printed and bound in the United States of America.

Library of Congress Cataloging-in-Publication Data

Names: Corning, Peter A., 1935– editor. | Kauffman, Stuart A., editor. | Noble, Denis, 1936– editor. |
    Shapiro, James Alan, 1943– editor. | Vane-Wright, Richard Irwin, editor. | Pross, Addy, 1945– editor.
Title: Evolution "on purpose" : teleonomy in living systems / edited by Peter A. Corning,
    Stuart A. Kauffman, Denis Noble, James A. Shapiro, Richard I. Vane-Wright, and Addy Pross.
Description: Cambridge, Massachusetts : The MIT Press, [2023] | Series: Vienna series in
    theoretical biology | Includes bibliographical references and index.
Identifiers: LCCN 2022043536 (print) | LCCN 2022043537 (ebook) |
    ISBN 9780262546409 (paperback) | ISBN 9780262376020 (epub) | ISBN 9780262376013 (pdf)
Subjects: LCSH: Evolution (Biology)—Philosophy. | Teleology.
Classification: LCC QH360.5 .E967 2023 (print) | LCC QH360.5 (ebook) | DDC 576.801—dc23/eng/20221021
LC record available at https://lccn.loc.gov/2022043536
LC ebook record available at https://lccn.loc.gov/2022043537

10 9 8 7 6 5 4 3 2 1

Contents

Series Foreword   vii

Preface   ix

1   **Introduction**   1
Peter A. Corning, Stuart A. Kauffman, Denis Noble, James A. Shapiro,
Richard I. Vane-Wright, and Addy Pross

2   **Teleonomy in Evolution: "The Ghost in the Machine"**   11
Peter A. Corning

3   **Cellular Basis of Cognition in Evolution: From Protists and Fungi
Up to Animals, Plants, and Root-Fungal Networks**   33
František Baluška, William B. Miller Jr., and Arthur S. Reber

4   **Constructing "On Purpose": How Niche Construction Affects
Natural Selection**   59
Dominik Deffner

5   **Relational Agency: A New Ontology for Coevolving Systems**   79
Francis Heylighen

6   **Teleonomic Anticipatory Configurations in Biological Evolution:
The Downward Dynamical Nature of Goal-Directedness**   105
Abir U. Igamberdiev

7   **From Teleonomy to Mentally Driven Goal-Directed Behavior:
Evolutionary Considerations**   119
Eva Jablonka and Simona Ginsburg

8   **Beyond the Newtonian Paradigm: A Statistical Mechanics of
Emergence**   141
Stuart A. Kauffman and Andrea Roli

9   **On the Concept of Meaning in Biology**   161
Kalevi Kull

10   Collective Intelligence of Morphogenesis as a Teleonomic Process   175
      Michael Levin

11   Form, Function, Agency: Sources of Natural Purpose in
      Animal Evolution   199
      Stuart A. Newman

12   How Purposive Agency Became Banned from Evolutionary Biology   221
      Denis Noble and Raymond Noble

13   Goal Attributions in Biology: Objective Fact, Anthropomorphic Bias,
      or Valuable Heuristic?   237
      Samir Okasha

14   Toward the Physicalization of Biology: Seeking the Chemical Origin
      of Cognition   257
      Robert Pascal and Addy Pross

15   Evolutionary Change Is Naturally Biological and Purposeful   275
      James A. Shapiro

16   Agency, Teleonomy, Purpose, and Evolutionary Change in
      Plant Systems   299
      Anthony Trewavas

17   Agency, Goal Orientation, and Evolutionary Explanations   325
      Tobias Uller

18   Evolutionary Foundationalism: The Myth of the Chemical Given   341
      Denis M. Walsh

      Contributors   363
      Index   365

Series Foreword

Biology is a leading science in this century. As in all other sciences, progress in biology depends on the interrelations between empirical research, theory building, modeling, and societal context. But whereas molecular and experimental biology have evolved dramatically in recent years, generating a flood of highly detailed data, the integration of these results into useful theoretical frameworks has lagged behind. Driven largely by pragmatic and technical considerations, research in biology continues to be less guided by theory than seems indicated. By promoting the formulation and discussion of new theoretical concepts in the biosciences, this series intends to help fill important gaps in our understanding of some of the major open questions of biology, such as the origin and organization of organismal form, the relationship between development and evolution, and the biological bases of cognition and mind. Theoretical biology has important roots in the experimental tradition of early-twentieth-century Vienna. Paul Weiss and Ludwig von Bertalanffy were among the first to use the term *theoretical biology* in its modern sense. In their understanding the subject was not limited to mathematical formalization, as is often the case today, but extended to the conceptual foundations of biology. It is this commitment to a comprehensive and cross-disciplinary integration of theoretical concepts that the Vienna Series intends to emphasize. Today, theoretical biology has genetic, developmental, and evolutionary components—the central connective themes in modern biology— but it also includes relevant aspects of computational or systems biology and extends to the naturalistic philosophy of sciences. The Vienna Series grew out of theory-oriented workshops organized by the KLI, an international institute for the advanced study of natural complex systems. The KLI fosters research projects, workshops, book projects, and the journal *Biological Theory*, all devoted to aspects of theoretical biology, with an emphasis on—but not restriction to—integrating the developmental, evolutionary, and cognitive sciences. The series editors welcome suggestions for book projects in these domains.

Gerd B. Müller, Thomas Pradeu, and Katrin Schäfer

Preface

Peter A. Corning and Richard I. Vane-Wright

Although it is now widely accepted that living systems exhibit an internal teleology, or teleonomy, the full implications of this distinctive biological property have not been sufficiently explored. This volume addresses various aspects of this important phenomenon, including the origins and history of the concept, its scope and meaning, and many of the various ways in which teleonomy has influenced the evolutionary process.

The edited papers presented here result from a Linnean Society of London online conference, *Evolution "on Purpose": Teleonomy in Living Systems*, that took place on 28/29 June 2021. Many of the speakers at that event are contributors to this volume, and/or to a special issue of the *Biological Journal of the Linnean Society* (Vane-Wright & Corning, 2023). The 2021 conference, and these proceedings, have largely been inspired by the Third Way of Evolution Group, led by Denis Noble, James A. Shapiro, and Raju Pookottil; and a Royal Society/British Academy meeting, "New Trends in Evolutionary Biology: Biological, Philosophical and Social Science Perspectives," held in London during November 2016 (see Bateson et al., 2017; also Kull, 2016).

Acknowledgments

Clearly, *Evolution "On Purpose"* is, in many ways, a collective effort. However, in addition to our contributors and other academic peers, the co-editors also wish to express their gratitude to other colleagues who have assisted in making this publication possible—notably Anne-Marie Bono, Kate Elwell, and Roger Wood at the MIT Press and the production team at Westchester Publishing Services led by Madhulika Jain with Mona Tiwary. We also thank Arc Indexing Inc. (Florida) for their care in preparing the index.

References

Bateson, P., Cartwright, N., Dupré, J., Laland, K., & Noble, D. (2017). New trends in evolutionary biology: Biological, philosophical and social science perspectives. *Interface Focus*, 7, 20170051. http://dx.doi.org/10.1098/rsfs.2017.0051

Kull, K. (2016). What kind of evolutionary biology suits cultural research? *Sign Systems Studies*, *44* (4), 634–647. https://ojs.utlib.ee/index.php/sss/article/view/SSS.2016.44.4.09

The Third Way. (n.d.). *Evolution in the era of genomics and epigenomics.* https://www.thethirdwayofevolution.com

Vane-Wright, R. I. & Corning, P. A. (eds.). (2023, in press). Teleonomy in living systems. *Biological Journal of the Linnean Society.*

1 Introduction

Peter A. Corning, Stuart A. Kauffman, Denis Noble, James A. Shapiro, Richard I. Vane-Wright, and Addy Pross

The so-called modern synthesis in evolutionary biology, which dates back to biologist Julian Huxley's landmark volume, *Evolution: The Modern Synthesis* (1942), assigns a central role in evolution to the genes and their supposedly random, incremental changes (mutations) over time. This has been the standard model in evolutionary biology for more than 80 years, and it was reinforced in the public mind by biologist Richard Dawkins' best-selling popular book, *The Selfish Gene* (1989/1976). As recently as 2014, there was a spirited defense of the modern synthesis by a prominent group of biologists (see Wray et al., 2014; but see also Laland et al., 2015).

There has also been a major effort in recent years by Massimo Pigliucci, Gerd Müller, and a number of their colleagues to update evolutionary theory with what has been called the extended evolutionary synthesis (EES) (Pigliucci & Müller, 2010; Laland et al., 2014, 2015; see also Danchin et al., 2011; Wade, 2011; Laubichler & Renn, 2015). Rather than replace the underlying principles and assumptions of the modern synthesis, these theorists would modify and supplement them to accommodate the recent work in developmental biology, epigenetic inheritance, genomics, multilevel selection, niche construction theory, and the like. In their view, these constructive processes "share the responsibility for the direction and rate of evolution" with the classical model (Laland et al., 2015).

Here we seek to contribute to this vexed theoretical issue. We hope to show that teleonomy (evolved biological purposiveness), in its many forms and facets, has played a major—often critical—part in shaping the trajectory of evolution over time. This suggests that a more inclusive theoretical synthesis is required (as in Noble, 2017a: fig. 6), in order to encompass all of the many diverse and important causal influences in the history of life on Earth.

In chapter 2, Peter Corning traces the origin and the meaning of the term *teleonomy*. It was coined by the biologist Colin Pittendrigh (1958) in connection with the landmark 1957 conference and subsequent edited volume on the role of behavior in evolution (Roe & Simpson, 1958). Pittendrigh, and most other theorists at that time, applied the term both to behavior and to morphology.

A few words are in order here about the term *teleonomy*, which we believe represents a basic property of living systems. To borrow the words of the Nobel biologist Jacques Monod (1971, 9), teleonomy is "one of the fundamental characteristics common to all living beings without exception: that of being objects endowed with a purpose or project, which at the same time they show in their structure and execute through their performances." Unfortunately, in the twentieth century, the original definition of teleonomy was qualified by making "the purpose or project" of an organism simply the working-out of natural selection acting on random genetic variations over time (see chapter 12). Teleonomy was relegated merely to apparent purposiveness, or as Mayr (1974, 98) put it, the "seemingly goal-directed behavior of organisms." Still today, Wikipedia tells us that teleonomy means only "the quality of *apparent* purposefulness" (emphasis added). But this is not correct. Even Mayr contradicted it in other writings—see chapter 2. Moreover, Krieger (1998) and Ayala (1998) concluded that Mayr's definition was tautological.

Corning goes on in his chapter (chapter 2) to survey some of the many developments in evolutionary biology over the years that highlight the role of teleonomy in evolution. Among other things, there is growing evidence, championed especially by biologist Lynn Margulis (1970, 1993, 1998; see also Margulis & Fester, 1991), that symbiosis—cooperative relationships between organisms of different species with complementary capabilities—is a widespread phenomenon in the natural world, and that Mereschkowsky's "symbiogenesis" (Kowallik & Martin, 2021) has played a major role in shaping the evolutionary trajectory over time (see also Margulis & Sagan, 2002; Sapp, 1994).

Another challenge to the modern synthesis is the discovery (e.g. Hilario & Gogarten, 1993) that single-celled prokaryotes are profligate sharers of genetic material via "horizontal" (or lateral) gene transmission and do not strictly follow the pattern of competition and Mendelian "vertical" inheritance from parent to offspring, as assumed under the modern synthesis. Thus, cooperative phenomena of various kinds, which are portrayed as being highly constrained and problematic under the modern synthesis, are now seen to play an important causal role in living systems, and in evolution (see Corning, 2018; Shapiro, 2022a).

The rise of evolutionary developmental biology (or evo-devo) has also produced serious challenges to the modern synthesis, especially the extensive work on morphological development and "phenotypic plasticity" (Müller & Newman, 2003; West-Eberhard, 2003, 2005a, 2005b). As West-Eberhard has shown, this is an important source of innovation in evolution.

There is also the burgeoning evidence that the genome is in fact a "two-way, read-write system," as microbiologist James Shapiro (2011, 2013, 2022b) characterizes it. Recent progress in microbiology has shown that an overwhelming majority of DNA changes in the genome are the result of what Shapiro (2011, 2022b) calls "natural genetic engineering" and the influence of internal regulatory and control networks, not random mutations and incremental "additive" selection. In fact, rapid genome alteration and restructuring can be achieved by a variety of mobile DNA "modules"—transposons (McClintock & Moore, 1987), integrons, CRISPRs, retroposons, variable antigen determinants, and more (Craig, 2002; Noble, 2006, 2011, 2013, 2017b; Sapp, 2009; Koonin, 2009, 2011, 2016; Shapiro, 2011, 2013; Craig et al., 2015). As Shapiro (2011, 2) emphasizes, "The capacity of living organisms to alter their own heredity is undeniable." (See also Shapiro, 2022b.)

A further challenge to a gene-centered model of evolution is our growing appreciation of the fact that what was long considered to be irrelevant "junk DNA"—because it was repetitive, was noncoding for proteins, and therefore presumably not subject to natural selection—in fact plays an important role in shaping genome evolution, epigenetic development, and gene regulation. Thus, mobile elements, part of the repetitive DNA, are major actors as mutagenic agents in evolving genomic regulatory circuits and other adaptive new configurations in the genome (Morris, 2012; Adelman & Egan, 2017; Mattick, 2018). Repetitive DNA also accounts for a large fraction of the so-called "noncoding" genome regions that serve as templates for transcription of "noncoding" lncRNAs, which carry out high-level functions in eukaryotic physiology and genome regulation. Long lncRNAs nucleate many kinds of multimolecular complexes in both the nucleus and cytoplasm. The lncRNAs have also been shown to act as regulators of complex traits, such as pregnancy, pluripotency, embryonic development, and nervous system connectivity. It is also interesting to note that repetitive DNA is the most "volatile" part of the genome in evolution, and its abundance tracks organismal complexity better than the protein-coding sequences (Liu, Mattick, & Taft, 2013). As Shapiro has noted, our understanding of genomes has fundamentally changed; we know they encode much more than proteins; they have an amazingly broad range of functionalities.

Likewise, it has long been appreciated that "microevolution" at the level of individual traits may have a very different pattern of causation from "macroevolution"—systematic changes in populations, species, and lineages over time (e.g., Eldredge & Gould, 1972; Mayr, 2001). However, this categorical distinction can be breached by hybridization between species. We now know that hybridization is ubiquitous—both in plants and in animals—and that it plays a significant role in creating new species (Shapiro, 2022a).

Perhaps most significant theoretically, there has been a growing recognition that the "purposeful" (teleonomic) behavior of living organisms themselves has had a major influence in shaping the course of evolution over time. Some contemporary theorists have adopted the concept of "agency" to characterize this defining biological characteristic (e.g., Kauffman, 2000; Walsh, 2015). Others have adopted Humberto Maturana and Francisco Varela's concept of "autopoiesis" (e.g., Capra & Luisi, 2014). However, the basic idea of the organism as an autonomous, self-organized and self-directed agent can be traced back at least to Lamarck, who first proposed that changes in an animal's "habits," stimulated by environmental changes, have been a primary source of evolutionary change over time (Lamarck, 1984/1809). (Even Darwin was open to Lamarck's idea and mentioned it no less than 12 times in *The Origin of Species*, 1968/1859. See also Noble, 2019.)

A "Darwinized" version of Lamarck's insight, called "organic selection theory" made a brief appearance at the end of the nineteenth century (Baldwin, 1896). The basic idea was that purposeful behavioral changes could alter the selective context for natural selection. However, this was soon overwhelmed and supplanted by "mutation theory" and the later work in population genetics that led to the modern synthesis (see Corning, 2014).

Organic selection theory, and the idea of behavior as an influence in evolutionary change, was tentatively reintroduced by the paleontologist George Gaylord Simpson (1953) under the neologism of the Baldwin effect. However, he portrayed it as being of only minor significance in evolution. A turning point came with an important set of conferences and a subsequent edited volume called *Behavior and Evolution* (Roe & Simpson, 1958). One of the conference attendees, biologist Ernst Mayr (1960), concluded in an important follow-up essay on the subject: "It is now quite evident that . . . the evolutionary changes that result from adaptive shifts are often initiated by a change in behavior, to be followed secondarily by a change in structure. . . . Changes of evolutionary significance are rarely, except on the cellular level, the direct results of mutation pressure. . . . The selection pressure in favor of the structural modification is greatly increased by a shift to a new ecological niche, by the acquisition of a new habit, or by both." Mayr characterized these Lamarckian behavioral innovations as the "pacemakers" of evolution (although he did not refer to Lamarck, of course).

Many other theorists over the years have also drawn attention to the role of behavior in evolution, as Corning notes, including Conrad Waddington, Jean Piaget, Jacques Monod, Edward Stuart Russell, Lynn Margulis, Jan Sapp, Henry Plotkin, Patrick Bateson, Eva Jablonka, Bruce Weber, David Depew, Terrence Deacon, Denis Noble, Kevin Laland, John Odling-Smee, Denis Walsh, and Peter Richerson. Equally important, the idea of "purposeful" adaptive changes has now been extended to the genome and to ontogeny (see Goodman, 1998; Shapiro, 2011, 2022b; Noble, 2017b).

Perhaps the most significant example of the role of behavior in evolution is the rise of humankind. As described in detail in Corning (2018), there have been three keys to our ancestors' extraordinary success over time: close social cooperation, adaptive innovations, and functional synergy. (Culture and cultural evolution have also played an important role, an insight that goes back to Darwin's *The Descent of Man*, 1871; see also Henrich, 2016; Richerson et al., 2021; Whiten, 2021). Some anthropologists have also invoked the idea of culture as a "collective brain" (see also Corning & Szathmáry, 2015).

There is much evidence that these adaptive social behaviors and technological innovations in human evolution preceded by many generations the anatomical changes that paleoanthropologists have used to define the major phases of our evolution as a species. In other words, over the course of several million years, the human species in effect invented itself through an entrepreneurial process involving gradual cultural innovations that changed our ancestors' relationship to the environment—and to one another. And these changes, in turn, led to the natural selection of supportive anatomical and psychological traits. As biologist Jonathan Kingdon (1993) put it in the title of his insightful book, we are the *Self-Made Man* (see also Boyd & Richerson, 2009; Boyd, Richerson, & Henrich, 2011, 2013). Corning concludes that teleonomy is a core property of living systems at all levels and, in innumerable ways, it has greatly influenced the evolution of life on Earth (see Corning, 2014, 2018, 2019, 2020, 2022).

In this volume, some 23 other theorists also explore the role of teleonomy in living systems, and in the evolutionary process. In chapter 3, on the "Cellular Basis of Cognition in Evolution," František Baluška, William B. Miller Jr., and Arthur S. Reber stress that all living organisms utilize a form of sentience, and that the emergence of eukaryotes and more complex living systems from prokaryotes required the development of a unitary cognitive system. In chapter 4, "Constructing 'On Purpose,'" Dominik Deffner reviews the literature on niche construction theory, which refers to a class of examples where organisms "actively" and "purposefully" choose and modify their environments. Deffner also develops a model that suggests how this influences natural selection. In chapter 5 on "Relational Agency," Francis Heylighen proposes that we replace the traditional object-based ontology. The basic "model," he says, should include "condition-action rules, challenges, agents, reaction networks, and chemical organizations." The fundamental evolutionary mechanism is agencies and reactions mutually adapting so as to form self-maintaining organizations.

Abir U. Igamberdiev, in chapter 6, writes about "Teleonomic Anticipatory Configurations in Biological Evolution," which he says are rooted in internal "models" and produce goal-directed behaviors. Eva Jablonka and Simona Ginsberg, in chapter 7, explore the evolutionary transition "From Teleonomy to Mentally Driven" behavior in evolution, most notably in humankind, where what they term an "unlimited associative learning" (UAL) system was the basis for the evolution of goals driven by rational design and abstract values. In chapter 8, "Beyond the Newtonian Paradigm," Stuart Kauffman and Andrea Roli argue provocatively that the traditional laws of physics are unable to explain the emergent evolution of living systems and that a new statistical mechanics will be required to account for a dynamic process that involves novelty, natural selection, and heritability—and teleonomy. Kalevi Kull, in chapter 9, proposes the surprising idea that "meaning" (an idea that goes back to Jakob von Uexküll and the origin of the science of semiotics) in all its forms has played an important part in evolution.

In chapter 10, on "Collective Intelligence of Morphogenesis as a Teleonomic Process," Michael Levin calls teleonomy a "lynchpin concept" that helps address questions around evolvability, biological plasticity, and basal cognition. Stuart Newman, in chapter 11 on the "Sources of Natural Purpose in Animal Evolution," challenges the "standard" (traditional) assumptions about the process of complexification in biological evolution, presenting evidence that it has been "more directional and purposeful" than the "opportunistic, random-search-based scenario suggests." He sees purpose as a game-changer in evolutionary theory. Likewise, Denis Noble and Raymond Noble, in their chapter (12) on "How Purposive Agency Became Banned from Evolutionary Biology," recount how Julian Huxley, contrary to the long-standing myth about his paradigm-defining 1942 book, *Evolution: The Modern Synthesis*, actually recognized a wide range of influences, even purposive factors. It became simplified by others, or "hardened" (in Stephen Jay Gould's term), into the gene-centered model that most biologists are familiar with only much later, in the 1950s and 1960s. The Nobles assert that "purposive agency needs to be restored in the study of living organisms."

Philosopher of science Samir Okasha, who has written extensively on the concept of "agency" in evolution, dissects the three different ways in which biologists use the concept of a "goal or endpoint" with respect to an organismic activity or process in his chapter (13) on "Goal Attributions in Biology." It arises in three different biological contexts, Okasha suggests. He concludes that the third one, "treating behavior-functions [in living organisms] as organismic goals" is indeed appropriate, given the fact that all living organisms must seek to survive and reproduce. However, according to Okasha, this presupposes that organisms exhibit a "unity of purpose." In chapter 14, "Toward the Physicalization of Biology" (on the search for a chemical origin of cognition), Robert Pascal and Addy Pross suggest that the recently proposed chemical domain of dynamic kinetic chemistry, and its underlying stability kind, dynamic kinetic stability (DKS), may provide a physico-chemical basis for understanding the different behavioral characteristics of living and nonliving things (Pross, 2016; Pascal & Pross, 2015; Pascal et al., 2013), including that of cognition. The DKS state is critically dependent on its environment for energy and material support, and this might lead to cognitive functions ("perception, memory, learning, decision-making, information transfer") with its associated teleonomic character. Crucially, that physical character suggests that teleonomy in biology may be physically tractable after all, and that central biological attributes may yet be reducible to physics and chemistry.

Microbiologist James A. Shapiro, who has published pioneering books over the years on mobile genetic elements, "natural genetic engineering," bacterial multicellularity, and the two-way "read-write genome" in evolution, provides documentation in chapter 15 of some of the extensive evidence for the thesis that "evolutionary change is naturally biological and purposive." Beginning with a review of Barbara McClintock's paradigm-shattering work in the 1940s on transposable controlling elements (transposons), which at the time was widely rejected or ignored (although she was later awarded the Nobel Prize), Shapiro then provides in tabular form a comprehensive review of the work over many years on genomic innovations that could be attributed to mobile DNA elements, as well as the work on how environmental stresses trigger increased mobile DNA activity. He notes that "mobile DNA elements and other biological engines of genome change constitute essential survival tools in a dynamic ecology, and that is why they are ubiquitous in living organisms." He concludes: "Recognition of the cognitive aspect of the evolutionary process opens a range of hitherto forbidden questions for scientific exploration. . . . It may well be the case that living organisms [which] are *incapable* of actively modifying their genomes would be doomed to extinction" (italics added). If that inference turns out to be correct, then genome self-modification, or re-writing, would be another essential goal-oriented function to incorporate into our definition of what it means to be alive.

Anthony Trewavas, in chapter 16, summarizes his many publications which have documented cognitive processes and decision making in plants. Chapter 17, by Tobias Uller, "Agency, Goal Orientation, and Evolutionary Explanations," presents the "radical proposal," as he puts it, "that organismal goals themselves can be explanatory for evolutionary change." Such naturalistic, teleonomic explanations can fill "an explanatory gap left by causal explanations" and encourage theories and models that allow for organismal activities "that originate as a result of their internal organisation." Finally, in chapter 18, Denis Walsh, who has been a leading advocate for the concept/role of agency in evolution, provides a detailed comparison between what he calls the "foundationalism" of the modern synthesis and the "contact phenomena" that are a crucial "interface" between living organisms and their environments. He concludes that the modern synthesis is incapable of accounting for "inheritance, development, innovation, and adaptively biased change." An agency theory must be included in our explanation of the evolutionary process.

In sum, what the contributors to this volume have collectively shown is that living systems exhibit/demonstrate an evolved, means–ends purposiveness (teleonomy), in a myriad of different ways. This arises from, and is necessitated by, the fact that all living organisms are contingent dynamic (and kinetic) systems that must actively seek to survive and reproduce in their many different, often changing environments. Their "agency" derives from this unavoidable "struggle for existence"—in Darwin's famous characterization. Teleonomy in living systems is not, after all, only "apparent." It is a fundamental fact of life.

References

Adelman, K., & Egan, E. (2017). Non-coding RNA: More uses for genomic junk. *Nature, 543* (7644), 183–185. https://doi.org/10.1038/543183a

Ayala, F. J. (1998). Teleological explanations *versus* teleology. *History and Philosophy of the Life Sciences, 20,* 41–50.

Baldwin, J. M. (1896). A new factor in evolution. *American Naturalist, 30,* 441–456, 536–553.

Boyd, R., & Richerson, P. J. (2009). Culture and the evolution of human cooperation. *Philosophical Transactions of the Royal Society of London Series B, Biological Sciences, 364,* 3281–3288. https://doi.org/10.1098/rstb.2009.0134

Boyd, R., Richerson, P. J., & Henrich, J. (2011). The cultural niche: Why social learning is essential for human adaptation. *Proceedings of the National Academy of Sciences of the United States of America, 108,* 10918–10925. https://doi.org/10.1073/pnas.1100290108

Boyd, R., Richerson, P. J., & Henrich, J. (2013). Cultural evolution of technology: Facts and theories. In P. J. Richerson & M. Christiansen (Eds.), *Cultural evolution: Society, technology, language, and religion* (pp. 119–142). Cambridge, MA: MIT Press.

Capra, F., & Luisi, P. L. (2014). *The systems view of life: A unifying vision.* Cambridge: Cambridge University Press.

Corning, P. A. (2014). Evolution "on purpose": How behaviour has shaped the evolutionary process. *Biological Journal of the Linnean Society, 112*, 242–260. https://doi.org/10.1111/bij.12061

Corning, P. A. (2018). *Synergistic selection: How cooperation has shaped evolution and the rise of humankind.* Singapore: World Scientific.

Corning, P. A. (2019). Teleonomy and the proximate-ultimate distinction revisited. *Biological Journal of the Linnean Society, 127* (4), 912–916. https://doi.org/10.1093/biolinnean/blz087

Corning, P. A. (2020). Beyond the modern synthesis: A framework for a more inclusive biological synthesis. *Progress in Biophysics and Molecular Biology, 153*, 5–12. https://doi.org/10.1016/j.pbiomolbio.2020.02.002

Corning, P. A. (2022). A systems theory of biological evolution. *BioSystems, 214* (April 2022), No. 104583.

Corning, P. A., & Szathmáry, E. (2015). 'Synergistic selection': A Darwinian frame for the evolution of complexity. *Journal of Theoretical Biology, 371*, 45–58.

Craig, N. L. (2002). *Mobile DNA II.* Washington, DC: American Society for Microbiology Press.

Craig, N. L., Chandler, M., Gellert, M., Lambowitz, A., & Rice, P. A. (Eds.). (2015). *Mobile DNA III.* Washington, DC: American Society for Microbiology.

Danchin, E., Charmantier, A., Champagne, F. A., Mesoudi, A., Pujol, B., & Blanchet, S. (2011). Beyond DNA: Integrating inclusive inheritance into an extended theory of evolution. *Nature Review, Genetics, 12*, 475–486. https://doi.org/10.1038/nrg3028

Darwin, C. R. (1968/1859). *On the origin of species by means of natural selection, or the preservation of favoured races in the struggle for life.* Baltimore, MD: Penguin.

Darwin, C. R. (1871). *The descent of man and selection in relation to sex* (2nd ed.). London: Charles Murray.

Dawkins, R. (1989/1976). *The selfish gene.* New York: Oxford University Press.

Eldredge, N., & Gould, S. J. (1972). Punctuated equilibria: An alternative to phyletic gradualism. In T. J. M. Schopf (Ed.), *Models in paleobiology* (pp. 82–115). San Francisco, CA: Freeman, Cooper.

Goodman, M. F. (1998). Purposeful mutations. *Nature, 395*(6699), 221–223.

Henrich, J. (2016). *The secret of our success: How culture is driving human evolution, domesticating our species, and making us smarter.* Princeton, N.J.: Princeton University Press.

Hilario, E., & Gogarten, J. P. (1993) Horizontal gene transfer of ATPase genes—the tree of life becomes a net of life. *BioSystems, 31*, 111–119. https://doi.org/10.1016/0303-2647(93)90038-E

Huxley, J. S. (1942). *Evolution: The modern synthesis.* New York: Harper & Row.

Kauffman, S. A. (2000). *Investigations.* New York: Oxford University Press.

Kingdon, J. (1993). *Self-made man: Human evolution from Eden to extinction?* New York: John Wiley & Sons.

Koonin, E. V. (2009). The Origin at 150: Is a new evolutionary synthesis in sight? *Trends in Genetics, 25*(11), 473–475. https://doi.org/10.1016/j.tig.2009.09.007

Koonin, E. V. (2011). *The logic of chance: The nature and origin of biological evolution.* Upper Saddle River, NJ: FT Press Science.

Koonin, E. V. (2016). Viruses and mobile elements as drivers of evolutionary transitions. *Philosophical Transactions of the Royal Society of London, Ser. B 371*, 20150442. https://doi.org/10.1098/rstb.2015.0442

Kowallik, K. V., & Martin, W. F. (2021). The origin of symbiogenesis: An annotated English translation of Mereschkowsky's 1910 paper on the theory of two plasma lineages. *BioSystems, 199*, 104281.

Krieger, G. J. (1998). Transmogrifying teleological talk? *History and Philosophy of the Life Sciences, 20*, 3–34.

Laland, K. N., Uller, T., Feldman, M. W., Sterelny, K., Müller, G. B., Moczek, A., Jablonka, E., & Odling-Smee, J. (2014). Does evolutionary theory need a rethink? (Yes, urgently.). *Nature, 514*(7521), 161–164. https://doi.org/10.1038/514161a

Laland, K. N., Uller, T., Feldman, M. W., Sterelny, K., Müller, G. B., Moczek, A., Jablonka, E., & Odling-Smee, J. (2015). The extended evolutionary synthesis: Its structure, assumptions and predictions. *Philosophical Transactions of the Royal Society London, Ser. B 282*, 20151019. https://doi.org/10.1098/rspb.2015.1019

Lamarck, J.-B. (1984/1809). *Zoological philosophy: An exposition with regard to the natural history of animals* (H. Elliot, trans.). Chicago: University of Chicago Press.

Laubichler, M. D., & Renn, J. (2015). Extended evolution: A conceptual framework for integrating regulatory networks and niche construction. *Journal of Experimental Zoology, 324B*, 565–577. https://doi.org/10.1002/jez.b.22631

Liu, G., Mattick, J. S., & Taft, R. J. (2013). A meta-analysis of the genomic and transcriptomic composition of complex life. *Cell Cycle, 12*(3), 2061–2072. https://doi.org/10.4161/cc.25134

Margulis, L. (1970). *Origin of eukaryotic cells*. New Haven, CT: Yale University Press.

Margulis, L. (1993). *Symbiosis in cell evolution* (2nd ed.). New York: W. H. Freeman.

Margulis, L. (1998). *Symbiotic planet: A new look at evolution*. New York: Basic Books.

Margulis, L., & Fester, R. (Eds.). (1991). *Symbiosis as a source of evolutionary innovation: Speciation and morphogenesis*. Cambridge, MA: MIT Press.

Margulis, L., & Sagan, D. (2002). *Acquiring genomes: A theory of the origins of species*. New York: Basic Books.

Mattick, J. S. (2018). The state of long non-coding RNA biology. *Noncoding RNA, 4*(3), 17. https://doi.org/10.3390/ncrna4030017

Mayr, E. (1960). The emergence of evolutionary novelties. In S. Tax (Ed.), *Evolution after Darwin* (Vol. 1, pp. 349–380). Chicago: University of Chicago Press.

Mayr, E. (1974). Teleological and teleonomic: A new analysis. In R. S. Cohen & M. W. Wartofsky (Eds.), *Boston studies in the philosophy of science* (vol. 14, pp. 91–117). Boston: Reidel.

Mayr, E. (2001). *What evolution is*. New York: Basic Books.

McClintock, B., & Moore, J. A. (Eds.). (1987). *The discovery and characterization of transposable elements: The collected papers of Barbara McClintock*. New York: Garland Publishers.

Monod, J. (1971). *Chance and necessity* (2nd ed.) (A. Wainhouse, trans.). New York: Alfred A. Knopf.

Morris, K. V. (Ed.). (2012). *Non-coding RNAs and epigenetic regulation of gene expression: Drivers of natural selection*. Poole, UK: Caister Academic Press.

Müller, G. B., & Newman, S. A. (Eds.). (2003). *Origination of organismal form: Beyond the gene in developmental and evolutionary biology*. Cambridge, MA: MIT Press.

Noble, D. (2006). *The music of life: Biology beyond the genes*. Oxford, UK: Oxford University Press.

Noble, D. (2011). Neo-Darwinism, the modern synthesis and selfish genes: Are they of use in physiology? *Journal of Physiology, 589*(5), 1007–1015. https://doi.org/10.1113/jphysiol.2010.201384

Noble, D. (2013). Physiology is rocking the foundations of evolutionary biology. *Experimental Physiology, 98*(8), 1235–1243. https://doi.org/10.1113/expphysiol.2012.071134

Noble, D. (2017a). Evolution viewed from physics, physiology and medicine. *Interface Focus, 7*, 20160159. http://dx.doi.org/10.1098/rsfs.2016.0159

Noble, D. (2017b). *Dance to the tune of life: Biological relativity*. Cambridge: Cambridge University Press.

Noble, D. (2019). Exosomes, gemmules, pangenesis, and Darwin. In L. R. Edelstein, J. R., Smythies, P. J. Quesenberry, & D. Noble (Eds.), *Exosomes: A clinical compendium* (pp. 487–501). Amsterdam: Elsevier.

Pascal, R., & Pross, A. (2015). Stability and its manifestation in the chemical and biological worlds. *Chemical Communications, 51*(90), 16160 -16165. doi.org/10.1039/c5cc06260h

Pascal, R., Pross, A., & Sutherland, J. D. (2013). Towards an evolutionary theory of the origin of life based on kinetics and thermodynamics. *Open Biology, 3*(11), 130156. doi.org/10.1098/rsob.130156

Pigliucci, M., & Müller, G. B. (Eds.). (2010). *Evolution: The extended synthesis*. Cambridge, MA: MIT Press.

Pittendrigh, C. S. (1958). Adaptation, natural selection and behavior. In A. Roe and G. G. Simpson (Eds.), *Behavior and evolution* (pp. 390–416). New Haven, CT: Yale University Press.

Pross, A. (2016). *What is life? How chemistry becomes biology*, 2nd ed. Oxford, UK: Oxford University Press.

Richerson, P. J., Gavrilets, S., & de Waal, F. B. M. (2021). Modern theories of human evolution foreshadowed by Darwin's *Descent of Man*. *Science, 372*(6544), eaba3776. https://doi.org/10.1126/science.aba3776

Roe, A., & Simpson, G. G. (Eds.). (1958). *Behavior and evolution*. New Haven, CT: Yale University Press.

Sapp, J. A. (1994). *Evolution by association: A history of symbiosis*. New York: Oxford University Press.

Sapp, J. A. (2009). *The new foundations of evolution, on the tree of life*. Oxford, UK: Oxford University Press.

Shapiro, J. A. (2011). *Evolution: A view from the 21st century*. Upper Saddle River, NJ: FT Press Science.

Shapiro, J. A. (2013). How life changes itself: The read-write (RW) genome. *Physics Life Reviews, 10*, 287–323. https://doi.org/10.1016/j.plrev.2013.07.001

Shapiro, J. A. (2022a, online). Engines of innovation: Biological origins of genome evolution. *Biological Journal of the Linnean Society*. blac041. https://doi.org/10.1093/biolinnean/blac041

Shapiro, J. A. (2022b). *Evolution: A view from the 21st century. Fortified,* 2nd ed. *Why evolution works as well as it does*. Chicago: Cognition Press.

Simpson, G. G. (1953). The Baldwin effect. *Evolution, 2*, 110–117. https://doi.org/10.1111/j.1558-5646.1953.tb00069

Wade, M. J. (2011). The neo-modern synthesis: The confluence of new data and explanatory concepts. *Bioscience*, *61*(5), 407–408. https://doi.org/10.1525/bio.2011.61.5.10

Walsh, D. M. (2015). *Organisms, agency, and evolution*. Cambridge: Cambridge University Press.

West-Eberhard, M. J. (2003). *Developmental plasticity and evolution*. Oxford, UK: Oxford University Press.

West-Eberhard, M. J. (2005a). Developmental plasticity and the origin of species differences. *Proceedings of the National Academy of Sciences. United States of America*, *102*(Suppl. 1), 6543–6549. https://doi.org/10.1073/pnas.0501844102

West-Eberhard, M. J. (2005b). Phenotypic accommodation: Adaptive innovation due to developmental plasticity. *Journal of Experimental Zoology B, Molecular and Developmental Evolution*, *304*(6), 610–618. https://doi.org/10.1002/jez.b.21071

Whiten, A. (2021). The burgeoning reach of animal culture. *Science*, *372*(6537), eabe65. https.//doi.org/10.1126/science.abe6514

Wray, G. A., Hoekstra, H. E., Futuyma, D. J., Lenski, R. E., Mackay, T. F. C., Schluter, D., & Strassman, J. E. (2014). Does evolutionary theory need a rethink? (No, all is well). *Nature*, *514*(7521), 161–164. https://doi.org/10.1038/514161a

2 Teleonomy in Evolution: "The Ghost in the Machine"

Peter A. Corning

Natural selection is not a mechanism; it is a consequential happening. Teleonomy is a product—and a cause.
—Peter A. Corning

Overview

Although it is now widely accepted that living systems exhibit an evolved purposiveness, or *teleonomy*, the theoretical implications of this distinctive biological property have yet to be fully explored. Here I briefly discuss the origins and history of the concept, along with its scope and meaning and some of its many forms and facets. I also attempt to clarify the often-misunderstood concept of natural selection. However, I focus especially on the causal role of purposeful behaviors in shaping natural selection, and on how teleonomy and functional synergy (combined or co-operative effects of various kinds) have together influenced the rise of biological complexity in the natural world. An important example of this causal dynamic is the evolution of humankind, which the zoologist Jonathan Kingdon, in his book-length study (1993), characterized as the "self-made man." Teleonomy is ultimately one of the most determinative properties of living systems.

2.1 Introduction

The Ghost in the Machine is the title of a provocative book by the polymath and famed twentieth-century novelist Arthur Koestler (1967), in which he disputed the then-fashionable view, often attributed to Descartes, that the human mind is a dualistic, nonmaterial entity. Koestler argued that, on the contrary, the mind is embedded in, and is a product of, the natural world. (Koestler's ironic title was borrowed from the philosopher Gilbert Ryle [1949].)

I have revived this distinctive title to underscore the cardinal fact that teleonomy in evolution is not simply a product of natural selection. It is also an important cause of natural selection and has been a major shaping influence over time in biological evolution. Natural selection is not an exogenous force or a "mechanism." It is an outcome of the relationships and interactions between purposeful living organisms—agents, if you will—and their lived-in environments, inclusive of other organisms. Here I explore the concept of teleonomy and seek to illuminate its influence in shaping natural selection and the trajectory of evolution.[1]

2.2 Defining Teleonomy

The term *teleonomy* was originally coined by the biologist Colin Pittendrigh in connection with the landmark 1957 conference on behavior in evolution (Roe & Simpson, 1958). Pittendrigh was seeking to draw a contrast between an "external" teleology (Aristotelian or religious) and the "internal" purposiveness and goal-directedness of living systems, which are products of the evolutionary process and of natural selection. As the eminent twentieth-century biologist Theodosius Dobzhansky later explained (using the equivalent phrase "internal, or natural, teleology"):

Purposefulness, or teleology, does not exist in nonliving nature. It is universal in the living world. It would make no sense to talk of the purposiveness or adaptation of stars, mountains, or the laws of physics. Adaptedness of living beings is too obvious to be overlooked. . . . Living beings have an *internal*, or natural, teleology. Organisms, from the smallest bacterium to man, arise from similar organisms by ordered growth and development. Their internal teleology has accumulated in the history of their lineage. On the assumption that all existing life is derived from one primordial ancestor, the internal teleology of an organism is the outcome of approximately three and a half billion years of organic evolution. . . . Internal teleology is not a static property of life. Its advances and recessions can be observed, sometimes induced experimentally, and analyzed scientifically like other biological phenomena (Dobzhansky et al., 1977, 95–96).

Many other theorists over the years have expressed similar views, as Samir Okasha has documented in his book-length study, *Agents and Goals in Evolution* (2018; see also Walsh, 2015). For instance, the Nobel biologist Jacques Monod (1971, 9) concluded that "one of the most fundamental characteristics common to all living things [is] that of being endowed with a project, or a purpose." Likewise, the biologist Ernst Mayr, one of the founding fathers of the so-called modern synthesis in evolutionary biology, wrote "goal directed behavior . . . is extremely widespread in the natural world; most activity connected with migration, food-getting, courtship, ontogeny, and all phases of reproduction is characterized by such goal orientation" (Mayr, 1988, 45).

2.3 Differing Definitions

Over the years, many theorists have interpreted teleonomy broadly. Pittendrigh (1958) himself characterized it as a "fundamental property" and defining feature of all biological phenomena, including behavior. Similarly, Monod, in his influential book *Chance and Necessity*, concluded: "All the structures, all the performances, all the activities contributing to the essential project [of life] will hence be called 'teleonomic'. . . . It is the very definition of living beings" (Monod, 1971, 9, 14). Monod pointed to the central nervous system as an example.

However, Mayr (1974), in his classic essay "Teleological and Teleonomic: A New Analysis," strongly opposed such a broad definition. Mayr framed teleonomy as requiring a preexisting goal and "something material" that guides and controls a "process" to a "determinable end." In living organisms, he said, this a priori goal entails a "program"—an analogy Mayr borrowed from computers. It is the teleonomic program that is responsible for directing the process of developing a phenotype and its behavior, although an "open program" (as Mayr called it) allows for the influence of learning and experience (and "disturbances"). To illustrate, Mayr alluded to the science of cybernetics, or goal-directed control systems. He also insisted that a teleonomic program—an obvious euphemism for the genome—could only have a one-way flow of information, and that developmental influences are highly restricted. "The inheritance of acquired characters becomes quite unthinkable."

Mayr was adamant that it was inappropriate to attribute purposiveness to the process of evolution itself, or to the influence of natural selection, and he opposed applying the term *teleonomy* to any "static" biological system (meaning the structural elements of an organism). He cited the central nervous system as a negative example, thus implicitly contradicting Monod. Mayr also insisted that "it is misleading and quite inadmissible to designate such broadly generalized concepts as survival or reproductive success as definite and specified goals. Teleonomy does not exist outside the ultimately determinative influence of DNA and the genetic 'program.'"

In other words, Mayr was an apostle of the gene-centered, one-way, bottom-up evolutionary paradigm, commonly referred to as neo-Darwinism, that predominated at the time in evolutionary theory, and he seemed to exclude what he called "proximate" causes from exerting a direct influence on "ultimate" causes (natural selection and evolution). In 1963, Mayr wrote: "The proponents of the synthetic theory maintain that all evolution is due to the accumulation of small genetic changes, guided by natural selection" (Mayr, 1963, 586). Indeed, in an earlier paper, Mayr (1961) had identified only two categories of what he called legitimate "evolutionary causes": "genetic causes" and "ecological causes."

2.4 Moving On

However, there was a ghost in Mayr's machine. In fact, there were several ghosts. Mayr's adherence to what the Nobel biologist Francis Crick (1970) termed the "central dogma" of neo-Darwinian evolutionary theory, and his radical separation of proximate and ultimate causation, is no longer tenable (Laland, Sterelny et al., 2011; Laland, Odling-Smee et al., 2013; see also Calcott, 2013a, 2013b; Corning, 2019, 2020).

Among other things, there is growing evidence, championed especially by biologist Lynn Margulis (1970, 1993, 1998; Margulis & Fester, 1991), that symbiosis—cooperative relationships between organisms of different species with complementary capabilities—is a widespread phenomenon in the natural world, and that "symbiogenesis" has played a major causal role in shaping the evolutionary trajectory over time (see also Margulis & Sagan, 2002; Sapp, 1994, 2009; Gontier, 2007; Carrapiço, 2010; Archibald, 2014). Symbiogenesis theory shifts the locus of innovation away from "random" changes in genes, genomes, and the "classical" model of natural selection to the "purposeful" behavioral actions of the phenotypes and their functional consequences.[2]

An even greater challenge to neo-Darwinism arose with the discovery that single-celled prokaryotes are profligate sharers of genetic material via "horizontal" (or lateral) gene transmission and do not strictly follow the pattern of competition and Mendelian ("vertical") inheritance from parent to offspring, as the modern synthesis assumes (Sapp, 2009; Koonin,

2011; Crisp et al., 2015). As the biologist Eugene Koonin (2009) concluded, all the central tenets of the modern synthesis break down with prokaryotes and the findings of comparative genomics. It has been argued that the prokaryote world can best be described as a single, vast, interconnected gene pool.

The rise of evolutionary developmental biology (evo-devo for short) has also produced serious challenges to the modern synthesis, including the discovery that there are many deep homologies and highly conserved structural gene complexes in the genome (some of which are universal in living systems), along with the extensive work on morphological development and "phenotypic plasticity" (Müller & Newman, 2003; West-Eberhard, 2003, 2005a, 2005b; Koonin, 2011; Bateson & Gluckman, 2011). There is also the burgeoning evidence that the genome is in fact a "two-way read-write system," as the biologist James Shapiro (2011, 2013, 2022a, 2022b) characterizes it. The extensive and rapidly increasing evidence of epigenetic inheritance (changes in the phenotype that are transmitted to the germ plasm in the next generation) also falsifies the one-way, gene-centered theory (see also Jablonka, 2013; Jablonka & Raz, 2009; Jablonka & Lamb, 2014; Noble, 2012, 2013, 2015, 2016, 2018; Walsh, 2015; Huneman & Walsh, 2017).

Recent progress in microbiology has shown that an overwhelming majority of DNA changes in the genome are the product of internal regulatory and control networks, not random mutations and incremental "additive" selection. There is also much evidence of various biases in mutational processes (Stoltzfus, 2019). Indeed, rapid genome alteration and restructuring can be achieved by a variety of mobile DNA "modules"—transposons (McClintock & Moore, 1987), integrons, CRISPRs, retroposons, variable antigen determinants, and more (Craig, 2002; Craig et al., 2015; Sapp, 2009; Shapiro, 2011, 2013, 2022a, 2022b; Koonin, 2011, 2016; Noble, 2016). It is now also apparent that individual cells have a variety of internal regulatory and control capabilities that can significantly influence cell development and the phenotype. More significant, they may even provide feedback that modifies the genome and affects subsequent generations (Pan & Zhang, 2009; Gladyshev & Arkhipova, 2011; Koonin, 2011; Shapiro, 1991, 2011, 2022a, 2022b; Noble, 2006, 2011, 2016, 2018). Particularly significant are the discoveries related to the influence of exosomes, which resemble Darwin's speculative ideas of pangenesis and internal migratory "gemmules" (Darwin's term) in reproduction, as Noble (2019) has pointed out. Exosomes also clearly violate the so-called Weismann barrier (the assumption that genetic change is only a one-way process).

As Shapiro (2011, 2) emphasizes, "The capacity of living organisms to alter their own heredity is undeniable. Our current ideas about evolution have to incorporate this basic fact of life." Shapiro cites some 32 different examples of what he refers to as "natural genetic engineering," including immune system responses, chromosomal rearrangements, diversity-generating retroelements, the actions of transposons, genome restructuring, whole genome duplication, and symbiotic DNA integration (see also Shapiro, 1988, 2013, 2022a, 2022b). Likewise, Jablonka and Lamb (2014) identify four distinct "Lamarckian" modes of inheritance: (1) directed adaptive mutations, (2) the inheritance of characteristics acquired during development and the lifetime of the individual, (3) behavioral inheritance through social learning, and (4) language-based information transmission. All this prompted biologist Kevin Laland and his colleagues to publish two major critiques of Mayr's proximate-ultimate dichotomy (Laland, Sterelny et al., 2011; Laland, Odling-Smee et al., 2013). These critics argue that proximate and ultimate causes are interpenetrated and that the one-way causal model associated with the modern synthesis and neo-Darwinism should be replaced with one that recognizes a major role for "reciprocal causation" (see also Calcott, 2013a, 2013b).

2.5 Teleonomy and Natural Selection

To fully appreciate the causal role of teleonomic influences in evolution, it might be helpful to revisit the concept of natural selection. The neo-Darwinian definition of natural selection has always tended to be narrow, gene-centered, and circular. Evolution is defined as "a change in gene frequencies" in a given "deme," or breeding population, and natural selection is defined as a "mechanism" which produces changes in gene frequencies. As the biologist John H. Campbell put it in a review: "Changes in the frequencies of alleles by natural selection *are* evolution" (Campbell, 1994, 86).

By implication, it follows that mutations and related molecular-level changes—subject to the "approval" of natural selection—are the only important sources of novelty in evolution. Natural selection is in turn represented as being an external "mechanism" or "force" out there in the environment somewhere. Thus, Linnen and Hoekstra (2009, 155) write: "A complete understanding of the role of natural selection in driving evolutionary change requires estimates of the strength of selection acting in the wild" (rather like the wind conditions?).

The tendency to reify natural selection goes back to Darwin himself. In the first edition of *The Origin of Species* (Darwin, 1968/1859), he famously wrote: "Natural selection is daily and hourly scrutinizing, throughout the world, every variation, even the slightest; rejecting that which is bad, preserving and adding up all that is good; silently and insensibly working, whenever and wherever opportunity offers, at the improvement of each organic being in relation to its organic and inorganic conditions of life." Only in a later edition did Darwin add the all-important clarifying phrase: "It may metaphorically be said . . .".

Natural selection is not in fact a mechanism, or a force; it is a consequential happening. It does not *do* anything; nothing is ever actively selected (although sexual selection and artificial selection are special cases). Nor can the sources of causation be localized either within an organism or externally in the environment. As Darwin himself conceded, "natural selection" is metaphor—an umbrella term that identifies a fundamental aspect of the evolutionary process. The ground-zero premise of evolutionary biology is that life is, in essence, a "survival enterprise." Living organisms are inherently contingent, dynamic phenomena that must actively seek to survive and reproduce. This existential problem requires that they must be goal-directed in an immediate, proximate sense. (Of course, what we call "goal-directedness" may be our observation of the combined effect of many separate biological characteristics and traits that "act together" to achieve a predictable combined result: survival and reproduction.)

Thus, *natural selection* refers to whatever functionally significant factors are responsible in a given context for causing differential survival and reproduction. The well-known behaviorist psychologist B. F. Skinner (1981) called it "selection by consequences." Properly conceptualized, these causal "factors" are intensely interactional and relational; they are defined by both the organism(s) and their environment(s).

This important point can be illustrated with a classic example: "industrial melanism" (Kettlewell, 1955, 1973). Until the Industrial Revolution, a "cryptic" (light-colored) variety of the peppered moth (*Biston betularia* f. 'typica') predominated in the English countryside over a rare, darker "melanic" form (*Biston betularia* f. 'carbonaria'). The mottled wing coloration of the 'typica' form provided effective camouflage from avian predators as the moths rested on the trunks and branches of lichen-encrusted trees. The darker, melanic form did not share this advantage. But as industrial soot progressively blackened the trees in areas close to expanding industrial cities, the relative frequency of the two forms was reversed; the birds began to prey more heavily on the now more visible cryptic moths.

So, the question is, where in this example was natural selection "located"? The short answer is that natural selection encompassed the entire *configuration* of factors that combined to influence differential survival and reproduction. In this case, an alteration in the relationship between the coloration of the trees and the wing pigmentation of the moths, as a consequence of industrial pollution, was an important proximate factor. But this factor was important only because of the inflexible resting behavior of the moths and the feeding habits and perceptual abilities of the birds. Had the moths been subject only to insect-eating bats that use a kind of "sonar" to catch insects on the wing, rather than a visual detection system, the change in background coloration would not have been significant. Nor would it have been significant were there not genetically based differences in wing coloration between the two varieties that were available for "differential selection."[3]

Accordingly, the ongoing survival challenge for living systems (Darwin called it "the struggle for existence") imposes a potential constraint on every aspect of the process.[4] Every feature, or trait, of a given organism can be viewed in terms of its relationship (for better or worse) to this fundamental, built-in, inescapable problem. Thus, natural selection differentially favors proximate functional "means" over time that serve the ultimate biological "end" of survival and reproduction. Indeed, the very term *adaptation* is commonly defined as a feature that advances some process or deals with some challenge related to survival and reproduction.

The neo-Darwinian definition also tends to equate natural selection and evolution with genetic changes, rather than viewing evolution more expansively as a multileveled process in which genes, other molecular factors, genomes, developmental ("epigenetic") influences, mature phenotypes, and the natural environment interact with one another and evolve together in a dynamic relationship of mutual and reciprocal causation, including (in the current jargon) "upward" causation, "downward" causation, and even "horizontal" causation—for example, in predator-prey interactions or between symbionts. The rise of multilevel selection theory in biology during the past three decades has also been helpful as a corrective to the gene-centric model (see especially D. S. Wilson, 1997; see also Okasha, 2006; Traulsen & Nowak, 2006).

Another corrective has been the extensive work on "niche construction theory" (Odling-Smee et al., 1996, 2003; Laland et al., 1999), as well as the growing research literature on the role of cultural influences in evolution, culminating in humankind (Kingdon, 1993; Boyd & Richerson, 2005, 2009; Richerson & Boyd, 2005; Corning, 2005, 2014, 2018; Laland et al., 2010; Henrich, 2016; Laland, 2017; Whiten, 2021; Richerson et al., 2021).

Another way of framing it is that evolution involves four distinct categories of functional variation: (1) molecular-genetic variation, (2) phenotypic variation (inclusive of developmental, physiological, and behavioral variations), (3) ecological (environmental) variation, and (4) differential survival and reproduction as an outcome of the specific organism-environment relationships and interactions in a given context at a given time. Furthermore, the causal arrows between and among these domains can go in both directions.

Thus, many things, at many different levels, may be responsible for bringing about changes in an organism-environment relationship and differential survival. It could be a functionally significant mutation, a chromosomal transposition, a change in the physical environment that affects development, or a change in one species that affects another species, or it could be a change in behavior that results in a new organism-environment relationship. In fact, a whole sequence of changes may ripple through a pattern of relationships. For instance, a climate change might alter the ecology, which might prompt a behavioral shift to a new habitat, which might encourage an alteration in nutritional habits and precipitate changes in the interactions among different species, resulting ultimately in the differential survival and reproduction of organisms with differing morphological characters and the genes that support them. As Stuart Kauffman (2019, 137) points out, "New species literally create niches for yet further new species."

An in vivo illustration of this causal dynamic can be found in the long-running research program in the Galápagos Islands among "Darwin's finches" (Grant & Grant, 1979, 1989, 1993, 2002; Weiner, 1994). It is well known that birds often use their beaks as tools, and that their beaks tend to be specialized for whatever food sources are available in a given environment. In the Galápagos Islands, the zoologist Peter Grant and his colleague/wife Rosemary have observed many changes over the years among its (originally) 14 closely related bird species in response to environmental changes. During drought periods, for instance, small seeds become scarce and the most abundant food sources consist of much larger, tougher seeds that must be cracked open to get at their kernels. Birds with larger, stronger beaks have a functional advantage, and this is the proximate cause of their differential survival during a drought.

In sum, natural selection is focused on the functional causes of differential survival and reproduction, and is agnostic about how and why this may occur in any given context. Contrary to Mayr, the survival imperative can indeed be posited as an overarching goal in living systems (without any scare quotes or "as ifs"), inclusive of the proximate teleonomic phenomena that are, in fact, causal influences in natural selection. The basic unit of analysis in this alternative paradigm is not the genes but interdependent living "systems" and their parts—along with their external "affordances" and dependencies (Rosen, 1970, 1991; Bateson, 2004, 2005; Corning, 2005, 2018; Noble, 2006, 2016; Capra & Luisi, 2014; Walsh, 2015; Okasha, 2018; Kauffman, 2019). Some theorists (e.g., Gilbert et al., 2012) have adopted the term *holobiont* to characterize this frame shift. A living system represents a "combination of labor" with an overarching vocation, a means-ends teleonomy. As Eldredge (2004, 16) noted many years ago, "According to the popular press and television, everything, every bit of animal behavior, really boils down to passing genes along to the next generation. Forgotten in the process is the simple fact that animals need to eat simply to live." (Indeed, their often-helpless, newborn offspring may also need to be fed and protected.)

The late Patrick Bateson (2013) illustrated this alternative paradigm with an analogy. The recipe for a biscuit/cookie is rather like the genome in living organisms. It represents a set of instructions for how to make an end-product. A shopper who buys a biscuit/cookie selects the "phenotype"—the end-product, not the recipe. So, if the recipe survives and the number of cookies multiplies over time, it is only because shoppers like the end-product and are willing to purchase more of them.

2.6 Natural Selection and Lamarckism

Some contemporary theorists have adopted the concept of "agency" to characterize this defining biological characteristic (e.g., Kauffman, 2000; Walsh, 2015; Okasha, 2018). Other theorists have adopted Humberto Maturana's and Francisco Varela's concept of "autopoiesis" (e.g., Capra & Luisi, 2014). *Agency* is a term that is utilized in biology to characterize the ability of a living system to act as an autonomous, self-directed agent—to vary its morphology, its behavior, and its environment "purposefully" in relation to external or internal (physiological) conditions and goals. When a hungry fox chases an evasive hare, both are exercising agency—not God's will, or a philosophical concept, but an evolved capability for meeting their needs and coping with challenges/threats in their environments. Agency in living systems is a product of the evolutionary "trial and success" process, as Dobzhansky put it.[5] (For more on agency, see endnote 5.)

For the record, the importance of the organism as a self-organized and self-directed agent in evolution can be traced back at least to Jean Baptiste de Lamarck, who proposed that changes in an animal's "habits," stimulated by environmental changes, have been a primary source of evolutionary change over time. Lamarck (1984/1809, 114) wrote: "It is not the organs . . . of an animal's body that have given rise to its special habits and faculties; but it is, on the contrary, its habits, mode of life and environment that have over the course of time controlled . . . the faculties which it possesses." Darwin was receptive to Lamarck's idea, calling it the "use and disuse of parts," and mentioned it no less than 12 times in *The Origin of Species* (1968/1859). (Conversely, late in life, Lamarck embraced a precursor of Darwin's natural selection idea. See Corning, 2018, 70.)

To repeat what was said earlier, a Darwinized version of Lamarck's insight, called "organic selection theory" made a brief appearance at the end of the nineteenth century. The basic idea was that purposeful behavioral changes could alter the selective context for natural selection. However, this insight was soon overwhelmed and supplanted by "mutation theory" and the later work that led to the modern synthesis (see Corning, 2014). The idea that behavior is an influence in evolutionary change was tentatively reintroduced by the paleontologist George Gaylord Simpson (1953) under the neologism of the Baldwin effect, although he portrayed it as being of only minor significance in evolution.

A turning point came with the major conferences and an edited volume on *Behavior and Evolution* (Roe & Simpson, 1958). In a landmark follow-up essay on the subject, Mayr (1960) concluded: "It is now quite evident that . . . the evolutionary changes that result from adaptive shifts are often initiated by a change in behavior, to be followed secondarily by a change in structure. . . . Changes of evolutionary significance are rarely, except on the cellular level, the direct results of mutation pressure. . . . The selection pressure in favor of the structural modification is greatly increased by a shift to a new ecological niche, by the acquisition of a new habit, or by both." Mayr did not mention Lamarck, but he characterized these Lamarckian behavioral innovations as the "pacemakers" of evolution. (Mayr thus seemed to be contradicting himself in his 1961 article in the journal *Science*.)

2.7 Natural Selection and Behavior

In fact, the "purposeful" behavior of living organisms has had a major influence in shaping natural selection and the trajectory of evolution over time. It could be said—with

Dobzhansky—that the behavior of living organisms exhibits an internal, or natural, teleology. This term highlights the fact that evolved teleonomic processes and systems can exert a significant causal influence on the properties and actions of living systems, both in themselves and in others. Some theorists speak of "agency"; others of "autopoiesis" (or self-maintenance); still others of "cybernetic" (feedback-driven) goal-directedness. Indeed, even the neologism of Stephen Jay Gould and Elisabeth Vrba (1982), "exaptations" (traits that evolved in relation to one function, like feathers in dinosaurs, that were later co-opted for another use) implies behavioral shifts as a cause of evolutionary changes. I characterize the "ultimate" evolutionary consequences of this dynamic as "teleonomic selection."

A well-documented illustration involves the remarkable tool-using behavior of the so-called woodpecker finch. *Cactospiza pallidus* is one of the many species of highly unusual birds, first discovered by Darwin, that have evolved in the Galápagos Islands, probably from a single immigrant species of mainland ancestors. Although *C. pallidus* was not actually observed by Darwin, subsequent researchers have found that the woodpecker finch occupies a niche that is normally occupied on the mainland by conventional woodpeckers. However, as any beginning biology student knows, *C. pallidus* has achieved its unique adaptation in a highly unusual way. Instead of excavating trees with its beak and tongue alone, as the mainland woodpecker does, *C. pallidus* skillfully uses cactus spines or small twigs held lengthwise in its beak to probe beneath the bark. When it succeeds in dislodging an insect larva, it will quickly drop its digging tool, or else deftly tuck the tool between its claws long enough to devour the prey. Members of this species have also been observed carefully selecting digging "tools" of the right size, shape, and strength and carrying them from tree to tree (Lack, 1947; Weiner, 1994).

What is most significant about this distinctive behavior, for our purpose, is the "downward" effect it had on natural selection and the genome of *C. pallidus*. The mainland woodpecker's feeding strategy is in part dependent on the fact that its ancestors evolved an extremely long, probing tongue. But *C. pallidus* has no such "structural" modification. In other words, the invention of a digging tool enabled the woodpecker finch to circumvent the requirement for an otherwise necessary morphological change. This behavioral "workaround" in effect provided both a facilitator and a selective shield, or mask. (For a more recent example of social behavior as a facilitator of genetic change, see Shell et al., 2021; see also Richerson et al., 2021; Whiten, 2021.)

It is also frequently the case that the teleonomic behavioral choices of one species can become the instrument of natural selection in another species. One example among many can be found in the rainforest of the Olympic National Park, in the state of Washington, where there is intense competition among the towering evergreen trees (western hemlock, Sitka spruce, Douglas fir, and western cedar) inside a crowded forest canopy. Hemlocks produce by far the most seeds and are the best adapted to growing in the low sunlight conditions of the park. However, it is the Sitka spruce that dominates, and the reason is that the abundant Roosevelt elk in the park feed heavily on young hemlock trees and do not feed on the Sitka spruce. In other words, the food preferences of the elk are the "proximate cause" of differential survival (natural selection) between the hemlock and spruce trees (Warren, 2010).

2.8   Mind in Evolution

Over the past half century, the research on learning and innovation by living organisms—from "smart bacteria" to human-tutored apes and playful dolphins—has grown to cataract proportions. (Indeed, there is now so much of it that some excellent earlier work is being overlooked and forgotten.) The examples are almost endless: worms, fruit flies, honeybees, guppies, stickleback fish, ravens, various songbirds, hens, rats, gorillas, chimpanzees, elephants, dolphins, whales, and many others. In the index to their book on *Animal Traditions*, Avital and Jablonka (2000) list well over 200 different species (see also Whiten, 2021).

We now know that primitive *E. coli* bacteria, slime molds, *Drosophila* flies, ants, bees, flatworms, laboratory mice, pigeons, guppies, cuttlefish, octopuses, dolphins, gorillas, and chimpanzees, among many species, can learn novel responses to novel conditions, via "classical" and "operant" conditioning.

Our respect for the "cognitive" abilities of various animals also continues to grow (see de Waal, 2016). Innumerable studies have documented that many species are capable of sophisticated cost-benefit calculations, sometimes involving several variables, including the perceived risks, energetic costs, time expenditures, nutrient quality, resource alternatives, relative abundance, and more. Animals are constantly required to make "decisions" about habitats, foraging, food options, travel routes, nest sites, even mates. Many of these decisions are under tight genetic control, with "pre-programmed" selection criteria. But many more are also, at least in part, the product of experience, trial-and-error learning, observation and even, perhaps, some insight learning (Corning, 2014, 2018). One classic illustration is ethologist Bernd Heinrich's experiments in which naïve ravens in captivity quickly learned to use their beaks and claws to pull up "fishing lines" hung from their roosts in order to capture the food rewards attached at the ends (Heinrich, 1995). (Heinrich's 1999 book, *The Mind of the Raven*, provides extensive evidence for the mental abilities of these remarkable birds.)

Indeed, even plants make "decisions." In the marine alga *Fucus*, for example, biologists Simon Gilroy and Anthony Trewavas (2001) have found that at least 17 environmental conditions can be "sensed," and the information that it collects is then either summed or integrated synergistically as appropriate. Gilroy and Trewavas conclude: "What is required of plant-cell signal-transduction studies . . . is to account for 'intelligent' decision-making; computation of the right choice among close alternatives" (Gilroy & Trewavas, 2001, 307; see also Trewavas, 2014; Gilroy & Trewavas, 2022).

A striking example of "decision-making" in trees can be seen in the so-called whistling thorn acacia (*Vachellia drepanolobium*) in East Africa, which produces hollow galls that provide accommodations and nutrients for symbiotic ants. The ants repay the trees by providing additional defenses against browsers (such as elephants and giraffes) with their stingers and formic acid emissions. However, when the trees are not subject to browsers, they may kick out their symbionts and collapse their galls to conserve scarce water and nutrients (Palmer et al., 2008).

Especially important theoretically are the many forms of social learning through "stimulus enhancement," "contagion effects," "emulation," and even some "teaching." Social learning has been documented in many species of animals, from rats to bats, to lions and elephants, as well as some birds and fishes and, of course, domestic dogs. For instance, red-wing blackbirds, which readily colonize new habitats, are especially prone to acquire new food habits—or food aversions—from watching other birds (Weigl & Hanson, 1980). Pigeons can learn specific food-getting skills from other pigeons (Palameta & LeFebvre, 1985). Domestic cats, when denied the ability to observe conspecifics, will learn certain tasks much more slowly or not at all (John et al., 1968). And, in a controlled laboratory study, naïve ground squirrels (*Tamiasciurus hudsonicus*) that were allowed to observe an experienced squirrel feed on hickory nuts were able to learn the same trick in half the time it took for unenlightened animals (cited in Byrne, 1995, 58). True imitation (including the learning of motor skills) has also been observed in (among others) gorillas (peeling wild celery to get at the pith), rats (pressing a joystick for food rewards), African grey parrots (vocalizations and gestures), chimpanzees (nut-cracking with an anvil and a stone or wooden hammer), and bottlenose dolphins (many behaviors, including grooming, sleeping postures, even mimicking the divers that scrape the observation windows of their pools, down to the sounds made by the divers' breathing apparatus) (see Corning, 2014).

Not surprisingly, the most potent cognitive skills have been found in social mammals, especially the great apes. They display intentional behavior, planning, social coordination, understanding of cause and effect, anticipation, generalization, even deception. Primatologists Richard Byrne and Andrew Whiten, in their two important edited volumes on the subject, referred to it as "Machiavellian intelligence" (Byrne & Whiten, 1988; Whiten & Byrne, 1997; see also Gibson & Ingold, 1993; de Waal, 2016; Whiten, 2021). Cognitive skills and social learning have provided a powerful means—which humankind has greatly enhanced—for accumulating, dispersing, and perpetuating novel adaptations without waiting for slower-acting genetic changes to occur.

Tool-use is an especially significant and widespread category of adaptive behavior in the natural world—from insects to insectivores and omnivores—and it is utilized for a wide variety of purposes. As Edward O. Wilson (1975) pointed out in his comprehensive survey and synthesis, *Sociobiology*, tools provide a means for quantum jumps in behavioral invention, and in the ability of living organisms to manipulate their environments. Tool-use results in otherwise unattainable behavioral outcomes (synergies) (Wilson, 1975, 172; see also Beck, 1980; McGrew, 1992; Wrangham et al., 1994).

Finally, it is important to take note of the role of "culture" and cultural transmission in evolutionary change. The evidence continues to mount that cultural influences are present in many species—from primates to cetaceans, birds, fish, whales, even some insects (see Whiten, 2021). Of course, culture has played an especially important role in shaping human evolution (see Foley, 1995; Corning, 2005, 2018; Boyd & Richerson, 2005; Boyd et al., 2011; Foley & Gamble, 2011; Henrich, 2016; Richerson et al., 2021; Waring & Wood, 2021; Whiten, 2021).

Biologist Richard Dawkins, in his legendary popular book, *The Selfish Gene*, famously characterized humankind and other living systems as "survival machines—robot vehicles that are blindly programmed to preserve the selfish molecules known as genes" (Dawkins, 1989/1976, ix). We now know that this is seriously misleading. Arguably, it is the other way around; the genes have evolved in the service of living organisms, for the most part, and the exceptions prove the rule (see Shapiro, 2011, 2022a; Baverstock, 2021; Noble & Noble, 2021).

2.9   The Synergism Hypothesis

A major theoretical issue in mainstream evolutionary biology over the past two decades has concerned the rise of complexity in nature, and a search has been underway for "a Grand Unified Theory"—as biologist Daniel McShea (2015) characterized it—that is consistent with Darwin's great vision. McShea aspired to "some single principle or some small set of principles" that could explain the evolutionary trend toward greater complexity.

Likewise, biologist Deborah Gordon (2007) noted that "perhaps there can be a general theory of complex systems, but we don't have one yet."

As it happens, such a theory does already exist. It was first proposed in *The Synergism Hypothesis: A Theory of Progressive Evolution* (Corning, 1983), and it involves an economic (or perhaps bioeconomic) theory of complexity. Simply stated, cooperative interactions of various kinds, however they may occur, can produce novel combined effects—*synergies*—with functional advantages that may, in turn, become direct causes of natural selection. The focus of the synergism hypothesis is on the favorable selection of synergistic "wholes" and the combination of genes that produces these wholes. The parts (and their genes) that produce these synergies may, in effect, become interdependent units of evolutionary change.

In other words, the synergism hypothesis is a theory about the unique combined effects produced by the relationships between things. I refer to it as holistic Darwinism because it is entirely consistent with natural selection theory, properly understood. It is the functional (economic) benefits associated with various kinds of synergistic effects in any given context that are the underlying causes of cooperative relationships—and of complex organization—in the natural world. The synergy produced by the whole provides the proximate functional payoffs that may differentially favor the survival and reproduction of the parts, and their genes (see Corning, 1983, 2003, 2005, 2012, 2018).

It should also be obvious that these synergies can very often be quantified in various ways. A legendary example, among many others (see Corning, 2018), is the way in which emperor penguins huddle closely together in large colonies, sometimes numbering in the tens of thousands, to share heat during the bitterly cold Antarctic winter. In so doing, they are able to reduce their individual energy expenditures by 20 to 50 percent, depending upon where they are in the huddle and the wind direction and speed (Le Maho, 1977).

The biologists John Maynard Smith and Eörs Szathmáry (1995, 1999), in their path-breaking works on the "major transitions" in evolution, came to the same conclusion independently about the causal role of synergy in evolution—although they graciously acknowledged the priority of my 1983 book in one of their two books on the subject. They applied their version of the synergism hypothesis specifically to the problem of explaining the emergence of new levels of biological organization over time (see also Corning & Szathmáry, 2015). Maynard Smith (1982) also proposed the concept of synergistic selection as (in effect) a subcategory of natural selection. He illustrated with a formal mathematical model that included a term for "non-additive" benefits (when $2 + 2 = 5$). The idea is also distilled in the catchphrase "the whole is greater than the sum of its parts," which traces back to the *Metaphysics* of Aristotle (ca. 350 BC, Book H, 1045, 8–10). Synergistic selection refers to the many contexts in nature where two or more genes/genomes/individuals have a shared fate; their combined effects are functionally interdependent.

Thus, cooperative phenomena of various kinds, which are portrayed as being highly constrained and problematic under the predominately competitive assumptions of the modern synthesis, are now seen to play an important causal role in living systems, and in evolution. Indeed, the forest ecologist Suzanne Simard and her colleagues have shown that even trees can form cooperative communities bound together by networks of underground mycorrhizal fungi that facilitate information exchange and the sharing of various nutrients and chemicals (Simard et al., 2012; see also the wide-ranging popular treatment, *The Hidden Life of Trees*, by the forester Peter Wohllben, 2016).

Biologist Richard Michod (1999) has asserted that "cooperation is now seen as the primary creative force behind ever greater levels of complexity and organization in all of biology"; Martin Nowak (2011) has called cooperation "the master architect of evolution." However, it is not cooperation per se that has been the "creative force" or the "architect." Rather, it is the unique combined effects (the beneficial synergies) produced by cooperation. Synergies of various kinds have been a prodigious source of evolutionary novelties and the common underlying cause of cooperation and increased complexity in evolution over time (Corning, 1983, 2005, 2012, 2018).

Although it may seem like backwards logic, the thesis is that functional synergy is the cause of cooperation and complexity in living systems, not the other way around. To repeat, the synergism hypothesis is basically an economic theory of emergent complexity, and it applies equally to biological and cultural evolution, most notably in humankind. Indeed, it appears that social cooperation has played a key role in our own evolution as a species, and that synergy is the basic reason why we cooperate. In a very real sense, we have invented ourselves.

2.10   The "Self-Made Man"

As described in detail in my book *Synergistic Selection: How Cooperation Has Shaped Evolution and the Rise of Humankind* (Corning, 2018), there have been three keys to our ancestors' extraordinary success over time: close social cooperation, adaptive innovations, and functional synergy. (Cultural evolution has also played an important role, an insight that goes back to Darwin's *The Descent of Man*, 1871; see also Klein, 1999; Klein & Edgar, 2002; Richerson et al., 2021.) Our remote bipedal ancestors, the australopithecines, were relatively small (about three feet tall) and slow-moving. They would not have survived the harsh physical challenges involved in living on the ground, nor would they have held their own against the many large predators in their East African environment in those days—such as the pack-hunting *Palhyaena*—without foraging together in closely cooperative groups and defending themselves collectively with the tools that they had invented for procuring food, and for self-defense (probably digging sticks that doubled as clubs, and perhaps thrown rocks). The result was a game-changing synergy: cooperative outcomes that could not otherwise have been achieved.

The other major transitions in the multimillion-year history of our evolution as a species followed this same basic formula. Cooperation and innovation were the underlying themes, and the synergies that were produced (the economic benefits) were the reason why our ancestors cooperated, and why they survived. Thus, the emergence of the much larger and bigger-brained *Homo erectus* some two million years ago was a product of a synergistic joint venture: namely, the hunting of big game animals in closely cooperating groups with the aid of an array of potent new tools—finely balanced throwing spears, hand axes, cutting tools, carriers, and (eventually) fire and cooking. Not to mention (quite likely) sequestered home bases, midwifery, and the first baby-sitting cooperatives. It was a collective survival enterprise—a superorganism, like leaf cutter ants—and it was sustained by multiple synergies. True, the many large game animals in their environment provided a major "ecological opportunity," but our ancestors devised the means for exploiting it.

The final emergence of modern humankind, perhaps as early as 300,000 years ago, represented a further elaboration of this collective survival strategy; it was novel economic synergies that enabled the evolution of much larger groups. Each "tribe" was, in effect, a coalition of many biological families that was sustained by a sophisticated array of new technologies—shelters, clothing, food processing, food preservation and storage techniques, and much else. Especially important were the more efficient new hunting and gathering tools, like spear throwers (which greatly increased their range and accuracy), bows and arrows, nets, traps, and a variety of fishing techniques. Indeed, culture itself (including spoken language) became a powerful engine of cumulative evolutionary change. Our collective survival enterprise—our superorganism—became an autocatalytic engine of growth and innovation (and environmental disruption) as synergy begat more synergy. (Some anthropologists have invoked the idea of culture as a "collective brain." See Richerson et al., 2021; Whiten, 2021.)

Most importantly for our purpose, there is much evidence that these adaptive social behaviors and technological innovations preceded by many generations the anatomical changes that paleoanthropologists have used to define the major phases of our evolution as a species. In other words, over the course of several million years, the human species in effect invented itself through an entrepreneurial process involving gradual cultural innovations that changed our ancestors' relationship to the environment—and to one another. These changes, in turn, led to the natural selection of supportive anatomical and psychological traits. As biologist Jonathan Kingdon (1993) put it in the title of his insightful book with this theme, we are the *Self-Made Man* (see also Boyd, Richerson, & Henrich, 2011, 2013; Richerson et al., 2021).

2.11   "The Sentient Symphony"

The evolution of humankind is undoubtedly the most striking example of how teleonomy has exerted a shaping influence in biological evolution, but a case can be made that teleonomy was also involved in many of the great turning points and transitions in the history of life on Earth, including the earliest colonization of the seafloor, the emergence of the eukaryotes, the migration of life forms from the oceans onto the land, the rise of multicellular organisms, the development of land plants and trees, the origin of fish, birds, and mammals, the invention of social organization, the division of labor (task specialization), and more.

Teleonomy is also an implicit (though unspoken) influence in connection with many other familiar terms, I would argue, including "symbiogenesis," "organic selection theory," evolutionary "pacemakers," the "Baldwin effect," "major transitions theory," "niche construction theory," "gene-culture coevolution theory," "natural genetic engineering," many examples of "semiosis," and, recently, the concept of "agency" in evolution. These terms all suggest the role of purposive behavior. A radically different view of evolution has been emerging in this century. We now know that living systems actively shape their own evolution, in various ways. The distinguished geneticist James Shapiro (2011, 2022a, 2022b) shows us that the "selfish gene" model (and the so-called modern synthesis) is quite wrong (see also Noble & Noble, 2021).[6]

Biologist Lynn Margulis and Dorion Sagan, in their popular book *What Is Life?* (1995), characterized evolution as a "sentient symphony." What is most significant about the behavior of living organisms, they asserted, is their ability to make choices and act upon them.

(For the record, C. H. Waddington mounted a very similar argument back in the 1950s, as did Conwy Lloyd Morgan in the 1890s.) Margulis and Sagan, allowing themselves a bit of poetic license, wrote: "At even the most primordial level, living seems to entail sensation, choosing, mind" (Margulis & Sagan, 1995, 180).

Natural selection, over time, has been both a cause of this purposiveness and an outcome (inclusive of many outcomes that were both unintended and indirect). The purposiveness of living systems, culminating in humankind has, I would argue, thus exerted a major influence over the trajectory of the evolutionary process. As Denis Noble (2016) has argued, following West-Eberhard (2003), the genes have often been the followers rather than the leaders in evolutionary change. Teleonomy thus transcends the laws of physics and defies deterministic, predictive mathematical modeling (see Kauffman and Roli in chapter 8). It is, I believe, one of the most unique and ultimately determinative properties of living systems.

Notes

1. This is an original paper, but some basic ideas/concepts have been drawn from earlier publications by this author.

2. The basic idea of symbiogenesis, and even the term itself, traces back to a school of nineteenth- and early twentieth-century Russian botanists, including A. S. Famintsyn (1907a, b; 1918), Konstantin Mereschkovsky (1909, 1920), and B. M. Kozo-Polyansky (1924, 1932), but their pioneering work was not generally known to Western scientists until recent decades.

3. It should also be noted that subsequent challenges to Kettlewell's methods and the validity of his findings were resolved when a British geneticist, Mike Majerus (2009), undertook a study that confirmed the original results. Also see Hurley and Montgomery's "Peppered Moths & Melanism" (2009).

4. Darwin used this phrase, or a close alternative, more than 40 times in *The Origin of Species*. See https://simple .wikipedia.org/wiki/Struggle_for_existence#:~:text=The%20struggle%20for%20existence%20is,chapter%20 3%20of%20the%20Origin.

5. The term *agency* was imported into biology from the social sciences and philosophy, and it is entangled with theories of "mind," human cognition, intentional behavior, rationality, rational choice theory, and artificial intelligence, among other things. However, there have been some useful efforts to sort all this out for biologists. Walsh (2015), for instance, stresses that agency in biology refers to the goal-directed behavior of living organisms: their ability to pursue goals and to respond appropriately to conditions in their environments. Agency is fundamentally an "ecological phenomenon," he says, and he identifies three key properties of biological agency: (1) goals, (2) "affordances" (which are determined by both the organism and its environment) and (3) the organism's "repertoire" of behavioral responses. Okasha (2018) likewise identifies three rationales for applying the term *agency* in biology: (1) goal-directed activities in organisms with a "unified" goal, (2) behavioral flexibility, and (3) traits that are adaptations (serving intermediate "sub-purposes" related to the overarching goal).

I would add to this the following points: Because life is a contingent phenomenon, living organisms must actively pursue opportunities (resources) in their environments and must be able to avoid or cope with challenges and threats of various kinds. Agency is thus an evolved capability that enables a living system to respond to variability, and changing conditions, in relation to needed resources and challenges/threats in its environment. (Mobility in an organism also greatly increases this challenge, needless to say.) Agency in living systems requires: (1) the detection or "perception" of variations in internal and external conditions; (2) the ability to discriminate among these perceptions ("information"); (3) the ability to purposefully vary behavior, or actions; and (4) "control"— or the ability to link information with actions (cf., the cybernetic model of goal-oriented, "feedback"-driven behavior).

Agency is not dependent upon having a "brain." It can be based upon simple decision rules. However, its effectiveness can be greatly enhanced by being able to draw upon prior learning and memory, along with in situ cognitive and problem-solving skills. Agency will be favored by natural selection in relation to the degree of variability, and novelty, in the opportunities and threats in any given environmental context. But it is also a costly trait. It requires energy and functionally specialized biomass that must be built and maintained over time. Therefore, it will atrophy, or will not evolve at all, in conditions where it is not clearly advantageous for survival and reproduction. Examples of these points can be found in such diverse living entities as macrophages, bacteria at hydrothermal vents, slime molds, sea floor sponges, land plants, insects, fish, birds, and mammals.

6. I must disagree with those who seek to account for, and measure, teleonomy in terms of some discrete unit of information. Information is not a thing; it is the capacity to control the capacity to do work. Its salience/

effectiveness is determined by the user (see Corning, 2007). What I refer to as "control information" should properly be defined as a latent potential: "the capacity (know how) to control the acquisition, disposition and utilization of matter/energy in 'purposive' (cybernetic) processes in living systems." Consider the apparent paradox that even the absence of something can provide important information for a living organism—like the absence of water, or of prey, or of a mate. Indeed, the same "unit" of coded "information"—say, a binary bit—can have very different meanings in different contexts and for different potential users.

References

Archibald, J. (2014). *One plus one equals one: Symbiosis and the evolution of complex life*. Oxford, UK: Oxford University Press.

Aristotle. (1961/ca. 350 BC). *The metaphysics*. Cambridge, MA: Harvard University Press.

Avital, E., & Jablonka, E. (2000). *Animal traditions: Behavioral inheritance in evolution*. Cambridge: Cambridge University Press.

Bateson, P. P. G. (2004). The active role of behaviour in evolution. *Biology and Philosophy, 19*, 283–298.

Bateson, P. P. G. (2005). The return of the whole organism. *Journal of Bioscience, 30*, 31–39.

Bateson, P. P. G. (2013). Evolution, epigenetics and cooperation. *Journal of Bioscience, 38*(4), 1–10.

Bateson, P. P. G., & Gluckman, P. (2011). *Plasticity, robustness, development, and evolution*. Cambridge: Cambridge University Press.

Baverstock, K. (2021). The gene: An appraisal. *Progress in Biophysics and Molecular Biology, 164*, 46–62. https://doi.org/10.1016/j.pbiomolbio.2021.04.005

Beck, B. B. (1980). *Animal tool behavior*. New York: Garland Press.

Boyd, R., & Richerson, P. J. (2005). *The origin and evolution of cultures*. Oxford, UK: Oxford University Press.

Boyd, R., & Richerson, P. J. (2009). Culture and the evolution of human cooperation. *Philosophical Transactions of the Royal Society of London Series B, Biological Sciences, 364*, 3281–3288.

Boyd, R., Richerson, P. J., & Henrich, J. (2011). The cultural niche: Why social learning is essential for human adaptation. *Proceedings of the National Academy of Sciences of the U.S.A., 108*, 0918–10925.

Boyd, R., Richerson, P. J., & Henrich, J. (2013). Cultural evolution of technology: Facts and theories. In P. J. Richerson & M. Christiansen (Eds.), *Cultural evolution: Society, technology, language, and religion* (pp. 119–142). Cambridge, MA: MIT Press.

Byrne, R. W. (1995). *The thinking ape: Evolutionary origins of intelligence*. Oxford, UK: Oxford University Press.

Byrne, R. W., & Whiten, A. (1988). *Machiavellian intelligence: Social expertise and the evolution of intellect in monkeys, apes and humans*. Oxford, UK: Oxford University Press.

Calcott, B. (2013a). Why the proximate–ultimate distinction is misleading, and why it matters for understanding the evolution of cooperation. In K. Sterelny, R. Joyce, B. Calcott, & B. Fraser (Eds.), *Cooperation and its evolution* (pp. 249–263). Cambridge, MA: MIT Press.

Calcott, B. (2013b). Why how and why aren't enough: More problems with Mayr's proximate-ultimate distinction. *Biology and Philosophy, 28*(5), 767–780.

Campbell, J. H. (1994). Organisms create evolution. In J. H. Campbell & J. W. Schopf (Eds.), *Creative Evolution?!* (pp. 85–102). Boston, MA: Jones & Bartlett.

Capra, F., & Luisi, P. L. (2014). *The systems view of life: A unifying vision*. Cambridge: Cambridge University Press.

Carrapiço, F. (2010). How symbiogenic is evolution? *Theoretical Bioscience, 129*, 135–139.

Corning, P. A. (1983). *The synergism hypothesis: A theory of progressive evolution*. New York: McGraw-Hill.

Corning, P. A. (2003). *Nature's magic: Synergy in evolution and the fate of humankind*. Cambridge: Cambridge University Press.

Corning, P. A. (2005). *Holistic Darwinism: Synergy, cybernetics and the bioeconomics of evolution*. Chicago: University of Chicago Press.

Corning, P. A. (2007). Control information theory: The 'missing link' in the science of cybernetics. *Systems Research and Behavioral Science, 24*, 297–311.

Corning, P. A. (2012). Rotating the Necker cube: A bioeconomic approach to cooperation and the causal role of synergy in evolution. *Journal of Bioeconomics, 15*(2), 171–193. https://doi.org/10.1007/s10818-012-9142-4

Corning, P. A. (2014). Evolution 'on purpose': How behaviour has shaped the evolutionary process. *Biological Journal of the Linnean Society, 112*, 242–260.

Corning, P. A. (2018). *Synergistic selection: How cooperation has shaped evolution and the rise of humankind.* Singapore: World Scientific.

Corning, P. A. (2019). Teleonomy and the proximate-ultimate distinction revisited. *Biological Journal of the Linnean Society, 127*(4), 912–916. https://doi.org/10.1093/biolinnean/blz087

Corning, P. A. (2020). Beyond the modern synthesis: A framework for a more inclusive biological synthesis. *Progress in Biophysics and Molecular Biology, 153,* 5–12. https://doi.org/10.1016/j.pbiomolbio.2020.02.002

Corning, P. A., & Szathmáry, E. (2015). 'Synergistic Selection': A Darwinian frame for the evolution of complexity. *Journal of Theoretical Biology, 371,* 45–58.

Craig, N. L. (2002). *Mobile DNA II.* Washington, DC: American Society for Microbiology Press.

Craig, N. L., Chandler, M., Gellert, A., Lambowitz, P. A., & Rice, P. A. (Eds.). (2015). *Mobile DNA III.* Washington, DC: American Society for Microbiology Press.

Crick, F. (1970). Central dogma of molecular biology. *Nature, 227*(5258), 561–563.

Crisp, A., Boschetti, C., Perry, M., Tunnacliffe, A., & Micklem, G. (2015). Expression of multiple horizontally acquired genes is a hallmark of both vertebrate and invertebrate genomes. *Genome Biology, 16,* 50. https://doi.org/10.1186/s13059-015-0607-3

Darwin, C. R. (1968/1859). *On the origin of species by means of natural selection, or the preservation of favoured races in the struggle for life.* Baltimore, MD: Penguin.

Darwin, C. R. (1871). *The descent of man, and selection in relation to sex.* London: John Murray.

Dawkins, R. (1989/1976). *The selfish gene.* New York: Oxford University Press.

de Waal, F. (2016). *Are we smart enough to know how smart animals are?* New York: W. W. Norton.

Dobzhansky, T., Ayala, F. J., Stebbins, G. L., & Valentine, J. W. (Eds.). (1977). *Evolution.* San Francisco, CA: Freeman.

Eldredge, N. (2004). *Why we do it. Rethinking sex and the selfish gene.* New York: Norton.

Famintsyn, A. S. (1907a). Concerning the role of symbiosis in the evolution of organisms. *Mémoirs Academy of Science, Ser. 8, Physical-Mathematical Division, 20*(3), 1–14.

Famintsyn, A. S. (1907b). Concerning the role of symbiosis in the evolution of organisms. *Transactions of the St. Petersburg Society of Natural Science, 38*(1), Minutes of Session, 4, 141–143.

Famintsyn, A. S. (1918). What is going on with lichens? *Nature* (April–May), 266–282.

Foley, R. (1995). *Humans before humanity: An evolutionary perspective.* Oxford, UK: Blackwell.

Foley, R., & Gamble, C. (2011). The ecology of social transitions in human evolution. *Philosophical Transactions of the Royal Society of London Series B, Biological Sciences, 364,* 3267–3279.

Gibson, R., & Ingold, T. (Eds.). (1993). *Tools, language, and cognition in human evolution.* Cambridge: Cambridge University Press.

Gilbert, S. F., Sapp, J., & Tauber, A. I. (2012). A symbiotic view of life: We have never been individuals. *Quarterly Review of Biology, 87*(4), 325–341.

Gilroy, S., & Trewavas, A. (2001). Signal processing and transduction in plant cells: The end and the beginning. *Nature Reviews (Molecular Cell Biology), 2,* 307–314.

Gilroy, S., & Trewavas, A. (2022). Agency, teleonomy and signal transduction in plant systems. *Biological Journal of the Linnean Society,* blac021. https://doi.org/10.1093/biolinnean/blac021

Gladyshev, E. A., & Arkhipova, I. R. (2011). A widespread class of reverse transcriptase-related cellular genes. *Proceedings of the National Academy of Sciences of the U.S.A., 108*(51), 20311–20316. https://doi.org/10.1073/pnas.1100266108

Gontier, N. (2007). Universal symbiosis: An alternative to universal selectionist accounts of evolution. *Symbiosis, 44,* 167–181.

Gordon, D. M. (2007). Control without hierarchy. *Nature, 446,* 143.

Gould, S. J., & Vrba, E. S. (1982). Exaptation—a missing term in the science of form. *Paleobiology, 8*(1), 4–15. https://doi.org/10.1017/S0094837300004310

Grant, B. R., & Grant, P. R. (1979). Darwin's finches: Population variation and sympatric speciation. *Proceedings of the National Academy of Sciences of the U.S.A., 76,* 2359–2363.

Grant, B. R., & Grant, P. R. (1989). Natural selection in a population of Darwin's finches. *American Naturalist, 133,* 377–393.

Grant, B. R., &. Grant, P. R. (1993). Evolution of Darwin's finches caused by a rare climatic event. *Proceedings of the Royal Society of London Series B, Biological Sciences, 251,* 111–117.

Grant, P. R., & Grant, B. R. (2002). Adaptive radiation of Darwin's finches. *American Scientist, 90,* 130–139.

Heinrich, B. (1995). An experimental investigation of insight in common ravens (*Corvus corax*). *Auk, 112*, 994–1003.

Heinrich, B. (1999). *The mind of the raven: Investigations and adventures with wolf-birds*. New York: Harper Collins.

Henrich, J. (2016). *The secret of our success: How culture is driving human evolution, domesticating our species, and making us smarter*. Princeton, NJ: Princeton University Press.

Huneman, P., & Walsh, D. M. (2017). *Challenging the modern synthesis: Adaptation, development, and inheritance*. New York: Oxford University Press.

Hurley, C., & Montgomery, S. (2009). Peppered moths & melanism. https://darwin200.christs.cam.ac.uk /melanism-moths

Jablonka, E. (2013). Epigenetic inheritance and plasticity: The responsive germline. *Progress in Biophysics and Molecular Biology, 111*, 99–107. https://doi.org/10.1016/j.pbiomolbio.2012.08.014

Jablonka, E., & Lamb, M. J. (2014). *Evolution in four dimensions: Genetic, epigenetic, behavioral, and symbolic variation in the history of life* (rev. ed.). Cambridge, MA: MIT Press.

Jablonka, E., & Raz, G. (2009). Transgenerational epigenetic inheritance: Prevalence, mechanisms, and implications for the study of heredity and evolution. *Quarterly Review of Biology, 84*(2), 131–176. https://doi.org/10 .1086/598822

John, E. R., Chesler, P., Bartlett, F., & Victor, I. (1968). Observation learning in cats. *Science, 159*, 1489–1491.

Kauffman, S. A. (2000). *Investigations*. New York: Oxford University Press.

Kauffman, S. A. (2019). *A world beyond physics: The emergence and evolution of life*. New York: Oxford University Press.

Kettlewell, H. B. D. (1955). Selection experiments on industrial melanism in the Lepidoptera. *Heredity, 9*, 323–342.

Kettlewell, H. B. D. (1973). *The evolution of melanism: The study of a recurring necessity*. Oxford, UK: Oxford University Press.

Kingdon, J. (1993). *Self-made man: Human evolution from Eden to extinction?* New York: John Wiley.

Klein, R. G. (1999). *The human career: Human biological and cultural origins* (2nd ed.). Chicago: University of Chicago Press.

Klein, R. G., & Edgar, B. (2002). *The dawn of human culture*. New York: John Wiley & Sons.

Koestler, A. (1967). *The ghost in the machine*. New York: Macmillan.

Koonin, E. V. (2009). The Origin at 150: Is a new evolutionary synthesis in sight? *Trends in Genetics, 25*(11), 473–475. https://doi.org.10.1016/j.tig.2009.09.007

Koonin, E. V. (2011). *The logic of chance: The nature and origin of biological evolution*. Upper Saddle River, NJ: FT Press, Science.

Koonin, E. V. (2016). Viruses and mobile elements as drivers of evolutionary transitions. *Philosophical Transactions of the Royal Society of London, Series B, 371*, 20150442. https://doi.org/10.1098/rstb.2015.0442

Kozo-Polyansky, B. M. (1924). *A new principle of biology. Essay on the theory of symbiogenesis* [in Russian]. Voronezh.

Kozo-Polyansky, B. M. (1932). *Introduction to Darwinism* [in Russian]. Voronezh.

Lack, D. L. (1961/1947). *Darwin's finches*. New York: Harper & Row.

Laland, K. N. (2017). *Darwin's unfinished symphony: How culture made the human mind*. Princeton, NJ: Princeton University Press.

Laland, K. N., Odling-Smee, F. J., & Feldman, M. V. (1999). Evolutionary consequences of niche construction and their implications for ecology. *Proceedings of the National Academy of Sciences of the United States of America, 96*, 10242–10247.

Laland, K. N., Odling-Smee, F. J., Hoppitt, W., & Uller, T. (2013). More on how and why: Cause and effect in biology revisited. *Biology and Philosophy, 28*(5), 719–745.

Laland, K. N., Odling-Smee, F. J., & Myles, S. (2010). How culture shaped the human genome: Bringing genetics and the human sciences together. *Nature Reviews, Genetics, 11*, 137–148.

Laland, K. N., Sterelny, K., Odling-Smee, F. J., Hoppitt, W., & Uller, T. (2011). Cause and effect in biology revisited: Is Mayr's proximate-ultimate dichotomy still useful? *Science, 334*, 1512–1516.

Lamarck, J.-B. (1984/1809). *Zoological philosophy: An exposition with regard to the natural history of animals* (H. Elliot, Trans.). Chicago: University of Chicago Press.

Le Maho, Y. (1977). The emperor penguin: A strategy to live and breed in the cold. *American Scientist, 65*, 680–693.

Linnen, C. R., & Hoekstra, H. E. (2009). Measuring natural selection on genotypes and phenotypes in the wild. *Cold Spring Harbor Symposia on Quantitative Biology, 74*, 155–168. https://doi.org/10.1101/sqb.2009.74.045

Majerus, M. E. N. (2009). Industrial melanism in the peppered moth, *Biston betularia*: An excellent teaching example of Darwinian evolution in action. *Evolution: Education and Outreach, 2*(1), 63–74. https://doi.org/10.1007/s12052-008-0107-y

Margulis, L. (1970). *Origin of eukaryotic cells.* New Haven, CT: Yale University Press.

Margulis, L. (1993). *Symbiosis in cell evolution* (2nd ed.). New York: W. H. Freeman.

Margulis, L. (1998). *Symbiotic planet: A new look at evolution.* New York: Basic Books.

Margulis, L., & Fester, R. (Eds.). (1991). *Symbiosis as a source of evolutionary innovation: Speciation and morphogenesis.* Cambridge, MA: MIT Press.

Margulis, L., & Sagan, D. (1995). *What is life?* New York: Simon & Schuster.

Margulis, L., & Sagan, D. (2002). *Acquiring genomes: A theory of the origins of species.* New York: Basic Books.

Maynard Smith, J. (1982). The evolution of social behavior—a classification of models. In The King's College Sociobiology Group (eds.), *Current problems in sociobiology* (pp.28–44). Cambridge: Cambridge University Press.

Maynard Smith, J., & Szathmáry, E. (1995). *The major transitions in evolution.* Oxford, UK: Freeman Press.

Maynard Smith, J., & Szathmáry, E. (1999). *The origins of life: From the birth of life to the origin of language.* Oxford, UK: Oxford University Press.

Mayr, E. (1960). The emergence of evolutionary novelties. In S. Tax (ed.), *Evolution after Darwin* (Vol. 1, pp. 349–380). Chicago: University of Chicago Press.

Mayr, E. (1961). Cause and effect in biology—kinds of causes, predictability, and teleology are viewed by a practicing biologist. *Science, 134*(348), 1501–1506. https://doi.org/10.1126/science. 134.3489.1501

Mayr, E. (1963). *Animal species and evolution.* Cambridge, MA: Harvard University Press.

Mayr, E. (1974). Teleological and teleonomic: A new analysis. In R. S. Cohen & M. W. Wartofsky (Eds.). *Boston Studies in the Philosophy of Science* (Vol. 14, pp. 91–117). Boston, MA: Reidel.

Mayr, E. (1988). *Towards a new philosophy of biology.* Cambridge, MA: Harvard University Press.

McClintock, B., & Moore, J. A. (Eds.). (1987). *The discovery and characterization of transposable elements: The collected papers of Barbara McClintock.* New York: Garland Publishers.

McGrew, W. C. (1992). *Chimpanzee material culture: Implications for human evolution.* Cambridge: Cambridge University Press.

McShea, D. W. (2015). Bernd Rosslenbroich: On the origin of autonomy: A new look at the major transitions [book review]. *Biology and Philosophy, 30*(3), 439–446. https://doi.org/10.1007/s10539-0159474-2

Mereschkovsky, K. C. (1909). The theory of two plasms as the foundation of symbiogenesis, A new doctrine about the origins of organisms [in Russian]. *Proceedings of the Imperial Kazan University [USSR], 12*, 1–102.

Mereschkovsky, K. C. (1920). La plante considérée comme un complexe symbiotique. *Societé des Sciences Naturelles de l'Ouest de la France, Bulletin 6*, 17–98.

Michod, R. E. (1999). *Darwinian dynamics, evolutionary transitions in fitness and individuality.* Princeton, NJ: Princeton University Press.

Monod, J. (1971). *Chance and necessity* (A. Wainhouse, trans.). New York: Alfred A. Knopf.

Müller, G. B., & Newman, S. A. (Eds.) (2003). *Origination of organismal form: Beyond the gene in developmental and evolutionary biology.* Cambridge, MA: MIT Press.

Noble, D. (2006). *The music of life: Biology beyond the genes.* Oxford, UK: Oxford University Press.

Noble, D. (2011). Neo-Darwinism, the modern synthesis and selfish genes: Are they of use in physiology? *Journal of Physiology, 589*(5), 1007–1015. https://doi.org/10.111/jphysiol.2010.201384

Noble, D. (2012). A theory of biological relativity: No privileged level of causation. *Interface Focus, 2*, 55–64.

Noble, D. (2013). Physiology is rocking the foundations of evolutionary biology. *Experimental Physiology, 98*(8), 1235–1243. https://doi.org/10.1113/expphysiol.2012.071134

Noble, D. (2015). Evolution beyond neo-Darwinism: A new conceptual framework. *Journal of Experimental Biology, 218, pt. 1*, 7–13. https://doi.org/10.1242/jeb.106310

Noble, D. (2016). *Dance to the tune of life: Biological relativity.* Cambridge: Cambridge University Press.

Noble, D. (2018). Central dogma or central debate? *Physiology, 33*, 246–249. https://doi.org/10.1152/physiol.00017.2018

Noble, D. (2019). Exosomes, gemmules, pangenesis, and Darwin. In L. R. Edelstein, J. R. Smythies, P. J. Quesenberry, & D. Noble (Eds.). *Exosomes: A clinical compendium* (pp. 487–501). Amsterdam: Elsevier.

Noble, D., & Noble, R. (2021). Origins and demise of selfish gene theory. *Theoretical Biology Forum.*

Nowak, M. A. (2011). *Super cooperators: Altruism, evolution and why we need each other to succeed* (with R. Highfield). New York: Free Press.

Odling-Smee, F. J., Laland, K. N., & Feldman, M. W. (1996). Niche construction. *American Naturalist, 147,* 641–648.

Odling-Smee, F. J., Laland, K. N., & Feldman, M. W. (2003). *Niche construction: The neglected process in evolution.* Princeton, NJ: Princeton University Press.

Okasha, S. (2006). *Evolution and the levels of selection.* Oxford, UK: Oxford University Press.

Okasha, S. (2018). *Agents and goals in evolution.* Oxford, UK: Oxford University Press.

Palameta, B., & Lefebvre, L. K. (1985). The social transmission of a food-finding technique in pigeons: What is learned? *Animal Behaviour, 33,* 892–896.

Palmer, T. M., Stanton, M. L., Young, T. P., Goheen, J. R., Pringle, R. M., & Karban, R. (2008). Breakdown of an ant–plant mutualism follows the loss of large herbivores from an African savanna. *Science, 319*(5860), 192–195. https://doi.org/10.1126%2Fscience.1151579

Pan, D., & Zhang, L. (2009). Burst of young retrogenes and independent retrogene formation in mammals. *PloS One, 4*(3), e5040. https://doi.org/10.1371/journal.pone.0005040

Pittendrigh, C. S. (1958). Adaptation, natural selection and behavior. In A. Roe and G. G. Simpson (Eds.). *Behavior and evolution* (pp. 390–416). New Haven, CT: Yale University Press.

Richerson, P. J., & Boyd, R. (2005). *Not by genes alone: How culture transformed human evolution.* Chicago: University of Chicago Press.

Richerson, P. J., Gavrilets, S., & de Waal, F. B. M. (2021). Modern theories of human evolution foreshadowed by Darwin's *Descent of Man. Science, 372,* eaba3776 (2021). https://doi.org/10.1126/science.aba3776

Roe, A., & Simpson, G. G. (Eds.). (1958). *Behavior and evolution.* New Haven, CT: Yale University Press.

Rosen, R. (1970). *Dynamical systems theory in biology.* New York: Wiley Interscience.

Rosen, R. (1991). *Life itself: A comprehensive inquiry into the nature, origin, and fabrication of life.* New York: Columbia University Press.

Ryle, G. (1949). *The concept of mind.* New York: Hutchinson's University Library.

Sapp, J. (1994). *Evolution by association: A history of symbiosis.* New York: Oxford University Press.

Sapp, J. 2009. *The new foundations of evolution, on the tree of life.* Oxford, UK: Oxford University Press.

Shapiro, J. A. (1988). Bacteria as multicellular organisms. *Scientific American, 258,* 82–89.

Shapiro, J. A. (1991). Genomes as smart systems. *Genetica, 84,* 3–4.

Shapiro, J. A. (2011). *Evolution: A view from the 21st century.* Upper Saddle River, NJ: FT Science Press.

Shapiro, J. A. (2013). How life changes itself: The read-write (rw) genome. *Physics of Life Reviews, 10,* 287–323. https://doi.org/10.1016/j.plrev.2013.07.001

Shapiro, J. A. (2022a). Engines of innovation: Biological origins of genome evolution. *Biological Journal of the Linnean Society,* blac041. https://doi.org/10.1093/biolinnean/blac041

Shapiro, J. A. (2022b). *Evolution: A view from the 21st century. Fortified,* 2nd ed. Chicago: Cognition Press.

Shell, W. A., Steffen, M. A., Pare, H. K., Seetharam, A. S., Severin, A. J., Toth, A. L., & Rehan, S. M. (2021). Sociality sculpts similar patterns of molecular evolution in two independently evolved lineages of eusocial bees. *Communications Biology, 4,* 253. https://doi.org/10.1038/s42003-021-01770-6

Skinner, B. F. (1981). Selection by consequences. *Science, 213,* 501–504.

Simard, S. W., Beiler, K. J., Bingham, M. A., Deslippe, J. R., Philip, L. J., & Teste, F. P. (2012). Mycorrhizal networks: Mechanisms, ecology and modelling. *Fungal Biology Reviews, 26,* 39–60. https://doi.org/10.1016/j.fbr.2012.01.001

Simpson, G. G. (1953). The Baldwin effect. *Evolution, 2,* 110–117. https://doi.org/10.1111/j.1558-5646.1953.tb00069

Stoltzfus, A. (2019). Understanding bias in the introduction of variation as an evolutionary cause. In T. Uller and K. N. Laland (Eds.), *Evolutionary causation: Biological and philosophical reflections* (pp. 29–61). Cambridge, MA: MIT Press.

Traulsen, A., & Nowak, M. A. (2006). Evolution of cooperation by multilevel selection. *Proceedings of the National Academy of Sciences of the United States of America, 103,* 10952–10955.

Trewavas, A. (2014). *Plant behaviour and intelligence.* Oxford, UK: Oxford University Press.

Walsh, D. M. (2015). Organisms, agency, and evolution. Cambridge: Cambridge University Press.

Waring, T. M., & Wood, Z. T. 2021. Long-term gene–culture coevolution and the human evolutionary transition. *Proceedings of the Royal Society B, 288*, 20210538. https://doi.org/10.1098/rspb.2021.053

Warren, H. C. (2010). *Olympic: The story behind the scenery*. Wickenburg, AZ: KC Publications.

Weigl, P. D., & Hanson, E. V. (1980). Observational learning and the feeding behavior of the red squirrel *Tamiasciurus hudsonicus*: The ontogeny of optimization. *Ecology, 61*, 213–218.

Weiner, J. 1994. *The beak of the finch*. New York: Vintage Books.

West-Eberhard, M. J. (2003). *Developmental plasticity and evolution*. Oxford, UK: Oxford University Press.

West-Eberhard, M. J. (2005a). Developmental plasticity and the origin of species differences. *Proceedings of the National Academy of Sciences of the United States of America, 102*(Suppl. 1), 6543–6549. https://doi.org/10.1073/pnas.0501844102

West-Eberhard, M. J. (2005b). Phenotypic accommodation: Adaptive innovation due to phenotypic plasticity. *Journal of Experimental Zoology, 304B*, 610–618.

Whiten, A. (2021). The burgeoning reach of animal culture. *Science, 372*(6537), https://doi.org/10.1126/science.abe6514

Whiten, A. & Byrne, R. W. (1997). *Machiavellian intelligence II: Extensions and evaluations*. Cambridge: Cambridge University Press.

Wilson, D. S. (1997). Introduction: Multilevel selection theory comes of age. *American Naturalist, 150*(Suppl.), S1–S4.

Wilson, E. O. 1975. *Sociobiology: The new synthesis*. Cambridge, MA: Harvard University Press.

Wohllben, P. (2016). *The hidden life of trees: What they feel, how they communicate*. Vancouver, Canada: Greystone.

Wrangham, R. W., McGrew, W. C., de Waal, F. B. M., & Heltne, P. G. (Eds.). (1994). *Chimpanzee cultures*. Cambridge, MA: Harvard University Press.

3 Cellular Basis of Cognition and Evolution: From Protists and Fungi Up to Animals, Plants, and Root-Fungal Networks

František Baluška, William B. Miller Jr., and Arthur S. Reber

Overview

Cellular life on Earth started some four billion years ago, taking nearly two billion years to evolve from protocells to the first prokaryotic cells, including ancient archaea and bacteria that resemble modern counterparts within those cellular domains. More complex eukaryotic cells evolved via symbiotic interactions of prokaryotic cells, representing multicellular *cells within a cell* chimera-like cellular consortia. The first eukaryotic cells emerged some 2–1.5 billion years ago, which implies that it took nearly two billion years to get from prokaryotic to eukaryotic cells. Our cellular basis of consciousness (CBC) model states that all living cells utilize cellular sentience to survive and evolve. We argue that the prolonged timeline to evolve eukaryotic cells from prokaryotic cells was necessitated by the complex level of evolutionary novelty required to assemble unitary consciousness from several formerly independent prokaryotic versions of cellular consciousness, as successive orders of cognition. Prokaryotic sensitivity to anesthetics is now presented in further defense of this model. Once an initiating eukaryotic threshold of cognition was attained, eukaryotic evolution (based on its novel eukaryotic version of cellular sentience and cognition) proceeded relatively rapidly alongside an active unicellular sphere, including a huge diversity of protozoa and other protists that has thrived and evolved until our present day. Some 0.8 billion years ago, and on several occasions, colonial protists invented the multicellular forms that evolved into fungi, animals, and plants, emerging first in the sea and later also on land. Cellular cognition enabled multicellularity and permitted its successful continuous evolution toward the higher level of cohesive cellular complexities exhibited in multicellular organisms, with symbiotic fungal–plant/tree roots networks representing one of its most extensively integrated forms.

3.1 Cognitive and Evolutionary Biology: Problems Caused by Inverted Reasoning

In all sciences, the optimal strategy is to start your analysis with the most simple systems and only then to move on to more complex ones. Unfortunately, life sciences started with complex systems (humans) and only later began to investigate less complicated, interrelated systems such as unicellular eukaryotes including protists, or archaea and bacteria. This unhappy situation is causing problems and misunderstandings in the current attempts to explore the cognitive aspects of archaea, bacteria, protozoa, slime molds, and plants. Fundamental biological phenomena like sentience, learning, memory, and cognition (among others) are still reserved only for humans.

Any attempts to expand these basic and fundamental biological concepts to other organisms were, and still are, dismissed as examples of anthropomorphism. In 2009, we have characterized this situation as an "upside-down" biology (Baluška & Mancuso, 2009). In our biophysical model of cellular sentience, a crucial role is played by the excitable plasma membrane and dynamic cytoskeleton. The plasma membrane provides all cells with a sheltered space, allowing exotic biophysical phenomena based on charged ions, reactive oxygen species, and bioelectric as well as biomagnetic phenomena. Cellular sentience and cognition allow all organisms to accumulate biological order and knowledge during biological evolution (Baluška et al., 2022a, 2022b). Evolutionary development is creative not only through either mutations or natural selection but also—and mainly—through the linked cognitive activities and preferences of individual organisms (Miller et al., 2020a, 2020b; Niemann, 2021; Noble & Noble, 2021; Baluška et al., 2022a, 2022b; Corning, 2022).

3.2   Cognitive Cells as Basic Units of Life: Emergence of Eukaryotic Cells

Evolution of the first competent cells can be considered for the first niche construction event (Torday, 2016). Ever since the first cells evolved from hypothetical proto-cells, the maintenance of life has been based on cellular cognition, allowing the survival of cells and their evolution to their current levels of diversity as both unicellular and multicellular organisms. Our understanding of biological evolution suggests that life started approximately four billion years ago with ancient archaea and cyanobacteria inventing photosynthesis some 3.4 billion years ago (Fournier et al., 2021). Together with cellular aerobic respiration, these fundamental photosynthetic inventions are based on free electrons moving through chains of electron donor and acceptor complexes allowing cells to obtain enough energy to sustain life. For the first two billion years, all life evolved in the form of prokaryotic cells represented by ancient archaea and bacteria, some of which joined forces through endosymbiotic merger to generate the eukaryotic cell some 2–1.5 billion years ago (Bengtson et al., 2017; Porter, 2020). Since then, unicellular life on Earth has been based on complex interactions among archaea, bacteria, and eukaryotic cellular organisms identified collectively as protists. Consequently, the evolution of the complex eukaryotic cell (*cells within a cell* situation) is derived from simpler prokaryotic cells and it would necessarily have depended, at least in part, on their features and faculties on any path toward higher levels of complex cellular integration. Why it took almost two billion years to evolve the first eukaryotic cells out of ancient archaea and bacteria, whereas the first multicellular organisms emerged relatively soon after the emergence of very first eukaryotic cells, has remained a mystery (Lane, 2015a, 2015b; Mikhailovsky & Gordon, 2018; Gee, 2021). Pertinent to any understanding, it must be recognized that the eukaryotic cell is a multicellular assembly of "cells within a cell" where nucleus and microtubules represent parts of a more encompassing cell body (Baluška, Volkman, & Barlow, 1997, 2004a, 2004b; Baluška & Lyons, 2021).

This long evolutionary period can best be explained based on a CBC perspective (Reber, 2019; Baluška & Reber, 2019, 2021a, 2021b; Reber & Baluška, 2021). Contrary to some views (Lane, 2015a; Gee, 2021), this long time gap is not a "boring billions," but rather is a logical consequence of the complexities of the cellular sentience (Reber, 2019). It is argued that the merger of ancient sentient cells represented the solution to the demanding problem of assembling composite consciousness out of two different types of cellular sentience. Each of those merged cells was based on different cytoskeletal systems. The larger host cell was based on the actin cytoskeleton whereas the smaller guest cell relied on a microtubular cytoskeleton (Baluška, Volkman, & Barlow, 1997, 2004a). According to the CBC concept, cellular consciousness is generated via the plasma membrane populated by ion channels, receptors, sensors, and the dynamic cytoskeletal polymers that drive diverse bioelectric vectorial processes across membranes (Baluška & Mancuso, 2019; Baluška & Reber, 2019, 2021a, 2021b; Baluška, Miller, & Reber, 2021). Thus, merging and fully integrating these cells by relying on two different cytoskeletal polymers for supporting cellular sentience and motility represents a truly fundamental and singular leap in cellular evolution (Mikhailovsky & Gordon, 2018; Baluška & Lyons, 2021). Indeed, it happened only once some two billion years ago. Consequently, all eukaryotic organisms share the same predecessor and are truly monophyletic (Lane, 2015a; Lyons, 2020; Gee, 2021).

The organismal identities of the originating host and guest cells that generated the first eukaryotic cell (Baluška, Volkman, & Barlow, 2004a, 2004b) are still obscure, but current research suggests that ancient archaea represent the most relevant candidate for the host cell type (Baluška & Lyons, 2018, 2021). The recently found and characterized Asgard archaea superphylum assembles a eukaryotic-like dynamic actin cytoskeleton (Lindås et al., 2017; Akıl & Robinson, 2018; Akıl et al., 2020; Stairs & Ettema, 2020) and also generates cellular protrusions which are often long and branching (Imachi et al., 2020; Schleper & Sousa, 2020; Dance, 2021). Such protrusions might be very relevant for symbiotic interactions with any smaller partnering cell/organisms (Baluška & Lyons, 2021). Moreover, the Asgard archaea express subunits of the endosomal sorting complex required for a transport (ESCRT) complex which is closely related to the eukaryotic ESCRT complex (Zaremba-Niedzwiedzka et al., 2017); further, the Asgard *Vps4* protein complements the *vps4* null mutant of budding yeast (Lu et al., 2020). ESCRT-mediated extracellular vesicles released from archaea participate in gene transfer and nutrient cycling (Liu et al. 2021). Besides an actin cytoskeleton and ESCRT complex, Asgard archaea have other eukaryotic features such as TRAPP complexes and several coatomer proteins, including COPII complexes (Zaremba-Niedzwiedzka et al., 2017; York, 2017). All these co-existing features strongly suggest that ancient relatives of the Asgard archaea represented the putative host cell at the beginning of eukaryotic cell evolution (Baluška & Lyons, 2021). The identity of the original eukaryotic guest cell remains puzzling, although the key to an exploration of that origin might be based on contractile centrins and microtubular features.

3.3 Ancient Origins of Eukaryotic Cognition

Ever since the first eukaryotic cells capable of autonomous life were established, new types of complex organisms started their own branching evolution, achieving a perpetual place across the planet alongside archaea and bacteria. In order to survive and evolve, they deployed their newly formed eukaryotic cell sentience, which not only served well for unicellular life forms but also allowed them to embark on social multicellular life some 0.8–0.6 billion years ago (King, 2004; Sebé-Pedrós et al., 2017; Ros-Rocher et al., 2021). Unicellular protozoa and other protists are one of the most abundant and diverse life forms, occupying all the available niches of the Earth's biosphere (Sleigh, 1989; Finlay, 2002; Foissner & Hawksworth, 2009; Esteban & Fenchel, 2020). Ever since Leeuwenhoeck's landmark paper "Letter on the Protozoa" (Leeuwenhoeck, 1677) was published, our knowledge of these organisms has increased dramatically (Sleigh, 1989; Finlay & Esteban, 2001; Esteban et al., 2015; Fenchel et al., 2019; Esteban & Fenchel, 2020). Beginning with Ernst Haeckel's nineteenth-century observations, it has been apparent that unicellular protozoa use sophisticated intelligence and cognition to mount proper behavioral responses to environmental challenges (Haeckel, 1872, 1878; Romanes, 1883; Binet, 1887; Verworn, 1889; Jennings, 1906; Schloegel & Schmidgen, 2002). Haeckel introduced the term *protists*, meaning "the first organisms," and he considered them to be endowed with individuality, behavior, agency, and sentience (Haeckel, 1872, 1878; Schloegel & Schmidgen, 2002). This initial psycho-physiological analysis of protozoa was displaced with the onset of behaviorism and Loeb's mechanistic biology (Loeb, 1912).

Although behaviorism is outdated now (Searle, 2013), mechanistic thinking still dominates much of contemporary biology (Nicholson, 2013, 2014, 2019), posing barriers to the wider acceptance of the cellular basis of sentience (Reber, 2019; Baluška & Reber 2019, 2021a, 2021b; Reber & Baluška, 2021) and recognition of the cognitive behaviors of protozoa and other so-called "lower" organisms. Recently, new examples of complex protozoan behavior have been reported (Vogel & Dussutour, 2016; Tang & Marshall, 2018; Vallverdú et al., 2018; Coyle et al., 2019; Dexter et al., 2019; Marshall, 2019; Trinh et al., 2019; Wan, 2019; Coyle, 2020; Dussutour, 2021) that have decisively revived the earlier forgotten concept of protozoa as unicellular organisms with sufficient faculties to permit contingent self-directed behaviors (Haeckel, 1872, 1878; Schloegel & Schmidgen, 2002). More recently, this stance has received fresh support from the CBC concept of cellular sentience (Reber, 2019; Baluška & Reber, 2019, 2021a, 2021b; Reber & Baluška, 2021) and *accumulating proof of cellular cognition* (Baluška & Mancuso, 2009; Baluška & Levin, 2016; Baluška & Reber, 2019; Shapiro, 2021). We argue that the cellular sensitivity to anesthetics represents evidence of sentience at the cellular level, maintaining that it evolved first in unicellular organisms and was then retained in multicellular organisms. Further, we assert the protozoan basis of most cells in bodies of multicellular organism is further evidence.

3.4   Protozoan Basis of Multicellular Organisms

Before cell theory, some thinkers proposed that "little animals," as reported by Leeuwenhoeck in 1677, assemble together to form bodies of animals and humans. As discussed by Schloegel and Schmidgen (2002), Buffon, Needham, and Oken held this opinion. Moreover, Haeckel proposed that cells and protozoa possess psyche and represent individuals of the first order, capable of generating multicellular organisms (Haeckel, 1872, 1878; discussed in Schloegel & Schmidgen, 2002). Haeckel and Virchow concluded that cells are fundamental and primary organisms that build bodies of multicellular organisms as kinds of cellular states (Reynolds, 2007, 2008). This speculative proposal was never further elaborated, but new findings and discoveries indicate that it is relevant to renew this concept from the perspective of intelligent and sentient cells using cellular cognition to support their behaviors (Baluška & Mancuso, 2009; Baluška & Levin, 2016; Reber, 2019; Baluška & Reber, 2019, 2021a, 2021b; Reber & Baluška, 2021; Baluška & Lyons, 2021; Shapiro, 2021; Timsit & Grégoire, 2021).

Haeckel proposed gastrea as the first step toward animal multicellularity, a proposal that was updated by the Metchnikoff's phagocytella (Metchnikoff, 1886; Tauber, 2003). In the latter concept, flagellated cells assemble into a colony forming a proto-epithelium encircling a hollow space with internal cells (figure 3.1A). Nielsen termed this colony *choanoblastaea* (Nielsen, 2008), and Hyman proposed that these internal cells within a colony are germ cells (Hyman, 1940; Richter & King, 2013). Interestingly, colonies of the green alga *Volvox* are also formed by flagellated cells encircling internal germ cells (Kirk, 2005; Hallmann, 2011; Matt & Umen, 2016; figure 3.1B). In the next section, we briefly discuss cognition in protists as well as similarities between protists and the sexual gametes of plants and animals.

3.5   Cellular Cognition in Protozoa and Other Protists

All protozoa use their cellular sentience and intelligence to obtain sufficient nutrients, resolve stressful situations, and find suitable mating partners. As already noted, ever since Antony van Leeuwenhoek reported on them in 1677 (Leeuwenhoek, 1677; Lane, 2015b), protozoa

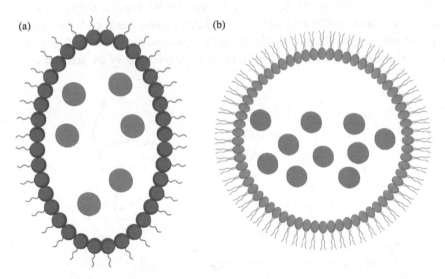

Figure 3.1
(a) Ur-epithelium formed by flagellated cells encircling internal germ cells. Internal spaces of these colonies are proposed to be populated by germ cells. This concept was proposed first by Ilya Metchnikoff as a phagocytella (Metchnikoff, 1886; see figure 1 in Tauber, 2003) and later evolved further by others (Hyman, 1940; Nielsen, 2008; Richter & King, 2013). Hypothetical urmetazoans closely resemble colonies of choanoflagellates (Nielsen, 2008; see also figure 1 in Richter & King, 2013) (b) A very similar situation is found in presently living colonies of *Volvox* algae (see also figure 1 in Kirk, 2005, and figure 2B in Matt & Umen, 2016).

have been a popular focus of scientific studies, and many scientists have concluded that their behavior must be based on primitive cellular sentience (Haeckel, 1872, 1878; Romanes, 1883; Binet, 1887; Verworn, 1889; Jennings, 1906; Schloegel & Schmidgen, 2002). There are two basic protozoa morphotypes: amoebae and ciliates; both include very intriguing examples of protozoan behavior. Among the more dramatic examples, amoebae are shape-shifters moving slowly via an actin-based cytoskeleton which drives protrusive pseudopodia used for hunting of bacteria for consumption via phagocytosis (Stockem et al., 1983; Koonce et al., 1986; Jeon, 1995; Hellstén & Roos, 1998; Tekle & Williams, 2016; Ecker & Kruse, 2021). Moving and hunting amoebae demonstrate associative conditioning as a sign of cognitive behavior (de la Fuente et al., 2019; Carrasco-Pujante et al., 2021; Schaap, 2021). Amoeboflagellates known as "vampire amoebae" locate prey algae, disintegrate a focal region of their cell wall, and suck out their protoplasts (Hess et al., 2012; Hess & Melkonian, 2013, 2014; Busch & Hess, 2016). Diverse ciliates use habituation and associative conditioning to support problem solving and guide their behavior (Jennings, 1906; Eisenstein, 1975; Applewhite, 1979; Hennessey et al., 1979; Tang & Marshall, 2018; Coyle et al., 2019; Wan, 2019; Dexter at al., 2019; Marshall, 2019; Trinh et al., 2019; Coyle, 2020; Gershman et al., 2021; Dussutour, 2021). The well-characterized avoidance and escape responses of *Stentor* serve as another excellent example of cellular cognition (figure 3.2) (Jennings, 1906; Dexter at al., 2019; Trinh et al., 2019). Another conspicuous example is the ciliate predator *Lacrymaria olor*, which shows sophisticated hunting behavior (Foissner, 1997; Coyle et al., 2019, Wan, 2019; Coyle, 2020). This hunting strategy, which is very unusual for unicellular organisms, is based on ultrafast extensions and retractions of slender necks supported by contractile cytoskeletal elements (figure 3.3). By a similar means, testate amoeba *Nebela vas* feeds on *Ironus* sp. nematodes (Yeates & Foissner, 1995) (figure 3.4). Further, testate amoeba *Difflugia tuberspinifera* actively hunt small rotifers using up to six contractile filiform pseudopods (Han et al., 2008). Testate amoeba *Cryptodifflugia operculata* apply a pack-hunting strategy to prey on much larger soil nematodes (Geisen et al., 2015). On the opposite side of the predator-prey

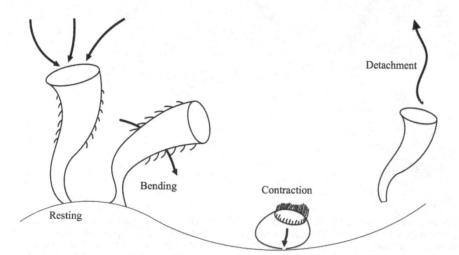

Figure 3.2
Avoidance response of ciliated protozoa *Stentor* (adapted from Jennings, 2006; see also Dexter at al., 2019; Trinh et al., 2019).

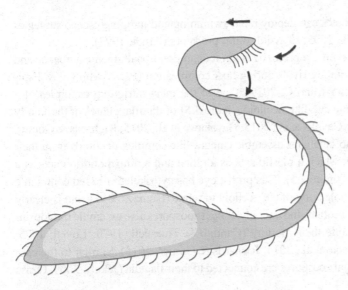

Figure 3.3
The hunting strategy of ciliate predator *Lacrymaria olor* based on ultrafast movements of their contractile necks
with the feeding apparatus (adapted from Foissner, 1997; see also Coyle et al., 2019; Wan, 2019; Coyle, 2020).

Figure 3.4
Testate amoeba *Nebela vas* feeding on *Ironus* sp. nematodes (adapted from figure 1 in Yeates & Foissner, 1995).

relations, some oligotrich ciliates can deploy rapid swimming and jumping escape strategies (Gilbert, 1994; Ueyama et al., 2005) to avoid rotifer predators (Arndt, 1993).

A further cognitive feature among protozoa is kin recognition in pathogenic amoebae and cooperation or competition among rivals during host colonization (Paz-Y-Miño-C & Espinosa, 2016; Espinosa & Paz-Y-Miño-C, 2019). One of the most intriguing examples of a protozoan sensory organ is the eye-like ocelloid (figure 3.5) of dinoflagellates of the family Warnowiaceae (Adler, 2015; Gavelis et al., 2015; Hayakawa et al., 2015; Richards & Gómes, 2015). These warnowiid dinoflagellates assemble camera-like complex ocelloids from their endosymbiotic organelles, in which a plastid acts as a retina and a mitochondrion acts as a cornea of a microbial "eye" (figure 3.5). This protist eye has cytoskeleton-based contractile appendages that allow the whole body of the ocelloid to rotate (Gómez, 2017), and is clearly a useful feature for predatory protists. Intriguingly, fungal zoospores also assemble rhodopsin-based ocelloids that help guide their motility (Cantino & Truesdell, 1970; Lovett, 1975; Saranak & Foster, 1997; Avelar et al., 2014; Richards & Gómes, 2015). Similar to the case of eyespots, ocelloids of fungal zoospores are connected to their flagella (Jékely, 2009; Ozasa et al., 2021).

The conspicuous ability of some protozoa to construct their shells using sand grains and other suitable construction materials such as algal coccoliths would be baffling if not ascribed to cognitive activity (Romanes, 1883; Verworn, 1889; Thomsen & Rasmussen, 2008; Hansell, 2011). Heron-Allen and Ford considered this protozoan ability to construct protective shells to be an expression of purpose and intelligence. Indeed, it is more. It is an explicit example of protozoan engineering which can only be accomplished by an intelligent agent (Heron-Allen, 1915; Ford, 2008; Hansell, 2011). The evolutionary success of these intelligent and cognitive amoebae is confirmed by fossils from the Lower Devonian Rhynie chert (Strullu-Derrien et al., 2019). Phylogenomics and morphological studies indicate the ancient status of

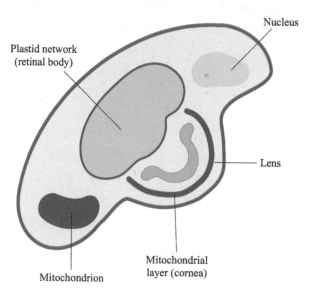

Figure 3.5
Eye-like ocelloid of warnowiid dinoflagellates based on symbiotic organelles (adapted from Richards & Gómes, 2015, and Gavelis et al., 2015).

testate amoebae as having diversified 790–730 million years ago (Lahr et al., 2019; Porter & Riedman, 2019). Moreover, this ancient protozoan cognition/sentience evolved further when protozoan cells joined forces to permit multicellular organisms such as fungi, plants, and animals. Notably, this same initiating multicellular cognitive sensibility still continues within modern multicellular organisms. We argue that immune and neural cells retain a protozoa-like nature with respect to their cognitive immune and neural functions. Further, this same pattern of integration of unicellular cognitive traits reiterates plant cognition in some aspects, best represented by the fungal-root supra-organismal symbiotic networks that will be further discussed.

3.6   Cellular Cognition in Animals and Humans

In animals including humans, two basic cell types loosely correspond to amoebae and flagellate / ciliate protozoa. Pertinently, in some tissues, these two basic cell types switch from one to the other during epithelial-mesenchymal or mesenchymal-epithelial transitions, typical within embryogenesis, development, and carcinogenesis (Lim & Thiery, 2012; Kölbl et al., 2016; Francou & Anderson, 2020; Amack, 2021; Davaine et al., 2021; Lachat et al., 2021; Ng-Blichfeldt & Röper, 2021). During epithelial-mesenchymal transitions, ciliated but non-motile epithelial cells transform into migratory amoeba-like mesenchymal cells. Amoeba-like macrophages contribute to intestinal epithelium homeostasis by inserting protrusions into epithelium affected by fungal toxins for their structural and functional repair (Chikina et al., 2020; Randolph, 2020). Similar macrophages also safeguard organ and bone homeostasis as well as repair (Sinder et al., 2015; Kang et al., 2020; Manole et al., 2021). Newly discovered nonimmune functions of macrophages also participate in the active protection of epithelia, aiding in the recognition of diverse gut microbiota and the detection of weak points in intestinal epithelial barriers requiring repair (Chiaranunt et al., 2021), and enabling cross-talk with other immune cells.

Interestingly too, choanoflagellate protozoa, the closest living relatives of animals, can accomplish flagellate-to-amoeboid switches of their morphology (Brunet et al., 2021; Velle & Fritz-Laylin, 2021). The typical choanoflagellate cellular architecture that resembles epithelial cells (figure 3.1) can undergo a fluid shift under threat to an ameboid phenotype, enabling myosin motility and thus permitting their escape from confinement. Also, interestingly, *Physarum* amoebae accomplish amoebo-flagellate transformation during their sexual cycle and starvation stress (Aldrich, 1968; Uyeda & Furuya, 1984, 1985; Ohta et al., 1991, 1992). As will be discussed below, this same phenomenon might relate to the fundamental differences between the male-female sex cells of all multicellular organisms, potentially having its origin in the emergence of the very first eukaryotic cells.

3.7   Protozoa-Like Oocytes and Sperm Cells

It is most intriguing that both plants and animals have dimorphic sexual cells with large nonmotile oocytes and smaller motile sperm cells, bearing a close resemblance to the putative cells that merged to form the first eukaryotic cell (Baluška & Lyons, 2021; Baluška, Volkman, & Barlow, 2004a; Baluška, Miller Jr., & Reber, 2022a, 2022b). The evolutionary origins of eukaryotic sex are still elusive, but, in contrast to previous views, all eukaryotes are considered to be sexual as an inherent attribute of eukaryotic life, having arisen only once (Goodenough & Heitman, 2014; Speijer, 2016; Speijer et al., 2016; Hofstatter & Lahr, 2019). Obviously then, eukaryotic sex emerged together with the first eukaryotes, which necessarily implies that is a direct outcome of the ancient archaeal merger that led to the very first eukaryotic cell (Baluška & Lyons, 2021). Ernst Haeckel reported that oocytes of the sponge *Olynthus* resemble amoebae and wandering cells of the immune system (Haeckel, 1878) (figure 3.6). Moreover, amoeba-like oocytes of *Calcera*, *Paraleucilla*, and *Geodia* sponges can feed on bacteria via phagocytosis (Sciscioli et al., 1994; Lanna & Klautau, 2010). The amoeba-like nature of oocytes in sponges is strong support for the endosymbiotic origin of the eukaryotic nucleus (Baluška & Lyons, 2021) as well as for the protozoan nature of cells in multicellular eukaryotic organisms. As we discuss later, the protozoan nature of multicellular gametes makes a very strong case for the protozoan origin of multicellular organisms. Obviously, sexual reproduction was not invented by eukaryotes, but rather is a consequence of the symbiotic origin of the very first eukaryotic cells (Baluška & Lyons, 2021).

In plants, all cells have prominent cell walls blurring their protist heritage. However, sperm cells of algae and lower plants, as well as gymnosperm trees such as cycads and *Ginkgo biloba* (Friedman, 1987), are flagellated and motile, thereby closely resembling sperm cells of animals (figure 3.7A,C) (Friedman, 1987; Southworth & Cresti, 1997; Renzaglia & Garbary, 2001; Norstog et al., 2004). During the later evolution of flowering plants, sperm cells lost their flagella and were delivered toward oocytes as passive passengers (Zhang et al., 2017), using tip-growing pollen tubes (figure 3.7A,B) that use signaling and communication for their navigation toward oocytes (Márton & Dresselhaus, 2010; Takeuchi & Higashiyama, 2011; Higashiyama & Takeuchi, 2015; Kanaoka, 2018). The further importance of the delivery of flowering plant sperm cells to oocytes via pollen tubes using neurotransmitter-based signaling between the male and female gametophytes is discussed below.

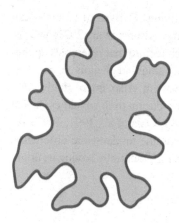

Figure 3.6
Oocytes of the sponge *Olynthus* resemble amoebae and wandering cells of the immune system (adapted from Haeckel, 1878).

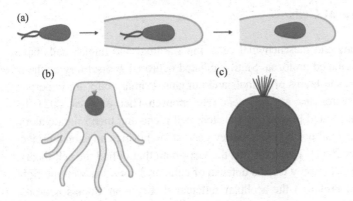

Figure 3.7
(a) Evolution of plant sperm cells from flagellated sex cells to generative cells within pollen tubes. (b) Branched pollen tube (the male gametophyte) of *Ginkgo biloba*, carrying flagellated sperm cell, and invading the female gametophyte (adapted from Friedman, 1987). (c) Flagellated sperm cells of gymnosperm cycad and ginkgo trees (adapted from Renzaglia & Garbary, 2001; Norstog et al., 2004).

3.8 Protozoa-Like Immune Cells of Animals and Humans

The innate immune system in animals and humans is based on numerous cell types, all of which closely resemble phagocytotic amoeba-like protozoa. This discovery by Ilya Metchnikoff was rather surprising and it earned him the Nobel Prize in physiology and medicine in 1908 (Tauber & Chernyak, 1991; Tauber, 2003; Gordon, 2016; Underhill et al., 2016). His discovery of mobile amoeba-like cells patrolling, fighting, and consuming foreign cells and invading pathogens as a type of cellular battle was derided as anthropomorphism. Consequently, he was considered controversial by the German scientific community, including Robert Koch and Paul Ehrlich, who advocated antibody-mediated immunity based on soluble macromolecules in blood plasma. In the end, both these originally antagonistic theories proved to be true, and Paul Ehrlich shared the Nobel Prize with Ilya Metchnikoff (Tauber, 2003; Kaufmann, 2008). The major role of immune cells is the recognition of nonself antigens from self-related antigens, placing immunity into the domain of cellular cognition (Cohen, 2000; Tauber, 2013). This knowledge has completely altered our views of the cellular basis of life and the major role of cellular cognition in the maintenance of tissue and body integrity in animals, including humans (Cohen, 2000; Lieff, 2020; Dettmer, 2021). The eureka-type discovery by Ilya Metchnikoff emanated from his zoological experiments with starfish at the Messina experimental station in 1882–1883. His early speculation about a cellular-phagocytic theory got encouragement from Rudolf Virchof, who was visiting the Messina station at the time of the discovery (Karnovsky, 1981; Chernyak & Tauber, 1988; Gordon, 2016; Vikhanski, 2016). In support of the protozoan nature of immune cells, as briefly discussed earlier in this chapter, both protozoa and immune cells are sensitive to anesthetics (Nunn, 1979; Eckenhoff, 2008; Sonner, 2008; Zhou et al., 2012; Stollings et al., 2016; Sedghi et al., 2017; Kelz & Mashour, 2019). Besides immune cells, neurons also show many analogous features, as discussed below, suggesting their protozoan origins.

3.9   Protozoa-Like Neural Cells

Besides amoeba-like immune and mesenchymal cells, most animal and human cells have primary cilia and resemble ciliated protozoa. Similar ciliated neuronal and sensory epithelia cells are found in ears, eyes, and brains of animals and humans. Primary cilia are important signaling hubs of neurons, astrocytes, and glia cells (Hasenpusch-Theil & Theil, 2021; Ki et al., 2021). Neurons are generated in the neuronal stem cell niche and then migrate along radial glial cells toward their final position where they generate dendritic and axonal projections (van den Ameele et al., 2014). The brain was the last organ that defied the cell theory. It was the genius of Santiago Ramón y Cajal's defense of cellular theory that won the fight against Camillo Golgi, who preferred the acellular reticular theory over Cajal's neuronal theory (Glickstein, 2006) which accorded special status to brain cells as neurons (Waldeyer-Hartz, 1891; Cimino, 1999; Jones, 1999). The neuron was considered to represent a very special cell which was part of the so-called neuron doctrine concept (Shepherd, 1991; Jones, 1999; Glickstein, 2006).

However, more recent studies have revealed that neurons are not exceptional as envisioned by the neuron doctrine (Bullock et al., 2005; Guillery, 2007); even plant cells share many so-called neuron-specific features (Baluška, 2010). Moreover, the evolutionary origins of neuronal cells are linked to ciliated protists, such as colonial choanoflagellates (Brunet et al., 2019; Dudin et al., 2019; Larson et al., 2020; Ros-Rocher et al., 2021) which needed the coordination of cilia (Jékely, 2011, 2021; Marinković et al., 2019; Arendt, 2021). Independent evolutionary origins and losses of neurons have been proposed (Ryan & Chiodin, 2015; Moroz & Kohn, 2016), and neurons have had their ancestry traced to ancient secretory cells of protists (Arendt, 2021; Jékely, 2021; Moroz, 2021). Recent advances in our understanding of nervous systems suggest that the enteric nervous system is evolutionarily older than the central nervous systems of brains (Spencer et al., 2021) and other organs, including heart and lungs that as additional organs maintain their own nervous systems (Kukanova & Mravec, 2006; Durães Campos et al., 2018; Kistemaker & Prakash, 2019; Fedele & Brand, 2020; Harper & Adams, 2021).

3.10   Cellular Cognition in Sexual Life of Plants

The sexual life of plants was discovered by Rudolf Jakob Camerarius in 1694 (Camerarius, 1694) and it was considered a scandal until some hundred years ago. The fight between plant sexualists and asexualists lasted more than 150 years (Taiz & Taiz, 2017) and represents a fine example of a paradigm shift in science (Kuhn, 1962). Typically, the victory of an unexpected paradigm is accepted silently by the mainstream. However, in 1988, the scientific journal *Plant Sexual Reproduction* was founded, and two years later a society devoted to its study was established (Willemse, 2008). The evolution of plant sexual cells is dramatically illustrated in the case of male gametes (figure 3.6). At first, these were freely motile using flagella (lower plants such as bryophytes and ferns); then, they were enclosed by invasive and tip-growing pollen tubes of gymnosperm trees (cycads and ginkgo); and finally, they became nonmotile within pollen tubes of all flowering plants (Southworth & Cresti, 1997; Renzaglia & Garbary, 2001; Márton & Dresselhaus, 2010; Higashiyama & Takeuchi, 2015;

Zhang et al., 2017). The navigation of pollen tubes is based on neurotransmitter-based (glutamate and GABA) cell-cell signaling and communication between the female and male (pollen tube) and female (pistil) gametophytes (Ma, 2003; Palanivelu et al., 2003; Yang, 2003; Michard et al., 2011; Wudick et al., 2018). It is significant to mention that plant-specific glutamate-receptor-like channels are also involved in sexual reproduction in mosses (Ortiz-Ramírez et al., 2017; Steinhorst & Kudla, 2017). In both instances, chemotaxis of flagellated sperm cells is guided by still unknown chemo-attractants activating moss glutamate-receptor-like channels PpGLR1 and PpGLR2.

3.11 From Cellular to Supracellular Consciousness: Extracellular Vesicles and Cell-Cell Channels

The cognitive unconscious is a concept, developed in human psychology, which refers to the cognitive processing of biologically relevant information and communication without direct access by the organismal consciousness (Reber, 1993; Reber & Allen, 2022). As every cell of multicellular organisms is gifted with its own cellular version of consciousness, the most urgent question is how cellular versions of consciousness merge to generate our unitary version of seamless multicellular consciousness. Unfortunately, our current knowledge in this respect is extremely limited, so only speculative proposals are possible. As recently discussed (Baluška & Reber, 2021a; Baluška, Miller, & Reber, 2021), ephaptic and senomic principles are relevant. Neurons in brains are coupled ephaptically (Anastassiou et al., 2011; Anastassiou & Koch, 2015; Stacey et al., 2015; Han et al., 2018) via ill-understood synchronous bio-electromagnetic fields (McFadden, 2002, 2020; Goldwyn & Rinzel, 2016; Martinez-Banaclocha, 2018, 2020; Chiang et al., 2019; Dickson, 2019; Keppler, 2021). But it is a puzzle how all neurons and other brain cells integrate into the unified consciousness we experience in our daily lives. Our recent proposal of a cellular senome concept (Baluška & Miller, 2018) and senomic cellular and supra-cellular fields is relevant in this respect (Miller et al., 2020a, 2020b). This proposal suggests that an aggregation of cellular information fields as individual sentience enables a shared attachment to information space-time, permitting the coordinate measurement of environmental cues that characterizes multicellular consciousness.

Here we further propose that extracellular vesicles released by cells self-assemble their own mini-senomic fields that move between cells and allow intercellular communication and synchronization (Cruz et al., 2018; Minakawa et al., 2021). In support of this role of extracellular vesicles, it has been shown that anesthetics inhibit the production and functional action of these vesicles (Liu et al., 2019; Buschmann et al., 2020; Wei et al., 2021). Further, there is another recently found structure that is potentially relevant for the formation of supracellular consciousness: a so-called tunneling nanotube (TNT), discovered in 2004 in cultured rat cells (Baluška, Hlavacka et al., 2004; Rustom et al., 2004). Following their initial discovery, TNTs have been found in diverse cell types and also in cells organized within intact tissues (Wang & Gerdes, 2012; Wittig et al., 2012; Abounit et al., 2016; Tardivel et al., 2016; Tóth et al., 2017; Souriant et al., 2019). TNTs are known to be induced by viruses and utilized for cell-cell transport (Jansens et al., 2020; Gánti et al., 2021). Importantly, TNTs provide neuron-astrocyte direct cytoplasmic pathways in the brain, allowing cell-cell transfer of pathological tau or prion proteins and even whole organelles, such as mitochondria (Agnati & Fuxe, 2014;

Chen & Cao, 2021; Rajasekaran & Witt, 2021; Senol et al., 2021; Wang et al., 2021). More-over, they also provide direct transport in retinal photoreceptors (Henderson & Zurzolo, 2021; Kalargyrou et al., 2021; Ortin-Martinez et al., 2021). Notably, both prokaryotic bacteria and archaea use similar TNTs for their communication (Matkó & Tóth, 2021), and plant cells are connected via analogous plant-specific plasmodesmata (Baluška, 2009; Baluška, Hlavacka et al., 2004; Baluška, Volkman, & Barlow, 2006; Brunkard & Zambryski, 2017), which can assemble between two different plants during plant parasitism (Fischer et al., 2021) Therefore, it is proposed that cell-cell channels and extracellular vesicles, representing truly archaic cel-lular structures, have perfect properties to support the advent and perpetuation of multicellular consciousness.

3.12   Fungal Networks for Supra-Organismal Cognition and Consciousness

As a further speculation, we propose that the integrated networks of plant roots and their symbiotic arbuscular mycorrhiza (AM) fungal networks represent another manifestation of multicellular consciousness as supracellular consciousness. This network integration is ancient, reaching back some 400 million years ago (Remy et al., 1994; Redecker, 2000; Field et al., 2015; Selosse et al., 2015; Hoysted et al., 2018). In intact forests that are always under environmental stresses, there are huge supra-organismal root-fungal networks, which together are characterized as the root-wide web (Simard, 2018, 2021; Sheldrake, 2020; Baluška & Mancuso, 2021). We argue that this further networked integration of individual supra-organismal root-fungal networks represents another prominent iteration of multicellular con-sciousness as supracellular supra-organismal consciousness. This bold thesis is supported by recent reports of rapid neuronal-like electric and chemical signals within these networks (Olsson & Hansson, 1995; Bais et al., 2005; Adamatzky, 2018; Schmieder et al., 2019; Volkov et al., 2019; Volkov & Shtessel, 2020). The concept of a fungal mind was recently proposed (Money, 2021) to explain their high sensitivities (Fricker et al., 2017; Nguyen et al., 2021), short-term memories, spatial recognition, and examples of cognitive fungal behavior and decision making (Aleklett & Boddy, 2021; Money, 2021). Recently, the Society for the Protection of Underground Networks (SPUN) got a $3.5 million donation from the Jeremy and Hannelore Grantham Environmental Trust to map these global and massive underground networks (https://tobykiers.com/spun). It is important that this project move briskly because these intact network sites are shrinking massively due to human impacts, especially agricul-ture and urbanization.

3.13   Conclusions and Outlook

3.13.1   Protozoan Nature of Sex Cells

Life evolved for the first two billion years in the form of unicellular organisms, remaining perpetually successful and abundant. Those unicellular organisms represent the primary living systems, not only due to their success as independent cellular domains, but especially because their features and faculties remain identifiably preserved in multicellular organisms, most particularly illustrated in the form of protozoa-like sexual cells in both plants and animals.

Consequently, current multicellular organisms represent secondary living systems insofar as all their cells are derived from protozoa-like sexual cells (gametes) after their sexual mergers.

3.13.2 Four Stages of Cellular Cognition

Biological cognition, as sentience, started with the first cells capable of autonomous life based on auto-replication and homeostasis. These most simple cellular organisms, currently represented by archaea and bacteria, are privileged with a first order of cellular cognition (FiOCC), likely based on their limiting excitable plasma membranes that support their cognitive cellular clocks based on reactive oxygen species (ROS) and excited free electrons (Baluška & Reber, 2021a). The eukaryotic cell is based on a second order of cellular cognition (SeOCC) which is supported by an initial aggregation of host and symbiotic cellular (acting as organelles) FiOCCs that are integrated via intracellular organismal synapses (Baluška & Mancuso, 2014). A third order of cellular cognition (ThOCC) is specific to eukaryotic multicellular organisms, as an aggregation and integration via extracellular cell-cell synapses, including immunological, neuronal, plant, and symbiotic synapses (Baluška, Volkman, & Menzel, 2005; Baluška & Mancuso, 2021). All three stages of cellular cognition are present in multicellular organisms, and are organized like Russian matryoshka dolls through their cellular and supra-cellular senomic fields (Baluška & Miller Jr., 2018; Miller Jr. et al., 2020a, 2020b). We propose a possible fourth order of cellular cognition (FoOCC), as a supracellular consciousness, permitting the effective integration of interlinked ThOCCs to better ensure their survival through coordinated responses to environmental stresses.

3.13.3 Sensitivity to Anesthetics as an Inherent Property of Unicellular Organisms

Similar to bacteria, protozoa and other protists are sensitive to anesthetics (Eckenhoff, 2008; Sonner, 2008; Zhou et al., 2012; Kelz & Mashour, 2019). Protozoa represent single eukaryotic cells as completely autonomous organisms. Therefore, they can be considered a primary form of eukaryotic organismal architecture. In contrast, all multicellular organisms are derived from unicellular protozoan organisms and should be therefore considered of a secondary nature. Nevertheless, all the multicellular organisms revert back to the primary protist-like stage during gametogenesis. Because both protozoa and germ cells are sensitive to anesthetics (Eckenhoff, 2008; Escher & Ford, 2020), it is proposed that this sensitivity to endogenous anesthetics evolved in ancient unicellular organisms to cope with severe stress via anesthesia-like dormancy states (Baluška et al., 2015). In fact, under severe stress all organisms, especially plants, generate and release substances with anesthetic properties (Baluška et al., 2015; Tsuchiya, 2017; Baluška & Yokawa, 2021). As anesthesia not only induces insensitivity, and detaches organisms from their environment, but also temporarily switches off higher orders of integrated cellular cognition, future studies of anesthetic effects will be useful to unravel the mysterious and elusive processes behind cellular cognition.

Acknowledgments

F. B. acknowledges support from the Stiftung Zukunft jetzt! (Munich, Germany). Figures were created by Felipe Yamashita via BioRender.com (accessed on 16 December 2021).

References

Abounit, S., J. W. Wu, K. Duff, et al. 2016. "Tunneling nanotubes: a possible highway in the spreading of tau and other prion-like proteins in neurodegenerative diseases." *Prion* 10:344–351.

Adamatzky, A. 2018. "On spiking behaviour of oyster fungi *Pleurotus djamor*." *Scientific Reports* 8:7873.

Adler, E. M. 2015. "Of ghrelin, cone cultivation, dinoflagellate eyes, and the cyanobacterial circadian clock." *Journal of General Physiology* 146:193–194.

Agnati, L. F., and K. Fuxe. 2014. "Extracellular-vesicle type of volume transmission and tunnelling-nanotube type of wiring transmission add a new dimension to brain neuro-glial networks." *Philosophical Transactions of the Royal Society London B: Biological Sciences* 369:20130505.

Akıl, C., and R. C. Robinson. 2018. "Genomes of Asgard archaea encode profilins that regulate actin." *Nature* 562:439–443.

Akıl, C., L. T. Tran, M. Orhant-Prioux, et al. 2020. "Insights into the evolution of regulated actin dynamics via characterization of primitive gelsolin/cofilin proteins from Asgard archaea." *Proceedings of the National Academy of Sciences of the U.S.A.* 117:19904–19913.

Aldrich, H. C. 1968. "The development of flagella in swarm cells of the myxomycete *Physarum flavicomum*." *Journal of General Microbiology* 50:217–222.

Aleklett, K., and L. Boddy. 2021. "Fungal behaviour: a new frontier in behavioural ecology." *Trends in Ecology and Evolution* 36:787–796.

Amack, J. D. 2021. "Cellular dynamics of EMT: lessons from live in vivo imaging of embryonic development." *Cell Communication and Signaling* 19:79.

Anastassiou, C. A., and C. Koch. 2015. "Ephaptic coupling to endogenous electric field activity: why bother?" *Current Opinion in Neurobiology* 31:95–103.

Anastassiou, C. A., R. Perin, H. Markram, and C. Koch. 2011. "Ephaptic coupling of cortical neurons." *Nature Neuroscience* 14:217–223.

Applewhite, P. B. 1979. "Learning in protozoa." *Biochemistry and Physiology of Protozoa* 1:341–355.

Arendt, D. 2021. "Elementary nervous systems." *Philosophical Transactions of the Royal Society London B: Biological Sciences* 376:20200347

Arndt, H. 1993. "Rotifers as predators on components of the microbial web (bacteria, heterotrophic flagellates, ciliates)." *Hydrobiologia* 255/256:231–246.

Avelar, G. M., R. I. Schumacher, P. A. Zaini, et al. 2014. "A rhodopsin-guanylyl cyclase gene fusion functions in visual perception in a fungus." *Current Biology* 24:1–7.

Bais, H. P., S. W. Park, T. L. Weir, et al. 2005. "How plants communicate using the underground information superhighway." *Trends in Plant Science* 9:26–32.

Baluška, F. 2009. "Cell-cell channels, viruses, and evolution: via infection, parasitism, and symbiosis toward higher levels of biological complexity." *Annals of the New York Academy of Sciences* 1178:106–119.

Baluška, F. 2010. "Recent surprising similarities between plant cells and neurons." *Plant Signaling & Behavior* 5:87–89.

Baluška, F., A. Hlavacka, D. Volkmann, and D. Menzel. 2004. "Getting connected: actin-based cell-to-cell channel in plants and animals." *Trends in Cell Biology* 14:404–408.

Baluška, F., and M. Levin. 2016. "On having no head: cognition throughout biological system." *Frontiers in Psychology* 7:902.

Baluška, F., and S. Lyons. 2018. "Energide-cell body as smallest unit of eukaryotic life." *Annals of Botany* 122:741–745.

Baluška, F., and S. Lyons. 2021. "Archaeal origins of eukaryotic cell and nucleus." *Biosystems* 203:10437.

Baluška, F., and S. Mancuso. 2009. "Deep evolutionary origins of neurobiology: turning the essence of 'neural' upside-down." *Communicative & Integrative Biology* 2:60–65.

Baluška, F., and S. Mancuso. 2014. "Synaptic view of eukaryotic cell." *International Journal of General Systems* 43:740–756.

Baluška, F., and S. Mancuso. 2019. "Actin cytoskeleton and action potentials: forgotten connections." In *The Cytoskeleton—Diverse Roles in a Plant's Life,* V. P. Sahi and F. Baluška (eds.), 63–84. Springer Nature Switzerland AG.

Baluška, F., and S. Mancuso. 2021. "Individuality, self and sociality of vascular plants." *Philosophical Transactions of the Royal Society London B: Biological Sciences* 376:20190760.

Baluška, F., and W. B. Miller Jr. 2018. "Senomic view of the cell: Senome versus genome." *Communicative & Integrative Biology* 11:1–9.

Baluška, F., W. B. Miller Jr., and A. S. Reber. 2021. "Biomolecular basis of cellular consciousness via subcellular nanobrains." *International Journal of Molecular Sciences* 22:2545.

Baluška, F., W. B. Miller Jr., and A. S. Reber. 2022a. "Cellular and evolutionary perspectives on organismal cognition: from unicellular to multicellular organisms." *Biological Journal of the Linnean Society*, blac005. https://doi.org/10.1093/biolinnean/blac005.

Baluška, F., W. B. Miller Jr., and A. S. Reber. 2022b. "Cellular sentience as the primary source of biological order and evolution." *BioSystems,* 218: 104694.

Baluška, F., and A. Reber. 2019. "Sentience and consciousness in single cells: how the first minds emerged in unicellular species." *BioEssays* 41:e1800229.

Baluška, F., and A. Reber. 2021a. "Senomic and ephaptic principles of cellular consciousness: the biomolecular basis for plant and animal sentience." *Journal of Consciousness Studies* 28:31–49.

Baluška, F., and A. Reber. 2021b. "CBC-Clock Theory of Life—Integration of cellular circadian clocks and cellular sentience is essential for cognitive basis of life." *BioEssays* 43:e2100121.

Baluška, F., D. Volkman, and P. W. Barlow. 1997. "Nuclear components with microtubule organizing properties in multicellular eukaryotes: functional and evolutionary considerations." *International Reviews of Cytology* 175:91–135.

Baluška, F., D. Volkman, and P. W. Barlow. 2004a. "Eukaryotic cells and their cell bodies: cell theory revisited." *Annals of Botany* 94:9–32.

Baluška, F., D. Volkman, and P. W. Barlow. 2004b. "Cell bodies in cage." *Nature* 428:371.

Baluška, F., D. Volkman, and P. W. Barlow. 2006. "Cell-cell channels and their implications for cell theory." In *Cell-Cell Channels,* F. Baluška et al. (eds.), 1–18. Berlin: Landes Bioscience—Springer Verlag.

Baluška, F., D. Volkman, and D. Menzel. 2005. "Plant synapses: actin-based domains for cell-to-cell communication." *Trends in Plant Science* 10:106–111.

Baluška, F., and K. Yokawa. 2021. "Anaesthetics and plants: from sensory systems to cognition-based adaptive behaviour." *Protoplasma* 258:449–454.

Baluška, F., K. Yokawa, S. Mancuso, and K. Baverstock. 2015. "Understanding of anesthesia. Why consciousness is essential for life and not based on genes." *Communicative and Integrative Biology* 9:e1238118.

Bengtson, S., T. Sallstedt, V. Belivanova, and M. Whitehouse. 2017. "Three-dimensional preservation of cellular and subcellular structures suggests 1.6 billion-year-old crown-group red algae." *PLoS Biology* 15:e2000735.

Binet, A. 1887. *The Psychic Life of Micro-Organisms.* London: Paul Kegan, Trench.

Brunet, T., M. Albert, W. Roman, et al. 2021. "A flagellate-to-amoeboid switch in the closest living relatives of animals." *eLife* 10:e61037.

Brunet, T., B. T. Larson, T. A. Linden, et al. 2019. "Light-regulated collective contractility in a multicellular choanoflagellate." *Science* 366:326–334.

Brunkard, J. O., and P. C. Zambryski. 2017. "Plasmodesmata enable multicellularity: new insights into their evolution, biogenesis, and functions in development and immunity." *Current Opinion in Plant Biology* 35:76–83.

Bullock, T. H., M. V. L. Bennett, D. Johnston, et al. 2005. "The neuron doctrine, redux." *Science* 310:791–793.

Busch, A., and S. Hess. 2016. "The cytoskeleton architecture of algivorous protoplast feeders (Viridiraptoridae, Rhizaria) indicates actin-guided perforation of prey cell wall." *Protist* 168:12–31.

Buschmann, D., F. Brandes, A. Lindemann, et al. 2020. "Propofol and sevoflurane differentially impact microRNAs in circulating extracellular vesicles during colorectal cancer resection." *Anesthesiology* 132:107–120.

Camerarius, R. J. 1694. *De Sexu Plantarum Epistola.* Tübingen: Martin Rommey.

Cantino, E. C., and L. C. Truesdell. 1970. "Organization and fine structure of the side body and its lipid sac in the zoospore of *Blastocladiella emersonii.*" *Mycologia* 62:548–567.

Carrasco-Pujante, J., C. Bringas, I. Malaina, et al. 2021. "Associative conditioning is a robust systemic behavior in unicellular organisms: an interspecies comparison." *Frontiers in Microbiology* 12:707086.

Chen, J., and J. Cao. 2021. "Astrocyte-to-neuron transportation of enhanced green fluorescent protein in cerebral cortex requires F-actin dependent tunneling nanotubes." *Scientific Reports* 11:16798.

Chernyak, L., and A. L. Tauber. 1988. "The birth of immunology: Metchnikoff the embryologist." *Cellular Immunology* 117:218–233.

Chiang, C.-C., R. S. Shivacharan, X. Wei, et al. 2019. "Slow periodic activity in the longitudinal hippocampal slice can self-propagate non-synaptically by a mechanism consistent with ephaptic coupling." *Journal of Physiology* 597:249–269.

Chiaranunt, P., S. L. Tai, L. Ngai, and A. Mortha. 2021. "Beyond immunity: underappreciated functions of intestinal macrophages." *Frontiers of Immunology* 12:49708.

Chikina, A. S., F. Nadalin, M. Maurin, and M. San-Roman. 2020. "Macrophages maintain epithelium integrity by limiting fungal product absorption." *Cell* 183:411–428.

Cimino, G. 1999. "Reticular theory versus neuron theory in the work of Camillo Golgi." *International Journal of History of Science* 36:431–472.

Cohen, I. R. 2000. *Tending Adam's Garden. Evolving the Cognitive Immune Self.* London: Elsevier Academic Press.

Corning, P. A. 2023. "Teleonomy in evolution." In *Evolution 'On Purpose': Teleonomy in Living Systems* (Vienna Series in Theoretical Biology), Corning, P., et al. (eds.). Cambridge, MA: MIT Press.

Coyle, S. M. 2020. "Ciliate behavior: blueprints for dynamic cell biology and microscale robotics." *Molecular Biology of the Cell* 31:2415–2420.

Coyle, S. M., E. M. Flaum, H. Li, et al. 2019. "Coupled active systems encode an emergent hunting behavior in the unicellular predator *Lacrymaria olor*." *Current Biology* 29:3838–3850.

Cruz, L., J. A. A. Romero, R. P. Iglesia, and M. H. Lopes. 2018. "Extracellular vesicles: decoding a new language for cellular communication in early embryonic development." *Frontiers in Cell and Developmental Biology* 6:94.

Dance, A. 2021. "The mysterious microbes that gave rise to complex life." *Nature* 593:328–330.

Davaine, C., E. Hadadi, W. Taylor, et al. 2021. "Inducing sequential cycles of epithelial-mesenchymal and mesenchymal-epithelial transitions in mammary epithelial cells." In *The Epithelial-to-Mesenchymal Transition: Methods and Protocols (Methods in Molecular Biology 2179)*, K. Campbell and E. Theveneau (eds.), 341–351.

de la Fuente, I. M., C. Bringas, I. Malaina, et al. 2019. "Evidence of conditioned behavior in amoebae." *Nature Communications* 10:3690.

Dettmer, P. 2021. *Immune—A Journey into the Mysterious System That Keeps You Alive.* New York: Random House.

Dexter, J. P., S. Prabakaran, and J. Gunawardena. 2019. "A complex hierarchy of avoidance behaviors in a single-cell eukaryote." *Current Biology* 29:4323–4329.

Dickson, C. T. 2019. "A jolt to the field: a self-generating and self-propagating, ephaptically mediated slow spontaneous network activity pattern in the hippocampus." *Journal of Physiology* 597:3.

Dudin, O., A. Ondracka, X. Grau-Bove, et al. 2019. "A unicellular relative of animals generates an epithelium-like cell layer by actomyosin-dependent cellularization." *eLife* 8:e49801.

Durães Campos, I., V. Pinto, N. Sousa, and V. H. Pereira. 2018. "A brain within the heart: a review on the intracardiac nervous system." *Journal of Molecular and Cellular Cardiology* 119:1–9.

Dussutour, A. 2021. "Learning in single cell organisms." *Biochemical and Biophysical Research Communications* 564:92–102.

Eckenhoff, R. G. 2008. "Why can all of biology be anesthetized?" *Anesthesia & Analgesia* 107:859–861.

Ecker, N., and K. Kruse. 2021. "Excitable actin dynamics and amoeboid cell migration." *PLoS ONE* 16:e0246311.

Eisenstein, E. M. 1975. *Aneural Organisms in Neurobiology.* New York: Plenum Press.

Escher, J., and L. D. Ford. 2020. "General anesthesia, germ cells and the missing heritability of autism: an urgent need for research." *Environmental Epigenetics* 6:dvaa007.

Espinosa, A., and G. Paz-Y-Miño-C. 2019. "Discrimination experiments in Entamoeba and evidence from other protists suggest pathogenic amebas cooperate with kin to colonize hosts and deter rivals." *Journal of Eukaryotic Microbiology* 66:354–368.

Esteban, G. F., and T. Fenchel. 2020. *Ecology of Protozoa. The Biology of Free-Living Phagotropic Protists.* Basel, Switzerland: Springer Nature Switzerland.

Esteban, G. F., B. J. Finlay, and A. Warren. 2015. "Free-living protozoa." In: *Ecology and General Biology. Thorp and Covich's Freshwater Invertebrates* (4th ed.), 113–132.

Fedele, L., and T. Brand. 2020. "The intrinsic cardiac nervous system and its role in cardiac pacemaking and conduction." *Journal of Cardiovascular Development and Disease* 7:54.

Fenchel, T., B. J. Finlay, and G. F. Esteban. 2019. "Cosmopolitan metapopulations?" *Protist* 170:314–318.

Field, K. J., S. Pressel, J. G. Duckett, et al. 2015. "Symbiotic options for the conquest of land." *Trends in Ecology and Evolution* 30:477–486.

Finlay, B. J. 2002. "Global dispersal of free-living microbial eukaryote species." *Science* 296:1061–1063.

Finlay, B. J., and G. F. Esteban. 2001. "Exploring Leeuwenhoek's legacy: the abundance and diversity of protozoa." *International Microbiology* 4:125–133.

Fischer, K., L. A.-M. Lachner, S. Olsen, et al. 2021. "The enigma of interspecific plasmodesmata: insight from parasitic plants." *Frontiers in Plant Science* 12:641924.

Foissner, W. 1997. "Faunistic and taxonomic studies on ciliates (Protozoa, Ciliophora) from clean rivers in Bavaria (Germany) with descriptions of new species and ecological notes." *Limnologica* 27:179–238.

Foissner, W., and D. L. Hawksworth. 2009. *Protist Diversity and Geographical Distribution.* Springer Verlag.

Ford, B. J. 2008. "Microscopical substantiation of intelligence in living cells." *Royal Microscopical Society* 12:6–20.

Fournier, G. P., K. R. Moore, L. T. Rangel, J. G. Payette, L. Momper, and T. Bosak. 2021. "The Archean origin of oxygenic photosynthesis and extant cyanobacterial lineages." *Proceedings of Royal Society B Biological Sciences* 288: 20210675.

Francou, A., and K. V. Anderson. 2020. "The epithelial-to-mesenchymal transition (EMT) in development and cancer." *Annual Reviews of Cancer Biology* 4:197–220.

Fricker, M. D., L. L. M. Heaton, N. S. Jones, and L. Boddy. 2017. "The mycelium as a network." *Microbiology Spectrum* 5:FUNK-0033–2017.

Friedman, W. E. 1987. "Growth and development of the male gametophyte of *Ginkgo biloba* within the ovule (in vivo)." *American Journal of Botany* 74:1797–1815.

Ganti, K., J. Han, B. Manicassamy, and A. C. Lowen. 2021. "Rab11a mediates cell-cell spread and reassortment of influenza A virus genomes via tunneling nanotubes." *PLoS Pathogens* 17:e1009321.

Gavelis, G. S., S. Hayakawa, R. A. White III, et al. 2015. "Eye-like ocelloids are built from different endosymbiotically acquired components." *Nature* 523:204–207.

Gee, H. 2021. *A (Very) Short History of Life on Earth. 4.6 Billions of Years in 12 Chapters.* London: Picador.

Geisen, S., J. Rosengarten, R. Koller, et al. 2015. "Pack hunting by a common soil amoeba on nematodes." *Environmental Microbiology* 17:4538–4546.

Gershman, S. J., P. E. Balbi, C. R. Gallistel, and J. Gunawardena. 2021. "Reconsidering the evidence for learning in single cells." *eLife* 10:e61907.

Gilbert, J. J. 1994. "Jumping behavior in the oligotrich ciliates *Strobilidium velox* and *Halteria grandinella*, and its significance as a defense against rotifer predators." *Microbial Ecology* 27:189–200.

Glickstein, M. 2006. "Golgi and Cajal: The neuron doctrine and the 100th anniversary of the 1906 Nobel Prize." *Current Biology* 16:R147–R151.

Goldwyn, J. H., and J. Rinzel. 2016. "Neuronal coupling by endogenous electric fields: cable theory and applications to coincidence detector neurons in the auditory brain stem." *Journal of Neurophysiology* 115: 2033–2051.

Gómez, F. 2017. "The function of the ocelloid and piston in the dinoflagellate *Erythropsidinium* (Gymnodiniales, Dinophyceae)." *Journal of Phycology* 53:629–641.

Goodenough, U., and J. Heitman. 2014. "Origins of eukaryotic sexual reproduction." *Cold Spring Harbor Perspectives in Biology* 6:a016154.

Gordon, S. 2016. "Phagocytosis: the legacy of Metchnikoff." *Cell* 166:1065–1068.

Guillery, R. W. 2007. "Relating the neuron doctrine to the cell theory. Should contemporary knowledge change our view of the neuron doctrine?" *Brain Research Reviews* 55:411–421.

Haeckel, E. 1872. *Monographie der Kalkschwämme.* Berlin: Georg Reimer Verloag.

Haeckel, E. 1878. *Zellseelen und Seelenzellen.* Leipzig: Alfred Kröner Verlag

Hallmann, A. 2011. "Evolution of reproductive development in the volvocine algae." *Sexual Plant Reproduction* 24:97–112.

Han, B.-P., T. Wang, Q.-Q. Lin, and H. J. Dumont. 2008. "Carnivory and active hunting by the planktonic testate amoeba *Difflugia tuberspinifera*." *Hydrobiologica* 596:197–201.

Han, K. S., C. Guo, L. Witter, et al. 2018. "Ephaptic coupling promotes synchronous firing of cerebellar Purkinje cells." *Neuron* 100:564–578.

Hansell, M. 2011. "Houses made by protists." *Current Biology* 21:R485-R487.

Harper, A. A., and D. J. Adams. 2021. "Electrical properties and synaptic transmission in mouse intracardiac ganglion neurons *in situ*." *Physiological Reports* 9:e15056.

Hasenpusch-Theil, K., and T. Theil. 2021. "The multifaceted roles of primary cilia in the development of the cerebral cortex." *Frontiers in Cell and Developmental Biology* 9:630161.

Hayakawa, S., Y. Takaku, J. S. Hwang, et al. 2015. "Function and evolutionary origin of unicellular camera-type eye structure." *PLoS ONE* 10:20118415.

Hellstén, M., and U.-P. Roos. 1998. "The actomyosin cytoskeleton of amoebae of the cellular slime molds *Acrasis rosea* and *Protostelium mycophaga*: structure, biochemical properties, and function." *Fungal Genetics and Biology* 24:123–145.

Henderson, J. M., and C. Zurzolo. 2021. "Seeing eye to eye: photoreceptors employ nanotube-like connections for material transfer." *EMBO Journal* 22:e109727.

Hennessey, T. M., W. B. Rucker, and C. G. McDiarmid. 1979. "Classical conditioning in Paramecia." *Animal Learning & Behavior* 7:417–423.

Heron-Allen, E. 1915. "A short statement upon the theory, and the phenomena of purpose and intelligence exhibited by the Protozoa as illustrated by selection and behaviour in the Foraminifera." *Journal of Royal Microscopy Society* 35:547–557.

Hess, S., and M. Melkonian. 2013. "The mystery of clade X: Orciraptor gen. nov. and Viridiraptor gen. nov. are highly specialised algivorous amoeboflagellates (Glissomonadida, Cercozoa)." *Protist* 164:706–747.

Hess, S., and M. Melkonian. 2014. "Ultrastructure of the algivorous amoeboflagellate Viridiraptor invadens (Glissomonadida, Cercozoa)." *Protist* 165:605–635.

Hess, S., N. Sausen, and M. Melkonian. 2012. "Shedding light on vampires: the phylogeny of vampyrellid amoebae revisited." *PloS ONE* 7:e31165.

Higashiyama, T., and H. Takeuchi. 2015. "The mechanism and key molecules involved in pollen tube guidance." *Annual Reviews of Plant Biology* 66:393–413.

Hofstatter, P. G., and D. J. G. Lahr. 2019. "All eukaryotes are sexual, unless proven otherwise. Many so-called asexuals present meiotic machinery and might be able to have sex." *BioEssays* 41:1800246.

Hoysted, G. A., J. Kowal, A. Jacob, et al. 2018. "A mycorrhizal revolution." *Current Opinion in Plant Biology* 44:1–6.

Hyman, L. H. 1940. *The Invertebrate: Protozoa through Ctenophora*. New York: McGraw-Hill.

Imachi, H., M. K. Nobu, N. Nakahara, et al. 2019. "Isolation of an archaeon at the prokaryote—eukaryote interface." *Nature* 577:519–525.

Jansens, R. J. J. J., A. Tishchenko, and H. W. Favoreel. 2020. "Bridging the gap: virus long-distance spread via tunneling nanotubes." *Journal of Virology* 94:e02120–19.

Jékely, G. 2009. "Evolution of phototaxis." *Philosophical Transactions of the Royal Society London B: Biological Sciences* 364:2795–2808.

Jékely, G. 2011. "Origin and early evolution of neural circuits for the control of ciliary locomotion." *Proceedings of Royal Society London B: Biological Sciences* 278:914–922.

Jékely, G. 2021. "The chemical brain hypothesis for the origin of nervous systems." *Philosophical Transactions of the Royal Society London B: Biological Sciences* 376:20190761.

Jennings, H. S. 1906. *Behavior of the Lower Organisms*. New York: Columbia University Press.

Jeon, K. W. 1995. "The large, free-living amoebae: wonderful cells for biological studies." *Journal of Eukaryotic Microbiology* 42:1–7.

Jones, E. G. 1999. "Golgi, Cajal and the Neuron Doctrine." *Journal of the History of the Neurosciences* 8:170–178.

Kalargyrou, A. A., M. Basche, A. Hare, et al. 2021. "Nanotube-like processes facilitate material transfer between photoreceptors." *EMBO Reports* 22:e53732.

Kanaoka, M. M. 2018. "Cell-cell communications and molecular mechanisms in plant sexual reproduction." *Journal of Plant Research* 131:37–47.

Kang, M., C.-C. Huang, Y. Lu, et al. 2020. "Bone regeneration is mediated by macrophage extracellular vesicles." *Bone* 141:115627

Karnovsky, M. L. 1981. "Metchnikoff in Messina: a century of studies on phagocytosis." *New England Journal of Medicine* 304:1178–1180.

Kaufmann, S. H. E. 2008. "Immunology's foundation: the 100-year anniversary of the Nobel Prize to Paul Ehrlich and Elie Metchnikoff." *Nature Immunology* 9:705–712.

Kelz, M. B., and G. A. Mashour. 2019. "The biology of general anesthesia from Paramecium to Primate." *Current Biology* 29:R1199–R1210.

Keppler, J. 2021. "Building blocks for the development of a self-consistent electromagnetic field theory of consciousness." *Frontiers in Human Neuroscience* 15:723415.

Ki, S. M., H. S. Jeong, and J. E. Lee. 2021. "Primary cilia in glial cells: an oasis in the journey to overcoming neurodegenerative diseases." *Frontiers in Neuroscience* 15:736888.

King, N. 2004. "The unicellular ancestry of animal development." *Developmental Cell* 7:313–325.

Kirk, D. L. 2005. A twelve-step program for evolving multicellularity and a division of labor. *BioEssays* 27:299–310.

Kistemaker, L. E. M., and Y. S. Prakash. 2019. "Airway innervation and plasticity in asthma." *Physiology* 34:283–298.

Koonce, M. P., U. Euteneuer, K. L. McDonald, et al. 1986. "Cytoskeletal architecture and motility in a giant freshwater amoeba, Reticulomyxa." *Cellular Motility and the Cytoskeleton* 6:521–533.

Kölbl, A. C., U. Jeschke, and U. Andergassen. 2016. "The significance of epithelial-to-mesenchymal transition for circulating tumor cells." *International Journal of Molecular Sciences* 17:1308.

Kuhn, T. S. 1962. *The Structure of Scientific Revolutions.* Chicago: University of Chicago Press.

Kukanova, B., and B. Mravec. 2006. "Complex intracardiac nervous system." *Bratislavske Lekarske Listy* 107:45–51.

Lachat, C., P. Peixoto, and E. Hervouet. 2021. "Epithelial to mesenchymal transition history: from embryonic development to cancers." *Biomolecules* 11:782.

Lahr, D. J. G., A. Kosakyan, E. Lara, E. A. D. Mitchell, L. Morais, A. L. Porfirio-Sousa, G. M. Ribeiro, A. K. Tice, T. Pánek, S. Kang, and M. W. Brown. 2019. "Phylogenomics and morphological reconstruction of Arcellinida testate amoebae highlight diversity of microbial eukaryotes in the neoproterozoic." *Current Biology* 29:991–1001.

Lane, N. 2015a. *The Vital Question: Energy, Evolution and the Origins of Complex Life.* New York: W. W. Norton.

Lane, N. 2015b. "The unseen world: reflections on Leeuwenhoek (1677) 'Concerning little animals'." *Philosophical Transactions of the Royal Society London* 370:20140344.

Lanna, E., and M. Klautau. 2010. "Oogenesis and spermatogenesis in *Paraleucilla magna* (Porifera, Calcarea)." *Zoomorphology* 129:249–261.

Larson, B. T., T. Ruiz-Herrero, S. Lee, et al. 2020. "Biophysical principles of choanoflagellate self-organization." *Proceedings of the National Academy of Sciences of the U.S.A.* 117:1303–1311.

Leeuwenhoeck, A. 1677. "Observations communicated to the publisher by Mr. Antony van Leewenhoeck concerning little animals by him observed in rain-well-sea and snow water, as also in water wherein pepper had lain infused." *Philosophical Transactions of the Royal Society London* 12:821–831.

Lieff, J. 2020. *The Secret Language of Cells: What Biological Conversations Tell Us About the Brain-Body Connection, the Future of Medicine, and Life Itself.* Dallas, TX: BenBella Books.

Lim, J., and J. P. Thiery. 2012. "Epithelial-mesenchymal transitions: insights from development." *Development* 139:3471–3486.

Lindås, A. C., K. Valegård, and T. J. G. Ettema. 2017. "Archaeal actin-family filament systems." *Subcellular Biochemistry* 84:379–392.

Liu, J., V. Cvirkaite-Krupovic, P. H. Commere, et al. 2021. "Archaeal extracellular vesicles are produced in an ESCRT-dependent manner and promote gene transfer and nutrient cycling in extreme environments." *ISME Journal* 15:2892–2905.

Liu, J., Y. Li, X. Xia, et al. 2019. "Propofol reduces microglia activation and neurotoxicity through inhibition of extracellular vesicle release." *Journal of Neuroimmunology* 333:476962.

Loeb, J. 1912. *The Mechanistic Conception of Life.* Chicago: University of Chicago Press.

Lovett, J. S. 1975. "Growth and differentiation of the water mold *Blastocladiella emersonii:* cytodifferentiation and the role of ribonucleic acid and protein synthesis." *Bacteriology Reviews* 39:345–404.

Lu, Z., T. Fu, Y. Liu, et al. 2020. "Coevolution of eukaryote-like Vps4 and ESCRT-III subunits in the Asgard archaea." *mBio* 11:00417–20.

Lyons, S. 2020. *From Cells to Organisms: Re-Envisioning Cell Theory.* Toronto, Canada: University of Toronto Press.

Ma, H. 2003. "Plant reproduction: GABA gradient, guidance and growth." *Current Biology* 13:R834-R836.

Manole, E., C. Niculite, J. M. Lambrescu, et al. 2021. "Macrophages and stem cells: two to tango for tissue repair?" *Biomolecules* 11:697.

Marinković, M., J. Berger, and G. Jékely. 2019. "Neuronal coordination of motile cilia in locomotion and feeding." *Proceedings of Royal Society B: Biological Sciences* 375:20190165.

Marshall, W. F. 2019. "Cellular cognition: sequential logic in a giant protist." *Current Biology* 29:R1303–R1305.

Martinez-Banaclocha, M. 2018. "Ephaptic coupling of cortical neurons: possible contribution of astroglial magnetic fields?" *Neuroscience* 370:37–45.

Martinez-Banaclocha, M. 2020. "Astroglial isopotentiality and calcium-associated biomagnetic field effects on cortical neuronal coupling." *Cells* 9:439.

Márton, M. L., and T. Dresselhaus. 2010. "Female gametophyte-controlled pollen tube guidance." *Biochemical Society Transactions* 38(2): 627–630.

Matkó, J., and E. A. Tóth. 2021. "Membrane nanotubes are ancient machinery for cell-to-cell communication and transport. Their interference with the immune system." *Biologia Futura* 72:25–36.

Matt, G., and J. Umen. 2016. "Volvox: a simple algal model for embryogenesis, morphogenesis and cellular differentiation." *Developmental Biology* 419:99–113.

McFadden, J. 2002. "Synchronous firing and its influence on the brain's electromagnetic field: evidence for an electromagnetic field theory of consciousness." *Journal of Consciousness Studies* 9:23–50.

McFadden, J. 2020. "Integrating information in the brain's EM field: the cemi field theory of consciousness." *Neuroscience of Consciousness* 2020:niaa016.

Metschnikoff, I. 1886. *Embryologische Studien an Medusen*. Vienna: A. Hölder.

Michard, E., P. T. Lima, F. Borges, et al. 2011. "Glutamate receptor-like genes form Ca^{2+} channels in pollen tubes and are regulated by pistil D-serine." *Science* 332:434–437.

Mikhailovsky, G., and R. Gordon. 2018. "Symbiosis: why was the transition from microbial prokaryotes to eukaryotic organisms a cosmic gigayear event?" In *Habitability of the Universe Before Earth. Astrobiology Exploring Life on Earth and Beyond*, R. Gordon and A. A. Sharov (eds.), 355–405. Cambridge, MA: Academic Press.

Miller, W. B. Jr., F. Baluška, and J. S. Torday. 2020a. "Cellular senomic measurements in cognition-based evolution." *Progress in Biophysics and Molecular Biology* 1562:20–33.

Miller, W. B. Jr., F. Baluška, and J. S. Torday. 2020b. "The N-space Episenome unifies cellular information space-time within cognition-based evolution." *Progress in Biophysics and Molecular Biology* 150:112–139.

Minakawa, T., T. Matoba, F. Ishidate, et al. 2021. "Extracellular vesicles synchronize cellular phenotypes of differentiating cells." *Journal of Extracellular Vesicles* 10:e12147.

Money, N. P. 2021. "Hyphal and mycelial consciousness: the concept of the fungal mind." *Fungal Biology* 125:257–259.

Moroz, L. L. 2021. "Multiple origins of neurons from secretory cells." *Frontiers in Cell and Developmental Biology* 9:669087.

Moroz, L. L., and A. B. Kohn. 2016. "Independent origins of neurons and synapses: insights from ctenophores." *Philosophical Transactions of the Royal Society London B: Biological Sciences* 371:20150041.

Ng-Blichfeldt, J.-P., and K. Röper. 2021. "Mesenchymal-to-epithelial transitions in development and cancer." In *The Epithelial-to Mesenchymal Transition: Methods and Protocols: Methods in Molecular Biology* 2179, K. Campbell and E. Theveneau (eds.), 43–62. Totowa, NJ: Humana Press.

Nguyen, T. A., S. Le, M. Lee, et al. 2021. "Fungal wound healing through instantaneous protoplasmic gelation." *Current Biology* 31:271–282.

Nicholson, D. J. 2013. "Organisms ≠ Machines." *Studies in History and Philosophy of Science Part C: Studies in History and Philosophy of Biological and Biomedical Sciences* 44:669–678.

Nicholson, D. J. 2014. "The machine conception of the organism in development and evolution: a critical analysis." *Studies in History and Philosophy of Science Part C: Studies in History and Philosophy of Biological and Biomedical Sciences* 48:162–174.

Nicholson, D. J. 2019. "Is the cell *really* a machine?" *Journal of Theoretical Biology* 477:108–126.

Nielsen, C. 2008. "Six major steps in animal evolution: are we derived sponge larvae?" *Evolution & Development* 10:241–257.

Niemann, H.-J. 2021. "Popper, Darwin, and evolution." In *Karl Popper's Science and Philosophy*, Z. Parusniková and D. Merritt (eds.), 231–256. Springer Nature, Switzerland AG.

Noble, D., and Noble, R. 2021. "Rehabilitation of Karl Popper's ideas on evolutionary biology and the nature of biological science." In *Karl Popper's Science and Philosophy*, Z. Parusniková and D. Merritt (eds.), 193–210. Springer Nature, Switzerland AG.

Norstog, K. J., E. M. Gifford, and D. W. Stevenson. 2004. "Comparative development of the spermatozoids of Cycads and *Gingko biloba*." *Botanical Reviews* 70:5–15.

Nunn, J. F. 1979. "Anesthesia and the leucocyte." *Acta Anaesthesiologica Belgica* 30:23–28.

Ohta, T., S. Kawano, and T. Kuroiwa. 1991. "Migration of the cell nucleus during the amoebo-flagellate transformation of *Physarum polycephalum* is mediated by an actin-generated force that acts on the centrosome." *Protoplasma* 163:114–124.

Ohta, T., S. Kawano, and T. Kuroiwa. 1992. "Fine structure of centrosome complex and its connection with cell nucleus in the slime mould, *Physarum polycephalum*." *Cytologia* 57:515–523.

Olsson, S., and B. S. Hansson. 1995. "Action potential-like activity found in fungal mycelia is sensitive to stimulation." *Naturwissenschaften* 82:30–31.

Ortin-Martinez, A., N. E. Yan, E. L. S. Tsai, et al. 2021. "Photoreceptor nanotubes mediate the *in vivo* exchange of intracellular material." *EMBO Journal* 40:e107264.

Ortiz-Ramírez, C., E. Michard, A. A. Simon, et al. 2017. "Glutamate receptor-like channels are essential for chemotaxis and reproduction in mosses." *Nature* 549:91–95.

Ozasa, K., H. Kang, S. Song, et al. 2021. "Regeneration of the eyespot and flagellum in *Euglena gracilis* during cell division." *Plants* 10:2004.

Palanivelu, R., L. Brass, A. F. Edlund, and D. Preuss. 2003. "Pollen tube growth and guidance is regulated by POP2, an Arabidopsis gene that controls GABA levels." *Cell* 114:47–59.

Paz-Y-Miño-C, G., and A. Espinosa. 2016. "Kin discrimination in Protists: from many cells to single cells and backwards." *Journal of Eukaryotic Microbiology* 63:367–377.

Porter, S. M. 2020. "Insights into eukaryogenesis from the fossil record." *Interface Focus* 10:20190105.

Porter, S. M., and L. A. Riedman. 2019. "Ancient fossilized amoebae find their home in the tree." *Current Biology* 29:R200–R223.

Rajasekaran, S., and S. N. Witt. 2021. "Trojan horses and tunneling nanotubes enable α-synuclein pathology to spread in Parkinson disease." *PLoS Biology* 19:e3001331.

Randolph, G. J. 2020. "Colonic macrophages combat fungal intoxication: Metchnikoff would be pleased." *Cell* 183:305–307.

Reber, A. S. 1993. *Implicit Learning and Tacit Knowledge: An Essay on the Cognitive Unconscious*. New York: Oxford University Press.

Reber, A. S. 2019. *The First Minds: Caterpillars, Karyotes, and Consciousness*. New York: Oxford University Press.

Reber, A. S., and R. Allen. (eds.). 2022. *The Cognitive Unconscious: The First Half-Century*. Oxford, UK: Oxford University Press. (In press).

Reber, A. S., and F. Baluška. 2021. "Cognition in some surprising places." *Biochemical and Biophysical Research Communications* 564:150–157.

Redecker, D. 2000. "Glomalean fungi from the Ordovician." *Science* 289:1920–1921.

Remy, W., T. N. Taylor, H. Hass, and H. Kerp. 1994. "Four hundred-million-year-old vesicular arbuscular mycorrhizae." *Proceedings of the National Academy of Sciences of the U.S.A.* 91:11841–11843.

Renzaglia, K. S., and D. J. Garbary. 2001. "Motile gametes of land plants: diversity, development, and evolution." *Critical Reviews of Plant Sciences* 20:107–213.

Reynolds, A. 2007. "The theory of the cell state and the question of cell autonomy in nineteenth and early twentieth-century biology." *Science in Context* 20:71–95.

Reynolds, A. 2008. "Ernst Haeckel and the theory of the cell state: remarks on the history of a bio-political metaphor." *History of Science* 46:123–152.

Richards, T. A., and S. L. Gómes. 2015. "Protistology: how to build a microbial eye." *Nature* 523:166–167.

Richter, D. J., and N. King. 2013. "The genomic and cellular foundations of animal origins." *Annual Reviews of Genetics* 47:509–537.

Romanes, G. J. 1883. *Mental Evolution in Animals*. London: Paul Kegan, Trench.

Ros-Rocher, N., A. Pérez-Posada, M. M. Leger, and I. Ruiz-Trillo. 2021. "The origin of animals: an ancestral reconstruction of the unicellular-to-multicellular transition." *Open Biology* 11:200359.

Rustom, A., R. Saffrich, I. Markovic, et al. 2004. "Nanotubular highways for intercellular organelle transport." *Science* 303:1007–1010.

Ryan, J. F., and M. Chiodin. 2015. "Where is my mind? How sponges and placozoans may have lost neural cell types." *Philosophical Transactions of the Royal Society London B: Biological Sciences* 370: 20150059.

Saranak, J., and K. W. Foster. 1997. "Rhodopsin guides fungal phototaxis." *Nature* 387:465–466.

Schaap, P. 2021. "From environmental sensing to developmental control: cognitive evolution in dictyostelid social amoebas." *Philosophical Transactions of the Royal Society London B: Biological Sciences* 376:20190756.

Schleper, C., and F. L. Sousa. 2020. "Meet the relatives of our cellular ancestor." *Nature* 577:478–479.

Schloegel, J. J., and H. Schmidgen. 2002. "General physiology, experimental psychology, and evolutionism. Unicellular organisms as objects of psychophysiological research, 1877–1918." *Isis* 93:614–645.

Schmieder, S. S., C. E. Stanley, A. Rzepiela, et al. 2019. "Bidirectional propagation of signals and nutrients in fungal networks via specialized hyphae." *Current Biology* 29:217–228.

Sciscioli, M., E. Lepore, M. Gherardi, and L. Scalera Liaci. 1994. "Transfer of symbiotic bacteria in the mature oocyte of *Geodia cydonium* (Porifera, Demospongiae): an ultrastructural study." *Cahiers de Biologie Marine* 35:471–478.

Searle, J. 2013. "Theory of mind and Darwin's legacy." *Proceedings of the National Academy of Sciences of the U.S.A.* 110(Suppl. 2):10343–10348.

Sebé-Pedrós, A., B. M. Degnan, and I. Iñaki Ruiz-Trillo. 2017. "The origin of metazoa: a unicellular perspective." *Nature Reviews Genetics* 18:498–512.

Sedghi, S., H. L. Kutscher, B. A. Davidson, and P. R. Knight. 2017. "Volatile anesthetics and immunity." *Immunological Investigations* 46:793–804.

Selosse, M. A., C. Strullu-Derrien, F. M. Martin, et al. 2015. "Plants, fungi and oomycetes: a 400-million years affair that shapes the biosphere." *New Phytologist* 206:501–506.

Senol, A. D., M. Samarani, S. Syan, et al. 2021. "α-synuclein fibrils subvert lysosome structure, and function for the propagation of protein misfolding between cells through tunneling nanotubes." *PLoS Biology* 19:e3001287.

Shapiro, J. A. 2021. "All living cells are cognitive." *Biochemical and Biophysical Research Communications* 564:134–149.

Sheldrake, M. 2020. *Entangled Life: How Fungi Make Our Worlds, Change Our Minds, and Shape Our Futures.* New York: Random House.

Shepherd, G. M. 1991. *Foundations of the Neuron Doctrine.* Oxford, UK: Oxford University Press.

Simard, S. 2018. "Mycorrhizal networks facilitate tree communication, learning and memory." In *Memory and Learning in Plants,* F. Baluška, M. Gagliano, and G. Witzany (eds.), 191–213. Springer Nature Switzerland AG.

Simard, S. 2021. *Finding the Mother Tree: Discovering the Wisdom of the Forest.* New York: Alfred A. Knopf.

Sinder, B. P., A. R. Pettit, and L. K. McCauley. 2015. "Macrophages: their emerging roles in bone." *Journal of Bone and Mineral Research* 30:2140–2149.

Sleigh, M. 1989. *Protozoa and Other Protists.* London: Edward Arnold.

Sonner, J. M. 2008. "A hypothesis on the origin and evolution of the response to inhaled anesthetics." *Anesthesia & Analgesia* 107:849–854.

Souriant, S., L. Balboa, M. Dupont, et al. 2019. "Tuberculosis exacerbates HIV-1 infection through IL-10/STAT3-dependent tunneling nanotube formation in macrophages." *Cell Reports* 26:3586–3599.

Southworth, D., and M. Cresti. 1997. "Comparison of flagellated and non-flagellated sperms in plants." *American Journal of Botany* 84:1301–1311.

Speijer, D. 2016. "What can we infer about the origin of sex in early eukaryotes?" *Philosophical Transactions of the Royal Society London B: Biological Sciences* 371:20150530.

Speijer, D., J. Lukeš, and M. Eliáš. 2016. "Sex is a ubiquitous, ancient, and inherent attribute of eukaryotic life." *Proceedings of the National Academy of Sciences of the U.S.A.* 112:8827–8834.

Spencer, N. J., L. Travis, L. Wiklendt, et al. 2021. "Long range synchronization within the enteric nervous system underlies propulsion along the large intestine in mice." *Communications Biology* 4:955.

Stacey, R. G., L. Hilbert, and T. Quail. 2015. "Computational study of synchrony in fields and microclusters of ephaptically coupled neurons." *Journal of Neurophysiology* 113:3229–3241.

Stairs, C. W., and T. J. G. Ettema. 2020. "The archaeal roots of the eukaryotic dynamic actin cytoskeleton." *Current Biology* 30:R521–R526.

Steinhorst, L., and J. Kudla. 2017. "Sexual attraction channelled in moss." *Nature* 549:35–36.

Stockem, W., H.-U. Hoffmann, and B. Gruber. 1983. "Dynamics of the cytoskeleton in *Amoeba proteus.* I. Redistribution of microinjected fluorescein-labeled actin during locomotion, immobilization and phagocytosis." *Cell and Tissue Research* 232:79–96.

Stollings, L. M., L.-J. Jia, P. Tang, et al. 2016. "Immune modulation by volatile anesthetics." *Anesthesiology* 125:399–411.

Strullu-Derrien, C., P. Kenrick, T. Goral, and A. H. Knoll. 2019. "Testate amoebae in the 407-million-year-old Rhynie Chert." *Current Biology* 29:461–467.

Taiz, L., and L. Taiz. 2017. *Flora Unveiled: The Discovery and Denial of Sex in Plants.* Oxford, UK: Oxford University Press.

Takeuchi, H., and T. Higashiyama. 2011. "Attraction of tip-growing pollen tubes by the female gametophyte." *Current Opinion in Plant Biology* 14:614–621.

Tang, S. K. Y., and W. F. Marshall. 2018. "Cell learning." *Current Biology* 28:R1180–R1184.

Tardivel, M., S. Bégard, L. Bousset, et al. 2016. "Tunneling nanotube (TNT)-mediated neuron-to neuron transfer of pathological Tau protein assemblies." *Acta Neuropathologica Communications* 4:117.

Tauber, A. I. 2003. "Metchnikoff and the phagocytosis theory." *Nature Reviews of Molecular and Cellular Biology* 4:897–901.

Tauber, A. I. 2013. "Immunology's theories of cognition." *History and Philosophy of Life Sciences* 35:239–264.

Tauber, A. I., and L. Chernyak. 1991. *Metchnikoff and the Origins of Immunology: From Metaphor to Theory.* New York: Oxford University Press.

Tekle, Y. I., and J. R. Williams. 2016. "Cytoskeletal architecture and its evolutionary significance in amoeboid eukaryotes and their mode of locomotion." *Royal Society Open Science* 3:160283.

Thomsen, E., and T. L. Rasmussen. 2008. "Coccolith-agglutinating foraminifera from the Early Cretaceous and how they constructed their tests." *Journal of Foraminifera Research* 38:193–214.

Timsit, Y., and S.-P. Grégoire. 2021. "Towards the idea of molecular brains." *International Journal of Molecular Sciences* 22:11868.

Torday, J. 2016. "The cell as the first niche construction." *Biology* 5:19.

Tóth, E. A., A. Oszvald, M. Péter, et al. 2017. "Nanotubes connecting B lymphocytes: high impact of differentiation-dependent lipid composition on their growth and mechanics." *Biochimica & Biophysica Acta* 1862:991–1000.

Trinh, M. K., M. T. Wayland, and S. Prabakaran. 2019. "Behavioural analysis of single-cell aneural ciliate, *Stentor roeseli*, using machine learning approaches." *Journal of Royal Society Interface* 16:20190410.

Tsuchiya, H. 2017. "Anesthetic agents of plant origin: a review of phytochemicals with anesthetic activity." *Molecules* 22:1369.

Ueyama, S., H. Katsumaru, T. Suzaki, and Y. Nakaoka. 2005. "*Halteria grandinella*: a rapid swimming ciliate with a high frequency of ciliary beating." *Cell Motility and Cytoskeleton* 60:214–221.

Underhill, D. M., S. Gordon, B-A. Imhof, et al. 2016. "Élie Metchnikoff (1845–1916): celebrating 100 years of cellular immunology and beyond." *Nature Reviews in Immunology* 16:651–656.

Uyeda, T. Q. P., and M. Furuya. 1984. "Cycloheximide insensitive amoebo-flagellate transformation in starved amoebae of *Physarum polycephalum*." *Development, Growth & Differentiation* 26:121–128.

Uyeda, T. Q. P., and M. Furuya. 1985. "Cytoskeletal changes visualized by fluorescence microscopy during amoeba-to-flagellate and flagellate-to-amoeba transformations in *Physarum polycephalum*." *Protoplasma* 126:221–236.

Vallverdú, J., O. Castro, R. Mayne, M. Talanov, M. Levin, F. Baluška, Y. Gunji, A. Dussutour, H. Zenil, and A. Adamatzky. 2018. "Slime mould: the fundamental mechanisms of biological cognition." *Biosystems* 165:57–70.

van den Ameele, J., L. Tiberi, P. Vanderhaeghen, and I. Espuny-Camacho. 2014. "Thinking out of the dish: what to learn about cortical development using pluripotent stem cells." *Trends in Neurosciences* 37:334–342.

Velle, K. B., and L. K. Fritz-Laylin. 2021. "Closest unicellular relatives of animals crawl when squeezed." *Current Biology* 31:R353–R355.

Verworn, M. 1889. *Psycho-Physiologische Protisten-Studien.* Jena: Gustav Fischer Verlag.

Vikhanski, L. 2016. *Immunity—How Metchnikoff Changed the Course of Modern Medicine.* Chicago: Chicago Review Press.

Vogel, D., and A. Dussutour. 2016. "Direct transfer of learned behaviour via cell fusion in non-neural organisms." *Philosophical Transactions of the Royal Society London B: Biological Sciences* 283:2016238.

Volkov, A. G., S. Toole, and M. WaMaina. 2019. "Electrical signal transmission in the plant-wide web." *Bioelectrochemistry* 129:70–78.

Volkov, A. G., and Y. B. Shtessel. 2020. "Underground electronic signal transmission between plants." *Communicative & Integrative Biology* 13:54–58.

Waldeyer-Hartz, H. W. G. 1891/2018. *Über einige neuere Forschungen im Gebiete der Anatomie des Zentralnervensystems.* Stuttgart, Germany: Deutscher Apotheker Verlag.

Wan, K. Y. 2019. "Ciliate biology—the graceful hunt of a shape-shifting predator." *Current Biology* 29:R1174–R1176.

Wang, F., X. Chen, H. Cheng, et al. 2021. "MICAL2PV suppresses the formation of tunneling nanotubes and modulates mitochondrial trafficking." *EMBO Reports* 22:e52006.

Wang, X., and H.-H. Gerdes. 2012. "Long-distance electrical coupling via tunneling nanotubes." *Biochimica & Biophysica Acta* 1818 (8):2082–2086.

Wei, K., C. Hollabaugh, J. Miller, et al. 2021. "Molecular cardioprotection and the role of exosomes: the future is not far away." *Journal of Cardiothoracic and Vascular Anesthesia* 35:780–785.

Willemse, M. T. M. 2008. "History and prospects of plant sexual reproduction congresses, the IASPRR and sexual plant reproduction." *Sexual Plant Reproduction* 21:89–97.

Wittig, D., X. Wang, C. Walter, et al. 2012. "Multi-level communication of human retinal pigment epithelial cells via tunneling nanotubes." *PLoS ONE* 7:e33195.

Wudick, M. M., M. T. Portes, E. Michard, et al. 2018. "CORNICHON sorting and regulation of GLR channels underlie pollen tube Ca^{2+} homeostasis." *Science* 360:533–536.

Yang, Z. 2003. "GABA, a new player in the plant mating game." *Developmental Cell* 5:185–186.

Yeates, G. W., and W. Foissner. 1995. "Testate amoebae as predators of nematodes." *Biology and Fertility of Soils* 20:1–7.

York, A. 2017. "Archaeal evolution: evolutionary insights from the Vikings." *Nature Reviews Microbiology* 15:65.

Zaremba-Niedzwiedzka, K., E. F. Caceres, J. H. Saw, et al. 2017. "Asgard archaea illuminate the origin of eukaryotic cellular complexity." *Nature* 541:353–358.

Zhang, J., Q. Huang, S. Zhong, et al. 2017. "Sperm cells are passive cargo of the pollen tube in plant fertilization a." *Nature Plant* 3:17079.

Zhou, M., H. Xia, Y. Xu, et al. 2012. "Anesthetic action of volatile anesthetics by using Paramecium as a model." *Journal of Huazhong University of Science and Technology—Medical Sciences* 32:410–414.

4 Constructing "On Purpose": How Niche Construction Affects Natural Selection

Dominik Deffner

Overview

Organisms actively and often purposefully modify and choose components of their local environments. This "niche construction" can affect ecological processes, and thus may change natural selection pressures and contribute to transgenerational transmission of information through ecological inheritance. Importantly, if that environmental regulation by organisms is systematic and directional, it can potentially impose biases on selection, and thereby influence evolutionary outcomes. Here I summarize and extend recent work that explores how the variability and strength of natural selection are affected by organisms that regulate the experienced environment through their activities (whether by constructing components of their local environments, such as nests, burrows, or pupal cases, or by choosing suitable resources). This work provides compelling evidence for reduced temporal variation and weaker selection in response to constructed compared to nonconstructed sources of selection and some evidence for reduced spatial variation in selection. For this chapter, I reanalyze the temporal selection gradient dataset compiled by Clark et al. (2020) using a Bayesian multilevel time-series model. This analysis reveals two distinct pathways for organisms to reduce the variation in selection that they experience depending on the degree of control they have over relevant factors in the selective environment. Overall, these findings suggest that organism-manufactured or chosen components of environments may have properties that are qualitatively different from other environmental features, providing a plausible route for biological agents to "co-direct" their own evolutionary trajectories.

4.1 Organism and Environment

Natural selection is probably Charles Darwin's most potent metaphor.[1] This is evidenced by the fact that the phrase has lost its original metaphorical meaning and is now widely understood as brute empirical fact. Inspired by the observation of how the artificial selection (here in the literal sense) of especially desirable traits by Victorian pigeon breeders can lead to dramatic changes in the birds' appearance, Darwin envisaged that nature also actively "selects" those trait variants that enable organisms to survive and reproduce (Theunissen, 2012). This active filtering done by the environment then results in adaptation and evolutionary change. Inspired by this original metaphor, the asymmetric relationship of passive organisms that are selected by independent environmental agents and adapt into predefined niches has since dominated evolutionary thought. George Williams, for instance, famously wrote that "adaptation is always asymmetrical; organisms adapt to their environment, never vice versa" (Williams, 1992). With the formulation of the modern synthesis and especially the emergence of the gene-centered view of evolution (Ågren, 2021), the conceptual focus shifted from changes in organismal traits to changes in allele frequencies—and selection was henceforth conceived of as a process of environmental filtering of randomly generated genetic variation. However, the idea of the asymmetrical relationship between organism and environment, founded in Darwin's metaphor, remained essentially unchanged, leaving aside complications such as coevolution and habitat selection (Sultan, 2015).

Although the focus on population-level changes in allele frequencies, which are "selected" by external, environmental forces, proved instrumental in establishing biology as a rigorous scientific discipline in the twentieth century, there were early advocates of a more reciprocal relationship between organism and environment. More generally, these authors argued for a more prominent role of ontogeny and development in evolution. In "Evolutionary Systems," Conrad Waddington (1959a) assumed four major factors in biological evolution. In addition to the genetic system and natural selection pressures, he introduced an epigenetic and an exploitive system. Waddington is now probably best known for his idea of the "epigenetic landscape" that describes the developmental pathways arising from gene–environment interactions. Another prominent idea, the canalization and genetic assimilation of acquired characteristics, is based on the observation that the development of certain traits, whose formation originally depended on the presence of certain environmental factors, can be stabilized through natural selection such that the traits would form even in the absence of external stimuli (Waddington, 1959b). Even if less well-known, his notion of an exploitive system was equally important and innovative. Waddington suggested that organisms are always surrounded by a vast range of potential environments and have to choose where to live and which elements to interact with. As a consequence, organisms necessarily modify the selective pressures they experience and co-direct their evolutionary pathway (Waddington, 1959a). In his own words: "It is the animal's behavior which to a considerable extent determines the nature of the environment to which it will submit itself and the character of the selective forces with which it will consent to wrestle. This 'feedback' or circularity in a relation between an animal and its environment is rather generally neglected in present-day evolutionary theorizing" (Waddington, 1975: 170).

Later, Harvard biologist Richard Lewontin elaborated on these ideas. In the traditional view, evolutionary change in organisms over time is seen as a function of both the organisms themselves and the environment, whereas environmental change is only a function of environments independently of organismal influences (Lewontin, 1983). Lewontin contrasts this view with his constructivist notion of mutual interdependence in which both organismal and environmental changes are functions of themselves and of each other. The organism, therefore, is "both the object of natural selection and the creator of the conditions of that selection" (Levins & Lewontin, 1985).

If organisms and environments are indeed reciprocally dependent on each other, as Waddington and Lewontin suggested, evolution can no longer be fully described as a process in which only organisms evolve in response to independently existing environments. Instead, Sultan (2015) suggested that the whole system of organism–environment interactions should be regarded as the fundamental unit of evolution, rather than the organism alone, a view closely related to what Susan Oyama and collaborators describe as "developmental systems" (Oyama, 2000; Oyama, Griffiths, & Gray, 2003). In the developmental systems view of evolution, differential replication of life cycles does not stem from the degree to which organisms match an external environment, but from the fact that life cycles that include certain interactions have a greater capacity to reconstruct themselves than life cycles without these interactions (Griffiths & Gray, 1994). Selective agents, in that view, are not independent environmental factors but ecological interactants in developmental systems that selectively enhance the replication of the entire system (Oyama, 2000). Selection, then, is not an objective external force acting on organisms but a statistical abstraction that describes how certain traits persist in dynamical interactions with an environment that organisms themselves (in part) control.

4.2   Niche Construction Theory: An Overview

Despite compelling theoretical arguments for a reciprocal relationship between organism and environment articulated during the twentieth century by Waddington, Lewontin, and others, these ideas did not result in a coherent research program that made them empirically tractable. The formulation of the niche construction perspective within evolutionary biology by Odling-Smee, Laland, and Feldman (Odling-Smee, 1988; Laland, Odling-Smee, & Feldman, 1996, 1999; Odling-Smee, Laland, & Feldman, 2003) provided such an integrated theoretical framework that has made organism-driven environmental modifications and their evolutionary consequences amenable to detailed theoretical and empirical investigation.

Niche construction refers to the process whereby organisms actively modify their own and/or each others' niches (Odling-Smee et al., 2003). The evolutionary niche of a population describes the sum of all natural selection pressures to which the population is exposed (Odling-Smee et al., 2003) and thus can only be specified in relation to this population. Examples of niche construction include the building of physical artifacts by animals (such as bird nests, beaver dams, termite mounds, insect leaf-ties, or football stadiums); the alteration of physical and chemical conditions by corals or earthworms; the creation of shade and the alteration of nutrient cycling by plants; and also consistent choices of animals, for mates, habitats (including flower sources among pollinators), and food (Odling-Smee et al., 2003). When such organism-mediated modifications of the environment alter natural selection pressures, they can result in evolution by niche construction (Matthews et al., 2014; Sultan, 2015).

Niche construction is not only ubiquitous in nature, as apparent from the various cases just mentioned, it can be regarded as a universal property of living systems (Odling-Smee et al., 2003). In line with the second law of thermodynamics, the entropy of any isolated system can only increase, except for minor random fluctuations. Put differently, the disorder of a system always increases over time, unless it can exchange energy with its environment. In his seminal 1944 book *What Is Life?*, physicist Erwin Schrödinger noted that life as a complex, self-organized system must resist increasing entropy because its structure and functionality will be endangered as soon as the entropy reaches a critical threshold (Schrödinger, 1944). Just to exist, organisms must exchange entropy with their world and thereby modify the physical and chemical composition of their surroundings. But unconstrained and random niche-constructing activities could not sustain life, as any far-from-equilibrium system cannot maintain its boundaries by chance alone, as illustrated by the famous demon in Clerk Maxwell's thought experiment from 1871 (see Odling-Smee et al., 2003). Therefore, niche construction must at least partly be guided by information, specifically by semantic information that relates to the fitness of a specific organism. This semantic information encompasses statistical regularities in the environment as well as modes of interacting with this environment in a way that satisfies the requirements of an organism and keeps it alive. Odling-Smee et al. (2003) state that semantic information can either be generated by the process of natural selection and stored in DNA sequences, or learned in the lifetime of an organism and represented in neuronal network activity or—at least in the case of humans—incorporated in cultural institutions and artifacts (see also Oyama, 2000, for concerns with common notions of information in biology).

Because niche construction is at least partly governed by semantic information, it is expected that organisms will disproportionately generate environmental states that are coherent with the constructing organism's demands and those of its descendants. At least in the short term and with respect to a key dimension, organism-driven changes in the environment are expected to be adaptive for the constructor (Odling-Smee et al., 2003; Laland, Odling-Smee, & Endler, 2017).

Cases of niche construction can be heuristically categorized along two distinct axes. First, organism-mediated modifications of the environment can be either *inceptive* or *counteractive*. *Inceptive niche construction* refers to all cases where organisms initiate a modification in an environmental factor and thereby change the relationship between their own features and their environment's factors. It can be obligate—for example, if a population is forced into a new habitat, as it is often the case in anthropogenic environments (see Alberti, 2015; Alberti et al., 2017)—or facultative, if a population opportunistically switches to a new habitat. Counteractive niche construction, in contrast, describes cases where organisms oppose or (wholly or partly) neutralize a change in an environmental factor that is already happening, thereby buffering the environmental variation they are experiencing. Counteractive niche construction often serves to limit environmental fluctuations to a specific range that is adaptive or at least tolerable for the organism. By building nests, for example, birds reduce the environmental variation in temperature and humidity that eggs experience, and thereby provide optimal conditions for the development of their offspring (Hansell, 2000, 2007). Second, organisms can change the selection pressures they experience by either *perturbation* or *relocation*. Perturbational niche construction occurs when organisms physically change their surroundings at a specific location and time by secreting chemicals, exploiting resources, or building artifacts. This term thus describes the causal impact organisms have on their world. Organisms engage in relocating niche construction when they actively move in space and choose habitats, with the consequence that they expose themselves to different environmental factors and selection pressures.

The environmental modifications created by organisms can sometimes outlast the original niche constructor and affect future generations of their own and other species. This "ecological inheritance" not only constitutes an additional mechanism for transmitting information across generations, and thus an important component in a broadened conception of inheritance (Danchin et al., 2011; Bonduriansky, 2012; Jablonka & Lamb, 2014), it also provides a plausible non-Lamarckian route by which acquired characteristics can influence evolution, namely by generating long-lasting modifications in the patterns of natural selection. Especially pronounced in humans, but also present in other species (Whiten, 2021), socially learned information represents a second system of transgenerational inheritance resulting in cultural evolution (Cavalli-Sforza & Feldman, 1981; Boyd & Richerson, 1985; Henrich & McElreath, 2003; Deffner & Kandler, 2019). Humans (and other social animals) are smart individually, but not nearly smart enough to overcome even basic adaptive problems on their own (Henrich & McElreath, 2003). Our ecological success relies not only on individual problem solving but also on our special abilities to learn from others. Such social learning allows our species to accumulate cultural information over time and develop well-adapted tools, beliefs, and institutions that are too complex for any single individual to invent on their own (Boyd & Richerson, 1985; Henrich & McElreath, 2003). In a process called cumulative cultural evolution, socially acquired information can build up over time and modify selective environments to an extent that is unprecedented in noncultural species (Laland, Odling-Smee, & Feldman, 2001; Laland, 2018). Besides the anthropological textbook examples of gene-culture coevolution—(1) selection for adult lactose tolerance in societies with a long history of farming and (2) selection for sickle-cell allele in communities where traditional yam agriculture provided favorable conditions for malaria-spreading mosquitoes (Holden & Mace, 2009; Feldman & Laland, 1996; Beja-Pereira et al., 2003)—mathematical geneticists have demonstrated that as much as 10% of the genome has experienced selective sweeps during the past 100,000 years. These events of fast and consistent selection especially affected genes involved in the immune system, digestion, and brain function, making cultural selection pressures at least highly plausible (Williamson et al., 2007; Laland, Odling-Smee, & Myles, 2010). Looking beyond humans, Whitehead et al. (2019) reviewed the evidence for gene–culture coevolution in other animals, especially birds, cetaceans, and primates. They found that socially learned traditions can both relax and intensify selection, create new selection pressures by changing ecology or behavior, favor adaptations (including in other species), and even shape genetic population structure and diversity.

Selective feedback due to niche construction has also been studied by two-locus population-genetic models (Laland et al., 1996, 1999; Odling-Smee et al., 2003). These models investigate the feedback dynamics between alleles influencing the extent of niche-constructing activities and alleles at a second locus where fitness depends on the amount of an environmental resource which is influenced by the first locus. These models show that niche construction can significantly alter evolutionary processes and create both inertia and acceleration. Evolutionary effects of niche construction can be substantially delayed in time because niche-constructing activities often do not become consequential until a critical threshold in the modified environmental resource is reached. In other scenarios, positive feedback may also lead to accelerated selective responses. When niche construction changes the direction of selection compared to environmental conditions unmodified by the organism, it can also lead to unexpected evolutionary outcomes such as fixation of otherwise disfavored alleles and promotion or prevention of stable polymorphisms (Odling-Smee et al., 2003). Going beyond only selection, Fogarty and Wade (2022) have recently developed quantitative-genetics models to partition the effects of niche construction on trait heritability and selection. They showed that niche construction can substantially alter the way phenotypes respond to selection and hence the pace of evolutionary change. These theoretical models show that niche construction can dramatically change evolutionary dynamics and lead to otherwise counterintuitive outcomes.

4.3   Niche Construction as Evolutionary Bias

It is now widely accepted that niche construction occurs in nature and can have important ecological and evolutionary consequences (Odling-Smee et al., 2003; Pelletier, Garant, & Hendry, 2009; Post & Palkovacs, 2009; Odling-Smee et al., 2013; Matthews et al., 2014; Scott-Phillips et al., 2014; Sultan, 2015). However, from a traditional point of view, changes in the environment brought about by organisms are not regarded as inherently different from other changes in the environment. Natural selection is still portrayed as the evolutionary process, while niche construction, and environmental change in general, are treated as mere background conditions for selection (see arguments of "sceptics" on gene-culture coevolution and lactose tolerance in Scott-Phillips et al., 2014). Futuyma (2017), for instance, claims: "Niche construction can influence or even cause the evolutionary process of natural selection, but it is not itself an evolutionary process, any more than a changing environment is." If *evolution* is defined as the change of allele frequencies over time (the textbook definition of evolution), it makes sense to identify only those processes as evolutionary processes that directly change allele frequencies. These include natural selection, along with drift, mutation, gene flow, and other phenomena known from population genetics. Although niche construction can directly change gene frequencies (e.g., through consistent prey choice), its main way to direct evolutionary change is through modifying the selective environment that the constructor and/or other species experience. Although this form of causal influence of niche construction is not generally recognized as an evolutionary process, broadened conceptions of evolutionary causation have been proposed. Endler (1986), for instance, formulated such a broader classification scheme, which involves a number of categories of evolutionary processes including "adaptive processes," "rate-determining processes," and "direction-determining processes." Such a conceptualization can incorporate all causal factors that are evolutionarily significant (such as niche construction) and not just those factors that *directly* change gene frequencies.

Following from that, Laland et al. (2017) argue that, in principle, it is possible to differentiate between standard natural selection (direct environmental influence on fitness) and organism-induced changes in environments (niche construction) leading to natural selection. They propose that the specific properties of niche construction should lead to consistent and detectable differences in the strength and temporal dynamics of selection generated by "constructed" and "autonomous" (i.e., nonconstructed) sources of selection. By autonomous or nonconstructed sources of selection, Laland et al. (2017) mean environmental conditions that are not influenced or only weakly influenced by the activities of organisms (see Sober, 1984, for his distinction between "source laws" and "consequence laws"). Niche construction is expected to initiate or modify patterns of selection in an orderly, directed, and sustained manner because of feedback loops that can lead to a self-reinforcing process of cause-effect relationships. Figure 4.1, adapted from Laland et al. (2017), visualizes the self-reinforcing nature of evolution by niche construction. Niche-constructing activities can change selective environments and thereby modify gene frequencies, not only in the constructor but also in

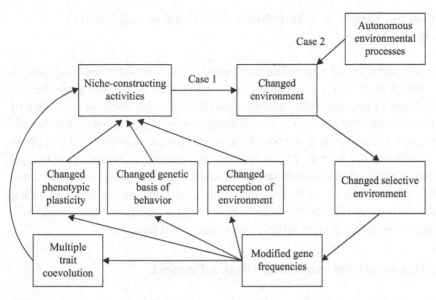

Figure 4.1
Illustration of the cause-effect relationships associated with evolution by niche construction. Through feedback loops between niche-constructing activities and selection imposed by organism-driven modifications in the environment, environmental changes stemming from the activities of organism (case 1) are expected to be qualitatively different from changes stemming from autonomous (i.e., nonconstructed) processes (case 2). In particular, there should be less variation in constructed than in nonconstructed sources of selection (adapted from Laland et al., 2017).

other populations that share the environmental resources affected by the niche-constructing activities. Via multiple causal pathways, these changes in gene frequencies can feed back and lead to changes in the niche-constructing activities themselves. This self-reinforcing nature of the cause–effect cycle associated with niche construction is assumed to lead to significantly less variation in the source of selection compared to systems without such feedback. This in turn is assumed to make selection stemming from constructed sources (Case 1) more reliable, consistent, and predictable than selection imposed by autonomous sources (Case 2). In this manner, niche construction is expected to impose a statistical bias on selection and to direct adaptive evolution.

Laland et al. (2017) also compare evolution by niche construction with artificial selection and posit that niche construction occupies the middle ground between artificial and natural selection. What is central about artificial selection is not the underlying intention or deliberate attempt to produce a certain outcome but the fact that breeders/experimentalists impose particularly consistent, reliable, and sustained selection on the respective organism. It is now widely accepted that multiple unintentional pathways led to initial domestication (Zeder, 2006, 2015), and it has been proposed that the emergence of domestication can be described as the establishment of co-evolutionary mutualisms between active niche construction by both humans and their plant/animal "partners" (Zeder, 2016). In fact, artificial selection can be regarded as an extreme form of niche construction. Similar to artificial selection, the self-reinforcing nature of niche construction is expected to result in especially predictable and consistent patterns of evolution.

4.4   Putting the Theory to a Test: Niche Construction and Natural Selection in the Wild

In principle, a straightforward way to determine whether organism-modified environments differ from other features of the environment in the selection pressures they generate is to compare measures of selection (e.g., selection gradients) deriving from the two kinds of environments. It is now well known that natural selection in wild populations is highly variable in its strength and consistency, and consideration of the source of selection might partly explain this variation. Clark et al. (2020) test the predictions that organism-constructed sources of selection that buffer environmental variation (i.e., regulatory or counteractive niche construction) will result in (i) reduced variation in selection gradients, including reduced variation between (a) years (temporal variation) and (b) locations (spatial variation); and (ii) weaker directional selection relative to nonconstructed sources.

4.5   Selection Gradients and Categorization Protocol

Based on previous data sets (Kingsolver et al., 2001; Siepielski, DiBattista, & Carlson, 2009; Siepielski et al., 2013) and performance of an extensive literature search, Clark et al. compiled large data sets of 1,045 temporally replicated selection gradients, 257 spatially replicated selection gradients, and a pooled data set of 1,230 selection gradients. Clark et al. (2020) focused solely on published studies that reported standardized linear selection gradients from quantitative traits in wild populations (*sensu* Lande & Arnold, 1983). Selection gradients were introduced as a standard metric to quantify the direct relationship between a phenotypic trait and fitness and to compare selection across different traits, fitness measures, and study systems. The selection gradient was originally defined as the partial derivative of (relative) fitness with respect to the mean value for a phenotypic trait; empirically, selection gradients can be estimated as the partial regression coefficients from a multiple regression where multiple phenotypic traits are included as predictors to model relative fitness as the response variable (Lande & Arnold, 1983). Clark et al. (2020) used selection gradients rather than selection differentials because selection differentials do not control for the influence of other, correlated traits, and therefore potentially confound selection arising from a constructed source with selection arising from nonconstructed sources (and vice versa).

After compilation of the database, these authors developed a robust and reliable protocol of categorizing cases as "constructed" or "non-constructed" (figure 4.2; see the appendix in Clark et al. [2020] for details). This is not straightforward because in many (perhaps most) traits and studies, there are likely influences of both organismal activity and autonomous environment. In simple terms, identified sources of selection were categorized as "constructed" if there were dominating environmental factors chosen or manufactured by organisms in a way that kept environmental parameters within suitable bounds. Examples included nests, burrows, leaf ties, and highly specialized prey choices. Conversely, environmental features were categorized as "nonconstructed" in all cases where the focal trait was not, or was only weakly, influenced by niche-constructing activities. Cases in which there was a clear and strong influence both of regulatory niche construction and of the autonomous environment

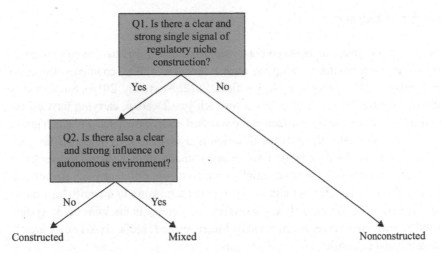

Figure 4.2
Categorization of niche construction. From figure A1 of Clark et al. (2020).

were categorized as "mixed," with the expectation that such cases should lie between purely constructed and nonconstructed cases. Note that Clark et al. (2020) report analyses where the three categories are treated separately as well as analyses where all cases involving constructed sources of selection (i.e., "constructed" and "mixed") were combined into one category. Most authors of the primary studies included in the database explicitly stated what they assumed to be the primary source(s) of selection. This was accepted by Clark et al. (2020), assuming author expertise regarding the study system. Two neutral coders independently coded the data blind to its actual value. Their coding exhibited high interobserver agreement, which was important because it demonstrates that the categorization procedure can be used to code traits in published studies as "constructed" and "nonconstructed" in a consistent and reliable fashion. To check the robustness of the conclusions to categorization error, the authors also conducted simulations that deliberately miscategorized 10% of the data. Reassuringly, this procedure did not substantially change any of the major findings.

As discussed above, active environmental regulation and habitat choice are both activities that function to preserve an adaptive match between the features of an organism and factors in its environments. For this reason, both have been subsumed within the wider category of "niche construction" (Odling-Smee et al., 2003; see Edelaar & Bolnick, 2019, for major differences). However, some prominent researchers—notably Richard Dawkins (2004)—have questioned the utility of such a broad conception, preferring to restrict niche construction to the construction of physical artifacts, or the activities of a focal species (see Scott-Phillips et al., 2014, for similar arguments). By repeating the analyses for broad and narrow definitions of niche construction, Clark et al. (2020) explored whether or not habitat modification and choice generated equivalent effects on selection. Likewise, by separately analyzing cases in which relevant environmental regulation was undertaken by the focal species or another species, and by a single or multiple species, they were able to evaluate whether and how these differences affect selection.

4.6   Statistical Analyses

To compare variation in selection between constructed and nonconstructed categories while taking into account variation due to sampling error, Clark et al. (2020) conducted Bayesian mixed-effects meta-analyses (Morrissey & Hadfield, 2012; Morrissey, 2016). Such models estimate the true unobserved selection gradient for each year/location, carrying forward the uncertainty from the original study encoded in the standard errors associated with each gradient. The model then estimates the average direction and magnitude of selection for each study and decomposes the total variation into variation among study systems and variation within study systems, such as temporal (spatial) variation across different years (locations). From these model parameters, the authors then computed a measure to quantify how much selection differs between two randomly chosen years or locations in the same study system and calculated a consistency score, as proposed by Morrissey and Hadfield (2012), separately for niche construction categories.

To overcome limitations of previous models, Clark et al. (2020) additionally developed a more general multilevel framework that explicitly models study-specific variances and thus allowed them to account for study differences (i.e., differences between species, traits, etc.) in temporal variation, not only in the direction and magnitude of selection. Finally, to investigate whether the mean strength of selection also differs in response to constructed and nonconstructed sources of selection, Clark et al. (2020) conducted additional random effects meta-analyses accounting for possible differences among species, studies, fitness measures, and phenotypic traits, all controlling for uncertainty of the single selection gradient estimates.

4.7   Main Findings

Their results confirmed that natural selection deriving from organism-constructed sources exhibits reduced temporal variation in selection gradients compared to nonconstructed sources (see figure 4.3, adapted from Clark et al., 2020, for temporal variation results using the combined category model on the top, the separate category model in the middle, and the double-hierarchical model on the bottom). These results hold for (i) broad (including animal choices) and narrow (physical artifacts only) definitions of niche construction, (ii) whether the focal species or another species engages in the niche construction, (iii) cases in which a single or multiple species engage in functionally equivalent niche construction, and (iv) are little affected by taxonomy. The fact that regulation through physical artifacts or by habitat selection, by a single or multiple species, or by the focal species or another species, generates similar impacts on selection supports the hypothesis that the functional equivalence of regulatory activity is often more important than the precise identity of the constructor in determining the selection that ensues. This helps to justify a broad definition of niche construction (Odling-Smee et al., 2003). The parallel effects on selection observed here do not imply that different forms of niche construction will always have identical ecological and evolutionary causes and consequences. For instance, habitat modification will typically generate more profound ecological and evolutionary consequences for other species than habitat choice. However, it is precisely because the parallels between these processes are not always self-

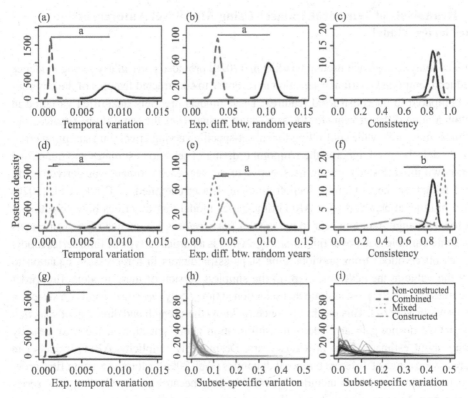

Figure 4.3
Posterior distributions for temporal variation using the combined category (constructed+mixed) model (A–C) and the separate category model (D–F) and a broad definition of niche construction. G–I: Marginal posterior probability distributions for the double-hierarchical model. Expected temporal variation ($1/\lambda$) was aggregated over all subsets (G) and standard deviations for each subset in the combined (H) and nonconstructed (I) categories. a = 99% credible intervals for the two distributions do not overlap; b = 95% credible intervals for the two distributions do not overlap (adapted from Clark et al., 2020).

evident to many researchers that the more general term *niche construction* can have explanatory utility and brevity.

Clark et al. (2020) also found evidence for reduced spatial variation in selection, suggesting that selection is more consistent across different locations in cases that are affected by organismal activities. These results were less robust, however, due to the substantially smaller sample size. Lastly, Clark et al. (2020) also found convincing evidence for reduced overall strength (i.e., reduced intensity) of selection deriving from organism-constructed sources compared to nonconstructed sources. While mixed cases (cases that were strongly influenced by both constructed and nonconstructed sources) showed only a slight reduction in the magnitude of selection, "purely" constructed cases were characterized by substantially weaker or almost absent selection. Summarizing, their results confirmed the predictions made by Laland et al. (2017) that natural (ecological) selection deriving from organism-constructed sources will exhibit reduced temporal and spatial variation in selection gradients and weaker selection compared to nonconstructed sources.

4.8   Reanalysis of Temporal Dataset Using Multilevel Autoregressive Time-Series Model

The meta-analytic approach used in Clark et al. (2020) provided particularly strong evidence for reduced temporal variation in selection in response to constructed features of the environment compared to nonconstructed features. These models precisely quantified variation in selection among different years for the same study system (i.e., the same study, species, trait and fitness measure), while controlling for subset-specific varying effects and sampling error. These models, however, ignored the temporal ordering of selection estimates and quantified variation within each study system irrespective of the year each estimate was derived from. Here, I reanalyze the temporal selection gradient dataset compiled by Clark et al. (2020) using a multilevel autoregressive AR(1) time-series model that describes how selection is expected to change from one year to the next. A *time series* is a sequence of measurements of the same variable made over time; *autoregression* is a particular type of time series model that uses observations from previous time steps as predictors in a regression equation to model the value at the next time step. In the simplest version of these models, first-order autoregression or AR(1), we assume that selection at time t only *directly* depends on selection in the previous year $t-1$. This means that once we know the strength and direction of selection this year, we do not gain any additional information about selection in the next year by learning about estimates from previous years. Despite this simplicity, AR(1) models can capture a wide range of different temporal dynamics (Wei, 2006). Moreover, using first-order autoregression allows one to include all available data, because some selection time series comprise only two consecutive years. The model is constructed as follows:

$$\hat{\beta}_{j,t} = \alpha_{k[j]} + b_{k[j]}\hat{\beta}_{j,t-1} + \varepsilon_{j,t}$$

where $\varepsilon_{j,t} \sim N(0, \sigma_{k[j]})$, and

$$\beta_{j,t} \sim N(\hat{\beta}_{j,t}, SE_{j,t}^2).$$

The observed selection gradient for each study system j in year t, $\beta_{j,t}$, is assumed to follow a normal distribution centered at the unobserved true value and with a variance calculated from the standard errors (SEs) reported in the literature. This estimated ("true") selection gradient, $\hat{\beta}_{j,t}$, is composed of the average selection in j's niche construction category k, the autoregressive effect of selection in the preceding year, $b_{k[j]}$, and a normally distributed noise term (with mean 0 and dispersion, $\sigma_{k[j]}$, that is specific to each niche construction category) which represents variation in selection that is not explained by the time series. Assuming stationarity, $b_{k[j]}$ can vary between -1 and 1. When $b_{k[j]}=1$, the expected direction and magnitude of selection stays constant over time; when $b_{k[j]}=0$, there is no temporal autocorrelation in selection and the process reduces to white noise; when $b_{k[j]}=-1$, there is a strong negative correlation across time such that strong positive selection in one year is likely followed by strong negative selection in the following year.

I account for dependencies arising from the nested data structure by including varying (or random) effects for all parameters, such that not only the average selection ($\alpha_{k[j]}$), but also the temporal stability ($b_{k[j]}$) and the residual variation ($\sigma_{k[j]}$) can vary among study systems. The model is implemented using the Hamiltonian Monte Carlo engine *Stan* (Stan Development Team, 2018; see Github repository for full model code and details: https://github.com/DominikDeffner/Niche-construction-time-series). Compared to the previous meta-analytic approach, this model allows one to track selection on a year-to-year basis and decompose observed time trends into variation that is due to internal dynamics of the system and external dynamics that cannot be explained by the time trend in selection.

Figure 4.4 shows posterior probability distributions for the two major model parameters, b_k (what I call "temporal stability") and σ_k ("residual variation"), separately for different niche construction categories. Residual variation—that is, year-to-year changes that are unexplained by the selection time series—is substantially lower for both constructed and mixed cases compared to nonconstructed cases. This suggests that niche constructing activities buffer external environmental factors that would otherwise influence selection in an undirected and unpredictable way. The results for temporal stability—that is, year-to-year covariation in selection—add further detail about the way niche construction can affect selection. In constructed cases where organisms more or less completely control the selective environment of a trait, there is little temporal autocorrelation among selection gradients compared to nonconstructed cases. This is because in such scenarios, selection is very weak and gradients tend to cluster around zero (compare results on average magnitude of selection in Clark et al., 2020). In mixed cases, in contrast, where organismal activities only partially influence the selective environment, the year-to-year autocorrelation in selection is higher compared to nonconstructed cases. So, although in such scenarios the buffering activities of organisms might not be strong enough to effectively remove consistent selection, they might still control them in a way that keeps selection relatively constant from one year to the next.

Based on the parameter estimates alone, it can be difficult to understand the temporal dynamics implied by the model. To visualize the range of implied selection time series, figure 4.5 shows 1,000 simulated time series (per niche construction category) of 50 years of selection based on fitted parameter estimates from the multilevel autoregressive time-series model. It is obvious that selection varies much more from year to year in nonconstructed cases (top) compared to mixed (middle) and constructed cases (bottom). Although

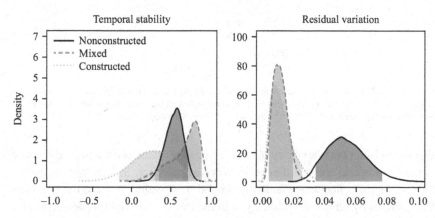

Figure 4.4
Marginal posterior probability distributions using the autoregressive time-series model for nonconstructed (solid, dark gray), mixed (dashed, mid gray), and constructed (dotted, pale gray) cases. The left plot shows estimates for the coefficient of the autoregressive effect, b_k, and thus represents the temporal stability of selection from one year to the next. The right plot shows estimates for the extrinsic, residual variation, σ_k, that is not explained by the selection time series.

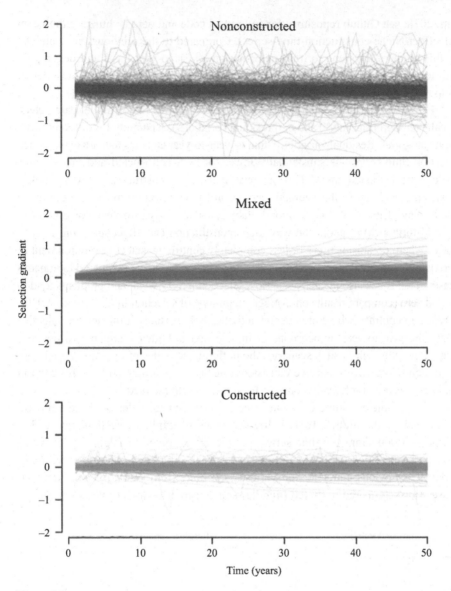

Figure 4.5
Simulated time series of 50 years of selection based on parameter estimates from multilevel autoregressive time-series model (nonconstructed on the top, mixed in the middle, constructed on the bottom). Each plot shows 1,000 time series based on 10 independent draws from the posterior for 100 randomly selected study systems.

there is smaller overall temporal variability in cases affected by niche construction, the simulated time series also reveal some differences. Whereas selection in response to purely constructed sources mostly fluctuates randomly around 0, selection in response to mixed sources is characterized by remarkable stability of even relatively strong selection from one year to the next. This confirms the interpretation that if organisms are in almost complete control of the selective environment they experience, directional selection can be virtually removed, whereas organisms that have only partial control of the environment are not able to remove directional selection but can keep it constant and thus predictable over time.

4.9   Summary and Outlook

The work summarized and extended in this chapter supports the view that organism-driven modifications of the environment are inherently different from other (nonconstructed or autonomous) changes in the environment and may generate systematically different evolutionary outcomes. Clark et al. (2020) provide the first meta-analytical study to demonstrate that niche construction substantially affects temporal (and spatial) variation as well as the average magnitude of natural selection in wild populations. These results add further weight to the argument that niche construction should be recognized as an important factor (or even an evolutionary process) that might impose a statistical bias on evolution in the wild. The idea that regulatory behavior might constrain evolution has sporadically appeared in the literature before. For instance, Bogert (1949) already suggested that reptiles can buffer selection through their behavior. More recently, studies exploring this Bogert effect report results that are broadly in line with the present findings (e.g., Huey, Hertz, & Sinervo, 2003; Marais & Chown, 2008; Stellatelli et al., 2018). Crucially, the findings reported in Clark et al. (2020) and extended here suggest that phenomena similar to the Bogert effect may be more general than previously appreciated. Organism-mediated regulatory activities buffering selection may extend beyond behavior to encompass buffering that is mediated by artifacts and other environmental resources as well. Similar effects may also extend beyond animals to any species that experiences environmental resources or conditions regulated by themselves or other organisms. It should be noted that the present findings show that regulatory niche construction affects selection in wild populations, but they do not prove that it imposes a statistical bias on evolution; still, a failure to find differences in selection in response to constructed versus nonconstructed sources would have undermined that view.

Although a rigorous and standardized protocol informed by the expertise of study authors and detailed knowledge of the respective study systems was used to classify cases according to the dominant source of selection, there inevitably remains some uncertainty about whether the identified environmental factor was indeed the dominant source of selection. Though much is known about natural selection in the wild, its average strength, and its spatial and temporal dynamics, surprisingly little is known about the ecological factors underlying phenotypic selection (Wade & Kalisz, 1990). MacColl (2011, p. 514) even notes concerning the ecological causes of evolution that "Darwin's assertion that 'We know not exactly what the checks are in even one single instance' is almost as true today as it was 150 years ago." To overcome this fundamental limitation, future studies of selection in the wild should record as many ecological variables as possible over a significant period of time. By relating these longitudinal ecological variables to variation in selection, it will become possible to shed more light on the ecological causes of evolution (Wade & Kalisz, 1990; MacColl, 2011). Additionally, we need new empirical selection studies that are specifically designed to test the predictions on effects of niche construction made by Laland et al. (2017) and others. For example, one could manipulate different nest properties and autonomous environmental variables at once and identify traits for which selection is primarily determined by autonomous or constructed sources, respectively. Thereby, it would be possible to directly quantify the relative importance of different factors going beyond the relatively coarse constructed-mixed-autonomous categorization scheme employed here. Given the enforced reliance on more circumstantial data to categorize studies and the low number of studies that fulfilled inclusion criteria (see Clark et al. 2020, for details), the reported analyses must be regarded as provisional rather than definitive.

Future research should investigate the effects of niche construction in a wider range of natural and experimental settings including different species, trait types, and forms of organism-driven modifications of the environment (note that for inceptive niche construction we would expect the opposite pattern of stronger selection compared to nonconstructed sources). This will help elucidate how far-reaching and pervasive the effects of niche construction and organismal agency on evolution really are. Complementing long-term field observations, detailed and controlled experiments on long-term selection in the lab or selective responses of domesticated species to artificial selection can clarify the exact mechanisms underlying niche-constructing activities and their effects on evolution. Ideally, it would be possible to link genetic variants associated with niche construction activities directly to genetic variants associated with target traits affected by niche construction. In the end, the usefulness and success of the niche construction concept will not be determined by its theoretical appeal but by the amount of fruitful research it is able to stimulate and the novel findings it may generate.

The findings of the present line of research have implications for areas extending beyond evolutionary biology. Domestication of plant and animal species constitutes a landmark in human evolution (Henrich, 2015; Zeder, 2015; Laland, 2018). As noted earlier, the process of initial domestication can be described as the establishment of co-evolutionary mutualisms between active niche construction by both humans and their plant/animal "partners" (Zeder, 2016). Therefore, more detailed knowledge about the strength and temporal dynamics of selection under niche construction can help explain the mechanisms underlying the idiosyncratic evolutionary pathways resulting in domestication. Relatedly, Hare (2017) introduced the "Human Self-Domestication Hypothesis," which portrays human evolution as a process in which selection for prosociality and against in-group aggression led to the evolution of a human "domestication syndrome." This suite of traits includes, among others, the early-emerging cooperative-communicative abilities that uniquely characterize our species. A similar self-domestication hypothesis involving sexual selection against male aggression has also been proposed for bonobo evolution (Hare, Wobber, & Wrangham, 2012). Thus, insights into the emergence of domestication and the role of constructed environments could also help explain how humans and other great apes evolved.

Another implication pertains to the evolution of social learning and culture. One central tenet in models on the adaptive value of culture is that the evolution of social learning depends on the level of environmental variation (Boyd & Richerson, 1985; Rogers, 1988; Hoppitt & Laland, 2013; Deffner & McElreath, 2022). If the environment varies too rapidly, selection is expected to favor individual (e.g., trial-and-error) learning because it provides up-to-date environmental information. If, however, the environment varies too slowly, genetic adaptation, the most reliable form of information transmission, will suffice to track such slow changes. Hence, it is often assumed that rates of environmental change must occupy a certain middle ground for reliance on social information to evolve. In light of the core finding that (counteractive) niche construction results in reduced temporal variation in selection, there is an intriguing possibility that humans (and possibly some other animals) themselves increasingly created the conditions that favored ever more social learning through their cultural niche-constructing activities (Laland et al., 2001; Laland & O'Brien, 2011), resulting in a self-reinforcing process (Laland, 2018). Such cultural niche construction that influences the predictability of the environment also has important implications for emerging levels of cultural diversity or complexity. For this reason, a population's relationship with its environment (i.e., its niche-constructing activities) must be considered alongside its effective population size (Deffner, Kandler, & Fogarty, 2022) in studies of cultural diversity and complexity (Fogarty & Creanza, 2017).

To summarize, if evolutionary processes are by definition restricted to processes that directly change allele frequencies, researchers might miss crucial details in the intricate causal webs underlying adaptation and evolution. Organisms do not merely modify the selection pressures they and other species experience: they modify them in an orderly, directed, and sustained manner resulting in detectable and surprisingly predictable selective outcomes. Therefore, broadening conventional conceptions of evolutionary causation and inheritance and incorporating the impact that organisms have on their selective environment will provide a richer and more detailed description of the causal network underlying selection and adaptive evolution.

Note

1. Note that sections 4.1–4.3 and 4.9. are partly based on D. Deffner's *Niche construction and the strength and temporal dynamics of natural selection in the wild* (unpublished MSc Thesis, University of St Andrews, UK, 2017).

References

Ågren, J. A. (2021). *The gene's eye view of evolution*. Oxford University Press.

Alberti, M. (2015). Eco-evolutionary dynamics in an urbanizing planet. *Trends in Ecology & Evolution, 30*(2), 114–126.

Alberti, M., Correa, C., Marzluff, J. M., Hendry, A. P., Palkovacs, E. P., Gotanda, K. M., & Zhou, Y. (2017). Global urban signatures of phenotypic change in animal and plant populations. *Proceedings of the National Academy of Sciences*, 201606034.

Beja-Pereira, A., Luikart, G., England, P. R., Bradley, D. G., Jann, O. C., Bertorelle, G., . . . & Erhardt, G. (2003). Gene-culture coevolution between cattle milk protein genes and human lactase genes. *Nature Genetics, 35*(4), 311.

Bogert, C. M. (1949). Thermoregulation in reptiles, a factor in evolution. *Evolution, 3*(3), 195–211.

Bonduriansky, R. (2012). Rethinking heredity, again. *Trends in Ecology & Evolution, 27*(6), 330–336.

Boyd, R., & Richerson, P. J. (1985). *Culture and the evolutionary process*. University of Chicago Press.

Cavalli-Sforza, L. L., & Feldman, M. W. (1981). *Cultural transmission and evolution: A quantitative approach* (No. 16). Princeton University Press.

Clark, A. D., Deffner, D., Laland, K., Odling-Smee, J., & Endler, J. (2020). Niche construction affects the variability and strength of natural selection. *The American Naturalist, 195*(1), 16–30.

Danchin, É., Charmantier, A., Champagne, F. A., Mesoudi, A., Pujol, B., & Blanchet, S. (2011). Beyond DNA: Integrating inclusive inheritance into an extended theory of evolution. *Nature Reviews Genetics, 12*(7), 475–486.

Dawkins, R. (2004). Extended phenotype—but not too extended. A reply to Laland, Turner and Jablonka. *Biology and Philosophy, 19*(3), 377–396.

Deffner, D., & Kandler, A. (2019). Trait specialization, innovation, and the evolution of culture in fluctuating environments. *Palgrave Communications, 5*(1), 1–10.

Deffner, D., Kandler, A., & Fogarty, L. (2022). Effective population size for culturally evolving traits. *PLoS Computational Biology, 18*(4), e1009430.

Deffner, D., & McElreath, R. (2022). When does selection favor learning from the old? Social Learning in age-structured populations. *PloS One, 17*(4), e0267204.

Edelaar, P., & Bolnick, D. I. (2019). Appreciating the multiple processes increasing individual or population fitness. *Trends in Ecology & Evolution, 34*(5), 435–446.

Endler, J. A. (1986). *Natural selection in the wild* (No. 21). Princeton University Press.

Feldman, M. W., & Laland, K. N. (1996). Gene-culture coevolutionary theory. *Trends in Ecology & Evolution*, *11*(11), 453–457.

Fogarty, L., & Creanza, N. (2017). The niche construction of cultural complexity: Interactions between innovations, population size and the environment. *Philosophical Transactions of the Royal Society B: Biological Sciences*, *372*(1735), 20160428.

Fogarty, L., & Wade, M. J. (2022). Niche construction in quantitative traits: heritability and response to selection. *Proceedings of the Royal Society B*, *289*(1976), 20220401.

Futuyma, D. J. (2017). Evolutionary biology today and the call for an extended synthesis. *Interface Focus*, *7*(5), 20160145.

Griffiths, P. E., & Gray, R. D. (1994). Developmental systems and evolutionary explanation. *The Journal of Philosophy*, *91*(6), 277–304.

Hansell, M. (2000). *Bird nests and construction behaviour*. Cambridge University Press.

Hansell, M. (2007). *Built by animals: The natural history of animal architecture*. Oxford University Press.

Hare, B. (2017). Survival of the friendliest: *Homo sapiens* evolved via selection for prosociality. *Annual Review of Psychology*, *68*(1), 155–186.

Hare, B., Wobber, V., & Wrangham, R. (2012). The self-domestication hypothesis: Evolution of bonobo psychology is due to selection against aggression. *Animal Behaviour*, *83*(3), 573–585.

Henrich, J. (2015). *The secret of our success: How culture is driving human evolution, domesticating our species, and making us smarter*. Princeton University Press.

Henrich, J., & McElreath, R. (2003). The evolution of cultural evolution. *Evolutionary Anthropology: Issues, News, and Reviews*, *12*(3), 123–135.

Holden, C., & Mace, R. (2009). Phylogenetic analysis of the evolution of lactose digestion in adults. *Human Biology*, *81*(5/6), 597–619.

Hoppitt, W., & Laland, K. N. (2013). *Social learning: An introduction to mechanisms, methods, and models*. Princeton University Press.

Huey, R. B., Hertz, P. E., & Sinervo, B. (2003). Behavioral drive versus behavioral inertia in evolution: A null model approach. *The American Naturalist*, *161*(3), 357–366.

Jablonka, E., & Lamb, M. J. (2014). *Evolution in four dimensions, revised edition: Genetic, epigenetic, behavioral, and symbolic variation in the history of life*. MIT Press.

Kingsolver, J. G., Hoekstra, H. E., Hoekstra, J. M., Berrigan, D., Vignieri, S. N., Hill, C. E., . . . & Beerli, P. (2001). The strength of phenotypic selection in natural populations. *The American Naturalist*, *157*(3), 245–261.

Laland, K. N. (2018). *Darwin's unfinished symphony*. Princeton University Press.

Laland, K. N., & O'Brien, M. J. (2011). Cultural niche construction: An introduction. *Biological Theory*, *6*(3), 191–202.

Laland, K. N., Odling-Smee, J., & Endler, J. (2017). Niche construction, sources of selection and trait coevolution. *Interface Focus*, *7*(5), 20160147.

Laland, K. N., Odling-Smee, F. J., & Feldman, M. W. (1996). The evolutionary consequences of niche construction: A theoretical investigation using two-locus theory. *Journal of Evolutionary Biology*, *9*(3), 293–316.

Laland, K. N., Odling-Smee, F. J., & Feldman, M. W. (1999). Evolutionary consequences of niche construction and their implications for ecology. *Proceedings of the National Academy of Sciences*, *96*(18), 10242–10247.

Laland, K. N., Odling-Smee, J., & Feldman, M. W. (2001). Cultural niche construction and human evolution. *Journal of Evolutionary Biology*, *14*(1), 22–33.

Laland, K. N., Odling-Smee, J., & Myles, S. (2010). How culture shaped the human genome: Bringing genetics and the human sciences together. *Nature Reviews Genetics*, *11*(2), 137–148.

Lande, R., & Arnold, S. J. (1983). The measurement of selection on correlated characters. *Evolution*, *37*(6), 1210–1226.

Levins, R., & Lewontin, R. C. (1985). *The dialectical biologist*. Harvard University Press.

Lewontin, R. C. 1983. Gene, organism, and environment. In D. S. Bendall (ed.), *Evolution from molecules to men*, pp. 273–285. Cambridge University Press.

MacColl, A. D. (2011). The ecological causes of evolution. *Trends in Ecology & Evolution*, *26*(10), 514–522.

Marais, E., & Chown, S. L. (2008). Beneficial acclimation and the Bogert effect. *Ecology Letters*, *11*(10), 1027–1036.

Matthews, B., De Meester, L., Jones, C. G., Ibelings, B. W., Bouma, T. J., Nuutinen, V., . . . & Odling-Smee, J. (2014). Under niche construction: An operational bridge between ecology, evolution, and ecosystem science. *Ecological Monographs*, *84*(2), 245–263.

Morrissey, M. B. (2016). Meta-analysis of magnitudes, differences and variation in evolutionary parameters. *Journal of Evolutionary Biology*, *29*(10), 1882–1904.

Morrissey, M. B., & Hadfield, J. D. (2012). Directional selection in temporally replicated studies is remarkably consistent. *Evolution*, *66*(2), 435–442.

Odling-Smee, F. J. (1988). Niche-constructing phenotypes. In H. C. Plotkin (ed.), *The role of behavior in evolution* (pp. 73–132). The MIT Press.

Odling-Smee, F. J., Laland, K. N., & Feldman, M. W. (2003). *Niche construction: the neglected process in evolution* (No. 37). Princeton University Press.

Odling-Smee, J., Erwin, D. H., Palkovacs, E. P., Feldman, M. W., & Laland, K. N. (2013). Niche construction theory: A practical guide for ecologists. *The Quarterly Review of Biology*, *88*(1), 3–28.

Oyama, S. (2000). *The ontogeny of information: Developmental systems and evolution*. Duke University Press.

Oyama, S., Griffiths, P. E., & Gray, R. D. (eds.). (2003). *Cycles of contingency: Developmental systems and evolution*. MIT Press.

Pelletier, F., Garant, D., & Hendry, A. P. (2009). Eco-evolutionary dynamics. *Philosophical Transactions of the Royal Society B: Biological Sciences*, *364*(1523), 1483–1489.

Post, D. M., & Palkovacs, E. P. (2009). Eco-evolutionary feedbacks in community and ecosystem ecology: Interactions between the ecological theatre and the evolutionary play. *Philosophical Transactions of the Royal Society of London B: Biological Sciences*, *364*(1523), 1629–1640.

Rogers, A. R. (1988). Does biology constrain culture? *American Anthropologist*, *90*(4), 819–831.

Schrödinger, E. (1944/1992). *What is life?* (reprint). Cambridge University Press.

Scott-Phillips, T. C., Laland, K. N., Shuker, D. M., Dickins, T. E., & West, S. A. (2014). The niche construction perspective: A critical appraisal. *Evolution*, *68*(5), 1231–1243.

Siepielski, A. M., DiBattista, J. D., & Carlson, S. M. (2009). It's about time: The temporal dynamics of phenotypic selection in the wild. *Ecology Letters*, *12*(11), 1261–1276.

Siepielski, A. M., Gotanda, K. M., Morrissey, M. B., Diamond, S. E., DiBattista, J. D., & Carlson, S. M. (2013). The spatial patterns of directional phenotypic selection. *Ecology Letters*, *16*(11), 1382–1392.

Sober, E. (1984). *The nature of selection*. MIT Press.

Stan Development Team. 2018. *Stan modeling language users guide and reference manual*, version 2.18.0. http://mc-stan.org

Stellatelli, O. A., Block, C., Villalba, A., Vega, L. E., Dajil, J. E., & Cruz, F. B. (2018). Behavioral compensation buffers body temperatures of two Liolaemus lizards under contrasting environments from the temperate pampas: A Bogert effect? *Ethology Ecology & Evolution*, *30*(4), 297–318.

Sultan, S. E. (2015). *Organism and environment: Ecological development, niche construction, and adaptation*. Oxford University Press.

Theunissen, B. (2012). Darwin and his pigeons: The analogy between artificial and natural selection revisited. *Journal of the History of Biology*, *45*(2), 179–212.

Waddington, C. H. (1959a). Evolutionary systems—animal and human. *Nature*, *183*(4676), 1634–1638.

Waddington, C. H. (1959b). Canalization of development and genetic assimilation of acquired characters. *Nature*, *183*(4676), 1654.

Waddington, C. H. (1975). *The evolution of an evolutionist*. Cornell University Press.

Wade, M. J., & Kalisz, S. (1990). The causes of natural selection. *Evolution*, *44*(8), 1947–1955.

Wei, W. W. (2006). Time series analysis. In *The Oxford handbook of quantitative methods in psychology*, vol. 2. Oxford University Press.

Whitehead, H., Laland, K. N., Rendell, L., Thorogood, R., & Whiten, A. (2019). The reach of gene–culture coevolution in animals. *Nature Communications*, *10*(1), 1–10.

Whiten, A. (2021). The burgeoning reach of animal culture. *Science*, *372*(6537), 1–7.

Williams, G. C. (1992). *Natural selection: Domains, levels, and challenges*. Oxford University Press.

Williamson, S. H., Hubisz, M. J., Clark, A. G., Payseur, B. A., Bustamante, C. D., & Nielsen, R. (2007). Localizing recent adaptive evolution in the human genome. *PLoS Genetics*, *3*(6), e90.

Zeder, M. A. (2006). Central questions in the domestication of plants and animals. *Evolutionary Anthropology: Issues, News, and Reviews*, *15*(3), 105–117.

Zeder, M. A. (2015). Core questions in domestication research. *Proceedings of the National Academy of Sciences*, *112*(11), 3191–3198.

Zeder, M. A. (2016). Domestication as a model system for niche construction theory. *Evolutionary Ecology*, *30*(2), 325–348.

5 Relational Agency: A New Ontology for Coevolving Systems

Francis Heylighen

Overview

A wide variety of approaches and mechanisms have been proposed to "extend" the neo-Darwinist theory of evolution, including self-organization, symbiogenesis, teleonomy, systems biology, and niche construction. These extensions share a focus on agents, networks, and processes rather than on independent, static units such as genes. To develop a new evolutionary synthesis, we therefore need to replace the traditional object-based ontology with one that is here called *relational agency*. This paper sketches the history of both object-based and relational worldviews, going back to their roots in animism and Greek philosophy. It then introduces the basic concepts of the relational agency model: condition-action rules, challenges, agents, reaction networks, and chemical organizations. These are illustrated with examples of self-contained ecosystems, genes, and cells. The fundamental evolutionary mechanism is that agencies and reactions mutually adapt so as to form a self-maintaining organization, in which everything consumed by one process is produced again by one or more other processes. Such autonomous organization defines a higher-level agency, which will similarly adapt, and thus become embedded in a network of relationships with other agencies.

5.1 Introduction

The dominant paradigm in biology since about the mid-twentieth century has been neo-Darwinism, also known as the "modern synthesis" of Darwinian evolution with genetics (Mayr & Provine, 1980). It assumes that all features of living systems can be explained by evolution at the level of the genes. More precisely, the theory reduces evolution to changes in the frequency of certain genes in the population under the influence of natural selection. In other words, the environment imposes a selective pressure or force of natural selection that increases the relative frequency of alleles that confer higher fitness. From the neo-Darwinian perspective, the only additional mechanism needed to explain evolution is random variation that creates new, potentially fitter genes, from which the best ones can then be selected in an ongoing process that generates organisms that are ever better adapted to their environment.

Neo-Darwinism has been very successful in proposing a simple, clear, and explicit mechanism that can explain a wide range of biological phenomena, many of which seemed difficult to explain otherwise—such as the precarious balance between selfish and cooperative behaviors (Dawkins, 2006). However, over the past decades its assumptions have drawn increasing criticism for being too reductionist, and in particular for neglecting a wide range of phenomena characterized by complex interactions. These interactions take place at, and in between, different levels of complexity, including molecules, genes, organelles, cells, tissues, organs, organisms, groups, ecosystems, and physical environments. Of these, neo-Darwinism only really focuses on genes and environments, while at best acknowledging the importance of organisms as the phenotypes, or gene-carrying "vehicles," that mediate between the two (Dawkins, 2006).

A wide variety of approaches have tried to complement or extend the evolutionary synthesis with some of these more complex processes (Fábregas-Tejeda & Vergara-Silva, 2018a; Laland et al., 2015; Pigliucci & Müller, 2010). There are too many of those to review in the space of this introductory section, so I will here just list the most influential ones.

Self-organization is a mechanism that explains how complex organization can emerge from local interactions without need for selection by the environment (Ashby, 1962; Camazine et al., 2003; Heylighen, 2001; Kauffman, 1993). *Systems biology* is an approach that focuses on the complex network of reactions, pathways, and cycles necessary to maintain a living system's metabolism (Palsson, 2015). *Coevolution* is the idea that the environment of an organism mostly consists of other organisms, each of which is adapting to the changes occurring in the others. Thus, they are pushing each other to evolve further and further. *Synergy* is the idea that the resulting relations between organisms are not solely competitive, but that they can be mutually beneficial, creating selection for synergetic relationships (Corning, 2003). *Symbiogenesis* is the formation of more complex organisms by the merging of independently evolved organisms that have developed some form of synergy (Agafonov, Negrobov, & Igamberdiev, 2021). *Niche construction* notes that organisms are not just adapting to a pre-existing environment, but that they themselves change the environment in which they live, thus affecting their own selection (Laland, Matthews, & Feldman, 2016). *Teleonomy* refers to the observation that development and behavior of organisms consist not just of passively reacting to the environment, but are actively directed toward remote goals, by compensating for perturbations that otherwise would make the system deviate from these goals (Mayr, 1974; Corning, 2023). *Evo-devo* is an approach that focuses on how the ontogenetic development of embryos into complex organisms both constrains and enables the organic forms that can evolve on the genetic level (Fábregas-Tejeda & Vergara-Silva, 2018b; Müller, 2007). *Multilevel selection* notes that selection takes place not only at the level of genes, but also at the levels of cells, colonies, multicellular organisms, and groups of organisms (Okasha, 2005; Wilson, 1997). The theory of *major evolutionary transitions* tries to understand how cooperative interactions between systems at one level, such as cells, can give rise to the emergence of a collective system at a higher level, such as a multicellular organism (Maynard Smith & Szathmáry, 1997; West et al., 2015).

While seemingly disparate, these different approaches have two assumptions in common. First, they are *relational* (Heylighen, 1990; Marmodoro & Yates, 2016). Instead of reducing biological systems to independent units, such as genes or individuals, they investigate how these units interact and thus form part of a network of interdependencies that may give rise to a higher-order system. Second, they emphasize autonomous action or *agency*. Instead of seeing organisms as passively undergoing the forces of natural selection, they investigate how systems and processes actively seek or produce fit, synergetic arrangements, while counteracting environmental influences that push them away from these preferred states. Putting the two assumptions together leads me to formulate a new, synthetic perspective that I will call *relational agency*. Simply formulated, this is an approach that sees the world as a network of interacting agencies rather than as a collection of independent objects subjected to external forces. In the remainder of this paper, I will try to formulate some foundational concepts and principles for this worldview.

Yet, to really understand the revolutionary implications of this perspective, we first need to distinguish it from the more traditional object-based worldview, whose ontology Walsh has summarized as "objectcy" (Walsh, 2018). This ontology is implicit not only in neo-Darwinism, but in most modern scientific theories. As I hope to show, these implicit assumptions about the nature of reality make it intrinsically difficult to go beyond the traditional perspective and develop a deep insight into complex interaction. Understanding them better is a necessary first step in developing an alternative ontology that transcends the limitations of the object-based view.

5.2   The Ontology of Objects

The modern scientific worldview originated and developed within the European culture of the last 2,500 years, the roots of which go back to Greek philosophy. Parmenides, supported by his student Zeno, argued that change ("Becoming") is an illusion, and that only eternal "Being" is real. Inspired by the abstract structures of geometry, Plato postulated that the ever-varying phenomena we perceive are merely imperfect reflections of absolute, unchanging forms. Although his student Aristotle was more interested in concrete phenomena, he kept a focus on their invariant essences. This resulted in an ontology that sees the world as consisting of objects having properties that are either essential (i.e., defining the category to which the object belongs) or accidental. For example, an object in the category of "humans" has essential properties, such as having a heart and being able to think, and accidental properties, such as being tired or being located in a particular place.

The next step in the development of the object-based worldview, due to thinkers such as Galileo, Descartes, and Newton, was the insight that properties can be measured: that is, determined in a precise, objective manner as *quantities*. The essential properties remain invariant, but the accidental ones, such as temperature, position, or velocity, can vary over the course of time. The corresponding variables define a state space or phase space. Changes in the values of these variables are described by dynamical equations, which express the effect of different forces, such as gravitation, on the objects. Thus, all change can be reduced to the deterministic, time-parameterized trajectory of a collection of invariant objects through their state space (Heylighen, Cilliers, & Gershenson, 2007).

This assumption is the foundation of the Newtonian, mechanistic paradigm that has dominated science at least until the beginning of the twentieth century—when it was put into question by developments such as relativity theory, quantum mechanics, and especially quantum field theory. However, most of the assumptions of this paradigm are still implicitly held by scientists and laypeople alike when they think about how the universe is structured. That is in part because the ontology of objects appears so simple, concrete, and intuitive, in part because it is so dominant in the Western, scientific worldview.

Nevertheless, there have always been alternative worldviews that focused on processes and relations rather than on invariant objects. These can be found in pre-Socratic and non-Western philosophies, such as those of Heraclitus, Daoism, and Buddhism, and in certain ideas proposed by Western philosophers such as Leibniz, Spinoza, Whitehead, Teilhard de Chardin, and Bergson. Yet, the success of mechanistic science in explanation, prediction, and application was so great that these alternatives were largely ignored. Nevertheless, as our insight into complex, evolving systems advances, the shortcomings of mechanicism and its object ontology become ever more apparent (Heylighen, Cilliers, & Gershenson, 2007). To better understand these, we must first formulate the underlying assumptions more explicitly.

The ontology of objects assumes that all phenomena we observe can be reduced to objects and their movement in space. Objects are separate, invariant pieces composed of some inert *substance*, which we call *matter*. Matter may be repositioned, divided, or reshaped, but it cannot be destroyed or created. It just stays there, passively waiting until some force moves it into a different spatial arrangement. By assembling elementary pieces of matter, which we call *particles*, into different forms or shapes, different objects can be created. But the objects are intrinsically just as passive, inert, or insensitive as the matter that constitutes them. Unless some force acts on them, they remain as they are: rigid, invariant, unchanging.

Objects are distinguished by their location in space. No two objects can occupy the same space: some minimal distance must separate them. Therefore, distinct objects are intrinsically independent; you can move the one without affecting the others. Objects can be identified by their form, that is, the geometry of the matter that constitutes them. Form and position determine their properties. Therefore, properties are intrinsic, absolute, and objective. They do not depend on the presence or absence of other objects, including those constituting what we call an observer, environment, or context.

Complex objects (*systems*) that are constituted of smaller objects (*parts*) inherit their properties from these parts. There are no emergent properties; the whole does not have anything that is more than the sum of its parts. Therefore, to understand the behavior of the whole, it is sufficient to accurately observe the changes in the properties of its parts: that is, in the variables that describe the system. These changes are driven by external forces obeying natural laws. For example, the fundamental object of neo-Darwinist theory is a population. Its variable properties are the frequencies of different genes. Changes in these frequencies are driven by the force of natural selection, here conceived of as adaptation to a given, external environment.

Up to this point, the ontology of objects is perfectly consistent. However, if we wish to introduce the observing subject—the human individual who is perceiving and manipulating these objects—then we are confronted with a paradox. If we assume that this observer can reason and freely decide how to act on these objects, then this observer cannot itself be an inert object. The observing subject must have some form of autonomous agency. This includes a sensitivity to the outside world, some form of motive or goal, and the ability to manipulate and relate to these objects it is observing. Descartes resolved this paradox by postulating that human subjects are constituted out of the very different substance of *mind*,

Table 5.1
Difference between objects and agencies

	Objects	Agencies
Sensitivity	inert	sensitive
Activity	passive	active
Dynamics	mechanical	goal-directed
Shape	rigid	variable
Positioning	localized	localized or distributed
Constituents	material particles	processes
Existence	independent	relational

which exists independently of the *matter* that constitutes objects. This ontology is known as *dualism*, because it assumes two separate realms: mind and matter.

Though intuitively plausible, from a scientific perspective dualism is a highly unsatisfactory ontology, which creates many more questions and paradoxes than it resolves (Heylighen & Beigi, 2018). These largely center on the seemingly unsolvable mind-body problem: how can the nonphysical mind affect, and be affected by, the inert matter of the body, the activity of which is already fully determined by physical forces? Moreover, dualism is incompatible with an evolutionary worldview, which sees the human mind as gradually emerging from much simpler animal, unicellular, and inorganic forms of organization.

In earlier work (Heylighen & Beigi, 2018), I have proposed a radical resolution to this conundrum, by interpreting both objects and subjects as special cases of the more general category of *agents*: that is, phenomena that act on other phenomena. The remainder of this paper spells out some of the implications of this ontology for understanding biological organization and its evolution, by interpreting not only organisms, but also their physical components as well as the ecosystems they collectively form, as interrelated agencies. Table 5.1 provides a first sketch of how agencies differ from objects, a summary that will be elaborated in the rest of the paper.

5.3　Animism: The Origin of Relational Agency

The roots of the worldview that I will call *relational agency* are actually much older than those of the objectcy worldview. The worldview of hunter-gatherers has been characterized as *animism* (Harvey, 2014). That means that they see the phenomena around them as animated: that is, as having agency. These phenomena of course include the animals they hunt, live with, or fear, but also plants, rivers, clouds, and potentially even rocks. To survive and thrive in their local environment, they need to establish good relationships with these other agents. That requires getting to know their habits, capabilities, and preferences, while paying them due respect and trying to keep on good terms with them. They do not see themselves as separate from nature, but as merely one type of agents among many, taking part in a complex network of relationships.

According to the traditional view, animism is a naïve, irrational projection of a human-like mind or "spirit" into physical objects, thus attributing supernatural properties to these objects.

However, a more recent perspective in anthropology (Astor-Aguilera & Harvey, 2018; Bird-David, 1999; Ingold, 2006) interprets animism in a more pragmatic manner, as a practical and adaptive way of understanding the complex ecosystem in which hunter-gatherers live. Supernatural attributions, in this view, are misinterpretations made by old-school anthropologists who were biased by modern religions with their dualist ontology, in which spirits are conceived as disembodied, ghostlike entities separate from the physical objects they may inhabit. The Latin root of the word *spirit* (as well as the "anima" that gave rise to animism), however, simply means "breath." Breathing is what distinguishes a living, animated body from a dead one. Thus, in the more recent interpretation, animists are merely giving due respect to the aliveness of the world that surrounds them.

For hunter-gatherers, that aliveness, animation, or agency is not limited to biological organisms. A river, for example, is an active, self-organizing process that adapts its flow pattern to rainfall, landslides, or mud accretions, while persisting in its goal of discharging its water into a lake or sea. Even a rock may adapt to circumstances, such as eroding under the effect of frost, becoming slippery in the rain, or providing a substrate for animals to hide in its holes. Thus, a rock too is active: the processes it takes part in are merely much slower than those of animals and plants (Abram, 2010). Therefore, it may be seen as animated by some "spirit of the rock" whose habits and preferences it would be good to know. That would be useful if, for example, you regularly need to seek shelter underneath that rock, where you may otherwise be surprised by changed conditions, such as burrowing animals or leaking water.

Such detailed knowledge and understanding of natural agencies seem to have helped our hunter-gatherer ancestors to survive and thrive for millions of years before the advent of civilization. In the paleolithic environment in which humans evolved, animistic thinking appears to have been an adaptive strategy. It would therefore be shortsighted to dismiss animism as merely a prescientific misconception. A more sensible approach would be to ask *why* we eventually felt the need to shift from animism to our present object-based worldview.

5.4   The Origin of the Object Ontology

The anthropologist and philosopher David Abram has argued that this shift followed the adoption of alphabetic writing, first by the ancient Hebrews, then by the Greeks (Abram, 2012). Expressing ideas in the form of abstract symbols registered in an enduring document separates these ideas from the concrete, variable phenomena to which they refer. Moreover, the shape of letters and words has no relation with the shapes of these phenomena—another disconnect. Furthermore, the words registered on a page are intended to be read at a different time—that is, in a different context—than the one in which they were written. Therefore, their meaning (the phenomena they refer to) is supposed to be invariant across changes in context (Heylighen & Dewaele, 2002). Thus, it stands on its own, independent of other phenomena. According to Abram (2012), this is how written words became the prototype of the eternal, absolute "forms" postulated by Plato to be the fundamental constituents of reality. These in turn led Aristotle to conceive of objects as the material embodiments of these invariant and independent essences.

It is this context-independence of written descriptions that allowed knowledge to be recorded for the long term and to be disseminated widely without losing its meaning (Heylighen, 1999). Thus, for example, ancient Greek treatises about astronomy, geometry, and philosophy could still be read and understood two thousand years later in the very different context of the European Renaissance, providing an impetus for the scientific revolution. The further development of mathematical formalisms, such as Cartesian coordinates, calculus, and algebra, helped scientists to describe the properties of objects increasingly precisely and objectively, making abstraction of the observing subject and the context of observation. The resulting formalization allowed knowledge to be expressed in the form of universal "laws of nature" that are independent of space, time, or observer, thus forming the foundation of the classical scientific worldview (Heylighen, 2012).

Such precise, context-independent description also facilitated the development and spread of technology, because it allowed engineers to write down a full specification of the components and mechanisms of the objects they designed. Yet, even before formal descriptions, the object-based way of thinking had been stimulated by our use of an ever-larger array of artifacts, such as furniture, pottery, and tools. Indeed, unlike natural systems, artifacts are designed to be mere instruments—that is, passive objects that can be dependably manipulated by their human users—without deploying any agency of their own. You would not want your table to move to a different room because it prefers the view there! Neither would you want your hammer to decide it is better to swing sideways when you would like it to swing downward.

As people spent more and more time inside artificial environments, such as houses, cities, and offices, their interactions with the environment were increasingly restricted to the perception and handling of such passive objects. This stands in sharp contrast with the natural environment inhabited by hunter-gatherers and early farmers, where most phenomena, such as plants, animals, forests, rivers, and clouds, are intrinsically animated. Those that do behave like inert objects, such as pebbles or sticks, appear to be the exception rather than the rule. In our present environment, in contrast, we are surrounded by passive objects, and nonhuman agencies have become the exception. As a result, we have great difficulty conceiving of and experiencing nature as fundamentally active and relational.

Thus, it seems as if the development of civilization, and especially of the symbolic systems supporting science and technology, has alienated humans not just from their natural environment, but also from the animistic awareness that makes us intuitively understand and feel part of that living nature (Charlton, 2007). Abram and others (Abram, 2010; Harvey, 2019) have argued that our actions have severely upset the ecological balance—exhausting resources, polluting rivers, and driving many species to extinction—in part because we see nature as a mere collection of objects to be manipulated, rather than as a network of agencies in which we are intimately involved through a variety of relationships.

5.5   A Renaissance of Relational Agency?

Is there a way to recover the relational understanding needed for a deep ecological consciousness, without losing the benefits of objective, scientific description? That would require a conceptual framework and formal language able to accurately represent agencies and their relations, without reducing them to independent objects. The core message of this paper is that the elements of such a description are at hand: they merely have to be further clarified, elaborated, and integrated. Let us illustrate this development with some recent approaches.

One example of a presently popular formalism is *agent-based modeling* (Bonabeau, 2002; Macal & North, 2010). This pragmatic method is used to simulate the often-complex interactions between autonomous agents, and thus come to understand the emerging patterns of their joint evolution. Agents here are conceived as any entities that act on each other and the environment they share. These can be organisms, cells, molecules, firms, robots, or people. Computer simulation, unlike written language, can represent actual processes, showing us in real time how actions result in further actions. Such *multiagent simulations* (Ferber, Gutknecht, & Michel, 2004; Macal & North, 2010) are commonly used to illustrate a theoretical perspective known as *complex adaptive systems*. Typical examples of such systems are ecosystems, markets, and societies. Their dynamics combine individual variation and selection, adaptation, competition, cooperation, and collective self-organization (Hartvigsen, Kinzig, & Peterson, 1998; Holland, 2012; Miller & Page, 2007).

A related conceptual framework, *actor-network theory*, originates in sociology (Latour, 1996). It describes societal evolutions, such as the development of nuclear energy, as resulting from the interaction between and among very diverse "actors." These include human individuals, organizations, institutions, technologies, and even physical resources such as uranium, while forming a network of mutual dependencies. In the humanities, a similar perspective has inspired a new philosophical stance known as *posthumanism*. Posthumanists argue that there is no essential difference or separation between humans and nonhuman agents—such as animals, plants, or robots—and that, when reasoning and deciding about our common future, we humans should take into account our relations with these other agents (Barad, 2003; Nayar, 2018).

This brief sample of recently developed approaches illustrates the potential of relational agency to become not just a philosophical perspective, but a scientific method for investigating complex, evolving systems. Nevertheless, these existing approaches remain as yet fragmentary, disparate, and incoherent, originating from widely divergent viewpoints, disciplines, and theoretical frameworks. The same applies to the different "extensions" of evolutionary theory that we reviewed, to domains such as ecology, development, and molecular biology. In order to unify these different approaches, we need a conceptual framework and language that can deal with relations and agencies at all levels, from molecules to organisms, ecosystems, and societies. In the remainder of this paper, I propose a number of concepts and modeling methods that may serve as a foundation for such an integrated theory of relational agency.

5.6   The Ontology of Actions

The ontology of objects assumes that there are elementary objects, called "particles," out of which all more complex objects—and therefore the whole of reality—are constituted. Similarly, the ontology of relational agency assumes that there are elementary processes, which I will call *actions* or *reactions*, that form the basic constituents of reality (Heylighen, 2011; Heylighen & Beigi, 2018; Turchin, 1993).

A rationale for the primacy of processes over matter can be found in quantum field theory (Bickhard, 2011; Kuhlmann, 2000). Quantum mechanics has already made clear that observing some phenomenon, such as the position of a particle, is an action that necessarily affects the phenomenon being observed: there is no observation without interaction. Moreover, in general the result of that observation is indeterminate before the observation is made: the action of observing in a real sense *creates* the property being observed, through a process known as the collapse of the wave function (Heylighen, 2021; Tumulka, 2006). For example, before the observation a particle such as an electron typically does not have a precise position in space, but immediately afterward it does. More generally, quantum mechanics tells us that microscopic objects such as particles in general do not have objective, determinate properties, but that such properties are (temporarily) generated through interaction (Barad, 2003).

Quantum field theory adds that the objects (particles) themselves do not have a permanent existence, but that they too can be created or destroyed through interactions, such as nuclear reactions. Particles can even be generated by fluctuations of the vacuum (i.e., out of nothing), albeit that in this case their existence is so transient that they are called "virtual" (Milonni, 2013).

At a larger scale, the molecules that constitute living organisms are similarly ephemeral, being both produced and broken down by the chemical reactions that constitute the organism's metabolism. Cells and organelles in the body too are in a constant flux, being broken down by processes such as apoptosis and autophagy, while new ones are grown through cell division and from stem cells. Similar creation-destruction processes can be found at the level of ecosystems, where relations of predation, symbiosis, and reproduction between organisms join with meteorological and geological forces to produce a constantly changing landscape of forests and rivers, mountains and meadows. Even the planet is in a constant flux, as magma moves up, down, and sideways under its mantle, moving continents and opening up volcano-studded rifts. And the sun and stars that irradiate Earth are merely boiling cauldrons of nuclear reactions generating new elements in their cores, while releasing immense amounts of energy.

Against Parmenides, we may conclude that at all levels, from particles to stars, change ("Becoming") is more fundamental than permanence ("Being") (Prigogine, 1980), and that the stability of macroscopic objects is only apparent, ready to be dissolved when we more closely observe the microscopic processes that constitute them. That is why contemporary science requires a formal language that treats change as primary, and stability as derivative.

5.7 Reactions, Actions, and Agencies

Possibly the simplest way to represent an elementary process may be called a *condition-action rule* (Heylighen & Beigi, 2018). It notes that given some condition X, an action will take place that produces the new condition Y. This can be interpreted as a transformation from the situation X to the new situation Y, or as a causation linking the cause X to its effect Y. It can be expressed more briefly as "*if X, then Y*" or in symbols:

$$X \rightarrow Y$$

Conditions here are any states of affairs whose presence can be distinguished from their absence. Examples of conditions are the presence of various types or species of particles, chemical compounds, living organisms, habitats, geological features, or meteorological circumstances. An example is "*if* dark clouds are gathering, *then* it will rain."

These conditions can in general be decomposed into conjunctions of more elementary conditions, such as the joint presence of two or more circumstances, particles, chemicals, or biological species. For example, "*if* dark clouds are gathering *and* people are in the streets, *then* people will seek shelter." Adopting the notation used for reactions in physics and chemistry, I will denote conjunctions ("and") by the "+" operator:

$$a+b+\ldots \rightarrow e+f+\ldots$$

In this notation, the condition-action rules can be interpreted as *reactions*, in the sense that the conditions $\{a, b, \ldots\}$ on the left-hand side of the arrow "react" with each other in order to produce the novel conditions $\{e, f, \ldots\}$ on the right-hand side of the arrow. Such a reaction represents an elementary process that takes the elements on the left as its inputs and transforms them into the elements on the right as its outputs (Heylighen, Beigi, & Veloz, 2015).

Agencies (A) can now be defined in this framework as necessary conditions for the occurrence of a reaction—which, however, are not themselves affected by the reaction:

$$A+X \rightarrow A+Y$$

In chemistry, the function of A is the one of a *catalyst*: it enables the reaction that converts X into Y. Because A remains invariant during the reaction but has to be present in order for the reaction to take place, it can be seen as the agency of the conversion. The *reaction* among A, X, and Y can therefore be reinterpreted as an *action* performed by the *agency* A on condition X in order to produce condition Y. This can be represented in shorter notation as:

$$A: X \rightarrow Y$$

Note that while an agency remains invariant during the reactions it catalyzes, there in general exist other reactions that destroy (consume) or create (produce) that agency. Thus, while agencies can play the role reserved in the old ontology for objects or forces, they are neither inert nor invariant.

An agency will in general catalyze several reactions. This means that it will react to different conditions by different actions so as to produce different new conditions. For example:

$$A: X \rightarrow Y, Y \rightarrow Z, U \rightarrow Z$$

This set of actions triggered by A can be interpreted as a *dynamical system* that maps initial states (X, Y, U) onto subsequent states (Y, Z, Z) (Heylighen 2022; Sternberg 2010). Such a dynamical system typically has one or more *attractors*, which are (sets of) states to which the dynamics converge. This means that the processes catalyzed by the agency A will lead into the attractor, but that A does not provide any way out: once in the attractor, further actions can only produce states that are also in the attractor. An attractor is surrounded by a *basin of attraction*. This basin contains the states that converge to the attractor but are not part of the attractor itself. In the preceding example, Z is an attractor, and X, Y, and U are in its basin. This means that processes starting from these conditions necessarily end up in Z, without possibility of returning.

Such an attractor-centered dynamic can be interpreted as a model of *goal-directedness* or *teleonomy* (Heylighen, 2022; Heylighen & Beigi, 2018): the actions executed by agency A are directed toward the goal Z. Thus, Z constitutes the shared final state or "end" of processes starting from the different initial states X, Y, and U, a feature called *equifinality* (Lyman, 2004) or *plasticity*. Moreover, perturbations (i.e., processes not controlled by A)

that make a state deviate from its goal-directed trajectory will be counteracted by A so that they still end up in the attractor, at least as long as the deviation remains within the basin of that attractor. This is a characteristic of goal-directed systems that has been called *persistence* or *regulation* through negative feedback (Heylighen, 2022).

Thus, an agency can be seen as the enabler of a bundle of actions that act on the present situation so as to drive it toward a restricted set of attractor states. These define the implicit goal, "preference," or end of the agency. Although the reasoning that led to this definition is very abstract and general, the definition fits in with our intuitive sense of agency as the capability for goal-directed action. For example, bacteria exhibit chemotaxis, which means that they seek food by swimming up gradients of food molecules toward the source (Sourjik & Wingreen, 2012). The source (region with the highest concentration of food) functions here as the attractor of the chemotaxis dynamics. This dynamics is governed by a simple condition-action rule obeyed by the bacterium: *if* the concentration of food molecules does not increase, *then* change direction of movement. The net result is that the bacterium will be persistently moving in a direction where food concentration increases, until it reaches the highest concentration.

In this example, the agency is located within a spatially delimited system: the bacterial cell. We may call such a locally bounded agency an "agent." For example, the bacterium is the agent of its food-directed movement. An agent is in that respect similar to an object. However, there are also agencies that lack such a clear localization.

An example of an agency that is difficult to delimit is a thunderstorm. Its action consists in producing rain, wind, and lightning, while its "goal" is to discharge the water it carries onto the earth. In earlier times, this agency may have been conceived as a god of thunder and rain, such as Thor or Zeus. In present times, we try to understand it by means of very complicated meteorological models. But in any case, the impact of this agency on its surroundings is great, and therefore it is important to be able to anticipate and adapt to its actions. This requires an understanding of that agency's typical behavior. That knowledge includes rules such as that people walking outside during a thunderstorm will get wet, that tall structures may be hit by lightning, and that rivers may flood. An object-based ontology is totally unhelpful in this respect, because a thunderstorm is not an object in any practical sense of the word, while it is useless to try to reduce it to the more object-like molecules of water and air that it contains.

5.8   Agents and Challenges

From the point of view of an individual agent (or more general agency), the environment consists of conditions to which it may or may not react by performing an action. If the present condition is not a precondition for one of the reactions in the repertoire of condition-action rules that characterize the agent, then that agent will simply ignore that condition. For example, the presence of nitrogen in the air will be ignored by animals, because free nitrogen does not react with any of their metabolic processes. Therefore, they are insensitive to the amount of nitrogen present, and would not notice if that nitrogen were replaced by a different nonreactive gas, such as argon. In contrast, they are highly sensitive to the amount of oxygen in the air, because they need it for their respiration and to produce energy. Still, there are nitrogen-sensitive agencies, namely bacteria, that can absorb nitrogen from the air and transform it into ammonia or nitrates, which in turn can be used by plants as fertilizer.

Some conditions, such as the presence of oxygen, do trigger reactions from the agent, such as breathing. I will call such activating conditions *challenges*, because they incite or challenge the agent to act: that is, to change something about the situation (Heylighen, 2012, 2014). For a goal-directed agent, a challenge may be positive or negative. It is positive when it helps the agent to get closer to its goal. It is negative when it makes it more difficult for the agent to reach its goal. For example, for a bacterium the ultimate goal at which its actions are directed is survival and multiplication. The presence of food is a positive challenge, because consuming that food will help the bacterium achieve that goal. The presence of a toxin, however, is a negative challenge, because it makes it less likely for the bacterium to survive. Therefore, the bacterium will act so as to approach and ingest food and to evade toxins.

A positive challenge may be called an *affordance*, because it provides the agent with an opportunity to achieve some benefit (i.e., move closer to its goal). A negative challenge may be called a *disturbance*; that is, something that pushes the agent away from its goal-directed course of action. A challenge can also be neutral, in the sense that the action it elicits is neither one of approach nor one of avoidance, but of, say, exploration. In that case, I may call it a *diversion*: it incites a deviation from the present course of action that brings the agent neither closer to, nor farther away from, its goal.

That course of action is in general unpredictable because the environment will constantly throw up challenges that demand some novel action from the agent, so as to remain on course to the goal (Heylighen, 2012). The reason is that the environment consists of other agencies. These similarly react to challenges by actions that produce new conditions. These in turn may challenge one or more agent to act. Thus, we may say that *challenges propagate*: the reaction to a challenge by one agent will typically create a challenge for one or more other agents, potentially setting in motion a cascade of actions triggering further actions (Heylighen, 2014). That is because agents share a common environment, which functions as a medium interconnecting the different actions and agencies, thus providing a channel for (implicit) communication and coordination of actions (Heylighen, 2016). Let us investigate the resulting network of relationships.

5.9   Relations Between Actions and Agencies

Unlike objects, the reactions that form the elements of our new ontology are intrinsically coupled. That is because the conditions that constitute the inputs and outputs of a reaction always must occur in some other reaction as well. This is the principle of "the difference that makes a difference" (Bateson, 2000): for some condition to be distinguishable, it must incite a reaction that itself leads to a distinguishable condition. A condition that does not produce any distinguishable effect simply cannot be observed, either directly or indirectly. Therefore, we may as well assume that it does not exist. For example, it is in principle possible to reintroduce inert objects in the action ontology by means of the reaction $a \rightarrow a$ (the object a is always followed by the object a, without any change). But if we want to observe whether object a is present, we will need at least a reaction of the form: $a + \text{observation} \rightarrow a + \text{observation result}$. For example, the observation may consist in shining light on the object, while the observation result consists of the detection of light reflected back by the object.

Therefore, different reactions necessarily share certain conditions, in the sense that the inputs or outputs of one reaction must also be outputs or inputs of some other reaction. That means that reactions are mutually dependent. If, for example, one reaction *produces* the condition X, while another reaction *is triggered by* the condition X, then the first one will automatically be followed by the second one. Yet, relations can be more complicated. For example, one reaction may be triggered by $a+b$, whereas another is triggered by $a+c$. If now some third reaction produces a, then, depending on the presence or absence of b and c, none, one, or both of these initial reactions may be triggered. Thus, reactions form a complex network of interdependencies (Veloz & Razeto-Barry, 2017).

Some of these relationships will be synergetic, in the sense that two or more agencies or reactions together can produce more of the conditions or resources they all need to continue functioning than each of them on its own. Others will be characterized by conflict or friction, in the sense that the activity of the one will impede the continued activity of the other(s). An agency surrounded by synergetic agencies will be more successful in achieving its goals (ultimately survival and multiplication) than one surrounded by agencies that have a relation of friction with it. Therefore, natural selection will tend to favor agencies profiting from synergies and to eliminate agencies suffering from frictions. There will be a general trend for evolution to promote synergetic relationships among agents, while weakening relationships characterized by conflict or friction (Corning, 2003). However, in a complex network of relationships involving several agencies, as we find in ecosystems or biological networks, it is not always obvious what precisely constitutes synergy. Let us illustrate such mutual dependencies with a simple example of a self-contained ecosystem, and then elaborate on it by introducing variation and selection, so as to develop a relational perspective on evolution.

5.10 Self-Maintaining Systems

Consider a special type of aquarium that does not exchange any matter with the outside world. Such a hermetically sealed, transparent bowl, called an ecosphere, contains air, seawater, shrimps, algae, and bacteria (Schilthuizen, 2008). The organisms in such a bowl can survive for years without requiring any outside intervention. That is because their activities form a self-maintaining whole that can be summarized by the following reactions:

$$\rightarrow light$$

$$shrimps + algae + O_2 \rightarrow shrimps + waste + CO_2 + heat$$

$$bacteria + waste + O_2 \rightarrow bacteria + nutrients + CO_2 + heat$$

$$algae + light + CO_2 + nutrients \rightarrow 2\ algae + O_2$$

$$heat \rightarrow$$

These reactions describe how the shrimps consume algae and oxygen, while producing waste and carbon dioxide. The bacteria transform the waste into nutrients, such as nitrates and phosphates, that act as fertilizer for the algae. These nutrients, together with carbon dioxide, allow the algae to grow, while using light for photosynthesis and producing oxygen as a byproduct. The additional algae feed the shrimps, so that these can survive, and sustain the initial reaction.

This network of reactions includes three agencies: shrimps, algae, and bacteria. Their actions are perfectly complementary: each produces what the others consume, and consumes what the others produce. The only resource that has to come from outside the system to keep it going is light. Its only output is heat. That is because such a self-maintaining system, even when it recycles most of the products it uses, must be thermodynamically open in order not to degrade to its maximum entropy state in which everything decays.

It may seem strange that the amount of resources produced and consumed would remain perfectly balanced, so that the system maintains itself over the years. The explanation is that the different agencies will adjust their production and consumption of resources until they are mutually adapted. Suppose that initially there are too many algae for the amount of nutrients available. More algae means more food for the shrimps and therefore more waste and thus nutrients produced. Thus, the amount of algae will diminish by increased consumption, while the amount of nutrients they need will increase, until both are in balance. Suppose now that there are not enough algae initially. This may lead to the dying-off of some of the shrimps. The bacteria will decompose the dead shrimps into nutrients. Fewer shrimps means less consumption of algae. This, together with the additional nutrients, will allow the algae to multiply, until their number is in balance with that of the other components of the system.

More generally, every fluctuation up or down of one of the resources that make up this system will elicit feedback that is negative, that is, that moves in the opposite direction of the initial fluctuation. The reason is that in such a self-maintaining network of reactions the reduced availability of some resource will normally lead to a reduced consumption of that resource, while production of that resource does not immediately decrease, so that the net availability increases again. A similar negative feedback will reduce increased availability: an increase in a resource will lead to an increase in its consumption, but not in its production, so that net availability decreases. Thus, resource concentrations, though they can vary, generally do not deviate too far from their equilibrium values.

Such a closed ecosystem illustrates a key concept in reaction networks: a *(chemical) organization*. This concept was introduced by Peter Dittrich, thus founding an approach known as chemical organization theory (Dittrich & Fenizio, 2007; Heylighen, Beigi, & Veloz, 2015; Veloz et al., 2022). In this theory, an organization is defined as a network of reactions and resources (also called "molecules" or "species") that is *closed* and *self-maintaining. Closed* here means that no resources are produced by the reactions that are not already part of the network. *Self-maintaining* means that all resources that are consumed by some reactions are produced at least as much by other reactions, so that their total amount does not go to zero. In other words, the resources in a chemical organization are fully recycled and thus remain perpetually available.

Interestingly, it can be shown that the *attractors* of the dynamical system defined by the network of reactions are all chemical organizations (Peter & Dittrich, 2011). This means that the system tends to spontaneously settle in one of these organizations, and that once there, it will remain there. In other words, such self-maintaining, closed networks tend to self-organize. Self-organization, as observed long ago by the cybernetician Ashby, can be conceptualized as the mutual adaptation or alignment of the different subsystems within an overall system (Ashby, 1962; Heylighen, 2001). Let us illustrate the principle by imagining how one could create a new self-contained ecosystem from scratch.

5.11 Introducing Variation and Selection

Instead of buying a commercially available ecosphere aquarium, you might prefer to start one of your own. A simple method is to fill a big glass jar with a variety of organisms and materials collected in ponds, such as soil, pond water, water plants, small crustaceans, snails, worms, and insect larvae, and perhaps even some tadpoles or small fish. You then close the jar so that the water cannot evaporate, leaving some air at the top, keep it in a place with sufficient light, and wait. Typically, many of the organisms will die, because they lack the right conditions necessary for their survival, or because they do not find enough resources, such as food organisms, to sustain a thriving population. If all the members of a species have died, so that no new ones can be produced, then that species has been removed forever from the closed ecosystem. This is *natural selection* at work on the micro-scale of an ecosphere: that species was not adapted to its local environment.

Yet, some species are likely to survive. Initially, their populations may fluctuate wildly, but then overproduction of one species (e.g., overgrowth of algae profiting from the nutrients released by the decomposition of dead organisms) will tend to be suppressed by the corresponding population growth of another species that consumes them (e.g., snails eating algae). As this negative feedback sets in, fluctuations decrease, and the system reaches some degree of stability. Note that the resulting self-sustaining ecosystem may not be very interesting to look at: larger and more active organisms, such as fish, are less likely to find the necessary resources to survive in such a small volume. But, speaking from personal experience, some of the plants, microorganisms, and invertebrates will survive for years without requiring any outside intervention. These species are *fit*: adapted to the closed environment they collectively constitute.

In terms of chemical organization theory (COT), this means that they are produced as least as much as they are consumed. When the rate of the different reactions in a reaction network can be estimated, the network can be modeled as a dynamical system. That allows you to simulate the evolution of the ecosystem, by calculating how the concentrations of the different resources or species go up and down in the course of time. Such simulation will typically end in an attractor regime, where some of the concentrations have gone down to zero (elimination), while others remain positive (survival), though they may continue to fluctuate (Veloz et al., 2022; Heylighen, Beigi, & Veloz, 2015). Even without such a quantitative simulation, COT often allows you to determine algebraically which of the species will necessarily be eliminated—e.g., because the reactions producing that species are insufficient to compensate for the reactions consuming it—but also which assemblies of species are capable of forming a self-maintaining organization.

This shows that fitness can in principle be defined in a purely relational manner: as the ratio of production to consumption in a network of reactions connecting different resources and agencies. This implies that the fitness of an agency is context-dependent: it can be large or small depending on the presence or concentration of other agencies and resources. Moreover, the definition of fitness in a given context is ultimately circular or "bootstrapping": the fitness, and therefore presence, of these other agencies in turn depends on the presence or fitness of the first agency. That is because they are all part of a closed organization of mutually feeding reactions. For example, in the example of the ecosphere, the fitness of the shrimps depends on the fitness of the bacteria, because the shrimps need the bacteria to turn a potentially toxic buildup of waste into nutrients that will feed the algae that they consume. Vice versa, the bacteria need the shrimps to eat and digest the algae, turning them into waste that the bacteria can feed on.

Such mutual dependencies are usually conceived as cycles, in which some resource x is turned by some agency into y, after which one or more other agencies turn it back into x. But the reaction network formalism can handle much more complex mutual dependencies, where resources are produced again via a complex, branching network of reactions that turn combinations of resources into other combinations.

Let us further expand the example of the self-maintaining ecosystem. At the largest scale, the planet Earth is a closed ecosystem, which only receives sunlight as an input, while radiating heat into space. The number of species and resources making up this ecosystem is immensely larger than the one in our closed jar. Nevertheless, the network of reactions describing their mutual dependencies is not fundamentally different. Both are characterized by processes of mutual adaptation and overall self-organization that generate a self-maintaining, closed whole. At the coarse level of the biosphere, we even find a similar balance between oxygen-producing plants, carbon dioxide-producing animals, and nutrient-producing decomposers—although here we also need to take into account processes of recycling in the atmosphere, oceans, and lithosphere (e.g., carbon dioxide being trapped in sediments and chalk formations). This self-maintaining system at the planetary scale is known as Gaia (Rubin, Veloz, & Maldonado, 2021).

Perhaps the most striking difference at this scale is that we can expect frequent processes of *variation*: agencies undergoing a transformation that changes the way they react to certain conditions. For example, imagine that some species of bacteria undergoes a mutation that makes it resistant to a common antibiotic. This species was being produced by a reaction of the form:

$$\text{bacteria} + \text{host} \rightarrow 2 \text{ bacteria} + \text{ill\_host},$$

while being consumed, and thus held in check, by a reaction of the form:

$$\text{bacteria} + \text{ill\_host} + \text{antibiotic} \rightarrow \text{host}$$

The mutated form, bacteria*, however, is no longer consumed by the latter reaction, and thus may spread exponentially via the first reaction—unless some new reaction kicks in to consume it, such as:

$$\text{bacteria*} + \text{host} \rightarrow \text{host} + \text{antibodies} + \text{bacteria*} \rightarrow \text{host}$$

Variation creates new types or species of agencies, which may or may not fit in with the existing ecosystem. Unfit variations are simply eliminated. However, fit variations change the ecosystem itself, by initiating reactions that produce new challenges for other agencies. These may shift the balance between consumption and production of resources and agencies, potentially leading to the extinction of hitherto fit species. In addition, these changes may carve out a stable niche for the newly evolved species. Thus, a new variation resets, in a minor or (more rarely) major way, the dynamical system defined by the reaction network, inciting it to converge to a different attractor regime.

5.12 Cells and Genes as Agencies

After zooming out to the largest scale of the planetary ecosystem, I will now zoom in to the smallest biological scale, the one of a single-celled organism such as a bacterium. In this I am following the old dictum "As above, so below" (ascribed to the mythical sage Hermes Trismegistus), which suggests a fundamental similarity between macrocosm and microcosm.

Indeed, from a relational agency perspective, a living cell is not essentially different from an ecosphere. Both are to some degree transparent containers, filled with a diversity of interacting "molecules" or "species" that are continually being consumed and produced, together forming a self-maintaining, closed network of reactions. Thus, the overall system is *autopoietic*, that is, self-producing (Heylighen, Beigi, & Veloz, 2015; Mingers, 1994; Razeto-Barry, 2012). The complex, branching cycles of consumption and production reactions constitute the cell's *metabolism*. Although as yet we do not know all the reactions in a bacterial cell's metabolism, those parts we do know, such as *E. coli*'s sugar metabolism, are easily modeled as chemical organizations (Centler et al., 2007).

In the cell, the role of the resources is played by "passive" molecules, such as glucose, ATP, and oxygen, which are consumed and produced by reactions in order to harness energy or build components. The role of the agencies is played by the enzymes, which catalyze and thus enable most of these reactions. Yet, the enzymes themselves are the product of more complex gene-expressing processes, which read a coding sequence of DNA and translate it into the right enzyme. As long as the cell does not divide, the DNA itself is invariant, neither produced nor consumed. That allows it to function as a dependable memory for the cell, storing the "knowledge" about which actions (enzyme-catalyzed reactions) to perform in order to deal with various conditions (presence or absence of particular molecules in the cell) (Heylighen, Beigi, & Busseniers, 2022).

For example, an antibiotic-resistant bacterium may have acquired a piece of DNA that codes for an antibiotic-neutralizing enzyme. The entry of the antibiotic into the cell then functions as a triggering condition to express that stretch of DNA into the corresponding enzyme. This enables a reaction that consumes the antibiotic, transforming it into a molecule that is no longer toxic for the cell.

From this perspective, the genes in a cell are functionally similar to the different agencies in a self-maintaining ecosystem. The reactions they trigger must be coordinated or mutually adapted so that the one produces what the others consume, and the system as a whole remains in balance—while being quick to adjust to fluctuations in the concentrations of the different molecular "species," or to the entry of food molecules that must be consumed or toxins that must be neutralized. Thus, for a cell to survive, its genes must efficiently cooperate, forming a synergetic whole where everything needed by one process is produced by one or more other processes, and vice versa. Note that these processes generally include reactions that transport resources into the cell, and waste out of the cell. Therefore, the system is not closed in the thermodynamic sense, only in the organizational sense.

From a "selfish gene" perspective (Dawkins, 2006), an explanation of how all these agencies would have learned to cooperate is not obvious. From a relational agency perspective, however, inputs and outputs of the different agencies will eventually adjust to each other, albeit that some of the agencies may be eliminated in the process—as illustrated by the ecosystem in a jar where some of the organisms do not survive. The remaining ones can be viewed as cooperative, since it is through their ongoing contributions that the overall, auto-poietic system is maintained. Note that this conclusion is not in contradiction with the selfish gene perspective, which sees genes shared by an organism as being "in the same boat" (Dawkins, 2006) (surviving or being eliminated together), and therefore forced by natural selection to work together effectively. What the reaction network model adds is a mechanism clarifying precisely what "working together" means, and how such an arrangement can evolve.

What the cell example further adds to the one of a self-maintaining ecosystem is that the cellular organization produces its own boundary, the cell membrane. This membrane is semipermeable: it allows "food" molecules to enter, and "waste" molecules to exit, but otherwise protects the self-producing metabolism from external molecules that may interfere with its reactions. That makes the autopoietic organization more robust or resilient (Heylighen, Beigi, & Busseniers, 2022; Veloz et al., 2022), and therefore less dependent on outside conditions, that is, more autonomous. Its implicit goal is self-maintenance (and eventual multiplication). Therefore, we can view the cell as a whole as an agency in its own right, a "super-agency" consisting of genes and enzymes as "sub-agencies" that control its internal metabolism. This agency acts on its environment by transforming certain available inputs into outputs that change the environmental situation. If we use a unicellular alga as an example, we are back to the reaction that started this series of examples:

$$\text{alga} + \text{light} + CO_2 + \text{nutrients} \rightarrow 2 \text{ algae} + O_2$$

This shows that the relational agency ontology, with its formalization in terms of reaction networks and chemical organization theory, is applicable at all levels of life, from molecules to the planetary ecosystem.

5.13   An Extended Evolutionary Synthesis?

I have presented a first sketch of a new framework that can help us to understand complex, interacting processes and the systems they produce. This framework functions both on the conceptual level and on the level of formal descriptions that can be investigated mathematically and through computer simulation. Because these descriptions start from the very general concepts of conditions and actions, they do not a priori distinguish between microscopic processes (such as reactions between molecules) and macroscopic processes (such as interactions between biological species and their environments). That makes them in principle very widely applicable, and thus potentially able to unify disparate fields (Heylighen, Beigi, & Veloz, 2015; Veloz & Razeto-Barry, 2017), both inside and outside biology. Let us here examine whether such a framework could support an extended evolutionary synthesis, whose divergent strands I sketched in the introduction.

First, the relational agency framework embraces *self-organization* from the very beginning: following the lead of Ashby, it sees the self-organization of a complex system as the mutual adaptation of the components of that system so as to develop a self-maintaining network of relationships. Thus, self-organization provides the system with "order for free," as Kauffman has called it: that is, without need for selection by the environment to produce such order (Kauffman, 1993). However, we should note that self-organization does imply a form of *internal selection* (Heylighen, 1999), in the sense that agencies or resources that do not manage to adapt to the rest of the emerging organization are eliminated. Moreover, of all the possible dynamical arrangements of reactions, only the self-maintaining organizations survive for the long term. Thus, there is also a selection of the organization as a whole—even in situations where the impact of the environment is minimal, like in the case of the closed ecosphere where the only external influence is light. The resulting network of mutually supporting reactions forms the main focus of *systems biology*, an approach that examines how processes within an organism work together to keep the metabolism going.

From the perspective of relational agency, evolution is always *coevolution*: there is no a priori distinction between the system that is evolving and the environment to which the system has to adapt. Indeed, an environment itself consists of agencies that are trying to adapt to the environment created by all the other agencies. Thus, adaptation in the one will in general challenge the others to adapt as well, potentially triggering a cascade of propagating challenges and concomitant adaptations. These processes of adaptation are in general not passive, but active: an agency intervenes in its environment by reacting to specific inputs and producing specific outputs, thus changing the conditions for itself and others. These actions are not just mechanical responses to external forces, but directed at the internal *goal* of self-maintenance. The reason for this goal-directed behavior is that agencies whose actions were not effective in safeguarding the continuation of their organization have simply been eliminated by natural selection. Therefore, the remaining agencies are intrinsically *teleonomic*: their condition-action rules have been selected to produce a dynamical regime that converges from a wide range of initial conditions to the same attractor of self-maintenance (Heylighen, 2022; Mossio & Bich, 2017). The short-term effect of these actions is to neutralize perturbations that make the agency deviate from this attractor. The long-term effect is to reshape the environment so that environmental conditions make self-maintenance easier—in other words, to *construct a niche* for the organization.

This niche construction is accompanied by the establishment of dependable relationships with the other agencies in the coevolving network. These relations are ideally *synergetic*: the agencies mobilize more resources for their continuing self-maintenance together—that is, in interaction—than they would have done when acting on their own. This synergy incites self-maintenance at the level of the collective organization, where some agencies produce what others consume, and vice versa. Thus, the agencies live in *symbiosis*, being to some degree dependent on each other's activities. This symbiosis is not necessarily mutualistic, but can encompass parasite-host or predator-prey types of relationships, where one species grows at the expense of another species (as in the example of shrimp consuming algae). The synergy in a self-maintaining reaction network does not require consumption to be immediately repaid by production. The contribution of the consumer (e.g., shrimp) to the producer (e.g., algae) can be indirect, via a branching pathway that involves several intervening agencies (e.g., bacteria) and their products (e.g., nutrients).

When the system of relationships becomes so strong that an agency can no longer afford to survive without it, the self-maintaining network starts to behave like an individual agency rather than a group of collaborating agencies—a process known as "individuation." In this way, agencies become integrated into a super-agency. If this agency is localized within a clear boundary, it forms an agent. When such a super-agent is constituted of organisms that belong to different species, the process is called *symbiogenesis*: the generation of a new type of organism by the integration of symbiotically living organisms. When they belong to the same type or species, the process is called an *evolutionary transition*. For example, cells can develop more synergetic relations by specializing in complementary functions, such as somatic versus reproductive, thus forming a differentiated, multicellular organism (Heylighen, Beigi & Busseniers, 2022; Maynard Smith & Szathmáry, 1997). Since self-maintenance can succeed or fail for the super-agent as well as for its constituent agencies, there is now selection at both levels. Thus, there is a continuing pressure for *evolution at multiple levels* to produce condition-action rules that make the respective agencies fitter.

I have not as yet touched on the more complicated issue of ontogenetic *development*, the process during which the cells of a growing embryo differentiate into various tissues and organs. Kauffman's influential work on self-organization and selection in evolution attempted to understand the role of genetic regulatory networks in guiding this differentiation (Kauffman, 1993). He modeled these networks as random Boolean networks, because that made it easy to simulate their self-organizing dynamics. This dynamics has multiple attractors. These can be interpreted as different cell types characterized by different combinations of genes being "on" (expressed, active) or "off" (silent). Development can then be understood as a process in which different cells spread out across the different valleys of their epigenetic landscape (Baedke, 2013), so as to end up in their respective attractors.

It seems likely that random Boolean networks can be expressed in the same language of reaction networks that we have been using to describe elementary processes. This language is actually richer, because it can deal not only with Boolean (binary) variables, but also with concentrations expressed as real numbers. Therefore, it may well provide a more realistic model of the networks that are formed by genes, functioning as agencies, that regulate reactions and each other's activities via enzymes. There is no reason to assume that the attractor landscape formed by the different self-maintaining "chemical organizations" in such a reaction network would be any less rich than the one characterizing a random Boolean network. Thus, it seems as if the relational agency paradigm holds the potential to model the self-organizing evolution of not just cells and ecosystems, but of multicellular organisms and their development.

5.14   Conclusion

While the theory of evolution is fundamentally about change, it is remarkable how much the standard version of it, neo-Darwinism, is focused on static entities: genes and environments. In that respect, it is not different from other scientific theories of change, such as Newtonian mechanics. That is because all these theories ultimately derive from an ontology dating back to Aristotle, which sees invariant objects as the most fundamental constituents of reality (Walsh, 2018). Change, then, is nothing more than the movement of these objects through some abstract space defined by their variable properties. This movement is driven by external forces, such as gravity or natural selection. The advantage of such an approach is that objects, their properties, and the forces that drive them are easy to express in a precise, context-independent manner, allowing mathematical modeling.

Yet, there is an alternative ontology, of which the earliest incarnation is known as animism, but which I have here called *relational agency*. It sees the basic constituents of reality as elementary processes that define a network of interdependent agencies. Until recently, such process philosophy was hard to express in a precise manner. Therefore, it had little impact on scientific theory, which is still dominated by the object ontology. At present, however, approaches such as agent-based modeling and reaction networks have provided us with the beginnings of a formal language and a method for computer simulation. This puts us in a position where we can start to develop a precise, integrated theory of evolution that can deal with relational and agential processes.

An elementary reaction can be modeled as a condition-action rule, stating how a particular conjunction of conditions is transformed into some new conjunction of conditions. Such reactions form networks of causal dependencies, in the sense that the outcomes of certain reactions form the conditions or inputs that trigger further reactions. Some of these networks turn out to be self-maintaining, in the sense that everything consumed (removed) by some reaction in the network is produced again by some other reaction, thus keeping the assembly of reactions running. Such a closed, self-maintaining network has been defined mathematically as a "chemical organization" (Dittrich & Fenizio, 2007). It provides a simple, formal model for some of the most fundamental phenomena that we associate with life: self-organization, autopoiesis, cells, organisms, symbiotic assemblies, and ecosystems.

In each of these cases, we see a higher level of organization emerging from the mutual adaptation of processes at the lower level. Thus, a stable agency, such as a cell, is in fact constituted by a self-perpetuating flux of reactions consuming and producing resources, such as molecules. This perspective, which sees stable wholes materialize out of processes, upends the one of the object ontology, which sees stable objects as primary and processes as secondary. Moreover, the reaction model defines a self-maintaining organization as fundamentally *autonomous*, initiating its own goal-directed actions rather than passively undergoing the effect of external forces. Thus, these organizations are *teleonomic*, directed at the ultimate, intrinsic attractor of self-maintenance.

Of course, this does not mean that such organizations are independent of their environment. Even in the extreme case of the hermetically sealed ecosphere, thermodynamic constraints require an input of light and an output of heat. Thus, autonomous agencies must establish good relationships—dependable, synergetic exchanges of resources—with the other agencies that constitute their environment.

While chemical organization theory is well defined mathematically and computationally, its applications to biology are as yet limited. These include a number of preliminary, yet inspiring, models of the origin of life (Hordijk, Steel, & Dittrich, 2018), metabolism (Centler et al., 2007, 2008), endosymbiosis (Veloz & Flores, 2021), and ecology (Veloz, 2019). For the wider ontological framework of relational agency introduced in this paper, the applications are even sketchier, and by necessity laid out in very broad strokes. Still, I have argued that this framework may provide a foundation for the highly desired extended evolutionary synthesis that would replace the overly reductionist neo-Darwinist paradigm. This synthetic theory would broaden the assumed one-way effect of the environment on the selection of genes to a complex network of interactions among agencies at multiple levels, from molecules via genes, cells, organs, and organisms, to colonies, ecosystems, and ultimately the planet.

Although the complexity of such an integrated picture of the evolutionary process may seem daunting, I have tried to show with examples that the reaction network formalism allows one to start with very simple models, which can then be gradually refined and made more complex and realistic. Thanks to the power of computers, even complex models should remain manageable, in the sense that it would be possible to analyze which subnetworks of reactions form self-maintaining agencies, and which resources or species are likely to be eliminated (Centler et al., 2008). Developing such more realistic models will of course require much further research. Yet, I hope to have shown that, at the conceptual level at least, the ontology of relational agency already throws a new light on many old problems, such as self-organization, autopoiesis, symbiosis, teleonomy, and multilevel selection, thus offering us a first glimpse of an eventual synthetic theory. I hope it will not be considered presumptuous of me to see grandeur in this relational view of life, which not only builds further on the one of Darwin, but also extends its horizons in both downward and upward directions, from particles to planets.

Acknowledgments

This research was funded by the John Templeton Foundation as part of the project "The Origins of Goal-Directedness" (grant ID61733), under its research program on "The Science of Purpose." I thank my VUB colleagues collaborating on this project (Veloz et al., 2022), and in particular Shima Beigi, Tomas Veloz, and Evo Busseniers, for many inspiring discussions on the ideas presented here.

References

Abram, David. 2010. *Becoming Animal: An Earthly Cosmology*. Knopf Doubleday.

Abram, David. 2012. *The Spell of the Sensuous: Perception and Language in a More-Than-Human World*. Knopf Doubleday.

Agafonov, Vladimir A., Vladimir V. Negrobov, and Abir U. Igamberdiev. 2021. "Symbiogenesis as a Driving Force of Evolution: The Legacy of Boris Kozo-Polyansky." *Biosystems* 199 (January): 104302. https://doi.org/10.1016/j.biosystems.2020.104302

Ashby, W. R. 1962. "Principles of the Self-Organizing System." In *Principles of Self-Organization*, edited by H. von Foerster and G. W. Zopf, 255–78. Pergamon Press. http://csis.pace.edu/~marchese/CS396x/Computing/Ashby.pdf

Astor-Aguilera, Miguel, and Graham Harvey. 2018. *Rethinking Relations and Animism: Personhood and Materiality*. Routledge.

Baedke, Jan. 2013. "The Epigenetic Landscape in the Course of Time: Conrad Hal Waddington's Methodological Impact on the Life Sciences." *Studies in History and Philosophy of Science Part C: Studies in History and Philosophy of Biological and Biomedical Sciences* 44 (4, Part B): 756–73. https://doi.org/10.1016/j.shpsc.2013.06.001

Barad, Karen. 2003. "Posthumanist Performativity: Toward an Understanding of How Matter Comes to Matter." *Signs: Journal of Women in Culture and Society* 28 (3): 801–31.

Bateson, G. 2000. *Steps to an Ecology of Mind*. University of Chicago Press.

Bickhard, Mark H. 2011. "Some Consequences (and Enablings) of Process Metaphysics." *Axiomathes* 21 (1): 3–32. https://doi.org/10.1007/s10516-010-9130-z

Bird-David, Nurit. 1999. "'Animism' Revisited: Personhood, Environment, and Relational Epistemology." *Current Anthropology* 40 (S1): S67–91. https://doi.org/10.1086/200061

Bonabeau, E. 2002. "Agent-Based Modeling: Methods and Techniques for Simulating Human Systems." *Proceedings of the National Academy of Sciences* 99 (Suppl. 3): 7280–87. https://doi.org/10.1073/pnas.082080899

Camazine, S., J. L. Deneubourg, N. R. Franks, J. Sneyd, G. Theraula, and E. Bonabeau. 2003. *Self-Organization in Biological Systems*. Princeton University Press.

Centler, Florian, Pietro Speroni di Fenizio, Naoki Matsumaru, and Peter Dittrich. 2007. "Chemical Organizations in the Central Sugar Metabolism of *Escherichia Coli*." In *Mathematical Modeling of Biological Systems, Volume I*, 105–19. Springer. http://link.springer.com/chapter/10.1007/978-0-8176-4558-8_10

Centler, Florian, C. Kaleta, P. S. di Fenizio, and P. Dittrich. 2008. "Computing Chemical Organizations in Biological Networks." *Bioinformatics* 24 (14): 1611–18.

Charlton, Bruce G. 2007. "Alienation, Recovered Animism and Altered States of Consciousness." *Medical Hypotheses* 68 (4): 727–31. https://doi.org/10.1016/j.mehy.2006.11.004

Corning, Peter A. 2003. *Nature's Magic: Synergy in Evolution and the Fate of Humankind*. Cambridge University Press.

Corning, Peter A. 2023. "Teleonomy in Evolution: 'The Ghost in the Machine.'" In *Teleonomy in Evolution* (Vienna Series in Theoretical Biology). MIT Press.

Dawkins, Richard. 2006. *The Selfish Gene*. 3rd ed. Oxford University Press.

Dittrich, Peter, and Pietro Speroni di Fenizio. 2007. "Chemical Organisation Theory." *Bulletin of Mathematical Biology* 69 (4): 1199–1231. https://doi.org/10.1007/s11538-006-9130-8

Fábregas-Tejeda, Alejandro, and Francisco Vergara-Silva. 2018a. "Hierarchy Theory of Evolution and the Extended Evolutionary Synthesis: Some Epistemic Bridges, Some Conceptual Rifts." *Evolutionary Biology* 45 (2): 127–39. https://doi.org/10.1007/s11692-017-9438-3

Fábregas-Tejeda, Alejandro, and Francisco Vergara-Silva. 2018b. "The Emerging Structure of the Extended Evolutionary Synthesis: Where Does Evo-Devo Fit In?" *Theory in Biosciences* 137 (2): 169–84. https://doi.org/10.1007/s12064-018-0269-2

Ferber, J., O. Gutknecht, and F. Michel. 2004. "From Agents to Organizations: An Organizational View of Multi-Agent Systems." *Lecture Notes in Computer Science*, 214–30.

Hartvigsen, G., A. Kinzig, and G. Peterson. 1998. "Complex Adaptive Systems: Use and Analysis of Complex Adaptive Systems in Ecosystem Science: Overview of Special Section." *Ecosystems* 1 (5): 427–30.

Harvey, Graham. 2014. *The Handbook of Contemporary Animism*. Routledge.

Harvey, Graham. 2019. "Animism and Ecology: Participating in the World Community." *The Ecological Citizen* 3 (1): 79–84.

Heylighen, Francis. 1990. "Relational Closure: A Mathematical Concept for Distinction-Making and Complexity Analysis." In *Cybernetics and Systems '90*, edited by R. Trappl, 335–42. World Science.

Heylighen, Francis. 1999. "Advantages and Limitations of Formal Expression." *Foundations of Science* 4 (1): 25–56. https://doi.org/10.1023/A:1009686703349

Heylighen, Francis. 2001. "The Science of Self-Organization and Adaptivity." In *The Encyclopedia of Life Support Systems*, 5: 253–80. EOLSS Publishers. http://pespmc1.vub.ac.be/Papers/EOLSS-Self-Organiz.pdf

Heylighen, Francis. 2011. "Self-Organization of Complex, Intelligent Systems: An Action Ontology for Transdisciplinary Integration." *Integral Review*. http://pespmc1.vub.ac.be/Papers/ECCO-paradigm.pdf

Heylighen, Francis. 2012. "A Tale of Challenge, Adventure and Mystery: Towards an Agent-Based Unification of Narrative and Scientific Models of Behavior." ECCO Working Papers 2012–06. ECCO. Vrije Universiteit Brussel. http://pcp.vub.ac.be/Papers/TaleofAdventure.pdf

Heylighen, Francis. 2014. "Challenge Propagation: Towards a Theory of Distributed Intelligence and the Global Brain." *Spanda Journal*, V (2): 51–63.

Heylighen, Francis. 2016. "Stigmergy as a Universal Coordination Mechanism I: Definition and Components." *Cognitive Systems Research*, Special Issue of Cognitive Systems Research—Human-Human Stigmergy, 38 (June): 4–13. https://doi.org/10.1016/j.cogsys.2015.12.002

Heylighen, Francis. 2021. "Entanglement, Symmetry Breaking and Collapse: Correspondences Between Quantum and Self-Organizing Dynamics." *Foundations of Science*. https://doi.org/10.1007/s10699-021-09780-7

Heylighen, Francis. 2022. "The Meaning and Origin of Goal-Directedness: A Dynamical Systems Perspective." *Biological Journal of the Linnean Society*, blac060. https://doi.org/10.1093/biolinnean/blac060

Heylighen, Francis, and Shima Beigi. 2018. "Mind Outside Brain: A Radically Non-Dualist Foundation for Distributed Cognition." In *Socially Extended Epistemology*, edited by Andy Clark, Spyridon Orestis Palermos, and Duncan Pritchard, 59–86. Oxford University Press. https://doi.org/10.1093/oso/9780198801764.003.0005

Heylighen, Francis, Shima Beigi, and Evo Busseniers. 2022. The Role of Self-maintaining Resilient Reaction Networks in the Origin and Evolution of Life. *Biosystems*, 219. 104720. https://doi.org/10.1016/j.biosystems.2022.104720

Heylighen, Francis, Shima Beigi, and Tomas Veloz. 2015. "Chemical Organization Theory as a Modeling Framework for Self-Organization, Autopoiesis and Resilience." ECCO Working Papers 2015–01. http://pespmc1.vub.ac.be/Papers/COT-ApplicationSurvey-submit.pdf

Heylighen, Francis, P. Cilliers, and C. Gershenson. 2007. "Complexity and Philosophy." In *Complexity, Science and Society*, edited by J. Bogg and R. Geyer, 117–34. Radcliffe Publishing. https://doi.org/10.1002/(SICI)1099-0526(199807/08)3:6<12::AID-CPLX2>3.0.CO;2-0

Heylighen, Francis, and J. M. Dewaele. 2002. "Variation in the Contextuality of Language: An Empirical Measure." *Foundations of Science* 7 (3): 293–340. https://doi.org/10.1023/A:1019661126744

Holland, John H. 2012. *Signals and Boundaries: Building Blocks for Complex Adaptive Systems*. MIT Press.

Hordijk, Wim, Mike Steel, and Peter Dittrich. 2018. "Autocatalytic Sets and Chemical Organizations: Modeling Self-Sustaining Reaction Networks at the Origin of Life." *New Journal of Physics* 20 (1): 015011.

Ingold, Tim. 2006. "Rethinking the Animate, Re-Animating Thought." *Ethnos* 71 (1): 9–20. https://doi.org/10.1080/00141840600603111

Kauffman, Stuart A. 1993. *The Origins of Order: Self-Organization and Selection in Evolution*. Vol. 209. Oxford University Press.

Kuhlmann, Meinard. 2000. "Processes as Objects of Quantum Field Theory." *Faye et Al*, 365–88.

Laland, Kevin N., Blake Matthews, and Marcus W. Feldman. 2016. "An Introduction to Niche Construction Theory." *Evolutionary Ecology* 30 (2): 191–202. https://doi.org/10.1007/s10682-016-9821-z

Laland, Kevin N., Tobias Uller, Marcus W. Feldman, Kim Sterelny, Gerd B. Müller, Armin Moczek, Eva Jablonka, and John Odling-Smee. 2015. "The Extended Evolutionary Synthesis: Its Structure, Assumptions and Predictions." *Proceedings of the Royal Society B: Biological Sciences* 282 (1813): 20151019.

Latour, Bruno. 1996. "On Actor-Network Theory: A Few Clarifications." *Soziale Welt* 47 (4): 369–81.

Lyman, R. Lee. 2004. "The Concept of Equifinality in Taphonomy." *Journal of Taphonomy* 2 (1): 15–26.

Macal, C. M., and M. J. North. 2010. "Tutorial on Agent-Based Modelling and Simulation." *Journal of Simulation* 4 (3): 151–62.

Marmodoro, Anna, and David Yates. 2016. *The Metaphysics of Relations*. Oxford University Press.

Maynard Smith, John, and Eors Szathmáry. 1997. *The Major Transitions in Evolution*. Oxford University Press.

Mayr, Ernst. 1974. "Teleological and Teleonomic, a New Analysis." In *Methodological and Historical Essays in the Natural and Social Sciences*, edited by Robert S. Cohen and Marx W. Wartofsky, 91–117. Boston Studies in the Philosophy of Science, vol. 14. Springer Netherlands. https://doi.org/10.1007/978-94-010-2128-9_6

Mayr, Ernst, and William B. Provine. 1980. *The Evolutionary Synthesis*. Vol. 231. Harvard University Press.

Miller, John H., and Scott E. Page. 2007. *Complex Adaptive Systems: An Introduction to Computational Models of Social Life*. Princeton University Press.

Milonni, Peter W. 2013. *The Quantum Vacuum: An Introduction to Quantum Electrodynamics*. Academic Press.

Mingers, John. 1994. *Self-Producing Systems: Implications and Applications of Autopoiesis*. Springer Science & Business Media.

Mossio, Matteo, and Leonardo Bich. 2017. "What Makes Biological Organisation Teleological?" *Synthese* 194 (4): 1089–1114. https://doi.org/10.1007/s11229-014-0594-z

Müller, Gerd B. 2007. "Evo–Devo: Extending the Evolutionary Synthesis." *Nature Reviews Genetics* 8 (12): 943–49. https://doi.org/10.1038/nrg2219

Nayar, Prayod K. 2018. *Posthumanism*. John Wiley & Sons.

Okasha, Samir. 2005. "Multilevel Selection and the Major Transitions in Evolution." *Philosophy of Science* 72 (5): 1013–25. https://doi.org/10.1086/508102

Palsson, Bernhard. 2015. *Systems Biology*. Cambridge University Press.

Peter, Stephan, and Peter Dittrich. 2011. "On the Relation between Organizations and Limit Sets in Chemical Reaction Systems." *Advances in Complex Systems* 14 (1): 77–96. https://doi.org/10.1142/S0219525911002895

Pigliucci, Massimo, and Gerd B. Müller. 2010. "Elements of an Extended Evolutionary Synthesis." In *Evolution—the Extended Synthesis*. MIT Press. https://doi.org/10.7551/mitpress/9780262513678.003.0001

Prigogine, I. 1980. *From Being to Becoming: Time and Complexity in the Physical Sciences*. Freeman.

Razeto-Barry, Pablo. 2012. "Autopoiesis 40 Years Later. A Review and a Reformulation." *Origins of Life and Evolution of Biospheres* 42 (6): 543–67. https://doi.org/10.1007/s11084-012-9297-y

Rubin, Sergio, Tomas Veloz, and Pedro Maldonado. 2021. "Beyond Planetary-Scale Feedback Self-Regulation: Gaia as an Autopoietic System." *Biosystems* 199 (January): 104314. https://doi.org/10.1016/j.biosystems.2020.104314

Schilthuizen, Menno, Ed. 2008. "Life in Little Worlds." In *The Loom of Life: Unravelling Ecosystems*, 1–10. Springer. https://doi.org/10.1007/978-3-540-68058-1_1

Sourjik, Victor, and Ned S. Wingreen. 2012. "Responding to Chemical Gradients: Bacterial Chemotaxis." *Current Opinion in Cell Biology, Cell Regulation* 24 (2): 262–68. https://doi.org/10.1016/j.ceb.2011.11.008

Sternberg, Shlomo. 2010. *Dynamical Systems*. Courier Corporation.

Tumulka, Roderich. 2006. "On Spontaneous Wave Function Collapse and Quantum Field Theory." In *Proceedings of the Royal Society of London A: Mathematical, Physical and Engineering Sciences* 462: 1897–1908. The Royal Society.

Turchin, Valentin. 1993. "The Cybernetic Ontology of Action." *Kybernetes* 22 (2): 10–30. https://doi.org/10.1108/eb005960

Veloz, Tomas. 2019. "The Complexity–Stability Debate, Chemical Organization Theory, and the Identification of Non-Classical Structures in Ecology." *Foundations of Science* 25 (November), 1–15. https://doi.org/10.1007/s10699-019-09639-y

Veloz, Tomas, and Daniela Flores. 2021. "Toward Endosymbiosis Modeling Using Reaction Networks." *Soft Computing* 25 (3): 1–10.

Veloz, Tomas, Pedro Maldonado, Evo Busseniers, Alejandro Bassi, Shima Beigi, Marta Lenartowicz, and Francis Heylighen. 2022. "An Analytic Framework for Systems Resilience Based on Reaction Networks." *Complexity*, art. no. 9944562. https://researchportal.vub.be/en/publications/an-analytic-framework-for-systems-resilience-based-on-reaction-ne

Veloz, Tomas, and Pablo Razeto-Barry. 2017. "Reaction Networks as a Language for Systemic Modeling: Fundamentals and Examples." *Systems* 5 (1): 11. https://doi.org/10.3390/systems5010011

Walsh, Denis M. 2018. "Objectcy and Agency: Towards a Methodological Vitalism." In *Everything Flows*, edited by Daniel J. Nicholson and John Dupré, 167–85. Oxford University Press.

West, Stuart A., Roberta M. Fisher, Andy Gardner, and E. Toby Kiers. 2015. "Major Evolutionary Transitions in Individuality." *Proceedings of the National Academy of Sciences* 112 (33): 10112–19.

Wilson, David Sloan. 1997. "Altruism and Organism: Disentangling the Themes of Multilevel Selection Theory." *The American Naturalist* 150 (S1): S122–34. https://doi.org/10.1086/286053

6 Teleonomic Anticipatory Configurations in Biological Evolution: The Downward Dynamical Nature of Goal-Directedness

Abir U. Igamberdiev

Overview

The goal-directedness of biological systems is based in part on their anticipatory behavior grounded in an internal model corresponding to the system's embedded description. It is shaped through the achievement of robust self-maintaining configurations acting as attractors resistant to external and internal perturbations. In these configurations, the biological system achieves a condition of maximization of its power via synergistic effects and is able to perform the external work most efficiently. Based on this fundamental principle, the epigenetic trajectories are selected retrocausally, thus establishing the goal-directedness of the individual development. The subdivision of a system into the controlling "self" and the controlled "body" is a physical precondition of goal-directedness. The self-maintenance of autocatalytic sets achieved via feedback controls and error corrections becomes a prerequisite of the teleonomic organization characteristic for biological systems. A realization of novel complex self-maintained configurations becomes possible via the plasticity of the genetic system, which has the property of acquiring new meanings of the coding structures of the genome through the events of recombination, duplication, and alternative rearrangements. In evolution, teleonomic strategies are realized as the process of *codepoiesis*, the anticipatory genetic search that incorporates combinatorial changes into the internal structure of genetic systems. We discuss how the fundamental ideas of anticipation and goal-directedness can be incorporated into the foundations of theoretical biology and evolutionary theory.

6.1 Introduction: Anticipation and Retrocausality

The main property of living systems is the possession of an internal description that directs their dynamics and forms a prerequisite of their teleonomic anticipatory behavior. This idea appears as a reference frame in biology. It was definitively formulated by Aristotle, in particular in his treatise *De Anima* (ca. 350 BC), which represents a profound analysis of the fundamental self-determination of living processes. According to these views, a living object changes according to its internal determination and exists as a teleonomic state (defined as "entelechy" by Aristotle) subdivided into two parts, one being the entelechy as a possession of knowledge, and the other, the entelechy as an actual exercise of knowledge. In the framework of modem biology, the entelechy as a possession of knowledge is expressed through certain structures, such as the genetic code. It is present even when visible features of life (the actual exercise of knowledge) are absent (e.g., in dormant seeds), and it also undergoes a transition to future generations. Heredity, according to Aristotle, is not a transfer of ready-made forms, but the transfer of patterns (i.e., information), which acquire meanings in the whole system of developing organisms during the accomplishment of the actual exercise of knowledge (Stent, 1971).

The view of living systems as subdivided into low-energy and high-energy parts constitutes the basis of theoretical biology as it was formulated by Jakob von Uexküll (1909). This idea was further developed in the foundations of molecular biology. According to Pattee (2001), the evolutionary process is possible upon the separation of energy-degenerate rate-independent genetic symbols from the rate-dependent dynamics of construction that they control, which Pattee defined as the epistemic cut. Such control is achieved via the process of measurement in which the dynamical state is coded into the molecular symbols. These symbols represent the system of codes that are interpreted in the context of the whole biological organization (Barbieri, 2015). The process of internal measurement (Matsuno, 1995) includes the measurement of genotype by phenotype (Rosen, 1991, 1996), in which the dynamic coding structures are interpreted, rearranged, and reassembled, and which involves the dynamic part of the coding system represented by RNA (Witzany, 2016). The dynamic cycles create internal closure events, and thus provide an engine for creating novelty when the boundary conditions of the system foster the constraints that fundamentally change the phase space (Roli & Kauffman, 2020; Lehman & Kauffman, 2021). The principle of epistemic cut is viewed by Pattee (2001) as the basic foundation of life. In the dynamic cycle of information perception and processing described by Uexküll (1909) and reformulated by Barham (1996), low-energy and high-energy processes are united in a single dynamic system adapting to the environment. This becomes the basis for generation of the boundary between the object and the subject (Rosen, 1993). A biological system thus can anticipate its response to external stimuli, and, as Matsuno (2022) mentioned, any biological organism as the internal observer is retrocausal in identifying and feeding upon the necessary resources. This feature can be defined as teleonomy (Mayr, 1974; Corning, chapter 2 in this volume), representing the quality of apparent purposefulness and goal-directedness of the structures and functions in living organisms (Pittendrigh, 1958).

Living systems are autopoietic: that is, they continuously reproduce their own components to support their integrity (Varela et al., 1974; Heylighen, 2014). This process involves circular autocatalysis and, in Aristotelian terms, is closed to efficient causation (Rosen, 1991). This causal mechanism forms and supports the system's "self," which is perceived internally by the system as the set of qualia. Such systems, being circularly causal, behave teleonomically as they strive to achieve some goal state (Heylighen, 2014). In fact, these systems are robust to external influences, being able to return to their state of self-maintenance. Nevertheless, living systems are capable of evolution—the expansion of their goal-directedness—which corresponds to the expansion of their coding systems. Thus, autopoiesis, being the property of self-maintained systems, is complemented by the property of codepoiesis (Barbieri, 2015), which can be described metamathematically as Gödel numbering (Igamberdiev, 2021; see also Shelah & Strüngmann, 2021) and represents an expansion and assignment of new meanings to the elements of a coding system, the process that Heraclitus defined as *self-growing logos*. The same process is observed in the expansion of consciousness described by Gunji et al. (2017) as an inverse Bayesian inference, as well as in the evolution of language via the creation of metaphors (Barbieri, 2020).

The selective meaning of newly generated coding systems is tested at the level of phenotype. In the process of evolution, we observe an increase in phenotypic plasticity. In fact, the appearance of consciousness represents the phenomenon corresponding to the maximum of phenotypic plasticity. This phenomenon was captured by Baldwin (1896), who formulated the effect of the causal role of purposeful behaviors in shaping natural selection, and noted how this influenced the rise of biological complexity in evolution. Even the principle of natural selection can be viewed as a phenotype-first principle, despite its generally upward-causation nature. Teleonomy can be substantiated via the effects of the increase of phenotypic plasticity, which is manifested as a measurement of genotype by phenotype (Rosen, 1996) or as the principle of genetic assimilation formulated by Waddington (1959). The final successful "goal state" can also retrocausally (downward) shape the early stages of development; this idea was actively introduced by Shishkin (2018) and by Cherdantsev and Scobeyeva (2012). The capacity for genome self-modification (Shapiro, 2016) is a prerequisite for the downward causation in evolution in challenging conditions of ecological disruption. The teleonomic process itself is developed as a complexification phenomenon that cannot be reduced to random processes and involves a generation and interpretation of new coding systems (Igamberdiev, 2021). This, however, is not simply a backward causation but rather a "reticulate natural causation" that occurs between distinct and interacting ontological hierarchies (Gontier, 2010; Sukhoverkhov & Gontier, 2021). One of the examples of such reticulate causation is symbiogenesis (Agafonov et al., 2021; Corning, 2021), which represents one of the most important evolutionary mechanisms of complexification.

In this chapter, we analyze the operation of downward "teleonomic" causation in biology (Corning, 2022) in relation to the process of biological evolution. Biological entities as anticipatory systems possess certain attractors, which the system approaches during its individual development. These attractors are characterized by a flexibility that makes it possible to reach new teleonomic states that fit better to a changing environment during the continuous evolutionary process.

6.2   Maximization of Power of the System as an Anticipatory Attractor of Its Behavior

According to Niels Bohr (1933), life is consistent with, but undecidable from, physics and chemistry. However, biological organization refers to a specific physical organization that has its own unique features. It does not contradict the laws of thermodynamics, but has specific thermodynamic properties that distinguish it from inanimate matter. The *Funktionkreis* introduced by Jakob von Uexküll (1909) depicts the interconnection between the low-energy recognition of external signals and the high-energy activity of the operation of the biological system. Von Uexküll's concept describes how the self-supporting energy autonomy of living systems from local high-energy potentials is achieved via their internal metabolic structure by sensitivity to nonlocal low-energy fluxes. The *Funktionkreis* encompasses the coherence, adaptation, and interaction between the system (acting as a subject) and its environment (the *Umwelt* appearing as an objective external reality) as a purposeful whole (*"planmäßiges Ganzes"*).

Thus, biological dynamics includes both low-energy and high-energy processes separated through the epistemic cut. To function successfully, a biological organism should efficiently maximize its power via synergistic effects. This principle was formulated by Lotka (1922), who even suggested that it may constitute the fourth principle of energetics in open-system thermodynamics. A biological cell is an example of an open system (Odum & Pinkerton, 1955) which during self-organization maximizes power intake and energy transformation in order to reinforce production and efficiency (Odum, 1995). Individual development of the system continuously transforms the organization and energy-flow structure of the system in such a way that useful energy transformation becomes maximized. This results in an ability to prevail against disturbance through the autocatalytic feedbacks incorporated in its combined organization, the property defined as *ascendency* (Ulanowicz, 1997). According to Bauer (1935), a capacity for the increase in external work is the main characteristic feature of the evolutionary process.

The most efficient maximization of power of the system assumes its effective control by its subsystem that is efficiently shielded from energy flows and approaches zero entropy. This can be reached in highly ordered coherent states that can exist, in particular, inside macromolecular complexes (Matsuno & Paton, 2000; Igamberdiev, 2004, 2007). The appearance of DNA that stores the information, which is more flexibly exchanged at the level of RNA, became the efficient controlling event that made possible the reproduction and shaping of living systems as organizationally invariant entities. The internal quantum state (IQS) shielded from thermal fluctuations (Igamberdiev, 2004, 2007), with its most coherent part corresponding to perception and conscious activity, governs the rest of the body of a complex living system by sending the commands to it and ensuring that the heat machine of the organism operates with maximum power and efficiency.

The famous work of the founder of quantum mechanics Erwin Schrödinger, "*What Is Life*" (1944), which played an essential role in the establishment of modern molecular biology, successfully explores these principles. Among the points formulated by Schrödinger are these:

1. Aperiodic crystal: a small group of atoms produces orderly events. This became the basis of molecular understanding of DNA structure and genetic organization.

2. Extraction of order from the environment or feeding by negative entropy. This point became the basis of the nonequilibrium thermodynamics that is used for understanding the physical basis of life. It should be complemented by the principle that explains the existence of anticipatory attractors in the dynamics of biological systems.

3. The hereditary structure is largely withdrawn from the disorder of heat motion. This principle is less explored, but it represents the basis for understanding the fundamental physical nature of the epistemic cut essential for explaining life.

The first point explains that the aperiodic structure of the DNA molecule generates order which makes it a carrier of information, processed through transcription and translation. Although the principles of order generation by the aperiodic molecule cannot be reduced solely to the operation of the transcription-translation mechanism, this statement of Schrödinger became the founding principle of molecular genetics (Kauffman, 2020). It is valid for all processes involving the operation of various biological codes (e.g., those described by Barbieri [2015]), and explains the basic feature of information processing in biological systems.

In the second point, Schrödinger anticipated nonequilibrium thermodynamics. More broadly, he reformulated the principle of steady nonequilibrium originally formulated by Ervin Bauer (1935). The notion that the existing order displays the power of maintaining itself and of producing orderly events is more relevant for life than the "order from chaos," and assumes the self-maintenance of orderly events in accordance with the existence of self-supporting stable nonequilibrium structures operating as autocatalytic loops.

The third point formulated by Schrödinger has not yet been fully introduced into the paradigm of modern biology. Nernst's third law of thermodynamics as the basic principle of order in thermodynamics was often assumed to be irrelevant for living systems operating at ~300 K. Schrödinger's statement that the aperiodic crystal forming the hereditary substance is largely *withdrawn from the disorder of heat motion* means that the physical principle, which is similar to Nernst's law, is realized in biological systems. In other words, a living system contains an entangled state controlling its dynamics and shielded from heat motion.

These entangled states, at least at higher levels of organization, are experienced as qualia that allow perception of the dynamic actions in which the initial states are used to accomplish the resulting outputs. A prerequisite for this entangled shielded part of living systems is the principle of steady nonequilibrium state formulated by Ervin Bauer (1935). This means that the third law of thermodynamics is as essential as the second law for biological processes; however, it should be properly understood and adequately formulated to explain its importance for living systems. A special approximation to very low dissipation of energy via the maintenance of quantum coherence within a heat engine can be achieved in living systems, which can be seen as equivalent to the approximation to zero temperature (Igamberdiev, 1993, 2004).

It is claimed that super-cold states composed of the identical wave function (Bose-Einstein condensates) are impossible in living systems, but the shielded states in heat engines of living bodies possess similar physical properties (Khrennikov, 2022). Iosif Rapoport, in his book *Microgenetics* (1965), in an attempt to explain why the transmission of information in biological systems occurs with a very high degree of precision, suggested that only in living systems can the orderly state be reached at temperatures of ~300 K, and considered this property inherent in living systems as the new law of thermodynamics operating in biology. In fact, such orderly states can be described as having "effective" temperatures close to 0 K, and are reached in highly ordered coherent states existing inside macromolecular complexes (Matsuno & Paton, 2000). In a recent paper, Khrennikov (2022) distinguishes between the orderly coherent states described by Fröhlich (1983) realized at high temperatures, and the Bose-Einstein condensates appearing at ultralow temperatures. According to the third law of thermodynamics of Walther Nernst, the zero entropy state can be reached only in conditions of zero temperature (0 K), but in fact, low energy dissipation in orderly coherent states also refers to their ultralow "effective" temperatures (in the range of nano- and microkelvin) that may indicate a nature similar to orderly coherent Fröhlich states and Bose-Einstein condensates.

The quantum coherent state is limited by the minimum uncertainty condition, allowing for the provision of computation and information transfer with almost 100% efficiency, which is described by the model of quantum nondemolition measurement (Igamberdiev,

1993). The information based on specific recognitions triggering dynamical energy-driven processes appears to be nondigital, while the transfer of digital information is realized within functional hypercycles and corresponds to the operation of the genetic code. Almost 100% efficiency of information transfer in biological systems, not only at the level of genetic information but also in the enzymatic catalysis, can be described as a close-to-zero entropy state of macromolecular biological systems, as was suggested by Iosif Rapoport (1965).

Thus, the physical structure of a biological system incorporates two states, one being a low-energy coherent state corresponding to the internal knowledge of itself, and the other being a high-energy dynamic state corresponding to the dynamic action of the system. Their opposition constitutes the principle of epistemic cut and represents the basic foundation of living organization. The highly ordered coherent state constrains the release of energy into a few degrees of freedom in nonequilibrium processes (Kauffman, 2020). Achieving maximum power appears as a teleonomic attractor in the individual development of a biological system. It is possible because of the precise control achieved through low-dissipation mechanisms that provide precise reactions of the system to certain stimuli. We can conclude that the functional autocatalytic cycle of living processes is based on the two inseparable thermodynamic principles, one being the principle of minimum dissipation achieved in low-entropy ordered states controlling the system, and the other representing the maximum power principle of the system's actual realization. They both substantiate the teleonomic nature of biological systems.

6.3 Teleonomy in the Evolutionary Process

In the teleonomic evolutionary process, the emergence of a new level of organization is similar to creating a new formula; that is, it corresponds to the basic principles of codepoiesis suggested by Barbieri (2015). Although initially appearing to be noncausal, the meaning of goal-directedness can be incorporated in rational science by interposing a "predictive model" as a transducer between the time present and the time future. This idea can be found in the concept of Robert Rosen (1985, 1991), who emphasized that not only human models, but also the internal models that biological systems possess, can be anticipatory, revealing that inherent goal-directedness is built into their organization. While the anticipation phenomenon is connected with impredicativity (Nadin, 2010; Igamberdiev, 2015), goal-directedness introduces new computable principles into the organization of biosystems. The embedded description inherent in a system generates a new model of its behavior, which can be computed, but the system also possesses an internal flexibility which represents its noncomputable constituent. If the internal model encoded within the system does not provide a correct result, then the system can evolve due to the acquisition of new statements inside the embedded description that overcome limitations of the existing model. Matsuno (2014) defined this as a durability of living systems via their abstracting capacity. Whereas the durability in the physical world is achieved via the principle of inertia formulated by Galileo (who incorporated it into the basis of physics as its reference frame), the holding self-identity of a living system is achieved via abstracting its own durability as a class property out of different events to be processed. Thus the flexible system of codes, which was anticipated by Aristotle, became the reference frame for biological systems and the foundation of biology as a rational science.

Biological evolution involves goal-directed changes in the realization and interpretation of coding systems, which cannot be computed but provide the frames for structural transformations defined by the pre-existing organization and lead to partially predictable patterns of morphogenesis (Igamberdiev, 2014). In the process of evolution, nomothetic and random decisions appear as complementary. The existing structure constrains possible directions of future evolution; however, adaptive changes are formed via the range of randomly organized genome rearrangements and then memorialized in the new coding structures (*codepoiesis*) (Barbieri, 2015). Thus, codepoiesis imposes a "predictive model" as a transducer between the time present and the time future, and thus underlies goal-directedness.

New traits in evolution appear as "exaptations," which are not adaptations but rather their anticipations in the teleonomic sense; that is, the features available for useful co-optation by descendants (Gould & Vrba, 1982; Gould, 2000). These features become empowered by meaning in the course of evolution and serve as attractors in the course of temporal development of biological systems. Initially they may not have a functional role, but they acquire it in the course of evolution, becoming the fitting states. These states were analyzed in the nomothetic evolutionary theory of Leo Berg (1922), who defined this process as a phylogenetic acceleration or the precession of phylogeny by ontogeny. The opposite view was presented by Ernst Haeckel as a recapitulation or biogenetic law (Olsson et al., 2017). The recapitulation pattern refers to the evolutionary history, whereas the anticipation (precession) pattern refers to the goal-directedness of the evolutionary process. Both patterns are complementary within the continuous evolutionary process. Among the examples of precession of characters, the following events can be mentioned: the reappearance of young characteristics in the adult stages of evolutionarily younger organisms (paleontology); the possession by larvae of morphological and physiological characters of a higher organization, which vanish in the adult state (embryology); the occurrence, in the lower organized groups, of characters that are peculiar to the groups standing higher in the system (comparative anatomy).

In epigenetic inheritance, an essential and significant role is played by different RNA species. The main role of RNA consists of providing a dynamic flexibility of the genetic system, which cannot be achieved in the gene–protein dual system. The latter may generate only a strict evolution through mutations followed by selection that may result in a devastating self-destructing process under high environmental pressures, whereas the intermediate component of the genetic system (RNA) turns evolution into a sophisticated language game, in which the function of RNA consists of moderating incomplete identification between genes and functions and increasing the plasticity of this interacting interface (Witzany, 2016). This role of noncoding RNA in genome rearrangement can be successfully studied in ciliates. These unicellular organisms contain two nuclei: a small diploid micronucleus and a large polyploid macronucleus generated from the micronucleus by amplification of the genome and heavy editing. The micronucleus is a germline nucleus not expressing genes, whereas the macronucleus provides RNA for vegetative growth. The genome editing in these organisms occurs via the removal of noncoding DNA and sorting of the remaining fragments (Nowacki & Landweber, 2009). During the development of the somatic macronucleus, most of the germline is destroyed in ciliates via the fragmentation of their chromosomes, and the remaining fragments become assembled by inversion or permutation. In these processes, RNA templates guide DNA rearrangements and transport somatic mutations to the next generations. This all means that RNA sculpts genetic information. The guiding DNA assembly via maternal RNA templates means that RNA molecules participate in genome rearrangement (Nowacki et al.,

2010). The somatic ciliate genome is really an epigenome, formed through the templates and signals arising from previous generations. Ciliates possess a spatial separation between the edited and nonedited genomes, whereas multicellular organisms impose this separation mainly through the temporal sequences of editing and then inheriting the genetic information. It is more correct to define the process of epigenetic inheritance not as the inheritance of acquired traits but as a *teleonomic anticipatory epigenetic formatting of the genome based on its flexibility* (Igamberdiev, 2015). The latter definition emphasizes an active search for configurations that correspond to the changing fitness landscape. Thus, goal-directedness is achieved in the evolutionary process via the combinatorial events that acquire meaning through the perfection of the final reproductive state of biological organisms.

Besides the already mentioned example of ciliates, several other examples indicate that the mechanism of evolution via RNA-governed genome rearrangement represents a general phenomenon operating at different levels of organization (Koonin & Wolf, 2009, 2012). Such a type of evolution via sorting of genetic information according to molecular addresses was initially suggested by Efim Liberman (1972, 1979). In this process, sequences with fuzzy meaning become recruited for various functional roles, and the process of transfer of biological meaning between the genomes of selfish elements and hosts in the process of their coevolution serves various evolutionary goals (Koonin, 2016). Another such mechanism is RNA interference of the CRISPR-Cas type for defense against transposable elements and viruses. The epigenetic inheritance of resistance to exogenous nucleic acids via small interfering (si) RNA is operational throughout different biological kingdoms and results in high flexibility in the evolutionary adaptation toward thermodynamically optimized final states (Koonin, 2014). All of these events of horizontal gene transfer represent a form of the anticipatory teleonomic type of inheritance.

6.4   The Final State and Its Features

The final state which appears as an attractor in individual development and in the evolutionary process is associated with a biological system's synergistic maximization of power. This important thermodynamic principle, which has a teleonomical nature, suggests that biological evolution generally follows a path which favors the maximum useful energy flow transformation (Lotka, 1922). Ervin Bauer (1935) reformulated it as the principle of capability for the increase in external work in the evolutionary process. Later, Odum (1995) and Ulanowicz (1997) applied this principle to ecosystems by defining those systems' ability to prevail against disturbance through the autocatalytic feedbacks incorporated in their internal organization. According to these views, the development of biological systems continuously transforms the systems' organization and energy flow structure in such a way that useful energy transformation becomes maximized via the constrained release of energy that delays the production of entropy (Kauffman, 2020). This assumes that energy in biological evolution can be analyzed not in terms of the second law of thermodynamics but via such economic criteria as productivity, efficiency, and the costs and benefits ("profitability") of various mechanisms for capturing and utilizing energy to build biomass and do work (Corning, 2020b).

Simon Shnoll (1979), in his book *Physico-Chemical Factors of Biological Evolution*, developed similar views regarding the kinetic perfection of biosystems, which is expressed, in particular, in the rates of metabolic processes, in the spatial movement of organisms, and in their mechanical construction, hydrodynamic, and aerodynamic characteristics. Kinetic perfection determines the choice of unique trajectory in the multidimensional space of biological evolution. The kinetically successful transformation of the thermodynamic potential of biological systems results in their continuous complexification (Vanchurin et al., 2022). At earlier stages of evolution, the achievement of kinetic perfection is expressed in the refinement of the mechanisms of replication, transcription, and translation toward higher reliability. At later stages, overcoming the diffusion limitations of kinetic processes becomes important via the active maintenance of ionic gradients in cells supporting vectorized fluxes of energy and matter, the development of transport systems in multicellular organisms, morphological progress, and the mechanisms of active motion in space. The latter is controlled by a continuous perfecting of the nervous system, the operation of which is apparently based on holding long-range coherent states (Igamberdiev & Shklovskiy-Kordi, 2017; Khrennikov, 2022).

The principle of maximum useful energy flow transformation, by substantiating the kinetic perfection of biological systems as their evolutionary attractor, determines the durability of their nonequilibrium state and thus their competitiveness in their adaptation to environment. The hypercyclic structure that underlies these properties is based on fitting the slow nonequilibrium fluxes and the fast equilibration processes into the structure of metabolism (Igamberdiev & Kleczkowski, 2009). In fact, the equilibria of coenzyme nucleotides and substrates established in cells generate simple rules that provide optimal conditions for the nonequilibrium fluxes of major metabolic processes (Igamberdiev & Kleczkowski, 2019). The equilibrium "buffering" enzymatic reactions serve as control gates for the nonequilibrium flux through the "engine" enzymes providing the directed diffusion of substrates to their active sites and establishing the balance of the fluxes of load and consumption of metabolic components. The search for the best buffering configurations for engine enzymes represents an optimized strategy in developmental trajectories. Evolutionary optimization is achieved through the coordinated operation of buffering and engine enzymes, which determine the establishment of stable and organizationally invariant nonequilibrium states that organize the fluxes of energy spatially and temporally by controlling the rates of major metabolic fluxes that follow thermodynamically and kinetically defined computational principles.

The genetic system is formed from the total pool of nucleoside triphosphates, and DNA is metabolically isolated from the major metabolic pool of nucleotides because it uses deoxynucleotides, the synthesis of which is separately controlled. Nucleotides serve as metabolites (free coenzymes) of enzymes, and their association into nucleic acids generates matrices for the reproduction of enzymes themselves. Polymerization of cosubstrates into nucleic acids generates a self-referential set of arrows for the set of catalysts, resulting in the appearance of the digital information of the genetic code forming the internal programmable structure of biosystems (Igamberdiev, 1999). An understanding of the metabolic aspect of the organizational invariance parameter is important for drawing the equations linking RNA biosynthesis with the availability of nucleoside triphosphates and cofactors such as magnesium and potassium, concentrations of which are established upon equilibrium of nucleoside triphosphates. Even in a very general form, these equations would reflect Rosen's structure of an (M, R) system and could be expanded to include the actual rates of each process and its regulation. What is replicated is a functional component, not a material part as such. From this point of view, the distinction between genotype and phenotype is dynamic and flexible, so it is not possible to define what is primary and what is secondary. Evolution is based on the reestablishment of the parameter of organizational invariance rather than on independent and casual changes in the genotype.

The genetic system determines potentially infinite evolutionary unfolding through the recursive embedding of new coding systems. In this evolutionary unfolding process, previously nonformalized relations are converted into relations having an algorithmic nature. As a result, the internal description actually "holds" the potential possibilities of the whole system and allows the system to function according to its internal recursive constraints, and to transform these constraints in accordance with the range of potential changes of its environment (the epigenetic aspect) and with internal transformational principles of its evolution (the orthogenetic or nomogenetic aspect). An active combinatorial process of self-modification of information represents an internalized language game, which preconditions the teleonomic behavior of the system and the evolutionary transformation of that system. This goal-directedness is a natural phenomenon based on operational features that include negative feedback, feedforward links, equifinality, and flexibility toward external influences. Apparent goals in this development represent far-from-equilibrium attractors of a dynamical system (Busseniers et al., 2021; Heylighen 2014, 2021; see also Heylighen, chapter 5 in this volume), which can be reached despite possible deviations from the trajectory of development due to external perturbations. In a system reaching the equifinal teleonomic state, natural selection does not necessarily act as the factor driving evolution, but it may play an important role in the conditions of high selective pressure. Natural selection appears not as a cause of teleonomic goal-directedness but rather as being caused by teleonomy, as noted by Peter Corning (chapter 2 in this volume). Biological entities as anticipatory systems exploit goal-directedness as their fundamental principle of evolutionary change, in which the stable nonequilibrium states appear as natural attractors achieved and reestablished in the course of evolution.

References

Agafonov, V. A., Negrobov, V. V., & Igamberdiev, A. U. 2021. Symbiogenesis as a driving force of evolution: The legacy of Boris Kozo-Polyansky. *Biosystems 199*, 104302. https://doi.org/10.1016/j.biosystems.2020.104302

Aristotle. ca. 350 BC/2017. *De Anima [On the Soul]*. Penguin Classics.

Baldwin, J. M. 1896. A new factor in evolution. *American Naturalist 30*, 441–451 and 536–553.

Barbieri, M. 2015. *Code Biology: A New Science of Life*. Cham: Springer. https://doi.org/10.1007/978-3-319 -14535-8

Barbieri, M. 2020. The semantic theory of language. *Biosystems 190*, 104100. https://doi.org/10.1016/j.biosystems .2020.104100

Barham, J. 1996. A dynamical model of the meaning of information. *Biosystems 38*(2–3), 235–241. https://doi .org/10.1016/0303-2647(95)01596-5

Bauer, E. S. 1935/1982. *Theoretical Biology*. Budapest: Akadémiai Kiadó.

Berg, L. S. 1922/1969. *Nomogenesis: or, Evolution Determined by Law*. Cambridge, MA: MIT Press.

Bohr, N. 1933. Light and life. *Nature 308*, 421–423 and 456–459.

Busseniers, E., Veloz, T., & Heylighen, F. 2021. Goal directedness, chemical organizations, and cybernetic mechanisms. *Entropy (Basel) 23*(8), 1039. https://doi.org/10.3390/e23081039

Cherdantsev, V. G., & Scobeyeva, V. A. 2012. Morphogenetic origin of natural variation. *Biosystems 109*(3), 299–313. https://doi.org/10.1016/j.biosystems.2012.04.010

Corning, P. A. 2020a. Beyond the modern synthesis: A framework for a more inclusive biological synthesis. *Progress in Biophysics and Molecular Biology 153*, 5–12. https://doi.org/10.1016/j.pbiomolbio.2020.02.002

Corning, P. A. 2020b. "Thermoeconomics": Time to move beyond the second law. *Progress in Biophysics and Molecular Biology 158*, 57–65. https://doi.org/10.1016/j.pbiomolbio.2020.09.004

Corning, P. A. 2021. "How" vs. "Why" questions in symbiogenesis, and the causal role of synergy. *Biosystems 205*, 104417. https://doi.org/10.1016/j.biosystems.2021.104417

Corning, P. A. 2022. A systems theory of biological evolution. *Biosystems 214*, 104630. https://doi.org/10.1016/j.biosystems.2022.104630

Fröhlich, H. 1983. Evidence for coherent excitations in biological systems. *International Journal of Quantum Chemistry 23*, 1589–1595. https://doi.org/10.1002/qua.560230440

Gontier, N. 2010. Evolutionary epistemology as a scientific method: a new look upon the units and levels of evolution debate. *Theory in Biosciences 129*(2–3), 167–182. https://doi.org/10.1007/s12064-010-0085-9

Gould, S. J. 2000. Of coiled oysters and big brains: how to rescue the terminology of heterochrony, now gone astray. *Evolution & Development 2*(5), 241–248. https://doi.org/10.1046/j.1525-142x.2000.00067.x

Gould, S. J., & Vrba, E. S. 1982. Exaptation—a missing term in the science of form. *Paleobiology 8*, 4–15. https://doi.org/10.1017/S0094837300004310

Gunji, Y. P., Shinohara, S., Haruna, T., & Basios, V. 2017. Inverse Bayesian inference as a key of consciousness featuring a macroscopic quantum logical structure. *Biosystems 152*, 44–65. https://doi.org/10.1016/j.biosystems.2016.12.003

Heylighen, F. 2014. Cybernetic principles of aging and rejuvenation: the buffering-challenging strategy for life extension. *Current Aging Science 7*(1), 60–75. https://doi.org/10.2174/1874609807666140521095925

Heylighen, F. 2021. Entanglement, symmetry breaking and collapse: correspondences between quantum and self-organizing dynamics. *Foundations of Science* 1–29. https://doi.org/10.1007/s10699-021-09780-7

Igamberdiev, A. U. 1993. Quantum mechanical properties of biosystems: a framework for complexity, structural stability and transformations. *Biosystems 31*(1), 65–73. https://doi.org/10.1016/0303-2647(93)90018-8

Igamberdiev, A. U. 1999. Foundations of metabolic organization: coherence as a basis of computational properties in metabolic networks. *Biosystems 50*(1), 1–16. https://doi.org/10.1016/s0303-2647(98)00084-7

Igamberdiev, A. U. 2004. Quantum computation, non-demolition measurements, and reflective control in living systems. *Biosystems 77*(1–3), 47–56. https://doi.org/10.1016/j.biosystems.2004.04.001

Igamberdiev, A. U. 2007. Physical limits of computation and emergence of life. *Biosystems 90*(2), 340–349. https://doi.org/10.1016/j.biosystems.2006.09.037

Igamberdiev, A. U. 2014. Time rescaling and pattern formation in biological evolution. *Biosystems 123*, 19–26. https://doi.org/10.1016/j.biosystems.2014.03.002

Igamberdiev, A. U. 2015. Anticipatory dynamics of biological systems: from molecular quantum states to evolution. *International Journal of General Systems 44*(6), 631–641. https://doi.org/10.1080/03081079.2015.1032525

Igamberdiev, A. U. 2021. The drawbridge of nature: Evolutionary complexification as a generation and novel interpretation of coding systems. *Biosystems 207*, 104454. https://doi.org/10.1016/j.biosystems.2021.104454

Igamberdiev, A. U., & Kleczkowski, L. A. 2009. Metabolic systems maintain stable non-equilibrium via thermodynamic buffering. *Bioessays 31*(10), 1091–1099. https://doi.org/10.1002/bies.200900057

Igamberdiev, A. U., & Kleczkowski, L. A. 2019. Thermodynamic buffering, stable non-equilibrium and establishment of the computable structure of plant metabolism. *Progress in Biophysics and Molecular Biology 146*, 23–36. https://doi.org/10.1016/j.pbiomolbio.2018.11.005

Igamberdiev, A. U., & Shklovskiy-Kordi, N. E. 2017. The quantum basis of spatiotemporality in perception and consciousness. *Progress in Biophysics and Molecular Biology 130*(A), 15–25. https://doi.org/10.1016/j.pbiomolbio.2017.02.008

Kauffman, S. 2020. Answering Schrödinger's "What Is Life?" *Entropy (Basel) 22*(8), 815. https://doi.org/10.3390/e22080815

Khrennikov, A. 2022. Order stability via Fröhlich condensation in bio, eco, and social systems: The quantum-like approach. *Biosystems 212*, 104593. https://doi.org/10.1016/j.biosystems.2021.104593

Koonin, E. V. 2014. Calorie restriction à Lamarck. *Cell 158*(2), 237–238. https://doi.org/10.1016/j.cell.2014.07.004

Koonin, E. V. 2016. The meaning of biological information. *Philosophical Transactions of the Royal Society A: Mathematical, Physical and Engineering Sciences 374*(2063), 20150065. https://doi.org/10.1098/rsta.2015.0065

Koonin, E. V., & Wolf, Y. I. 2009. Is evolution Darwinian or/and Lamarckian? *Biology Direct 4*, 42. https://doi.org/10.1186/1745-6150-4-42

Koonin, E. V., & Wolf, Y. I. 2012. Evolution of microbes and viruses: A paradigm shift in evolutionary biology? *Frontiers in Cellular and Infection Microbiology 2*, 119. https://doi.org/10.3389/fcimb.2012.00119

Lehman, N. E., & Kauffman, S. A. 2021. Constraint closure drove major transitions in the origins of life. *Entropy (Basel) 23*(1), 105. https://doi.org/10.3390/e23010105

Liberman, E. A. 1972. Molecular computers in cells. I. General considerations and hypotheses. *Biofizika 17*, 932–943. PMID: 5086104.

Liberman, E. A. 1979. Analog-digital molecular cell computer. *Biosystems 11*(2–3), 111–124. https://doi.org/10.1016/0303-2647(79)90005-4

Lotka, A. J. 1922. Contribution to the energetics of evolution. *Proceedings of the National Academy of Sciences of the U.S.A. 8*(6), 147–151. https://doi.org/10.1073/pnas.8.6.147

Matsuno, K. 1995. Quantum and biological computation. *Biosystems 35*(2–3), 209–212. https://doi.org/10.1016/0303-2647(94)01516-a

Matsuno, K. 2014. Self-identities and durability of biosystems via their abstracting capacity. *Biosystems 120*, 31–34. https://doi.org/10.1016/j.biosystems.2014.04.006

Matsuno, K. 2022. Approaching biology through the retrocausal apex in physics. *Biosystems 213*, 104634. https://doi.org/10.1016/j.biosystems.2022.104634

Matsuno, K., & Paton, R. C. 2000. Is there a biology of quantum information? *Biosystems 55*(1–3), 39–46. https://doi.org/10.1016/s0303-2647(99)00081-7

Mayr, E. 1974. Teleological and teleonomic, a new analysis. In R. S. Cohen & M. W. Wartofsky (Eds.), *Methodological and Historical Essays in the Natural and Social Sciences* (pp. 91–117). Springer. https://doi.org/10.1007/978-94-010-2128-9_6

Nadin, M. 2010. Anticipation and dynamics: Rosen's anticipation in the perspective of time. *International Journal of General Systems 39*, 3–33. https://doi.org/10.1080/03081070903453685

Nowacki, M., & Landweber, L. F. 2009. Epigenetic inheritance in ciliates. *Current Opinion in Microbiology 12*, 638–643. https://doi.org/10.1016/j.mib.2009.09.012

Nowacki, M., Haye, J. E., Fang, W. W., Vijayan, V., & Landweber, L. F. 2010. RNA-mediated epigenetic regulation of DNA copy number. *Proceedings of the National Academy of Sciences of the U.S.A. 107*, 22140–22144. https://doi.org/10.1073/pnas.1012236107

Odum, H. T. 1995. Self-organization and maximum empower. In C. A. S. Hall (Ed.), *Maximum Power: The Ideas and Applications of H. T. Odum*. Colorado University Press.

Odum, H. T., & Pinkerton, R. C. 1955. Time's speed regulator: the optimum efficiency for maximum output in physical and biological systems. *American Journal of Science 43*, 331–343.

Olsson, L., Levit, G. S., & Hoßfeld, U. 2017. The "Biogenetic Law" in zoology: from Ernst Haeckel's formulation to current approaches. *Theory in Biosciences 136*(1–2), 19–29. https://doi.org/10.1007/s12064-017-0243-4

Pattee, H. H. 2001. The physics of symbols: Bridging the epistemic cut. *Biosystems 60*(1–3), 5–21. https://doi.org/10.1016/s0303-2647(01)00104-6

Pittendrigh, C. S. 1958. Adaptation, natural selection, and behavior. In A. Roe & G. G. Simpson (Eds.), *Behavior and Evolution* (pp. 390–416). Yale University Press.

Rapoport, I. A. 1965. *Microgenetics*. Nauka.

Roli, A., & Kauffman, S. A. 2020. Emergence of organisms. *Entropy (Basel) 22*(10), 1163. https://doi.org/10.3390/e22101163

Rosen, R. 1985. *Anticipatory Systems: Philosophical, Mathematical and Methodological Foundations*. Pergamon Press.

Rosen, R. 1991. *Life Itself: A Comprehensive Inquiry into the Nature, Origin, and Fabrication of Life*. Columbia University Press. ISBN 978–0231075657.

Rosen, R. 1993. Drawing the boundary between subject and object: Comments on the mind-brain problem. *Theoretical Medicine and Bioethics 14*(2), 89–100. https://doi.org/10.1007/BF00997269

Rosen, R. 1996. Biology and the measurement problem. *Journal of Computational Chemistry 20*(1), 95–100. https://doi.org/10.1016/s0097-8485(96)80011-8

Schrödinger, E. 1944/1993. *What Is Life? The Physical Aspect of the Living Cell*. Cambridge University Press.

Shapiro, J. A. 2016. The basic concept of the read-write genome: mini-review on cell-mediated DNA modification. *Biosystems 140*, 35–37. https://doi.org/10.1016/j.biosystems.2015.11.003

Shelah, S., & Strüngmann L. 2021. Infinite combinatorics in mathematical biology. *Biosystems 204*, 104392. https://doi.org/10.1016/j.biosystems.2021.104392

Shishkin, M. A. 2018. Evolution as a search for organizational equilibrium. *Biosystems 173*, 174–180. https://doi.org/10.1016/j.biosystems.2018.10.002

Shnoll, S. E. 1979. *Physico-chemical factors of biological evolution*. Nauka.

Stent, G. S. 1971. *Molecular Genetics: An Introductory Narrative*. W.H. Freeman.

Sukhoverkhov, A. V., & Gontier, N., 2021. Non-genetic inheritance: Evolution above the organismal level. *Biosystems 200*, 104325. https://doi.org/10.1016/j.biosystems.2020.104325

Uexküll, J. von. 1909. *Umwelt und Innenwelt der Tiere*. Springer.

Ulanowicz, R. E. 1997. *Ecology, the Ascendent Perspective*. Columbia University Press.

Vanchurin, V., Wolf, Y. I., Koonin, E. V., & Katsnelson, M. I. 2022. Thermodynamics of evolution and the origin of life. *Proceedings of the National Academy of Sciences of the U.S.A. 119*(6), e2120042119. https://doi.org/10.1073/pnas.2120042119

Varela, F. G., Maturana, H. R., & Uribe, R. 1974. Autopoiesis: the organization of living systems, its characterization and a model. *Current Molecular Biology Reports 5*(4), 187–196. https://doi.org/10.1016/0303-2647(74)90031-8

Waddington, C. H. 1959. Canalization of development and genetic assimilation of acquired characters. *Nature 183*, 1654–1655. https://doi.org/10.1038/1831654a0

Witzany, G. 2016. Crucial steps to life: From chemical reactions to code using agents. *Biosystems 140*, 49–57. https://doi.org/10.1016/j.biosystems.2015.12.007

7 From Teleonomy to Mentally Driven Goal-Directed Behavior: Evolutionary Considerations

Eva Jablonka and Simona Ginsburg

Overview

Goal-directed activities that do not depend on conscious will or preconceived design are referred to as *teleonomic behaviors*. In living systems, the evaluative systems that underlie teleonomic behavior sense deviations from homeostasis and employ adaptive plasticity mechanisms that restore, sustain, and boost survival and reproduction. By virtue of their sustainability, all living organisms have an intrinsic teleonomic organization, but some show, in addition, a new kind of goal-directed behavior (GDB) that supports and overlays it: behavior that is driven and controlled by subjectively felt passions and aversions. Such organisms, which can be said to act because they want or do not want to reach some goal, experience sensory inputs as percepts and appraise them and their own actions through feelings. We discuss and compare two approaches to minimal, desire-driven GDB, which suggest that learning was the driving force in the evolution of consciousness: the unlimited associative learning (UAL) model and the hedonic interface (HIT) model. We argue that UAL presents a more comprehensive evolutionary view of consciousness and that it was the basis for the evolutionary construction of another layer of GDB, that of attaining goals guided by imagination. During the evolution of humans, yet another layer was added: goals driven by rational design and abstract values that involve symbolic systems of representation and communication.

7.1 Introduction

Since Aristotle, goal-directedness has been recognized to be the hallmark of living organisms. Aristotle's notion of "soul" refers to the dynamic, embodied organization that makes a living being intentional in the intrinsic sense—that is, having goals that are constructed by its own internal dynamics rather than goals that are externally designed for it. He suggested that there are three "soul" levels: those of "plant, beast, and man" (*On the Soul*, 431b2–4 [1984]). The first, the "nutritive reproductive" soul of plants, has the goal of self-maintenance and reproduction; the goal of the second, the "sensitive" soul of animals, is the satisfaction of felt appetites and desires; the goal of the third, the "rational" soul of humans, is the satisfaction of abstract symbolic values such as "the good" or "the true" (see Ginsburg &

Jablonka, 2019, 2020 for discussions). With the addition of each successive soul level, the realm of goals expands: new categories of goals become available for animals in comparison to plants, and for humans in comparison to plants and animals.

Aristotle's approach was reformulated by later scholars who assumed that external intelligent design must drive the goal-directed processes of living, embryonic development, animal adaptive behavior, and the human capacity for rational reflection. Following the scientific revolution, causal mechanisms were seen as the only legitimate explanations in science, and even Immanuel Kant, who first conceptualized the notion of self-organization, assumed that mechanistic causes cannot explain goals and values: "For it is quite certain that in terms of merely mechanical principles of nature we cannot even adequately become familiar with, much less explain, organized beings and how they are internally possible" (Kant, 1790/1987: 282). The incommensurability between teleological and mechanistic explanatory frameworks was the "hard" problem of life.

Darwin's theory of evolution by natural selection provided a functional, historical-naturalistic rationale for the adaptive, intrinsically goal-directed (teleological)[1] design of living organisms. At the physiological level, Claude Bernard and subsequent generations of physiologists showed that living organisms sense deviations from an internal equilibrium state (later called a state of homeostasis) and employ mechanisms that restore, sustain, and boost that state, so all living organisms have an intrinsic organization and physiological mechanisms that satisfy the basic existential requirements of survival and reproduction. This intrinsic goal-directed organization, which does not involve conscious intention or rational design (the mental was assumed to be extrinsic; probably a Cartesian legacy), was called *teleonomic* by Pittendrigh (1958) to distance it from the external teleology of theologians or the incomprehensible life-force of vitalists (for discussions of the concept, see Corning, chapter 2 in this volume).

The idea that all living organisms, from bacteria to humans, can be studied as teleonomic systems without recourse to mental intention or rational design was a central assumption of twentieth-century biology. One influential articulation of this assumption was suggested by Nikolaas Tinbergen (Tinbergen, 1963). He proposed four explanatory frameworks (which he called "questions" or "causes")—phylogenetic (evolutionary), functional, developmental, and immediate (mechanistic)—as both necessary and sufficient for the comprehensive scientific study of all living teleonomic systems (Bateson & Laland, 2013). For example, when asked why a sunflower turned toward the sun, biologists provide accounts in terms of the phototactic sensory *mechanisms* involved, in terms of the *ontogeny* of these mechanisms and their supporting structures, in terms of the *current function* of the plant's behavior, and in terms of the *evolutionary history* of phototaxis. Higher-level goals and corresponding value systems, such as a mental motivation, serve no explanatory causal role in this case. But are these explanatory frameworks not only necessary, but also sufficient for the scientific study of sentient and rational organisms? For example, when asking why a seagull parent or a human parent aggressively protects its offspring, are the four "questions" sufficient? Tinbergen was well aware that mental motivation, which drives sentient animals and human behavior, may be involved in such cases, but he believed that it could not be scientifically studied and observed: "The ethologist does not want to deny the possible existence of subjective phenomena in animals, [but] he claims it is futile to present them as causes, since they cannot be observed by scientific methods" (Tinbergen, 1951, p. 5).

Along with other investigators of consciousness today, we claim that Tinbergen's fifth "cause" can be scientifically studied. We suggest that the evolution of conscious (perceiving and feeling) organisms involved the construction of a dynamic teleonomic system based on new kinds of values and goals. Conscious living organisms have learning-based beliefs about the predictive relations between perceived aspects of the world and about the predicted outcomes of their own actions. Their GDB is driven by intentions and desires: they act because they want or do not want to reach a represented goal. Unlike a water-deprived plant which grows toward a water source, a water-deprived dog *feels* thirsty and *desires* water (its represented goal) which will alleviate its feeling of thirst. Although both the plant and the dog satisfy the nonconscious (teleonomic) demand for restored water homeostasis, the dog, unlike the plant, has motivations: feelings, desires, and beliefs. Although conscious desires and feelings are intrinsic to the organism—they are the outcome of its internal dynamic organization—they are based on perceptual and affective *representations*, on an additional level of teleonomic organization that enables the conscious construction of beliefs, incentives, and goals which make such GDB possible.

In this chapter, we discuss the evolutionary transition from a teleonomic nonconscious system to a system that enables goal-directed actions that are based on subjective experiencing—on perceptions, desires, and felt evaluations of behavioral outcomes. Our main argument is that the evolution of consciousness entailed the evolution of an open-ended or "unlimited" form of associative learning (UAL), which was based on representations of predictive relations among perceived stimuli, actions, and outcomes. According to this view, the function of consciousness is to form a new realm of goals and values—goals that are guided by desires and intentions to satisfy felt needs. We compare this view to another theory of consciousness developed by Dickinson and Balleine (Balleine & Dickinson, 1998a; Dickinson & Balleine, 2000, 2010) that highlights the role of instrumental goal-directed learning, which we believe can be subsumed under the more general UAL model that we describe. We then briefly discuss the further evolution of conscious GDB: behavioral planning, guided by imagined goals; and human-specific GDB that is driven by rational reflection and publicly shared, abstract, symbolic values.

7.2   Unlimited Associative Learning and the Evolution of Consciousness

In previous publications we suggested that the evolution of consciousness involved the evolution of associative learning that required representations of predictive relations among stimuli, actions, and outcomes (Ginsburg & Jablonka, 2007a, 2007b, 2010a, 2019; Birch, Ginsburg & Jablonka, 2020, 2021). We endorsed William James' view of the function of consciousness as a "fighter for ends"—informing the organism about its internal state and endowing it with motivations to reach represented goals that satisfy its needs. Our functional-evolutionary approach was inspired by the methodological framework suggested by the Hungarian chemist Tibor Gánti for investigating the organizational dynamics of minimal life.

James (1890) suggested that the representation of goals is central to the process of subjective experiencing (or consciousness). He regarded the growth of neural complexity during evolution as leading to a dilemma: "A low brain does few things, and in doing them perfectly forfeits all other use. The performances of a high brain are like dice thrown forever on a table.

Unless they be loaded, what chance is there that the highest number will turn up oftener than the lowest?" (James, 1890, I: 139). Consciousness, which informs the organism about its physiological state, resolves this problem, "loading the dice" of the noisy neural activity of a complex brain. The stored value-laden memories of sensory and motor states, generated during past learning and present online updating, act as internal guides and selectors of new neural relations, new behaviors, and new ends, leading to the unitary, subjective, and intentional internal dynamic states that we recognize as subjective experiencing. The associative link between the representation of action and the goal (the outcome of the action) permits the prediction of the sensory consequences of the action (the representation of the action predicts the outcome) as well as the selection of the action that has led to the goal (the representation of the goal primes the representation of the action). As James saw it, the function of consciousness was to be "a *fighter for ends*, of which many, but for its presence, would not be ends at all" (James, 1890 [his emphasis]). Consciousness, subjectively experienced perceptions, motivations, and past and present affects load the neural "dice," enabling the organism to make decisions and reach goals.[2]

James, however, did not describe the neural architecture of consciousness, and did not point to the mechanisms that bring about the "loading of the dice." Although he accepted in very general terms that the evolution of the nervous system led to consciousness, he did not attempt to outline its evolutionary history. But because his view of the function of consciousness as a process defining a new realm of goals—a process based on relations among perception, memory, attention, emotions, and instincts—informed and guided our evolutionary approach to consciousness, this view was our point of departure for the study of consciousness (Ginsburg & Jablonka, 2007a, 2007b, 2010a).

We used evolutionary theory as our framework of investigation and focused on the evolutionary transition from preconscious organisms to the first positively conscious ones. We reasoned that if we can identify the evolutionary transition to a conscious mode of being and describe it in terms of the changes in the system's organization, we would be able to characterize the mechanisms and the architectural dynamics that constitute a minimal conscious system without being misled by later-evolved neural and behavioral associations and dissociations.

Our analysis followed Gánti's methodology for studying the origin of life (Gánti, 1987, 2003). Gánti started by compiling a list of capacities that are deemed jointly sufficient for life by most biologists: maintenance of a boundary, metabolism, stability, information storage, regulation of the internal milieu, growth, reproduction, and irreversible disintegration (death). He described a system of coupled mechanisms and processes that implement these capacities (the chemoton model) and identified an experimentally tractable marker of a minimal life system, a minimal living teleonomic system. The marker for sustainable minimal life which Gánti suggested is *unlimited heredity*: the capacity to form lineages that vary in open-ended ways from the initial system, so the number of possible different variants is vast (Gánti, 2003; Maynard Smith & Szathmáry [1995] further sharpened this concept). If we find a system with the capacity for unlimited heredity anywhere in the universe, we should be able to reconstruct or reverse-engineer on its basis the simplest teleonomic living system with all the properties listed by Gánti. On a planet like Earth, it would be something like a proto-cell.

We suggested that Gánti's methodology can be applied to the study of other teleological modes of being—the conscious and the rational modes—and called a capacity which requires that all the properties attributed to a particular mode of being are in place an *evolutionary transition marker* (ETM) (Ginsburg & Jablonka, 2015; Bronfman, Ginsburg, & Jablonka, 2016a, 2016b; Ginsburg & Jablonka, 2019, 2020; Birch, Ginsburg, & Jablonka, 2020, 2021). Following this methodology, we sought a consensus list of consciousness characteristics that can be characterized in neural, cognitive, behavioral, and phenomenological terms and that would be regarded as jointly sufficient for the simplest conceivable system to be deemed subjectively experiencing. We then identified the ETM of the transition to consciousness and suggested a functional architecture that can implement this ETM and fulfill the function of consciousness identified by James: enable GDB based on mental states. The consensus characteristics we list here are partially overlapping, suggesting that they form a functional system:

• Percept unification and differentiation: mapping and updating of sensory (exteroceptive, interoceptive, and proprioceptive) stimuli stemming from the body and the world as well as their relations. Perceiving objects as wholes made of discernible parts.

• Global accessibility and broadcast: information from perception, memory, and evaluative systems is integrated and broadcast back to these and other systems. The integration processes lead to the construction of cognitive maps that represent predictive relations between stimuli and their reinforcing outcomes, between actions and their predicted sensory outcomes, and between outcomes and their predicted value. The interactions among representations form a common neural space that contextualizes and updates incoming inputs, enabling comparison, discrimination, generalization, and prioritization of evaluations, which inform decision making.

• Temporal depth: the integration of percepts over time. A conscious organism must hold on to incoming information long enough for that information to become integrated and evaluated. It must have a "working memory." In James' words: "the practically cognized present is no knife-edge, but a saddle-back, with a certain breadth of its own on which we sit perched, and from which we look in two directions into time" (James, 1890, vol. I, p. 609). Temporal depth involves integration at a given time (for stationary objects or scenes) and may involve integration over time, when sequential relocations are perceived as movement.

• Flexible value attribution, desires, and goals: predictive perceptions and actions that have had specific rewarding or punishing outcome-value are subsequently treated as desired or loathed goals, to be reached or avoided. Outcome-values and the actions that lead to them can be flexibly changed when conditions change (e.g., what was previously punishing is now rewarding, and rewards can be devalued). Because many things with different valences impinge on the organism, the value system must allow the prioritization and ranking of concurrently encountered reinforcing stimuli.

• Developmental selection and selective attention: there are excluding and amplifying mechanisms for making some stimuli and actions more salient than others according to predictive evaluations based on present and past experience.

• Intentionality (aboutness): signals from the world, body, and their relations are not merely integrated (summated, subtracted, amplified) but are *mapped*, and these maps or representations are updated immediately.

• Embodiment and agency: conscious organisms are embodied and spontaneously active. They have object-oriented spatial cognition requiring freedom of movement, their spontaneous exploratory activities have intrinsic positive valence, and they are able to control their motor actions: some of their movements are voluntary. When the agency of the animal is compromised, the animal suffers.

• Self–other distinction from a point of view (a sense of self): there is a stable perspective from which the system constructs models of the world and body and responds to changes in them. Such a system is able to distinguish between a stimulus that is the result of its own action and an identical stimulus that is independent of the system's action. This is true not just for the outcome of reflex actions, but for outcomes of *learned* actions and outcomes that are independent of the organism's past actions. Such an organism, which can adjust its expectations when the action-outcome contingency changes, can have a sense of owning its actions and perceptions.

These characteristics are not an ad-hoc collection: they are functionally and causally related and construct a unified complex dynamic system. Most researchers would consider the possibility that an organism exhibiting all these capacities is subjectively experiencing to be plausible.

We identified a single capacity, an evolutionary transition marker, which enables the construction of a minimal system that displays all the characteristics we listed. This ETM, we suggested, is a domain-general, open-ended associative learning, which we called unlimited associative learning. UAL refers to an organism's ability to:

(i) Discriminate among *differently organized, novel, multifeatured* patterns of sensory stimuli and select among new compound action patterns (e.g., Couvillon & Bitterman, 1988; Mansur, Rodrigues, & Mota, 2018; Telles et al., 2017). This requires representations of predictive relations among world-stimuli, and between actions and outcomes (Dickinson & Balleine, 1994, 2000).

(ii) Learn about a predictive, novel, compound neutral stimulus or action even when there is a time gap between the presentation of the compound stimulus or action and its reinforcement (such learning is called "trace conditioning"; see, for example, Lucas, Deich, & Wasserman, 1981; Dickinson, Watt, & Griffiths, 1992; Bangasser et al., 2006; Moyer, Deyo, & Disterhoft, 2015; Rodríguez-Expósito et al., 2017).

(iii) Have a flexible value system that enables motivational tradeoffs, prioritizing different outcomes in a context-sensitive manner (Solms, 2021). The animal can alter the valence attributed to patterns of sensory stimuli and motor actions when conditions change. Because animals have to *learn* about the desirability of commodities (incentive learning), if desirability is changed (as in "outcome devaluation"; Holland & Rescorla, 1975; Adams & Dickinson, 1981; Mizunami, 2021), the animal can reevaluate the outcome by direct contact with it in the new state.

(iv) Use previously learned stimuli and actions as a basis for future learning, leading to the formation of chains of actions (e.g., Holland & Rescorla, 1975; Hussaini et al., 2007) and to categorizations and transfers (e.g., Benard, Stach, & Giurfa, 2006).

Unlimited associative learning requires that predictive associations be formed among sensory, motor, and value representations. We regard UAL as a good evolutionary transition marker for consciousness because it requires that all the capacities in the consensus list be in place. Unification and differentiation are needed to construct and update compound stimuli and discriminate between differently organized sensory maps; global accessibility is needed for integrating information from memory, value, sensory, and motor systems; integration over time is needed for trace conditioning; a flexible evaluation-affective system is needed to make incentive learning possible and enable trade-offs and second-order conditioning; selective attention is needed to pick relevant stimuli out from the background; intentionality is needed, since the system maps (or represents) patterns of stimuli, actions, and their relations, stores associative links, and enables categorization and generalization; embodied agency is needed for exploring the world and the consequences of one's own controllable action: learning about predictive relations among stimuli and between one's own actions and their outcomes; a stable point of view is needed to compare patterns of stimuli and actions and recognize world and body invariants; self-world registration is needed so that the organism will distinguish between current and learned own-action–dependent outcomes (and the stimuli that predict them) and outcomes that are independent of its own actions.

UAL is a system-property that requires hierarchical, recurrent associations among world, body, and prospective action-program representations; a dedicated memory subsystem that stores event representations; a dedicated evaluation subsystem that can assign valence to any compound input configuration and that enables context-sensitive prioritization; and a central association unit where all these subsystems interact and from which outputs that inform decision making are sent. A schematic toy model of UAL (figure 7.1) describes the components of the system and their associations.

The UAL architecture gives rise to consciousness through the recurrent lateral and bottom-up and top-down associations among the system's representations, which are encoded in the integrating and the association units. We call these active systemic states *categorizing sensory states* (CSSs), because they categorize, through their dynamics, input, action, and outcome. Based on the internal physiological state of the organism (e.g., food deprivation), current sensory inputs (exteroceptive, interoceptive, and proprioceptive) interact with innate scaffolds and memory traces associated with the attainment of a desired outcome (e.g., food), as well as with representations of action-outcomes and their current valence, leading to decision making. The overall sensory states constituted by these interactions closely correspond, we argue, to what we call mental states.

The UAL model is descriptive (a computational model of UAL has yet to be developed). It is compatible with models emphasizing information integration (Tononi et al., 2016), hierarchical recurrent interactions among neural maps (Lamme, 2020; Feinberg & Mallatt, 2016), and the active inference model of associative learning suggested by Pezzulo, Rigoli, and Friston (2015). It is also compatible with Damasio's (2021) and Panksepp's (2011) models and with the focus on feelings as forms of evaluation as emphasized by Packard and Delafield-Butt (2014). However, unlike Panksepp we do not see feelings as prior to cognitive processes of learning. Simple learning (through sensitization, habituation, and limited associative learning) preceded the evolution of mentality and of feelings, which are entangled with all cognitive processes and in our view are, like perceptions, the outcomes of learning evolution (Ginsburg & Jablonka, 2019).

Our model is most closely allied, however, to the global neural network (GNW) model developed by Changeux, Dehaene, and their colleagues. According to the GNW model, recurrent interactions among sensory, motor, memory value, and attentional processors

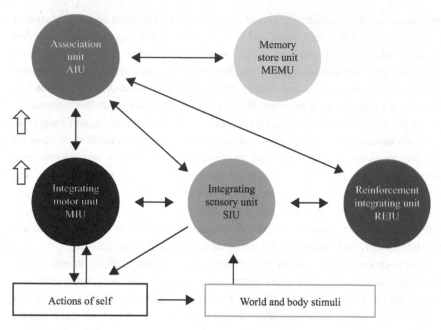

Figure 7.1
A general scheme of UAL cognitive architecture. Unlimited associative learning is hypothesized to depend on reciprocal re-entrant connections (depicted by double-headed arrows) between sensory (SIU), motor (MIU), reinforcement (value, affect [REIU]) and memory (MEMU) integrating processors, with a central association unit (AIU) at the core of the network. Hierarchical intervening levels are depicted by open arrows on the left. The units, including the AIU and MIU, can be distributed. Direct interactions of the sensory and motor units with the memory unit, as well as mappings of SIU-REIU and MIU-REIU relations and their interactions with MEMU at intervening hierarchical levels are not shown (see Ginsburg & Jablonka, 2019, for more details).

become integrated in a common workspace to form a unified, coherent representation of the world, broadcasting this content back to the input systems and onward to motor systems (Dehaene & Changeux, 2011; Mashour et al., 2020). The UAL model can be considered a minimal version of a GNW, which does not require dedicated attentional networks or the neural structures and processes supporting metacognitive tasks such as verbal representation, theory of mind, and full-blown episodic memory. In agreement with this minimal model, the capacity for UAL has been found in three animal phyla with very different brains: most vertebrates; some arthropods; and one group of mollusks, the cephalopods. Fossils of arthropods and vertebrate (fish) brains with structures supporting UAL first appeared during the Cambrian era, about 540 million years ago, and cephalopod mollusks evolved 250 million years later (Ginsburg & Jablonka, 2010b, 2019; see Feinberg & Mallatt, 2016, and Godfrey-Smith, 2020, for similar conclusions with regard to the origins of consciousness in these three groups).

7.3   Desired Goals: UAL and the Hedonic Interface Theory of Consciousness

The UAL model has several sets of predictions (Birch, Ginsburg, & Jablonka, 2020), but here we focus on behavioral predictions pertaining to the link between conscious awareness and success in UAL tasks. We predicted that experimental protocols such as backward masking, which selectively switch off conscious perception in humans but leave unconscious perception in place, will also selectively switch off UAL, while leaving more limited forms or aspects of learning in place. This means that humans and nonhuman animals will perform poorly in UAL tasks such as spatial learning, discrimination learning, trace conditioning, and reverse learning when the predictive perceptual cue is masked (subliminal) but will succeed in these tasks under normal (conscious) conditions. These learning tasks require that the animal be aware of a goal to be reached on the basis of predictive relations among cues, controllable actions, and outcomes. So far, experiments on human subjects (and few experiments on monkeys) support these predictions, although only some UAL tasks have been tested (for references see Birch, Ginsburg, & Jablonka, 2020, 2021).

A different and convergent approach, suggesting that GDB requires consciousness and that the evolution of GDB led to the emergence of consciousness, was developed by Dickinson and Balleine (hereafter referred to as D&B) who explored causal and incentive learning in rats (Balleine & Dickinson, 1998a, Dickinson & Balleine, 2000, 2010; Dickinson, 2012a). They defined GDB in terms of modifiable action-outcome and outcome-value representations: "an action is goal directed if its performance is mediated by the interaction of two representations: (1) a representation of the instrumental contingency between the action and the outcome, and (2) a representation of the outcome as a goal for the agent" (Dickinson & Balleine, 1994, p. 1). They pointed to similarities between patterns of sensitivity to action-outcome degradation in rats and humans: when the action-outcome contingency is relaxed and outcomes that were previously strongly dependent on action now also commonly appear independently of the action, the animal reduces the frequency of performing the action (reviewed in Dickinson & Balleine, 2000), a response that is interpreted in causal terms by humans ("the action no longer *causes* the outcome, so there is no need to act"). Rats and humans also show incentive learning based on *learned* outcome-value representations that inform the animal about the desirability of the outcome (interpreted in terms of desirability by reporting humans). This is in line with James' (1890) proposal that the desirability of any goal, including basic ones such as regarding mother's milk as delicious (for a baby mammal) or finding sitting on an egg wonderfully satisfying (for a hen), has to be learned through direct exposure to the goal-commodity.

D&B demonstrated that desirability is learned by manipulating the value of a learned outcome of an action. When a hungry rat learns that pressing a lever delivers food, the food is represented as a commodity that is greatly desired; when the desirability that had been originally learned when the rat was hungry is revalued (because the rat was fed to satiation with this food in conditions that were not contingent on its pressing the lever), the rat will press the lever less often when later presented with it[3] (reviewed in Dickinson & Balleine, 1994, 2000, 2010). Moreover, rodent action and human causal judgment show similar *illusions* under manipulations of the action-outcome contingency (for discussion see Dickinson & Balleine, 2000). These similarities suggest that GDB in rats and humans shares common cognitive and physiological processes, and that rats, like humans, are able to engage in causal learning. This conclusion was reinforced in a study by Blaisdell and colleagues (2006), which showed that rats can distinguish predictive relations between causal cues and outcomes that are independent of their actions and identical cues that are causally dependent on their own actions.

An animal that can distinguish between the consequences of its own actions and identical consequences that are independent of its own action can control its actions when faced with the opportunity to perform the action. It can choose to act, modify its action, or not to act. However, the ability to represent the causal relationship between action and outcome is not sufficient for adaptive behavior. This flexibility leads, in fact, to the problem of choice: the animal must not only be able to choose, but also must know what is worth choosing. In other words, the choice an animal makes about prospective action or inaction must be adaptive, but how does an animal know which choice is adaptive? An adaptive choice must lead to the satisfaction of the animal's current physiological needs, so it must be rooted in the animal's physiological state. Somehow, information about its own physiological state must become accessible to the animal and guide action-selection. Value must be assigned to a chosen outcome on the basis of relevant existential needs.

The hypothesis advanced by D&B is that evolution led to adaptive access to physiological states through the *feeling* of the hedonic value of the outcome. Feelings render the animal's consummatory behavior subjectively experienced, and lead to desire, which D&B define as "belief about the value of commodities" (Balleine & Dickinson, 1998a: 80). The function of feelings is thus to "inform" the organism about its current (sometimes latent) physiological state, something which is essential when the animal has many controllable action-options. As D&B put it: "It is only with the evolution of intentional control of goal-directed action that there is a function for the feelings and affective reactions elicited by motivationally significant events, that of grounding the assignment of value to the outcomes of action in biologically relevant processes" (Dickinson & Balleine, 2000: 200).

D&B described GDB in terms of interactions between two types of psychological systems. The first is a primitive learning system that is based on stimulus-response (S-R, also called the "reflex-machine" or the "beast-machine"), which enables animals to learn through simple Pavlovian or instrumental habit learning. Although this system enables adaptive behavior based on associative learning, this learning is limited in scope and flexibility. During the phylogeny of some animal groups, a second, more complex psychological system (which they call "cognitive"), enabling the formation of action-outcome (causal) representations that increase control of action, and an outcome-value association that enables learning about outcome desirability, evolved and was *added* to the reflex-machine. Animals that have such a cognitive system also have the primitive S-R system, and although the two systems interact they can be decoupled (Balleine & Dickinson, 1998a, 1998b; Dickinson & Balleine, 2000, 2010; Dickinson, 2012b). D&B suggest that it is the interaction of the direct motivational value of the S-R system with cognitive representations of action-outcome and outcome-value that renders the animal conscious of its bodily state and guides its choices, and they therefore call their theory the hedonic interface theory (HIT) of consciousness. In other words, and in their terms, D&B propose that the affect-free representation of desire became grounded in the S-R machine's consummatory reactions, and that this occurred through "the co-evolution of first order-conscious experience along with the cognitive control of goal-directed action" (Dickinson & Balleine, 2010: 78). As D&B recognize, it is not at all clear *why* the interface between the two systems leads to conscious feelings rather to an unconscious "informing," nor is it clear *how* this process of transformation occurs. However, the HIT model accords with the intuition that organism-level physiological reactions must lead to immediate and often binary decision and action—and feelings do just that.

Like D&B, we suggested that the function of consciousness is to control behavior through subjective feelings—to inform the animal about the desirability of a goal and motivate it to reach that goal. Moreover, in both approaches (UAL, GDB), the ability to learn in a certain way is indicative of consciousness, and therefore a model of such learning may capture some facets of the neural architecture underlying conscious states. At present, only humans, monkeys, rats, corvids, and starlings have been shown to be capable of GDB in D&B's sense (sensitivity to goal devaluation and incentive learning) and can therefore be assumed to be conscious according to HIT. However, we expect that GDB will be found to be far more common than hitherto found, overlapping with the distribution of UAL, which has been found in many animal taxa. Although corroborating tests using masking experiments that directly test the relation between conscious experience and learning have not been performed for GDB tasks, such experiments can be designed. If masking of predictive perceptual cues presented before action-outcome and outcome-value devaluations would render the animal much less sensitive to devaluation, this could more directly test the GDB–consciousness relation.

The focus of HIT on the function of consciousness is fundamental to a naturalistic-evolutionary view of consciousness, but the theory does not address and discuss other key aspects of consciousness. First, there is no discussion of sensory/perceptual contents (perceptual qualia) of the hedonic interface, and there is no reference to the sense of self, which is central to most theories of consciousness (e.g., Merker, 2007; Metzinger, 2007; Ginsburg & Jablonka, 2019; Damasio, 2021; Seth, 2021). Second, although the associative-cybernetic model of D&B (see especially Dickinson, 2012a, figure 3) captures some of the features of the dual psychology suggested by HIT, the developmental construction and the evolutionary history of the sensory, motor, memory, and value representations that constitute consciousness do not receive much attention. Third, a central assumption of HIT—that desire is a representation of a belief in the value of a goal and does not have a built-in affect—is debatable and requires critical discussion. Although desire/wanting has its origins in nonconscious exploratory behavior guided by primary motivational states, and "wanting" is distinct from the hedonic "liking" that characterizes the satisfaction of most physiological needs (Berridge, 2018), desire/wanting in UAL animals seems to have an *inherent* affective value that is related to what Panksepp called the basic emotion of seeking (Panksepp, 2011). In this view, the exploratory, arousal-based facet of desire is not an affect-free representation as HIT suggests, but also includes, in addition to the dimension of arousal, intrinsic valence and control/dominance dimensions (Bakker et al., 2014), which are apparent when animals are deprived of exploration and control (McMillan, 2020). For example, rats prefer signaled (preceded by a predicative cue signal) rather than unsignaled (and hence nonpredictable) electric shocks, even when the predicted shock is four to nine times longer and two to three times stronger than the unpredictable shock (Bassett & Buchanan-Smith, 2007). The inherent affective value of exploration and control (which is positive as long as the animal is not acutely stressed) is contextualized within the domain of learned hedonic values of specific outcomes, perceptions, and current physiological states. Although there may be pathologies that decouple the three constitutive aspects of desire, they are evolutionarily and developmentally coupled in minimally conscious animals.

Finally, the requirement for a dual-psychology architecture is not compelling. A unitary scheme such as that of UAL, according to which the breadth of learning had continuously increased through the evolutionary elaboration of simple Pavlovian and instrumental conditioning, seems more plausible from an evolutionary perspective. According to the unitary scheme, the key evolutionary changes that occurred as animals became able to contextualize their perceptions and actions were the addition of new hierarchical levels of integration—an increase in hierarchical depth—and the evolution of dedicated memory and evaluation systems.

The fact that conscious animals can, and often do, learn unconsciously is readily understood within the unitary UAL framework: such learning occurs when processing fails to reach high hierarchical levels of integration that contextualize incoming information and becomes insensitive to top-down signals. Resistance to devaluation of rewards as a result of overtraining (habit formation) can be attributed to greatly increased stabilization of mappings at lower levels that inhibits the ascent of signals to higher processing levels or renders them insensitive to top-down signals, whereas the failure in UAL tasks when predictive cues are masked can be interpreted in terms of failure of incoming signals to ascend to higher levels in the processing hierarchy. Blindsight, the result of a lesion to the V1 visual area, which leads to the unconscious encoding of visual stimuli, is another case where lower-level processing fails to reach higher levels of the cognitive hierarchy (because the lesion prevents further processing and recurrent interactions among levels within the visual processor's hierarchy). Hence, rather than two psychological systems as suggested by D&B, GDB can be explained in the framework of a single multilevel system, with increasingly more extensive mapping within and between modalities and between representations of action, perception, and value, which evolved from a primitive (nonconscious), less flexible and controllable learning system. Thus, consciousness emerged with the evolutionary elaboration and hierarchical extension of simpler, nonconscious associative learning, which included the evolution of new integrating sensory, motor, memory, and value neural structures that could map composite features and relations. A schematic representation of the architecture of a limited nonconscious learning system which evolved to an UAL architecture is depicted in figure 7.2, which points to both the shared functional units and to addition of hierarchical levels of control and of new dedicated memory and value units.

The view that the evolution of GDB involved the extension and sophistication of the basic architecture of associative learning rather than an interaction between two distinct systems was also suggested, from a different but complementary perspective, by Pezzulo, Rigoli, and Friston (2015). They interpret associative learning within an active inference framework, according to which organisms act to fulfill prior expectations that encode the evolutionarily constituted values of their physiological states. Their analysis of associative learning within this framework suggests that gradual and successive hierarchical contextualizations of sensorimotor constructs, which use the generative models that underpin active inference, led from reflexes to simple conditioning, and from simple conditioning to complex GDBs that include causal and incentive learning. They conclude that "in Active Inference, control is not dichotomized into two discrete systems, but viewed as distributed along a graded continuum going from the highest levels of abstract, prospective and conscious reasoning to more concrete, short-sighted unconscious levels of reasoning down to the arc reflex" (Pezzulo, Rigoli, & Friston, 2015: 24).

The evolution of UAL was relatively rapid in some lineages. It seems to have first evolved within the span of 50 million years, in vertebrates and arthropods during the Cambrian era, as suggested by Cambrian fossils with brain structures supporting UAL (Ginsburg &

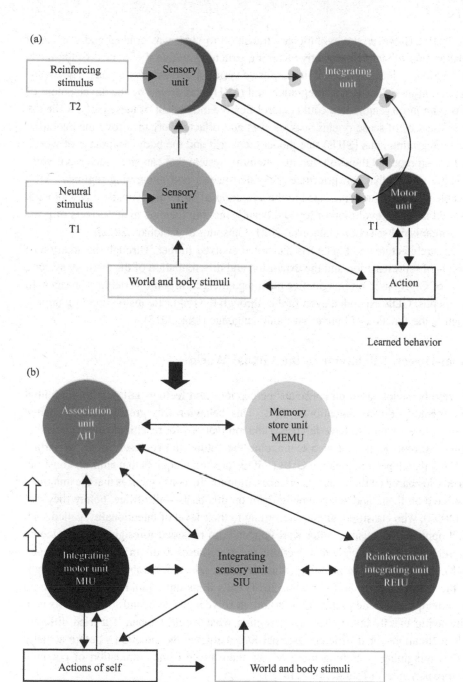

Figure 7.2
From limited, subliminal associative learning (top, A) to UAL (B). Note that memory in (A) is local (gray clouds at locations of association; there is no dedicated declarative memory system); reinforcement is local, too (dark gray crescent, representing the reinforcement value of the representation of the unconditional stimuli; there is no dedicated system that integrates and prioritizes evaluations). In (B), which reproduces figure 7.1, there are new hierarchical levels and new dedicated structures (reinforcement and memory) that implement UAL. The UAL functional architecture can be seen as an extension of the limited associative learning architecture in (A).

Jablonka, 2019). However, the evolutionary transition to UAL was gradual and was based on the elaboration of simple associative learning structures within the context of the intertwined evolutionary increase in body size, muscle systems, and sensory organs. The evolution of an action-modeling motor integrating unit (MIU) was driven by the development of body parts with large ganglia that could control flexible movement of these parts, while the increase in the size of sense organs such as eyes and olfactory organs drove the evolution of sensory integrating units (SIUs) that model the world and the body. Mappings of world, body, and action required the evolution of a memory system that can store such representations, a value system that can prioritize integrated needs, and integrating region/s (AIU) which enabled their interactions. It led to the evolution of suffering (Ginsburg & Jablonka, 2019) as well as to the evolution of joy and beauty, making feelings major drivers of post-Cambrian organismal selection (Jablonka, 2021, Ginsburg & Jablonka, 2022).

The UAL architecture and the GDBs it enabled evolved further. Through the addition of higher levels of representation and the expansion and differentiation of the memory system, a new type of GDB based on imaginative planning emerged in some vertebrate lineages. In the *Homo* genus, GDB expanded even further through the linguistic instruction of imagination, which is the function of human symbolic language (Dor, 2015).

7.4   Goal-Directed Behavior in the Virtual World

Goal-directed behavior based on conscious perceptions and feelings tallies with a minimal notion of mentally driven, intentional GDB. This behavior was greatly enriched when animals were able not only to have felt needs and motivations for reaching current goals that satisfy their current needs, but also to imagine the future and represent goals as virtual options. Fully developed imaginative GDB (IGDB for short) requires that animals construct virtual scenarios based on their past experiences, monitor the consequences that the imagined scenarios have on them, and select among them on this basis—all offline, before they act. Dennett (1997), who classified animals according to their level of intentionality, called such animals "Popperian organisms," after Karl Popper, who regarded foresight as the outcome of an internal selection principle that "permits our hypotheses to die in our stead" (Popper, 1979: 248). Such imaginative planning, which requires selection among offline representations, is the hallmark of what Tomasello (2014) calls thinking: "thinking occurs when an organism attempts, on some particular occasion, to solve a problem, and so to meet its goal not by behaving overtly but, rather, by imagining what would happen if it tried different actions in a situation—or if different external forces entered the situation—before actually acting. This imagining is nothing more or less than the 'off-line' simulation of potential perceptual experiences" (Tomasello, 2014: 9).

"Thoughtful" IGDB has been demonstrated in great apes and corvids (the number of species tested and the scope of testing has been very limited so far, so it is likely that more species capable of IGDB, as well as many IGDB gradations, will be found). The evolution of the capacity for IGDB involved enhanced top-down control of memory representations, greater differentiation of the hippocampus (the declarative memory hub, where episodic memory is encoded and reconstructed), and the elaboration of regulatory connections with executive and reward systems that enable self-control. An animal that can manifest IGDB has to be able to defer gratification, to "think" before acting, inhibiting not just reflexive responses but also, to a considerable extent, the feelings that elicit them. On the basis of their analysis of IGDB, Zacks, Ginsburg, and Jablonka (2022) suggested that planning-enabling IGDB may be considered a major transition in the evolution of cognition sensu Maynard Smith and Szathmáry (1995).[4]

The next great transition in the evolution of GDB was to human goal-directed behavior, which is different from other types of GDB in that it is driven by new type of values—symbolic values such as justice, freedom, virtue, and truth, which are beyond the emotional and intellectual horizons of other mammals. These kinds of goals are in the public symbolic domain, transcending the desires and goals of individuals, and are based on the ability of humans to communicate and reason about their planned actions and have extrinsic represen-tations of their goals through narratives, memorials, and art. This value system enabled rational design and extrinsic teleology, in the sense attributed by theologians to God.

The emergence of a new type of values led Ginsburg and Jablonka (2019) to suggest that the transition from IGDB to human symbol-imbued life is one of the three great teleological transitions in the history of life. However, although there is a qualitative difference between nonsymbolic and symbolic values, there is evolutionary continuity between the IGDB of nonhuman animals and the goal-directed, symbol-based behavior of humans. The connection between these two orders of things is central to Daniel Dor's theory of human language (Dor, 2015). According to Dor, language is a *technology for the instruction of imagination*: like some other mammals, the human world has a personal virtual aspect, but unlike other animals it also has a public aspect that humans can communicate about by employing a new com-munication technology—language.

According to Dor, language is based on a jointly identified set of conventional signs and norms of communication, which are constructed through social negotiations and cultural evolution. He suggests that ordered chains of words are used by speakers to intentionally and systematically instruct their listeners in the process of imagining their intended meaning. The private mental representations of the speaker are pared down to "skeletal" mutually agreed-upon concepts and protocols that are represented by conventional words and rules of communication. The listener uses the words of the speaker as scaffolds, raising past experi-ences from the listener's own memory and then reconstructing and recombining them to produce a novel, imagined experience.

Dor and Jablonka (2010) suggested that the evolution of language was driven by cumula-tive cultural evolution that had led to the partial genetic assimilation of the new communica-tion technology. Like other cultural practices such as literacy, which emerged and evolved culturally in parallel in several societies during the last 5,000 years, language underwent cumulative cultural evolution: it diversified, grew in size, and demarcated itself from the proto-symbolic (mimetic) communication system from which it grew (Donald, 1991). However, unlike literacy and other recent human technologies such as the cellphone, the lengthy, directional, cultural evolution of language drove the fixation of the genetic variations that supported it. The selection of the genetic supporting variations facilitated learning and opened up more learning opportunities, leading to positive feedbacks that accelerated the tempo and scope of language evolution.

Language enormously expanded the number of messages that can be exchanged and hence the realms of truth and falsity. It led to the ubiquity of lying but also to a notion of truth and of an objective world as well as to a reified representation of "self," that led to the feeling of "having" a self. This required an increased level of emotional and cognitive self-control, which is reflected in the new intellectual feelings of suspicion, doubt, and certainty. It also added a new dimension to the feeling of agency—a feeling of free will—which is based on the control of action-outcome of virtual causal relations in the world and in the private and public domains. It culminated in the ability to control the discourse about the external and social world, including the ability to veto social norms and interrogate causal relations and rational decrees (Dennett, 2017).

7.5   Some General Conclusions

All living beings, from bacteria to humans, are teleonomic systems, satisfying the basic goals of survival and reproduction through their intrinsic dynamics. Teleonomic behavior has evolved and led to what we recognize as the fully fledged intelligently designed goals of humans, who can do things not only because they are fitness-promoting or because they desire them, but because they have reasons for doing them.

The development of humans' ability to reason and to intelligently design was gradual and involved several evolutionary transitions. In this paper we highlighted the transition to consciousness, which has the function of enabling desire-guided GDB. This functional characterization of consciousness has been elaborated within the UAL and HIT models and can explain how the cognitive architecture that constitutes consciousness evolved. The detailed and explicit evolutionary scheme suggested by Ginsburg and Jablonka for the evolution of UAL favors successive hierarchical extensions of associatively learned representations within a unitary psychological scheme—a unitary scheme that is also suggested by the active inference model. The dual-psychology HIT model, though based on a convincing and well-argued functional characterization of consciousness, leaves many aspects of consciousness unaddressed and does not offer an explicit evolutionary scenario for the evolution of GDB. Comparative studies of GDB, at the behavioral and neural levels in different vertebrate classes, as well as in arthropods and cephalopods, are necessary for evaluating different models of desire-guided GDB.

A perspective that is based on an evolutionary approach to GDB forms the basis of our theory of consciousness. An explicit recognition of goal-directedness is required for the understanding of adaptive evolution and is of particular importance for addressing foundational evolutionary-origin questions: the origin/s of life, the origin/s of consciousness, the origin/s of imagination, and the origin of human rationality. In this paper we took a learning-cognitive approach to the study of the evolutionary origin of consciousness, but we would like to conclude by taking a more general perspective, that of *agency* as the point of departure for studying the evolution of goal-directed conscious behavior, because such a perspective complements and enriches the approach we have taken.

Spontaneous activity, the most basic expression of agency, is a biological primitive: all biological systems are inherently active. Furthermore, they are proactive, and not just reactive (Longo et al., 2015; Brembs, 2011). Exploratory, random, and semi-directed spontaneous activity occurs at all levels of biological organization. Examples are random and semi-random genetic mutations and epimutations, "noise" in biochemical and neural networks;

default-network activity in the brain; behavioral-motor exploration in moving organisms, and cultural variations. The stabilization of some of the effects of exploratory behavior during development and evolution is the basis of adaptive plasticity (West-Eberhard, 2003).

A major requirement for organisms that move through motor locomotion is the ability to distinguish between self-generated and world-generated stimuli. This is the basis of the difference between exafference and reafference in simple organisms: for example, the ability to inhibit the freezing reflex if the stimulus (e.g., friction) is the result of one's own action (e.g., crawling), while an identical stimulus which is independent of action (a world-stimulus elicited, for example, by another animal) elicits the freezing reflex. Such behaviors do not require consciousness (Merker, 2005; Crapse & Sommer, 2008), but their elaboration within a system of UAL, we argue, does. When the discrimination between world-generated and self-generated stimuli involves *learned* stimuli and controllable, learned action patterns (which require that causal relations are represented and contextualized), a primary conscious sense of self may be said to emerge: the animal recognizes itself as the *owner* and generator of its beliefs, desires, and evaluations. This feeling of agency is normally positively valenced and may be, as Panksepp suggested, a foundational emotional feeling (Panksepp, 2011). Many observations of control deprivation suggest, in both mammals and birds, that deprivation of seeking and deprivation of a sense of action-ownership and action-control have negative valence, leading to learned helplessness and depression, whereas relief of such deprivations has positive valence (McMillan, 2020). Comparative studies of the stress resulting from devaluation and degradation of control and agency in a variety of species (*all* vertebrate classes as well as some arthropods and cephalopods) will be important for understanding the biological bases of the relation between the sense of action-ownership and a sense of agency and control. And since the conscious sense of ownership of endogenously triggered, goal-directed behavior is a hallmark of voluntary action (Haggard, 2019), such studies can contribute to the understanding of volition.

The sense of self and agency were extended during the phylogeny of some lineages. Imaginative planning, exhibited in some mammalian and bird groups, enhances the control the animal has over its actions and emotions: when an animal plans, it must inhibit reflexive or learned responses by suppressing the expression of the feelings that drive them. This increased self-control enhances the sense of self, which is now self-represented as inhabiting the animal's present as well as its future.

In the human lineage, the further evolution of the sense of agency has led to morality and to the human-specific feeling of free will. Individuals in prelinguistic, *Homo erectus* societies underwent rich cultural evolution, and we may assume on the basis of the archeological record (as well as primate studies) that they had many social norms, and internalized the social decrees of their group. For these humans, the social norm—the "ought"—was imposed from without and from within. Social normative traditions were the context for the evolution of the social emotions of shame, guilt, embarrassment, and pride, which are all expressed in the uniquely human blush. The "ought" led to occasional internal conflicts of which the individual was often keenly aware, and which, we suggest, led to a nascent feeling of free will in humans. The later, culturally evolved ability to veto norms opened the "ought" itself to scrutiny and deliberation, enhancing the feeling of free will. Vetoing requires control over the power of the normative "ought": it requires that one does not just obey or disobey the ought, but obeys or disobeys for a reason, leading to systematic (philosophical, scientific) doubt, as well as to the willed rejection of the reflective doubt and to the self-conscious leap of faith (James, 1890).

Acknowledgments

We are grateful to Anthony Dickinson for his critical and constructive comments on section 7.3. The chapter is dedicated to the memory of Marion J. Lamb with whom many of these ideas were discussed.

Notes

1. We use "teleological" adjectivally here and elsewhere to refer to *any* goal-directed behavior: teleonomic, driven by feelings, and driven by planned calculation (human design), and to the evolutionary transitions to modes of being characterized by such behaviors. We avoid the *noun* teleology which, as Corning (2023) points out, is typically associated with extrinsic (godly) design.

2. Because consciousness opens up a new realm of goals, it may be more informative to talk about the goals of consciousness rather than its functions. For a discussion, see Ginsburg and Jablonka (2019, 2020).

3. This is tested under extinction, when pressing the lever results in no outcome (and hence no feedback), allowing the estimation of the effects of previous reevaluation.

4. This transition does not involve a new value system, and is therefore *not* a teleological transition. See Ginsburg and Jablonka (2019) for a discussion of different types of evolutionary transitions (ecological, intentional, informational, and teleological).

References

Adams, C. D., & Dickinson, A. (1981). Instrumental responding following reinforcer evaluation. *The Quarterly Journal of Experimental Psychology 33B*(2), 109–121.

Aristotle. (1984). "On the Soul." In J. Barnes (Ed.), J. A. Smith (Trans.), *The complete works of Aristotle. The revised Oxford translation*, Vol. I. Princeton, NJ: Princeton University Press.

Bakker, I., van der Voordt, T., Vink, P., et al. (2014). Pleasure, arousal, dominance: Mehrabian and Russell revisited. *Current Psychology 33*, 405–421. https://doi.org/10.1007/s12144-014-9219-4

Balleine, B. W., & Dickinson, A. (1998a). Consciousness—the interface between affect and cognition. In J. Cornwell (Ed.), *Consciousness and human identity*, 57–85. Oxford: Oxford University Press.

Balleine, B. W., & Dickinson, A. (1998b). Goal-directed instrumental action: contingency and incentive learning and their cortical substrates. *Neuropharmacology 37*(4–5), 407–419. PMID 9704982. https://doi.org/10.1016/S0028-3908(98)00033-1

Bangasser, D. A., Waxler, D. E., Santollo, J., & Shors, T. J. (2006). Trace conditioning and the hippocampus: the importance of contiguity. *Journal of Neuroscience 26*(34), 8702–8706. https://doi.org/10.1523/JNEUROSCI.1742-06.2006

Bassett, L., & Buchanan-Smith, H. M. (2007). Effects of predictability on the welfare of captive animals. *Applied Animal Behavior Science 102*, 223–245.

Bateson, P., & Laland, K. N. (2013). Tinbergen's four questions: an appreciation and an update. *Trends in Ecology and Evolution 28*(12), 712–718. https://doi.org/10.1016/j.tree.2013.09.013

Benard, J., Stach, S., & Giurfa, M. (2006). Categorization of visual stimuli in the honeybee *Apis mellifera*. *Animal Cognition 9*, 257–270. https://doi.org/10.1007/s10071-006-0032-9

Berridge, K. C. (2018). Evolving concepts of emotion and motivation. *Frontiers in Psychology 9*(647), 1–20.

Birch, J., Ginsburg, S., & Jablonka, E. (2020). Unlimited associative learning and the origins of consciousness: A primer and some predictions. *Biology and Philosophy 35*, S6. https://doi.org/10.1007/s10539-020-09772-0

Birch, J., Ginsburg, S., & Jablonka, E. (2021). The learning-consciousness connection. *Biology and Philosophy 36*, 49. https://doi.org/10.1007/s10539-021-09802-5

Blaisdell, A. P., Sawa, K., Leising, K. J., & Waldmann, M. R. (2006). Causal reasoning in rats. *Science 311*, 1020–1022.

Brembs, B. (2011). Towards a scientific concept of free will as a biological trait: spontaneous actions and decision-making in invertebrates. *Proceedings of the Royal Society B. 278*(1707), 930–939.

Bronfman, Z., Ginsburg, S., & Jablonka, E. (2016a). The evolutionary origins of consciousness: suggesting a transition marker. *Journal of Consciousness Studies 23*(9/10), 7–34.

Bronfman, Z. Ginsburg, S., & Jablonka, E. (2016b). The transition to minimal consciousness through the evolution of associative learning. *Frontiers in Psychology* 7, 1954. https://doi.org/10.3389/fpsyg.2016.01954

Corning, P. A. (2023). Teleonomy in evolution: "The ghost in the machine." In P. A. Corning, S. A. Kaufmann, D. Noble, J. A. Shapiro, R. I. Vane-Wright, and A. Pross (Eds.), *Evolution "on Purpose,"* 11–31. Cambridge, MA: MIT Press.

Couvillon, P. A., & Bitterman, M. E. (1988). Compound-component and conditional discrimination of colors and odors by honeybees: further tests of a continuity model. *Animal Learning and Behavior 16*, 67–74. https://doi.org/10.3758/BF03209045

Crapse, T. B., & Sommer, M. A. (2008). Corollary discharge across the animal kingdom. *Nature Reviews Neuroscience 9*, 587–600. https://doi.org/10.1038/nrn2457

Damasio, A. (2021). *Feeling and knowing: Making minds conscious.* New York: Pantheon.

Dehaene, S., & Changeux, J.-P. (2011). Experimental and theoretical approaches to conscious processing. *Neuron 70*(2), 200–227. https://doi.org/10.1016/j.neuron.2011.03.018

Dennett, D. C. (1997). *Kinds of minds: Towards an understanding of consciousness.* New York: Basic Books.

Dennett, D. C. (2017). *From bacteria to Bach and back: The evolution of minds.* New York: W. W. Norton.

Dickinson, A. (2012a). Associative learning and animal cognition. *Philosophical Transactions of the Royal Society of London Series B, Biological Sciences 367*, 2733–2742. PMID 22927572. https://doi.org/10.1098/rstb.2012.0220

Dickinson, A. (2012b). Why a rat is not a beast machine. *Frontiers of Consciousness: Chichele Lectures.* https://doi.org/10.1093/acprof:oso/9780199233151.003.0010

Dickinson, S., & Balleine, B. (1994). Motivational control of goal-directed action. *Animal Learning and Behavior 22*(1), 1–18.

Dickinson, A., & Balleine, B. W. (2000). Causal cognition and goal-directed action. In C. Heyes and L. Huber (Eds.), *Vienna series in theoretical biology: The evolution of cognition,* 185–204. Cambridge, MA: MIT Press.

Dickinson, A., & Balleine, B. (2010). Hedonics—the cognitive-motivational interface. In M. L. Kringelbach and A. C. Berridge (Eds.), *Pleasures of the brain,* 74–84. New York: Oxford University Press.

Dickinson, A., Watt, A., & Griffiths, W. J. H. (1992). Free-operant acquisition with delayed reinforcement. *The Quarterly Journal of Experimental Psychology* Section B, *45*(3), 241–258. https://doi.org/10.1080/14640749208401019

Donald, M. (1991). *The origins of the modern human mind.* Cambridge, MA: Cambridge University Press.

Dor, D. (2015). *The instruction of imagination: Language as a social communication technology.* Oxford, UK: Oxford University Press.

Dor, D., & Jablonka, E. (2010). Canalization and plasticity in the evolution of linguistic communication. In R. K. Larson, V. Deprez, and H. Yamakido (Eds.), *The evolution of human language,* 135–147. Cambridge: Cambridge University Press.

Feinberg, T. E., & Mallatt, J. (2016). *The ancient origins of consciousness: How the brain created experience.* Cambridge, MA: MIT Press.

Gánti, T. (1987). *The principle of life* (L. Vekerd, Trans.). Budapest, Hungary: Omikk.

Gánti, T. (2003). *The principles of life, with a commentary by James Griesemer and Eörs Szathmáry.* New York: Oxford University Press.

Ginsburg, S., & Jablonka, E. (2007a). The transition to experiencing: I. Limited learning and limited experiencing. *Biological Theory 2*, 218–230. https://doi.org/10.1162/biot.2007.2.3.218

Ginsburg, S., & Jablonka, E. (2007b). The transition to experiencing: II. The evolution of associative learning based on feelings. *Biological Theory 2*, 231–243. https://doi.org/10.1162/biot.2007.2.3.231

Ginsburg, S., & Jablonka, E. (2010a). Experiencing: A Jamesian approach. *Journal of Consciousness Studies 17*(5/6), 102–124.

Ginsburg, S., & Jablonka, E. (2010b). The evolution of associative learning: A factor in the Cambrian explosion. *Journal of Theoretical Biology 266*(1), 11–20. https://doi.org/10.1016/j.jtbi.2010.06.017

Ginsburg, S., & Jablonka, E. (2015). The teleological transitions in evolution: A Gántian view. *Journal of Theoretical Biology 381*, 55–60. https://doi.org/10.1016/j.jtbi.2015.04.007

Ginsburg, S., & Jablonka, E. (2019). *The evolution of the sensitive soul: Learning and the origins of consciousness.* Cambridge, MA: MIT Press.

Ginsburg, S., & Jablonka, E. (2020). Consciousness as a mode of being. *Journal of Consciousness Studies 27*(9–10), 148–162.

Ginsburg, S., & Jablonka, E. (2022). *Picturing the mind: Consciousness through the lens of evolution*. Cambridge, MA: MIT Press.

Godfrey-Smith, P. (2020). *Metazoa: Animal life and the birth of the mind*. London: William Collins.

Haggard, P. (2019). The neurocognitive bases of human volition. *Annual Review of Psychology 70*, 9–28. https://doi.org/10.1146/annurev-psych-010418-103348

Holland, P. C., and Rescorla, R. A. (1975). Second-order conditioning with food unconditioned stimulus. *Journal of Comparative Physiology and Psychology 88*(1), 459–467. https://doi.org/10.1037/h0076219

Hussaini, S. A., Komischke, B., Menzel, R., & Lachnit, H. (2007). Forward and backward second-order Pavlovian conditioning in honeybees. *Learning and Memory 14*, 678–683. https://doi.org/10.1101/lm.471307

Jablonka, E. (2021). Signs of consciousness? *Biosemiotics 14*, 25–29. https://doi.org/10.1007/s12304-021-09419-x

James, W. (1890). *The principles of psychology*. New York: Dover.

Kant, I. (1790/1987). *Critique of judgment* (W. S. Pulhar, Trans.). Indianapolis, IN: Hackett.

Lamme, V. (2020). Visual functions generating conscious seeing. *Frontiers in Psychology, 11*, 83. https://doi.org/10.3389/fpsyg.2020.00083

Longo, G., Montévil, M., Sonnenschein, C., & Soto, A. M. (2015). In search of principles for a theory of organisms. *Journal of Biosciences 40*(5), 955–968. https://doi.org/10.1007/s12038-015-9574-9

Lucas, G. A., Deich, J. D., & Wasserman, E. A. (1981). Trace autoshaping: acquisition, maintenance, and path dependence at long trace intervals. *Journal of Experimental Animal Behavior 36*, 61–74.

Mansur, B. E., Rodrigues, J. R. V., & Mota, T. (2018). Bimodal patterning discrimination in harnessed honey bees. *Frontiers in Psychology 9*, 1529. https://doi.org/10.3389/fpsyg.2018.01529

Mashour, G. A., Roelfsema, P., Changeux, J-P., and Dehaene, S. (2020). Conscious processing and the global neuronal workspace hypothesis. *Neuron 105*(5), 776–798. https://doi.org/10.1016/j.neuron.2020.01.026

Maynard Smith, J., & Szathmáry, E. (1995). *The major transitions in evolution*. Oxford, UK: Oxford University Press.

McMillan, F. D. (2020). The mental health and well-being benefits of personal control in animals. In F. D. McMillan (Ed.), *Mental health and well-being in animals* (2nd ed.). Wallingford, UK: CABI.

Merker, B. (2005). The liabilities of mobility: a selection pressure for the transition to consciousness in animal evolution. *Consciousness and Cognition 14*(1), 89–114. https://doi.org/10.1016/S1053-8100(03)00002-3

Merker, B. (2007). Consciousness without a cerebral cortex: a challenge for neuroscience and medicine. *Behavioral and Brain Science 30*(1), 63–134. https://doi.org/10.1017/S0140525X07000891

Metzinger, T. (2007). Self models. *Scholarpedia 2*(10), 4174. https://doi.org/10.4249/scholarpedia.4174

Mizunami, M. (2021). What is learned in Pavlovian conditioning in crickets? Revisiting the S-S and S-R learning theories. *Frontiers in Behavioral Neuroscience 15*, 661225. PMID: 34177477; PMCID: PMC8225941. https://doi.org/10.3389/fnbeh.2021.661225

Moyer, J. R., Deyo, R. A., & Disterhoft, J. F. (2015). Hippocampectomy disrupts trace eye blink conditioning in rabbits. *Behavioral Neuroscience 129*(4), 523–532. https://doi.org/10.1037/bne0000079

Packard, A., & Delafield-Butt, J. T. (2014). Feelings as agents of selection: putting Charles Darwin back into (extended neo-) Darwinism. *Biological Journal of the Linnean Society 112*(2), 332–353. https://doi.org/10.1111/bij.12225

Panksepp, J. (2011). Cross-species affective neuroscience decoding of the primal affective experiences of humans and related animals. *PLoS ONE 6*(9), e21236. https://doi.org/10.1371/journal.pone.0021236

Pezzulo, G., Rigoli, F., & Friston, K. (2015). Active inference, homeostatic regulation and adaptive behavioural control. *Progress in Neurobiology 134*, 17–35. Epub 2015 Sep 10. PMID: 26365173; PMCID: PMC4779150. https://doi.org/10.1016/j.pneurobio.2015.09.001

Pittendrigh, C. S. (1958). Adaptation, natural selection and behavior. In A. Roe and G. G. Simpson (Eds.), *Behavior and evolution*, 390–416. New Haven, CT: Yale University Press.

Popper, R. K. (1979). *Objective knowledge: An evolutionary approach*. London: Clarendon Press.

Rodríguez-Expósito, B., Gómez, A., Martín-Monzón, I., Reiriz, M., Rodríguez, F., & Salas, C. (2017). Goldfish hippocampal pallium is essential to associate temporally discontiguous events. *Neurobiology of Learning and Memory 139*, 128–134. https://doi.org/10.1016/j.nlm.2017.01.002

Seth, A. (2021). *Being you: A new science of consciousness*. London: Faber & Faber.

Solms, M. (2021). *The hidden spring: A journey to the source of consciousness*. London: Profile Books.

Telles, F. J., Corcobado, G., Trillo, A., & Rodríguez-Gironés, M. A. (2017). Multimodal cues provide redundant information for bumblebees when the stimulus is visually salient, but facilitate red target detection in a naturalistic background. *PLoS ONE 12*(9), e0184760. https://doi.org/10.1371/journal.pone.0184760

Tinbergen, N. (1951). *The study of instinct*. Oxford: Clarendon.

Tinbergen, N. (1963). On aims and methods of ethology. *Zeitschrift für Tierpsychologie 20*, 410–433.

Tomasello, M. (2014). *A natural history of human thinking*. Cambridge, MA: Harvard University Press.

Tononi, G., Boly, M., Massimini, M., & Koch, C. (2016). Integrated information theory: from consciousness to its physical substrate. *Nature Reviews Neuroscience 17*, 450–461.

West-Eberhard, M. J. (2003). *Developmental plasticity and evolution*. Oxford: Oxford University Press.

Zacks, O., Ginsburg, S., and Jablonka, E. (2022). The futures of the past. *Journal of Consciousness Studies*. 29(3–4), 29–61. https://doi.org/10.53765/20512201.29.3.029

8 Beyond the Newtonian Paradigm: A Statistical Mechanics of Emergence

Stuart A. Kauffman and Andrea Roli

Overview

Since Newton, all classical and quantum physics depends upon the "Newtonian paradigm." Here the relevant variables of the system are identified. For example, we identify the position and momentum of classical particles. Laws of motion in differential form connecting the variables are formulated. An example is Newton's three laws of motion and law of gravitation. The boundary conditions creating the phase space of all possible values of the variables are defined. Then, given any initial condition, the differential equations of motion are integrated to yield an entailed trajectory in the prestated and fixed phase space. It is fundamental to the Newtonian paradigm that the set of possibilities that constitute the phase space is always definable and fixed ahead of time.

All of this fails for the diachronic evolution of ever-new adaptations in our—or any—biosphere. The central reason is that living cells achieve constraint closure and construct themselves. With this, living cells, evolving via heritable variation and natural selection, adaptively construct new-in-the-universe possibilities. The new possibilities are opportunities for new adaptations thereafter seized by heritable variation and natural selection. Surprisingly, we can neither define nor deduce the evolving phase spaces ahead of time. The reason we cannot deduce the ever-evolving phase spaces of life is that we can use no mathematics based on set theory to do so. We can neither write nor solve differential equations for the diachronic evolution of ever-new adaptations in a biosphere.

These ever-new adaptations with ever-new relevant variables constitute the ever-changing phase space of evolving biospheres. Because of this, evolving biospheres are entirely outside the Newtonian paradigm.

One consequence is that for any universe such as ours with one or more evolving biospheres, there can be no final theory that entails all that comes to exist. The implications are large. We face a third major transition in science beyond the Pythagorean dream that "all is number," a view echoed by Newtonian physics.

In the face of this, we must give up deducing the diachronic evolution of the biosphere. However, all of physics, classical and quantum, applies to the analysis of existing life, a synchronic analysis.

But there is much more. We begin to better understand the emergent creativity of an evolving biosphere. Thus, we are on the edge of inventing a new physics-like statistical mechanics of emergence.

8.1   Introduction

Three centuries after Newton we are, we believe, at a third major transition in science. We hope to make clear the evidence and need for this transition, and the wide, unexpected landscape for new science that can be glimpsed.

We may attribute the first major transition to Newton, the invention of the differential and integral calculus, and the invention of classical physics. It is no understatement that Newton taught us how to think. Call this the "Newtonian Paradigm" (Smolin, 2013): First, find the relevant variables. In physics these are often position and momentum. Write laws of motion for these relevant variables in ordinary or partial differential deterministic equation form, or stochastic variants. Define ahead of time the boundary conditions, hence the phase space of all possible values of the relevant variables such as positions and momenta of particles of the system. For any initial condition of the relevant variables, integrate the laws of motion to obtain the entailed trajectory of the system in its phase space. It is fundamental to the Newtonian paradigm that we can and must always define the phase space ahead of time. For example, a clockwork universe that will unfold deterministically with a deistic god no longer able to work miracles. This clockwork universe renders "chance" merely epistemic—and it renders "mind" hapless at best.

The second major transition is nothing less than the reluctant discovery of the quantum of action in 1900 (Planck, 1901), thence the miracles of quantum mechanics and quantum field theory (Heisenberg, 1958; Feynman, 1998). Quantum theory, however, remains safely within the Newtonian paradigm with a prestated phase space, including Fock space; hence it has initial and boundary conditions, and the deterministic evolution of a probability distribution via the Schrödinger wave equation. Determinism is broken, on most interpretations of quantum mechanics, on the Born rule and von Neumann's projection postulate (Birkhoff and von Neumann, 1936). Among the most astonishing implications is spatial nonlocality (Einstein et al., 1935; Aspect et al., 1982).

The enormous power of the Newtonian paradigm can be found outside of physics. Ecology often considers a community of species linked by nonlinear dynamical equations of motion concerning the rate of reproduction of members of each species and the food web among the species. Integration of the equations in the predefined phase space of the relevant variables may exhibit limit cycles, multiple attractors, and other aspects of nonlinear dynamical systems (Svirezhev, 2008).

The foundational theory of microeconomics, competitive general equilibrium (CGE; Arrow and Debreu, 1954), is firmly within the Newtonian paradigm. CGE addresses the problem of the existence of an "equilibrium" vector of prices among a set of goods such that the supply and demand for all goods is balanced and "markets clear." Market clearing is the concept of "equilibrium." Consider the well-known supply and demand curve for a single good, such as bread. Supply and demand are inversely related to the price of the good. At the equilibrium price, supply equals demand and all supplied goods are sold. The market clears. The problem arises for two goods that are used together, such as bread and butter. If the price of butter goes up, the demand for bread will go down. Does an equilibrium pair of prices for bread and butter exist? For an arbitrarily large number of goods, will a vector of prices exist such that all markets for these goods clear? In short, does an equilibrium exist?

Arrow and Debreu solved the problem in 1954. We are to consider "all possible dated-contingent goods." An example of a dated-contingent good is "a kilogram of apples on your doorstep only if it rains in Shanghai on March 15th of this year." Next, we suppose that all the economic actors are infinitely rational and they also have probabilities, or expectations, with respect to all possible dated contingent goods. All agents also have their own utility function, or preferences. At the beginning of time, an auctioneer auctions off contracts for all possible dated-contingent goods. Contracts are let. In this setting Arrow and Debreu prove a fixed-point theorem in this continuum of dated-contingent goods showing that at least one vector of prices exists such that no matter how the future unfolds, all markets clear. It is a remarkable result. Competitive general equilibrium remains within the Newtonian paradigm. The prestated and also fixed phase space is the continuum of all possible dated-contingent goods.

CGE remains the foundation of microeconomics. There are familiar doubts about additional "sunspot" equilibria (Cass & Shell, 1983), and the implications of incomplete markets, incomplete knowledge, and other issues (Geanakoplos & Polemarchakis, 1986). We now wish to place ecology and CGE in a wider context. Ecology deals with a predefined set of species in a community. These provide the relevant variables, hence the predefined phase space. Over evolutionary time, species come and go. The set of species and their patterns of interactions themselves evolve. In the diachronic evolution of the biosphere, new adaptations emerge, existing adaptations vanish by extinction. Ecology can hope to be valid over time scales such that the species do not evolve relevant new features or lose relevant old ones. The issue we wish to raise, and the central question of this paper, asks whether we can predict or deduce the new relevant adaptive variables that arise and the old ones that vanish. Can we have well-founded expectations? We hope to demonstrate that the answer is "no."

The same issue arises with respect to CGE. We are asked to consider all possible dated-contingent goods and have well-formulated expectations with respect to them. But over the past 50,000 years of diachronic evolution of the econosphere, the number of goods has exploded from a few thousand to billions today. New goods arise, old goods vanish. The issue we again wish to raise, central to this paper, asks: Can we predict or deduce the new relevant economic variables that arise and the old ones that vanish? Can we have well-founded expectations? We hope again to demonstrate that the answer is "no."

If we cannot deduce the ever-changing phase space, it will be because we will be unable to write or solve equations of motion allowing deduction of those changing phase spaces. We will be outside of the Newtonian paradigm.

Life on earth has existed for almost four billion years, almost 30% of the lifetime of the universe. A failure of the Newtonian paradigm with respect to evolving life, let alone the evolving econosphere, will mean that major aspects of the cosmological evolution of the universe are outside of the Newtonian paradigm.

8.2   The Nondeducible Diachronic Evolution of the Biosphere

Life started on Earth about 3.7 billion years ago. The biosphere is the most complex system we know in the universe. The new central issue is that it really is not possible to deduce the diachronic evolution of our—or any—biosphere. The evolving biosphere is a propagating construction, not an entailed deduction (Longo et al., 2012; Montévil & Mossio, 2015; Kauffman, 2020).

The reasons at first seem strange (Kauffman & Roli, 2021):

1) The universe is not ergodic above the level of about 500 Daltons (Kauffman et al., 2020). The universe really will not make all possible complex molecules such as proteins 200 amino acids long in vastly longer than the lifetime of the universe (Kauffman, 2019; Cortês et al., 2022). Because the universe is not ergodic on time scales very much longer than the lifetime of the universe, it is true that most complex things will never get to exist.

2) Human hearts (very complex things weighing 300 grams and able to function to pump blood) exist in the universe. How can that be possible? The fundamental answer for why hearts exist in the universe is that life, based on physics, arose, evolved, and adapted in that universe over time. Living things have a special organization of nonequilibrium processes. Living things are Kantian wholes where the parts exist in the universe for and by means of the whole. Humans are Kantian wholes. We exist for and by means of our parts, such as hearts pumping blood, and kidneys purifying the blood in the loops of Henle making and excreting urine. Because we, as Kantian wholes, propagate our offspring, our sustaining parts, hearts, and kidneys are also propagated and evolve to function better. The "function" of the heart is to pump blood, not jiggle water in the pericardial sac. The function of a part is that subset of its causal properties that sustains the whole (Kauffman, 2019).

3) We cannot hope to account for the existence in the universe of a heart that can pump blood, or the loop of Henle in the kidney that can purify urine, without appeal to the function of these organs and their adaptive diachronic evolution by Darwin's heritable variation and natural selection. Selection is downward causation. Selection acts on the whole organism, not its evolving parts. What gets to exist in the evolving biosphere is that which was selected. The explanatory arrows point upward. The selection of the whole alters the parts.

4) In more detail, a Kantian whole has the property that the parts exist for and by means of the whole. A simple physical example is an existing 9-peptide collectively autocatalytic set (Kauffman, 2019; Ashkenasy et al., 2004). Here peptide 1 catalyzes a reaction forming a second copy of peptide 2 by ligating half fragments of peptide 2 into a second copy of peptide 2. The half fragments are "food" fed from an exogenous source. Similarly, peptide 2 catalyzes the formation of a second copy of peptide 3, and so on around a cycle in which peptide 9 catalyzes a second copy of peptide 1. The entire set of nine peptides is collectively autocatalytic. The set is a Kantian whole.

This collectively autocatalytic physical set has these properties:

1. It is collectively autocatalytic (Ashkenasy et al., 2004). No molecule catalyzes its own formation. Thus, this is a Kantian whole: the parts do exist for and by means of the whole.

2. The function of a part is that subset of its causal properties that sustains the whole. The function of peptide 1 is to catalyze the formation of a second copy of peptide 2. If, in doing so, the peptide jiggles the water in the Petri plate, that causal consequence is not the function of peptide 1.

3. The system achieves catalytic closure: All reactions requiring catalysis have catalysts within the same system.

4. The system achieves the newly recognized and powerful property of constraint closure (Montévil & Mossio, 2015). Thermodynamic work is the constrained release of energy into a few degrees of freedom (Atkins, 1984). These constraints constitute boundary conditions. The peptides in the nine-peptide collectively autocatalytic set are each a physical boundary condition that constrains the release of chemical energy. Each peptide binds the two substrates of the next peptide, thus lowering the activation barrier, thus chemical energy is released into a few degrees of freedom, and thermodynamic work is done to ligate the two fragments and construct the next peptide. Critically, the set of peptides constructs themselves, thus constructing the very constraints on the release of energy that constitutes the work by which they construct themselves! This is constraint closure (Montévil & Mossio, 2015; Kauffman, 2019, 2020).

Cells literally construct themselves. The evolving biosphere constructs itself. Automobiles do not construct themselves. We construct our artifacts. Living cells constitute a new class of matter and organization of process that is a new union of thermodynamic work, catalytic closure, and constraint closure (Montévil & Mossio, 2015). In a real sense this is the long sought "vital force," here rendered entirely nonmystical.

It is critical to emphasize that because living cells are open thermodynamic systems that construct themselves, they can and do construct ever-new boundary conditions that thereby create new-in-the-universe phase space possibilities that were not prestated (Kauffman, 2020). Not only do the boundary conditions change, but ever-new relevant variables emerge and constitute the new phase space. For example, with respect to the heart, systolic blood pressure, diastolic blood pressure, cardiac blood ejection volume, and blood oxygenation are among the now functionally relevant variables. Consider mimicry among butterflies. Good-tasting butterflies have evolved wing color patterns to mimic bad-tasting butterflies as camouflage to avoid predation by birds. The newly relevant variables for the butterflies involve the recognition capacities of the birds and the specific features of the bad-tasting butterflies. How are we to account for this without adaptive evolution of ever-novel functionalities?

5) The spontaneous emergence of life—of molecular Kantian wholes—in the evolving universe may well be an expected phase transition in complex chemical reaction networks. This body of theory and experiments is part of a theory of the origin of life on Earth and elsewhere that has developed over the past 50 years (Kauffman, 1971, 1986; Farmer et al., 1986; von Kiedrowski, 1986; Hordijk & Steel, 2004; Lincoln & Joyce, 2009; Vaidya et al., 2012; Lancet et al., 2018; Xavier et al., 2020). The central idea is a phase transition to collectively autocatalytic sets in sufficiently complex chemical reaction networks. Experimental collectively autocatalytic sets comprised of DNA, of RNA, and of peptides have been constructed. Most astonishingly, Xavier et al. (2020) analyzed Archaea and bacteria from before oxygen was in the atmosphere and found in each a small-molecule collectively autocatalytic set containing no polymers at all. No DNA, no RNA, no proteins.

Even more wonderfully, the small-molecule autocatalytic sets in Archaea and bacteria overlap in an intersection subset of 172 reactions and small molecules that is itself collectively autocatalytic. This strongly suggests that the smaller intersection subset was present in LUCA, the ancestor of Archaea and bacteria before the two kingdoms of life diverged. These are molecular fossils from more than two billion years ago.

These observations very strongly suggest that life arose as small-molecule collectively autocatalytic sets, very plausibly as the phase transition proposed. By four or five billion years ago the universe had cooked up a high diversity of small molecules, as seen in the Murchison meteorite formed with the solar system. This meteorite has tens of thousands of organic molecules (Kvenvolden et al., 1970). If such a diversity easily yields the spontaneous formation of small-molecule collectively autocatalytic sets, life is abundant among the solar systems in the universe.

On Earth, as observed (Xavier et al., 2020), an early formation of small-molecule collectively autocatalytic sets that synthesize amino acids and nucleotides may well have supported the subsequent formation of peptide-RNA autocatalytic sets that also evolved to catalyze the reactions already present in the small-molecule autocatalytic metabolism that sustained the emerging peptide-RNA system (Lehman & Kauffman, 2021). This emergence of early life would be followed by template replication and coding (Lehman & Kauffman, 2021). All this is now testable experimentally. Once such life arose, it was a Kantian whole achieving collective catalysis and constraint closure. Even without genes, such systems can evolve to some extent and so acquire new adaptations (Vasas et al., 2012). With the emergence of coding, that mystery of evolving life is fully formed on Earth.

The alternate and standard view of the origin of life posits the emergence of at least one template-replicating RNA sequence able to copy itself (Joyce, 2002). This has not yet been experimentally successful but may well become demonstrated. This theory faces the issue that ribonucleotides and polymers of RNA were hard to synthesize on the early Earth. Moreover, any such "nude" replicating RNA would have to evolve ribozymes to catalyze some connected metabolism that could sustain the RNA polymer system. This is easy to imagine. However, there seems to be no reason at all why such a connected small-molecule metabolism should itself be collectively autocatalytic. Why would such a de novo metabolism be able to reproduce itself? That new catalyzed metabolism was selected merely to sustain the RNA world system that uses it.

These facts now almost persuasively indicate that life arose as small-molecule collectively autocatalytic sets, perhaps widely in the universe. If life is widely distributed among the solar systems in the universe and that evolution is beyond the Newtonian paradigm, vast new domains of science must be created with respect to major aspects of the evolving universe.

6) Most adaptations in the evolution of the biosphere are "affordances," typically seized by heritable variation and natural selection. An example of an affordance (Gibson, 1966) is a horizontal surface that affords you a place to sit. Affordances are, in general, "the possible use by me of X to accomplish Y." "Accomplish" can occur without "mind," but by blind heritable variation and natural selection, as in the evolution of the heart and loop of Henle (Kauffman & Roli, 2021).

An affordance is not an independent feature of the world (Walsh, 2015). An affordance is in relation to the evolving organism for which it is an affordance to be seized or not by heritable variation and natural selection. Biological degrees of freedom are affordances, or relational opportunities available to evolving organisms.

7) Often in evolution adaptations emerge by co-opting the same organ for a new function. These are called Darwinian preadaptations or exaptations (Gould & Vrba, 1982).

Typical examples of such an affordance, or new Darwinian preadaptation, seized by heritable variation and natural selection include flight feathers, which evolved earlier for functions such as thermal insulation or as bristles but were co-opted for the new function of flight (Prum & Brush, 2002; Persons IV & Currie, 2015), and lens crystallins originated as enzymes (Barve & Wagner, 2013). A wonderful example is the evolution of the swim bladder that emerged in a lineage of fish (Kauffman, 2016). In this latter instance, the ratio of air and water in the swim bladder functions to assess neutral buoyancy in the water column. Paleontologists believe the swim bladder arose from the lungs of lungfish. Water got into some lung, making it now a sac filled with a mixture of air and water, so poised to evolve into a swim bladder. This is precisely finding a new use for the same initial "thing," the lung. A new function, neutral buoyancy in the water column, has emerged in the evolving biosphere. There is yet more: Once a swim bladder emerged, it became newly possible that a worm or bacterium might evolve to live in swim bladders. Natural selection presumably "worked" to craft a functioning swim bladder. But did natural selection "craft" the swim bladder such that it could become a new affordance, a new niche that might be seized by the worm or bacteria? No. Without selection achieving it, the evolving biosphere is creating the ever-new affordances, the ever-new niche possibilities, into which the biosphere evolves. The biosphere is constructing the very adjacent possible into which it enters (Kauffman, 2019).

8.3   The Insuperable Limits of Set Theory

We have established that in the evolution of the biosphere, ever-new phase spaces with new boundary conditions and new relevant variables arise that were not prestated. Thus, it is then essential to ask if those now-relevant variables might have been prestated. The surprising answer, we hope to show, is "No."

We cannot prestate the ever-new relevant variables because we can neither compute, predict, nor deduce ahead of time the coming into existence of new affordances and newly relevant variables seized by heritable variation and natural selection.

We cannot compute or deduce the new adaptive phase spaces because we cannot use set theory or any mathematics based on it to reliably and soundly model the evolutionary emergence of adaptations as seized affordances. The considerations are a bit unexpected and focus on the implications of biosphere evolution features for the foundations of set theory (Kauffman & Roli, 2021).

Although our argument concerns the case of affordances seized by evolution—and not cognitive ones—we believe an example from the tool-usage context may be greatly explicative. How many "uses" does a screwdriver have, alone or with other things, in London on March 22, 2021? *i.* Screw in a screw. *ii.* Open a can of paint. *iii.* Wedge a door closed. *iv.* Scrape putty off a window. *v.* As an *objet d'art*. *vi.* Tie to a stick and spear a fish. *vii.* Rent the spear to local fishermen and take 5% of the catch. *viii.* Lean the screwdriver against a wall, place plywood propped up by the screwdriver and use this to shelter a wet oil painting, and so on.

Is the number of uses of a screwdriver alone or with other things a specific number, say 11? No. Is the number of uses infinite? How would we know? The number of uses of a screwdriver now and over the next thousand years is *indefinite* or perhaps *unknown*. No one in 1690 could have used a screwdriver to short an electric connection. It is essential to recognize that we *cannot list all the possible uses of a screwdriver* (Kauffman, 2019), because not only can we not predict the possible future niches for the screwdriver, but also because the uses of a screwdriver also depend upon the user's goals and repertoire of actions (Walsh, 2015). The same considerations apply in general to any object: for example, to the uses of an engine block. It can be used to build an engine, as a chassis for a tractor, as a paperweight, to crack open coconuts against its sharp corners, and so on.

One may argue that we cannot list all the possible uses of an object by using our intuition, but we could do it by applying enumeration or deduction. This is not possible either. There are four mathematical ordering scales: nominal, partial order, interval, and ratio. The uses of an object are merely a nominal scale, therefore *there is no ordering relation between these uses*. Furthermore, in general a specific use of an object does not provide the basis for entailing another use. Hence, there is no *deductive relation* between the different uses of an object (e.g., it is not possible to deduce the use of an engine block to crack open coconuts from its use as a paperweight).

We believe it is apparent that these arguments hold also for the emergence of adaptations as seized affordances along the diachronic evolution of the biosphere: ever-new affordances appear, which are seized by evolution and shape ever-new niches and biological functions in an unpredictable way. Two main observations support this statement: first, the articulation of parts explanation (Kauffman, 1970), and second, the impredicative loop among affordances and agent's goals and actions (Walsh, 2015). In one sentence: biological evolution concerns constructing, not listing.

The implication of what we stated above is that *we cannot use set theory with respect to the diachronic emergence of new affordances seized by heritable variation and natural selection* (Kauffman & Roli, 2021). This concerns all the adaptations in the diachronic evolution of the biosphere.

A first axiom of set theory is the *axiom of extensionality*: "Two sets are identical if and only [if] they contain the same members" (Jech, 2006). But we cannot prove that the *unlistable* uses of a screwdriver are identical to the *unlistable* uses of an engine block, as we cannot prove, once and for all, the uses of object X. Therefore, no axiom of extensionality. Hence, no sound set theory can be formulated.

Worse, the implications also reach mathematical fields based on set theory. The *axiom of choice* (Moore, 2012), which comes into play whenever a choice function cannot be defined, cannot be applied. The axiom of choice is equivalent to "well ordering" (Potter, 2004), but an ordering among the unordered uses of X cannot exist. Therefore, we would reach a contradiction if we tried to postulate this in a formal description of the evolution of affordances.

A consequence of this argument is the impossibility of using numbers with respect to the emergence of novel functions in the evolving biosphere. One way to define numbers uses set theory (Russell & Whitehead, 1973). The number "0" is defined as the set of all sets each of which has zero elements. In our case this corresponds to "the set of all objects that have exactly 0 uses." Well, no, this cannot be grounded on objects in an evolving biosphere. The alternative approach to numbers is via Peano's axioms (Peano, 1889). These require a null set and a successor relation. But we have no null set. Furthermore, the different uses of X are unordered. We have no successor relation.

Therefore, with respect to all diachronically emerging adaptations via seizing affordances, no numbers. No integers, no rational numbers, no equations such as $2+3=5$. No equations, so no irrational numbers. No real line. No equations with variables. No imaginary numbers, no quaternions, no octonions. No Cartesian spaces. No vector spaces. No Hilbert spaces. No union and intersection of uses of X and uses of Y. No first-order logic. No combinatorics. No topology. No manifolds. No differential equations on manifolds.

Further, without an axiom of choice, we cannot integrate and take limits on the differential equations we cannot write.[1]

8.4   The Third Transition: We Are Beyond the Newtonian Paradigm

These facts mean that we are, surprised or not, at the third major transition in science. If we can neither write nor solve differential equations for the diachronic evolution of adaptations in the biosphere we cannot, in principle, prestate, compute, or deduce the relevant ever-new phase spaces of evolving biospheres. The evolving biosphere advances into the adjacent possible it creates, but we cannot deduce what is "in" that adjacent possible. Therefore, we do not know the sample space of the process, hence can neither define a probability measure nor define *random*. We truly have no well-founded expectations. With respect to the diachronic evolution of new adaptations, we are beyond the Newtonian paradigm.

The implications are very large. If we can write and solve no equations for the diachronic evolution of our or any biosphere and our evolving universe has at least one evolving biosphere, there can be no final theory that entails what comes to exist in the evolving universe. The famous equation destined for the T-shirt (Kaku, 2021), it now seems, does not exist.

This result is somewhat stunning at first, then perhaps not totally surprising. Gödel's first incompleteness theorem (Franzén, 2005) assures us that any consistent axiomatic system as rich as arithmetic has the property that, given the axioms and the inference rules, a statement exists such that it can neither be proved nor disproved inside the system. The nonprovable statement is itself generated algorithmically (Longo, 2019). If this algorithmically generated statement itself is added to the initial axioms, the new set of axioms again algorithmically generates statements whose truth cannot be either proved or disproved. *In short, Gödel's theorem, iterated, yields an open succession of ever-new axiom systems.* Gödel's theorem relies on self-reference.

The evolving biosphere instantiates Gödel's theorem, and even more. New adaptations, new uses of physical things such as molecules, as is true for the new uses of an engine block, cannot be deduced from the old uses. Also, and importantly, affordances are referential to the organism for which the affordance is relevant. Affordances are referential degrees of freedom, not independent features of the world. Thus, the referential new uses cannot be deduced as a theorem from knowledge of the properties and functions of the existing molecules and other physical properties of organisms prior to the new adaptation (Kauffman & Roli, 2021). Therefore, they are more than the analogue of algorithmically generated undecidable statements: they can be read as "If I get to exist in a new way for some time in the biosphere, my new existence cannot be deduced from the biosphere up to the present moment."

Like the ongoing generation of ever-richer axiom systems in Gödel's theorem by successive additions of new axioms, the generation of ever-new nondeducible adaptations by new uses among the indefinite possible uses of each physical thing in an evolving cell or multicellular organism successively says, "I get to exist in a new way in the biosphere that is not deducible."

Reluctant or not, we observe that the evolution of our or any biosphere is outside of the Newtonian paradigm. What are some implications?

1. There really can be no "final theory" that entails all that comes to exist in the evolving universe. The dream of such a final theory is magnificent and has been a driving motivation for superb science for centuries. Perhaps our arguments are wrong. If so, let them be vanquished.

2. The evolution of our or any biosphere in the universe is not only entailed by no law, but seems not even mathematizable by known techniques. Perhaps we can invent new mathematics.

3. If no law entails the evolution of biospheres and that evolution cannot even be mathematized, biological evolution is radically "free" to be and is vastly creative. Section 8.7 details some of the unexpected reasons for such ongoing creativity.

4. Most essentially, we really are at a third transition in science. The scale and meanings of this are quite unclear at present. Our universe is creative in ways we have not known. Somehow our understanding of the world will change.

8.5   Toward a Statistical Mechanics of Emergence

There is a pathway forward. A beginning point is to realize that, in fact, any physical (or other) object has indefinitely many uses that are not deducible from one another. The engine block really can be used as a paperweight and to crack open coconuts. Therefore, we must give up specific "properties" of objects and abstract an object as just that, an "object" or "thing." Given this step, we can think of "things" transforming to yield old or new "things." We can also think of "things" acting on and regulating the transformations among "things" to yield old or new "things."

From such a spare abstraction, a great deal can already be done. Loreto et al. (2016) formulated a modified urn model. Here one starts with two "things": for example, red and black balls in the urn. The process samples at random from the urn. If a new color, never seen before, is encountered, a ball with a new color is added to the urn. In this model, a single "thing" can only give rise to a single new "thing." "Things" are without properties save "color," which merely stands for "new thing." From this spare beginning, Loreto et al. derive Zipf's law and Heap's law (Loreto et al., 2016).

With others we are examining the theory of the "adjacent possible" (TAP) process, described by the following equation, eq. 1 (Steel et al., 2020):

$$M_{t+1} = M_t + \sum_{i=1}^{M_t} \alpha^i \binom{M_t}{i}, \quad 0 \le \alpha \le 1$$

In this process, there are at any time t, M_t "things." At each time step, any subset of the M_t things, 1,2,3, . . . , can be used to create a single new thing. At the next period of time, $t+1$, the system has the initial number of things, M_t, plus the new things created (see eq. 1). More specifically, the probability that any subset can be used to create a new thing decreases monotonically with the size of the subset, 1,2,3, . . . This is an entirely new equation. Because subsets up to size M_t can be used, as M_t increases this process has remarkable properties: If the process is started with only a few things, the number increases at a glacial pace then explodes suddenly. The continuous process reaches infinity in a finite time, and thus has a pole. The discrete process does not reach infinity, but explodes ever more rapidly (Steel et al., 2020).

The TAP process already seems to account for three features of many processes in the universe:

i. The increasing number of different molecular species, living species, and technological "tools" in the evolution of the universe, in the evolution of the biosphere, and in human technological evolution in the past 2.6 million years (Koppl et al., 2018; Steel et al., 2020).

ii. If we re-interpret M_t as the complexity of the most complex thing produced at time t, the same theory seems to account for the gradual then explosive diversification in molecular complexity, species complexity, and tool complexity over time. With respect to tools, 2.6 million years ago we had perhaps five to ten equally simple stone tools. Now we have billions ranging in complexity from needles to the International Space Station (Koppl et al., 2018).

iii. Every object created in the TAP process has one or more immediate ancestor objects and may have 0, 1, 2, or more immediate "children," then grandchildren and further descendants. TAP predicts a power law descent distribution with a slope of -1.1 to -1.35 depending upon parameters (Steel et al., 2020). The immediate and later descendants of a legal patent can be assessed by the citation of the parental patent. Remarkably, analysis of 3,000,000 patents in the U.S. Patent Office from 1835 to 2010 is a power law slope -1.30 (Koppl et al., 2021). Here the "objects" combined are not material at all, but ideas. Presumably, the descent distribution of actual technologies in the field is also a power law of the same slope.

A single theory, TAP, appears to explain three disparate phenomena (Koppl et al., 2021) suggesting that there may be something fundamentally correct about it.

iv. The TAP processes herald a *new fourth law of thermodynamics* in the nonergodic universe (Cortês et al., 2022). If we set $\alpha = 1.0$, the TAP process generates all the possibilities, TP. The total possible, TP, increases over time. If we set $\alpha < 1.0$, the TAP process generates the actualized possibilities, TA. The actualized possible, TA, a subset of the total possible TP, increases over time. The ratio of these, $R = TP/TA$, also increases over time: hence, the nonergodicity of the nonergodic system, R, increases over time. Thus, the localization of the system in its nonergodically expanding phase space, $1/R$, also becomes greater over time. The lower $1/R$, the greater the localization. *Over time, the universe creates an ever-tinier subset of what is now possible at the level of molecules, species and tools.* This new law plays a major role in the increasing complexity of the universe (Kauffman, 2022).

In this new fourth law, the increasing localization of systems is within their ever-expanding phase spaces. The relation of this new fourth law to the famous second law of thermodynamics and statistical mechanics in a prestated and fixed phase space where entropy always tends to increase and localization of the system in its fixed phase space tends to decrease, remains to be clarified (Kauffman, 2022).

It is hoped that the same abstract theory fits these three distinct phenomena. Both the Loreto-Strogatz urn model (Loreto et al., 2016) and the TAP process (Steel et al., 2020) are not ergodic. They are abstract representations of processes that reach into an unprestatable adjacent possible. In these two models, things transform to things. However, there is no notion of "function."

8.6   Kantian Wholes Provide the Missing Concept of Function

Any living cell or organism is a nonequilibrium physical system that is a Kantian whole which has the property that the parts exist for and by means of the whole. This provides a proper concept of *function*.

1) Any living cell or multicelled organism is a Kantian whole. Humans are Kantian wholes. We exist as complex things in the nonergodic universe above the level of 500 Daltons for and by means of our hearts and loops of Henle. Hearts and loops of Henle exist as complex things in the universe for and by means of us. Thus, we are Kantian wholes: our parts do exist for and by means of the us, the whole.

2) The function of a part is that subset of its causal properties that sustains the whole. The function of the heart is to pump blood, a process that sustains the whole organism of which the heart is a member. The existence of Kantian wholes as complex things in a nonergodic universe above 500 Daltons affords a clear meaning to the word *function*. Functions are real in the universe.

3) The system achieves catalytic or task closure: All reactions and tasks requiring fulfillment are fulfilled by components in the same system.

4) The system achieves the newly recognized and powerful property of constraint closure (Montévil & Mossio, 2015). Cells literally construct themselves.

8.7   The Evolution of Integrated Functionality: Emergent Creativity

We achieve a new understanding of the almost miraculous emergent self-construction and emergent coherent functional organization of processes in an evolving biosphere: There is no deductive relation between the different uses of any physical thing, such as a protein in a cell that can evolve to be used to catalyze a reaction, to carry a tension load, or to host a molecular motor on which it walks. Cells physically construct themselves.

Therefore, each molecule and structure in evolving cells and organisms in the biosphere stands *ever-available to be selected, alone or with other things, for indefinite adaptive new uses* such that myriad new adaptations and new physical things like new proteins arise all the time. The new uses are not open to deduction from the old uses.

Functional integration is always maintained, even as it transforms, because the functional evolution of the parts must always sustain the functioning Kantian whole upon which selection acts. Selection acting upon the whole determines what "gets to exist" for some time in the nonergodic biosphere. This is downward causation. The explanatory arrows do not point only downward (Weinberg, 1994).

The evolving biosphere really is a propagating adapting construction, not an entailed deduction. This is "sustained functional integrated emergence" in evolving Kantian wholes. It is the arrival of the fitter.

This is emergence. *Emergence is not engineering*. This radical emergence of a coevolving biosphere itself emerges only beyond the Newtonian paradigm. That we are at a third transition in science, beyond Newton's wonderful paradigm, is not a loss; rather, it is an invitation to participate in this magical emergence we have not ever seen before.

We hardly begin to understand this. An evolving biosphere is a self-constructing, functionally integrated blossoming emergence. This seems also to share a common ground with codependent origination (Laumakis, 2008).

An evolving biosphere is a propagating construction, not an entailed deduction. Hiding behind the equations we write, we cannot see the reality that they hide: the mystery of evolving life. We are of it, not above it.

8.8   Abstract Kantian Wholes as a Formalization of Functionality

An abstract representation of any Kantian whole includes both things transforming to things, and things regulating these transformations among things. This is a form of *digraph* (see figure 8.1). In general, "things" are represented by circles, while transformation among things, "reactions," are represented as dots. Every circle "thing" is connected to one or more transformation dots. Every transformation dot is connected to one or more circle "things." This is a digraph.[2]

In addition, the "thing regulating a specific transformation among things" is represented by an arrow from the thing circle to the transformation dot that the thing regulates. The result is a digraph augmented with 0, 1, or more arrows from each thing to any transformation it regulates.

Figure 8.1 is a typical example of an abstract Kantian whole. The Kantian whole has the property that the last step in the formation of each thing is positively regulated by one or more of the things in the set. In this abstract representation of a Kantian whole, the parts really get to exist for and by means of the whole. In a physical Kantian whole, the function of a part really is the subset of its indefinitely many causal properties that help sustain the whole.

Precisely because we have abstracted away any specific properties from a "thing," it can come to be used—hence function—in indefinitely many ways. The engine block, here abstracted from specific properties, can function as a paperweight and also function to crack open coconuts.

We have achieved an abstract model of the functional closure of a real physical Kantian whole, each of whose parts can come to function in ways that cannot be deduced from one another. Again, we achieve this precisely by abstracting away any specific properties of a thing, the transformations of things to things, and the way things can regulate the transformations. In this abstract representation, a thing can be a molecule, A transforming to B, and regulation can be catalysis of the transformation reaction by C. A thing can be a species in

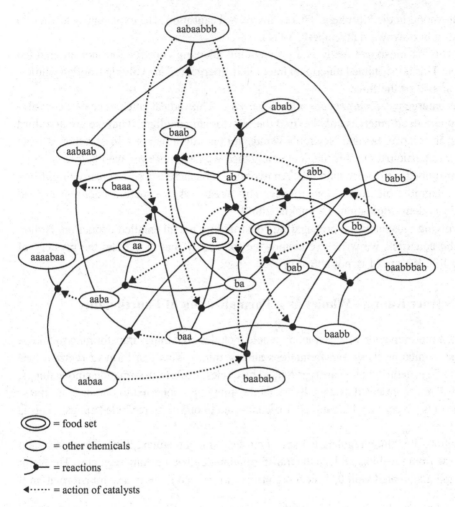

= food set

= other chemicals

= reactions

= action of catalysts

Figure 8.1
Example of a graph describing an autocatalytic set, taken from Farmer et al. (1986). Courtesy of the authors.

the process of surviving, and the things mediating this process of survival can be the things in the niche of the surviving species. The things can be goods sold in a market, regulated by relevant legal laws. A corporation is a Kantian whole embedded in the larger Kantian whole economic and legal world enabled by laws that constrain human activities into the specific human work that sustains the Kantian whole corporation and its enabling legal laws. By abstracting any properties from a thing, the indefinite actual uses of any physical thing can be captured. In short, the physical, legal, ideational character of a thing is irrelevant to the abstracted thing, its transformations, and the regulations of transformations among things by things.

If we are precluded from using set theory with respect to real things, stones, hammers, and enzymes, we are fully allowed here to use set theory with this fully syntactic model of things and their transformations.

8.9 A Statistical Mechanics of Emergence

A union of the TAP process and the evolution of Kantian wholes, TAP is an abstraction of nonergodic processes whereby one or more things can give rise to one new thing. A further step, closer to chemistry, is for one or more things to give rise to one or more things. To unite the TAP process with the evolution of functionally integrated Kantian wholes, we have merely to add to TAP that things can act on the transformation by which things yield things, to speed or slow the transformation: that is, to catalyze or inhibit the transformation. Most generally, let K things act with some probability, P, on any transformation, X. We can model the effect of the K things on this transformation by arbitrary Boolean functions on K inputs. We can explore different rules by which things come to act on transformations among things.

We here achieve for the first time an abstract union of the functionality of Kantian wholes and the nonergodic open transformation of things transforming to things by the TAP processes, or a generalization in which more than one thing can be produced in a transformation. This leads to the formation and evolution of abstract Kantian wholes with one another within the evolving TAP process as it creates an increasing diversity of things. The character of the things and transformations does not matter at all. Again, the things can be molecules reacting and forming a spray of new molecules (Scherer et al., 2017), and perhaps catalyzing or inhibiting those reactions. The things can be goods or services created out of input goods and transformed in factories, and other capital goods, into output goods (Cazzolla Gatti et al., 2020), or legal laws. The transformation can be carried out by human actions legally allowed or forbidden by extant evolving legal laws, as in human action in an economy.

There is a first hint of a statistical mechanics of emergence in nonergodic systems. Whether we consider an evolving chemical system in which molecules transform and regulate the transformations, an ecosystem of species creating and blocking niches for one another, or goods and services in an evolving economy creating and blocking market opportunities for one another, or the evolution of legal systems, we have a new set of conceptual—indeed, mathematical—tools.

New questions arise. Over time, how many abstract Kantian wholes emerge? What are the statistics of their sizes? Do they help or hinder one another? Do they coevolve? If so, what are the statistical structures of their coevolving fitness landscapes? Do those landscapes tend to asymptotic forms of criticality (Kauffman, 1995)? Genetic regulatory networks in cells and brains are dynamically critical (Beggs, 2008; Daniels et al., 2018; Villani et al., 2018). Do the Boolean functions in abstract Kantian wholes evolve to dynamical criticality with small attractors and a maximization of transfer entropy within and between the emergent Kantian wholes (Li et al., 2019)? Do the systems evolve to tune their own connectivity in some way? Why and how? How many "goods and services" do Kantian wholes exchange? Does any of this map to molecular and functional trading in microbial communities (Le & Wang, 2020)? Does it map to small ecosystems? To the entire evolving biosphere? Economy? Legal systems?

A bacterium is a Kantian whole. A eukaryotic cell contains mitochondria which themselves are Kantian wholes. Thus, a eukaryotic cell is a second-order Kantian whole enclosing a first-order Kantian whole. A multicelled organism is a third-order Kantian whole containing second-order Kantian wholes containing first-order Kantian wholes. The ecosystem in our guts and our cells is a fourth-order Kantian whole whose parts exist for and by means of the whole. Probably the entire biosphere is some form of high-order nested Kantian whole. Is an economy a nested set of Kantian wholes? With what emergent statistical regularities? Might it be possible to study abstract statistical properties of emergence of nested higher-order evolving Kantian wholes in a statistical mechanics of emergence?

8.10   Agency, Function, Purpose, and Teleonomy in Evolution

The very existence of Kantian wholes in the nonergodic universe above the level of atoms allows, as noted, a noncircular definition of *function*. The function of a part is that subset of its causal properties that sustains the Kantian whole. What of purpose? As Jacques Monod famously said in *Chance and Necessity* (Monod, 1971), the project of every cell is to create two cells. Living cells are molecular autonomous agents, able to reproduce, perform thermo-dynamic work cycles, and choose (Clayton & Kauffman, 2006). To choose, cells must sense the world, evaluate it as "good or bad for me," and act (Peil, 2014). This triad is central to the project of every cell and is the root of "value" (Peil, 2014). The same triad is the root of affordances: "goal," "affordance," and "means to seize the affordance" (Walsh, 2015). Here is where teleonomy comes into play: organisms act according to their values, and so their internal goals, and the opportunities they have. The tight, irreducible coupling between goals and affordances emphasizes the prominent role of organism purposiveness in the process of affordance-seizing exerted by natural selection, which is central for teleonomic selection (Corning, 2014, 2018). Furthermore, given that living organisms are nested Kantian wholes, teleonomy—and its impact on evolution—can be found at every level of organisms.

All this is central to the roles of consciousness in evolution. As we have seen, "seeing an affordance," such as seeing the use of an engine block to crack open coconuts, cannot be deduced. As we argue elsewhere (Kauffman & Roli, 2022), "seeing" of complex sequential affordances, as in jury-rigging (Jacob, 1977), cannot be achieved by a nonembodied Universal Turing machine or embodied robots, but can be achieved by humans. This suggests that mind is quantum and perhaps that quantum actualization underlies the consciousness that allows us to see complex affordances. Were this true, it would allow mind to have evolved and played its diverse roles as organisms mutually created more complex worlds with one another over the past 3.7 billion years.

8.11   Conclusions

The twenty-first century promises to be the century of biology. This of course embraces the explosion of biotechnology, an emergence of twenty-first-century medicine, and ever-deeper analysis of how cells and organisms that now exist function ("work") as physical systems at molecular, cellular, organism, and ecosystem levels. Here reliance on physics, chemistry, biophysics, biochemistry, and molecular biology is essential. The issues are massive in complexity and import. We are in the era of systems biology.

However, we confront the third major transformation in science, following Newtonian and quantum mechanics, the first two transformations. We are forced beyond the wonderful Newtonian paradigm. There really is no "Final Theory": the diachronic evolution of our or any biosphere is beyond entailing law and beyond any mathematics based on set theory.

There may well be other biospheres in the universe. Evolving biospheres are immensely creative in ways beyond our knowing or stating. We live forward in face of mystery. This implies that we humans are of nature, not above nature. Rather than a loss, this is, instead, an enormous invitation. We can try to understand in new ways how our or any biosphere, our global economy, and even our cultures diachronically construct themselves over billions, millions, and hundreds of thousands of years of unprestatable, nonentailed, ever-creative, nonergodic emergence. We are invited to construct a new statistical mechanics of emergence. We come to understand that we really are conscious agents. We are also invited to live responsibly in our shared biosphere.

Notes

1. Both the $(\varepsilon - \delta)$ formal definition of limits (Grabiner, 1983) and the one based on infinitesimals (Robinson, 2016) rely on set theory.

2. A similar formalism has also been introduced by Robert Rosen (1972; 1991).

References

Arrow, K., & Debreu, G. (1954). Existence of an equilibrium for a competitive economy. *Econometrica: Journal of the Econometric Society, 22*(3), 265–290.

Ashkenasy, G., Jagasia, R., Yadav, M., & Ghadiri, M. (2004). Design of a directed molecular network. *Proceedings of the National Academy of Sciences, 101*(30), 10872–10877.

Aspect, A., Dalibard, J., & Roger, G. (1982). Experimental test of Bell's inequalities using time-varying analyzers. *Physical Review Letters, 49*(25), 1804–1807.

Atkins, P. (1984). *The second law.* New York: Scientific American Library.

Barve, A., & Wagner, A. (2013). A latent capacity for evolutionary innovation through exaptation in metabolic systems. *Nature, 500*(7461), 203–206.

Beggs, J. (2008). The criticality hypothesis: how local cortical networks might optimize information processing. *Philosophical Transactions of the Royal Society A: Mathematical, Physical and Engineering Sciences, 366*(1864), 329–343.

Birkhoff, G., & von Neumann, J. (1936). The logic of quantum mechanics. *Annals of Mathematics, 37*(4), 823–843.

Cass, D., & Shell, K. (1983). Do sunspots matter? *Journal of Political Economy, 91*(2), 193–227.

Cazzolla Gatti, R., Koppl, R., Fath, B., Kauffman, S., Hordijk, W., & Ulanowicz, R. (2020). On the emergence of ecological and economic niches. *Journal of Bioeconomics, 22*(2), 99–127.

Clayton, P., & Kauffman, S. (2006). On Emergence, Agency, and Organization. *Biology and Philosophy, 21*(4), 501–521.

Corning, P. (2014). Evolution 'on purpose': How behaviour has shaped the evolutionary process. *Biological Journal of the Linnean Society, 112*(2), 242–260.

Corning, P. (2018). *Synergistic selection: How cooperation has shaped evolution and the rise of humankind.* Singapore: World Scientific.

Cortês, M., Kauffman, S., Liddle, A. R., & Smolin, L. (2022). Biocosmology: Biology from a cosmological perspective. *arXiv:2204.09379*. https://doi.org/10.48550/arXiv.2204.09379

Daniels, B. C., Kim, H., Moore, D., Zhou, S., Smith, H. B., Karas, B., Kauffman, S., & Walker, S. I. (2018). Criticality distinguishes the ensemble of biological regulatory networks. *Physical Review Letters, 121*(13), 138102.

Einstein, A., Podolsky, B., & Rosen, N. (1935). Can quantum-mechanical description of physical reality be considered complete? *Physical Review, 47*(10), 777.

Farmer, J., Kauffman, S., & Packard, N. (1986). Autocatalytic replication of polymers. *Physica D: Nonlinear Phenomena, 22*(1–3), 50–67.

Feynman, R. (1998). *Quantum electrodynamics.* Nashville, TN: Westview Press.

Franzén, T. (2005). *Gödel's theorem: An incomplete guide to its use and abuse.* Boca Raton, FL: CRC Press.

Geanakoplos, J., & Polemarchakis, H. (1986). Existence, regularity and constrained suboptimality of competitive allocations when the asset structure is incomplete. In W. Hell, R. Starr, and D. Starrett (Eds.), *Uncertainty, information and communication: Essays in honor of K. J. Arrow*, ch. 3, pp. 65–95. Cambridge: Cambridge University Press.

Gibson, J. (1966). *The senses considered as perceptual systems.* Boston: Houghton Mifflin.

Gould, S. J., & Vrba, E. S. (1982). Exaptation—a missing term in the science of form. *Paleobiology, 8*(1), 4–15.

Grabiner, J. (1983). Who gave you the epsilon? Cauchy and the origins of rigorous calculus. *The American Mathematical Monthly, 90*(3), 185–194.

Heisenberg, W. (1958). *Physics and philosophy: The revolution in modern science.* New York: Harper Torchbooks.

Hordijk, W., & Steel, M. (2004). Detecting autocatalytic, self-sustaining sets in chemical reaction systems. *Journal of Theoretical Biology, 227*(4), 451–461.

Jacob, F. (1977). Evolution and tinkering. *Science New Series, 196*(4295), 1161–1166.

Jech, T. (2006). *Set theory.* New York: Springer.

Joyce, G. (2002). The antiquity of RNA-based evolution. *Nature, 418*(6894), 214–221.

Kaku, M. (2021). *The God equation: The quest for a theory of everything.* New York: Doubleday.

Kauffman, S. (1970). Articulation of parts explanation in biology and the rational search for them. In *Topics in the Philosophy of Biology*, 245–263. Chicago: University of Chicago Press.

Kauffman, S. (1971). Cellular homeostasis, epigenesis and replication in randomly aggregated macromolecular systems. *Journal of Cybernetics, 1*(1), 71–96.

Kauffman, S. (1986). Autocatalytic sets of proteins. *Journal of Theoretical Biology, 119*(1), 1–24.

Kauffman, S. (1995). *At home in the universe: The search for the laws of self-organization and complexity.* Oxford: Oxford University Press.

Kauffman, S. (2016). *Humanity in a creative universe.* Oxford: Oxford University Press.

Kauffman, S. (2019). *A world beyond physics: The emergence and evolution of life.* Oxford: Oxford University Press.

Kauffman, S. (2020). Answering Schrödinger's "What Is Life?" *Entropy, 22*(8), 815.

Kauffman, S. (2022). Is there a fourth law for non-ergodic systems that do work to construct their expanding phase space? *Entropy, 24*(10), 1383.

Kauffman, S., Jelenfi, D., & Vattay, G. (2020). Theory of chemical evolution of molecule compositions in the universe, in the Miller–Urey experiment and the mass distribution of interstellar and intergalactic molecules. *Journal of Theoretical Biology, 486*, 110097.

Kauffman, S., & Roli, A. (2021). The world is not a theorem. *Entropy, 23*(11), 1467.

Kauffman, S., & Roli, A. (2022). What is consciousness? Artificial intelligence, real intelligence, quantum mind, and qualia. *Biological Journal of the Linnean Society*, blac092. https://doi.org/10.1093/biolinnean/blac092

Koppl, R., Devereaux, A., Herriot, J., & Kauffman, S. (2018). A simple combinatorial model of world economic history. *arXiv*:1811.04502. https://arxiv.org/abs/1811.04502

Koppl, R., Devereaux, A., Valverde, S., Solé, R., Kauffman, S., & Herriot, J. (2021). Explaining technology. *SSRN Papers.* https://papers.ssrn.com/sol3/papers.cfm?abstract_id=3856338

Kvenvolden, K., Lawless, J., Pering, K., Peterson, E., Flores, J., Ponnamperuma, C., Kaplan, I., & Moore, C. (1970). Evidence for extraterrestrial amino-acids and hydrocarbons in the Murchison meteorite. *Nature, 228*(5275), 923–926.

Lancet, D., Zidovetzki, R., & Markovitch, O. (2018). Systems protobiology: origin of life in lipid catalytic networks. *Journal of The Royal Society Interface, 15*(144), 20180159.

Laumakis, S. (2008). Interdependent arising. *Cambridge Introductions to Philosophy*, pp. 105–124. Cambridge: Cambridge University Press.

Le, M., & Wang, D. (2020). Structure and membership of gut microbial communities in multiple fish cryptic species under potential migratory effects. *Scientific Reports, 10*(1), 1–12.

Lehman, N., & Kauffman, S. (2021). Constraint closure drove major transitions in the origins of life. *Entropy, 23*(1), 105.

Li, M., Han, Y., Aburn, M., Breakspear, M., Poldrack, R., Shine, J., & Lizier, J. (2019). Transitions in information processing dynamics at the whole-brain network level are driven by alterations in neural gain. *PLoS Computational Biology, 15*(10), e1006957.

Lincoln, T., & Joyce, G. (2009). Self-sustained replication of an RNA enzyme. *Science, 323*(5918), 1229–1232.

Longo, G. (2019). Interfaces of incompleteness. In G. Minati, M. Abram, and E. Pessa (Eds.), *Systemics of incompleteness and quasi-systems*, pp. 3–55. New York: Springer.

Longo, G., Montévil, M., & Kauffman, S. (2012). No entailing laws, but enablement in the evolution of the biosphere. In *Proceedings of GECCO 2012—The 14th Genetic and Evolutionary Computation Conference*, pp. 1379–1392.

Loreto, V., Servedio, V., Strogatz, S., & Tria, F. (2016). Dynamics on expanding spaces: Modeling the emergence of novelties. In *Creativity and universality in language*, pp. 59–83. New York: Springer.

Monod, J. (1971). *Chance and necessity: An essay on the natural philosophy of modern biology*. New York: Alfred A. Knopf.

Montévil, M., & Mossio, M. (2015). Biological organisation as closure of constraints. *Journal of Theoretical Biology, 372*, 179–191.

Moore, G. (2012). *Zermelo's axiom of choice: Its origins, development, and influence*. North Chelmsford, MA: Courier Corporation.

Peano, G. (1889). *Arithmetices principia: Nova methodo exposita*. Turin, Italy: Fratres Bocca.

Peil, K. T. (2014). Emotion: The self-regulatory sense. *Global Advances in Health and Medicine, 3*(2), 80–108.

Persons IV, W., & Currie, P. (2015). Bristles before down: A new perspective on the functional origin of feathers. *Evolution, 69*(4), 857–862.

Planck, M. (1901). On the law of distribution of energy in the normal spectrum. *Annalen der physik, 4*, 553.

Potter, M. (2004). *Set theory and its philosophy: A critical introduction*. Oxford: Clarendon Press.

Prum, R., & Brush, A. (2002). The evolutionary origin and diversification of feathers. *The Quarterly Review of Biology, 77*(3), 261–295.

Robinson, A. (2016). *Non-standard analysis*. Princeton, NJ: Princeton University Press.

Rosen, R. (1972). Some relational cell models: The metabolism-repair systems. In *Foundations of mathematical biology*, vol. II, ch. 4, pp. 217–253. New York: Academic Press.

Rosen, R. (1991). *Life itself: A comprehensive inquiry into the nature, origin, and fabrication of life*. New York: Columbia University Press.

Russell, B., & Whitehead, A. (1973). *Principia mathematica*. Cambridge: Cambridge University Press.

Scherer, S., Wollrab, E., Codutti, L., Carlomagno, T., da Costa, S., Volkmer, A., Bronja, A., Schmitz, O., & Ott, A. (2017). Chemical analysis of a "Miller-Type" complex prebiotic broth—part II: Gas, oil, water and the oil/water-interface. *Origins of life and evolution of the biosphere: The journal of the International Society for the Study of the Origin of Life, 47*(4), 381–403.

Smolin, L. (2013). *Time reborn: From the crisis in physics to the future of the universe*. London: Penguin Books.

Steel, M., Hordijk, W., & Kauffman, S. (2020). Dynamics of a birth–death process based on combinatorial innovation. *Journal of Theoretical Biology, 491*, 110187.

Svirezhev, Y. (2008). Nonlinearities in mathematical ecology: Phenomena and models: Would we live in Volterra's world? *Ecological Modelling, 216*(2), 89–101.

Vaidya, N., Manapat, M., Chen, I., Xulvi-Brunet, R., Hayden, E., & Lehman, N. (2012). Spontaneous network formation among cooperative RNA replicators. *Nature, 491*(7422), 72–77.

Vasas, V., Fernando, C., Santos, M., Kauffman, S., & Szathmáry, E. (2012). Evolution before genes. *Biology Direct, 7*(1), 1–14.

Villani, M., La Rocca, L., Kauffman, S., & Serra, R. (2018). Dynamical criticality in gene regulatory networks. *Complexity, 18*, art. 5980636.

von Kiedrowski, G. (1986). A self-replicating hexadeoxynucleotide. *Angewandte Chemie International Edition in English, 25*(10), 932–935.

Walsh, D. (2015). *Organisms, agency, and evolution*. Cambridge: Cambridge University Press.

Weinberg, S. (1994). *Dreams of a final theory*. New York: Vintage.

Xavier, J., Hordijk, W., Kauffman, S., Steel, M., & Martin, W. (2020). Autocatalytic chemical networks at the origin of metabolism. *Proceedings of the Royal Society B, 287*(1922), 20192377.

9 On the Concept of Meaning in Biology

Kalevi Kull

Life can only be understood when one has acknowledged the importance of meaning.
—Jakob von Uexküll (1982/1940: 26)

Overview

This essay provides a provisional outline of biological approaches to the phenomenon of meaning. It includes brief descriptions of meaning construction based on function, purpose, survival, reproductive success, evolutionary history, code, signaling, communication, anticipation, and choice. It is argued that most of these approaches are not sufficient in demonstrating the meaning-making from the organism's own perspective. As concluded, the capacity of meaning-making is equivalent to the existence of *umwelt*.

9.1 Introduction

The terms *biological meaning* and *survival* were introduced into a common biological discourse almost together, in the 1860s, and their usage has been growing until the current century. This paralleled the critique of teleology and introduction of historical method into biology. The main focus of this essay is on the concept of meaning, ordinarily used in biology as a metaphor (that is, a concept which is not based on definition—cf. Reynolds 2022), but here viewed as being on its way toward becoming a scientific concept. Among other things, one may find an analogy between the topic of this article and William James' critique of Herbert Spencer's postulate of existence of survival interest (and its implications for primary mental phenomena) almost 150 years ago (James, 1878).

Survival, purpose, and meaning all have some teleological connotation. A general reconceptualization of teleology took place in the middle of the twentieth century as a result of cybernetic models, especially under the influence of the work by Arturo Rosenblueth, Norbert Wiener, and Julian Bigelow (1943; see also the comment by Taylor, 1950, and the response by Rosenblueth & Wiener, 1950). This also evoked the rapid growth of semiotics (for some historical links, see Halpern, 2005). The introduction of the concept of teleonomy by Colin Pittendrigh (1958), who studied chronobiological control mechanisms, occurred in the same period and context.

The concept of meaning has been difficult for biology. As viewed in the science of semiotics, meaning is neither a thing nor a process, but a relation. A biologist would ask: "Where can we find and identify relations if we study just things and processes? In what sense can meaning be real for an organism?" The problem seemed to be temporarily solved by introduction of an information approach in biology. However, there again, the quantification of information left the natural semantics out.

Among the contemporary developments in theoretical biology, one may observe a parallel and intertwined growth of epigenetic and biosemiotic studies. The latter is to a certain extent dependent on the former. Biosemiotic studies seek to understanding meaning-making in living systems; an important problem to solve is to find the necessary and sufficient conditions of biological meaning-making and to describe the types of such process(es).

What is meant by *meaning* or *meaning-relation* in living systems varies widely, however, with some overlap among alternative usages of the term. The term *meaning* does not have a fixed, generally agreed-upon definition, even in theoretical semiotics—ever since the classic work *Meaning of Meaning* by Ogden and Richards (1923), in which the concept was thoroughly analyzed. In semiotics, the concept of meaning has been separated into a series of concepts and usually replaced by more specific terms, such as denotatum, connotatum, designatum, sense, reference, signified, content, object of sign, interpretant, expression, value, affordance, semiotic fitting, and so on, so that it has sometimes been proposed not to use the term *meaning* (e.g., Morris, 1938: 43).

This essay reviews the common approaches to the concept of meaning (or to the conditions of prelinguistic meaningfulness) used in biology during the last hundred years. We exclude here the explicit applications of semiotic models in biology (like Hoffmeyer, 2008; Queiroz & El-Hani, 2006; Sonesson, 2006; Favareau, 2010; Emmeche & Kull, 2011; Maran et al., 2011; Kull & Favareau, 2022; and others) and philosophical analyses of the term (e.g., Millikan, 2009, and other works on teleosemantics). Instead, we focus on how some elements of semiotics spontaneously emerged from biological studies themselves. There exist several biological works on the origins of meaning in the living world: for instance, Canguilhem, 2008/1965; Tembrock, 1971; Goodwin, 1972; Prodi, 2021/1977; Kull, 1999; Cordelli & Galleni, 2003; Deacon, 2010, 2012; Stillwaggon Swan & Goldberg, 2010; Auletta, 2011; Kauffman, 2012; Winslow, 2014; Pharoah, 2020; among others. The background of these studies varies remarkably. But the scope of approaches to the problem of meaning in biology—meaning as it appears in living systems, or the forms of meaning in living systems as treated in biology—has not yet been systematically reviewed, as far as I know.

Why such a topic? Why is it interesting? It seems to be rather widely accepted in biology (at least in the biology termed as bioexceptionalism by Erik Peterson, 2022) that life is a process with the ability for meaning-making, in some sense. However, there is no agreement regarding in which form meanings exist in organisms or in their relationships. In many cases, the meaningfulness is described from a human perspective, not from that of a living being itself; or it is not explained how meaning can be real for an organism. It is also not clear which terms in biology have been used to refer to meaning.

In other words, if meaning-making is an attribute of life, then this implies that life is largely a semiotic phenomenon. The view that semiosis and life are coextensive was stated by Thomas Sebeok in the early 1980s. In a later formulation, he wrote: "the phenomenon that distinguishes life forms from inanimate objects is semiosis" (Sebeok, 2001: 1). The theory of life—theoretical biology—must therefore build its own semiotic foundation. And Conrad Waddington's conclusion that a paradigm similar to that of general linguistics (which is semiology, according to Ferdinand de Saussure), may become the theoretical basis for biology (Waddington, 1972: 289), can be interpreted in the same way. When introducing semiotics into biology, it is important to see how semiotics can be linked to the existing theoretical apparatuses in biology: that is, where (and why) meaning is hidden in biological theories or conceptual systems.

Thus, this article is not about semiotic theory, but about how biology itself has dealt with the problem of life's internal meaningfulness or meaning-creation. We are not building a model here that could cover the different approaches as special cases of meaning. This current work may rather serve as a preparation for such a further task. Instead of semiotics proper, we are going to disclose some cryptosemiotic approaches. Nevertheless, we need an overarching definition that would cover the approaches interpreted here as related to the concept of meaning, to somehow delimit our inquiry. Therefore, we define that a necessary (certainly not sufficient) condition for meaning is *the existence of mediated relation which is established by certain processes of life*. This definition will be used as a criterion for inclusion of an approach in the following list. Another necessary condition for meaning—that such relation *as a relation* (and not just a physical process) should be *real for the organism itself* (and not just for a human external observer)—will be added only in the later discussion section.

Our method is a simple conceptual-empirical typology, with some elements of heuristic typology (e.g., Bailey, 2005). Or, recalling the wonderful text "You know my method" by Thomas and Jean Sebeok (1981) from the Sebeoks' most productive period, it is retroduction, or abduction (as Charles Peirce used the term). As always in the case of taxonomies in the living realm, there exist intermediate forms, there are exceptions, and there can be alternative divisions into types or subtypes.

Relations that may be considered as a basis for biological meaning vary. According to our preliminary analysis, meaning in an organism's life, as described in biology, has been associated either with *function, purpose, survival, evolutionary past, coexistence, code, signaling, communication, anticipation*, or *umwelt*. In section 9.2 we look briefly into each of these approaches, recognizing that the list is not exhaustive and that the categories stem from existing traditions in biological discourse and not from any general theoretical model of life. To begin with, a simpler division into three main approaches is feasible.

9.2 Approaches to Meaning in Biology

In biological thought, three major approaches to the meaning of life phenomenon can be distinguished.

(a) *Functional*: meaning as a *functional correspondence*. An organism's features have meaning as adaptations to the environment, or as the means of some process of living, or as used for some purpose. This is a classical view, also of Lamarck and Darwin and non-Darwinians, widely used in empirical biology. Meaning is function. Functionality can be demonstrated without any evolutionary considerations.

(b) *Evolutionary*: meaning as a *relation to survival or to evolutionary experience*. Organisms' features have meaning as representing their fate—their reproductive success, evolutionary perspectives, or history. This is mostly a Darwinian view. Meaning is the relation to fitness.

(c) *Cognitive*: meaning as any *relation established in interpretation*. Meaning is a feature of organisms' umwelten, as these are fundamentally relational. This is Uexküll's view, among others. Meaning is the result of individual interpretation. Meaning is defined by the individual interpretant.

These three biological approaches to meaning are quite clearly distinguishable. However, when looking closer, several nuances and additional distinctions can be found that, on the one hand, show overlaps between the concepts, and, on the other hand, help to clarify the origins of proper semantics in the living world. The aim of our accounts in this article is to provide such a closer look, at least an introductory one.

9.2.1 Meaning from Function

It is common in biology to connect function, purpose, and meaning. Something has meaning due to its purpose, which is due to its function. "No function" implies "no use" implies "no meaning"—although function and purpose can be distinguished. In the case of function, an organ A works and results in a condition B; in the case of purpose, a goal B initiates the process A. The direction of influence in function and purpose is opposite.

Claus Emmeche, in his analyses of the concept of biological function, gives an example: "Saying that *cytochrome c* means something to the cell is the same as saying that it has a function" (Emmeche, 2002: 23). Meaning appears here, among other things, as a mereological aspect of a system: function is the function in relation to a certain condition of the whole (cell, organism, etc.).

In this sense, organisms' features have meaning also as functional adaptations—to the environment, or to other organs, and so on. This is the classical view of both of Lamarck and Darwin, later widely used in empirical biology. Here, meaning is function.

Caitrin Donovan (2019: 1) writes about biological function: "In biology, functions are attributed to the traits, behaviors, and parts of living things. A thing's function can refer to its purpose, a benefit it confers on an organism, or the causal role it contributes to a more complex system capacity." As Ulrich Krohs (2009) shows, function as related to design can be meaningful without reference to evolution.

When an organism recognizes an *affordance*, this indicates that it finds a way how to use it, or recognizes its function. Thus, the meaning from affordance would also belong to this class (Roli & Kauffman, 2020).

9.2.2 Meaning from Purpose

Assigning a function to something is very close to assigning a *purpose*, or *goal*, an end: that means finality, teleology. Such teleological discourse, which derives meaning from a future state and uses intermittently the terms *function*, *meaning*, and *purpose*, works well, until one is not posing a question about the origin (or reality) of goals. Of course, the question about the origin of purpose was raised. Darwin proposed the mechanism of natural selection for explanation of phenomena described as if organisms had a purpose. As a result, teleology, "the biologist's mistress" (see, e.g., Alexander, 2011), was driven away—but not completely.

For instance, botanist Edmund Sinnott writes in his book about the biology of purpose: "A remarkable fact about organic regulation, both developmental and physiological, is that, if the organism is prevented from reaching its norm or 'goal', in the ordinary way, it is resourceful and will attain this by a different method. [One may recall here the concept of equifinality by Hans Driesch.—K.K.] The end rather than the means seems to be the important thing" (Sinnott, 1950: 33).

Sinnott's account is remarkably concordant with the description provided by Terrence Deacon (2012: 270–271): the "fundamentally open and generic nature of living processes also means that they can additionally entrain and assimilate any number of intermediate supportive components and processes. This generic openness is what allows new functions and (in more complex organisms) new end-directed tendencies to evolve. By giving this general dynamical logic the name *teleodynamics*, we are highlighting this consequence-relative organization."

Existence of external purpose is generally disproved, while internal teleology (*teleonomy*) as redefined by Dobzhansky can be seen as a basis for biological meaning (Corning, 2014; on the history of the concept of teleonomy, see also Hennig, 2011).

9.2.3 Meaning from Survival

In attempting to explicate and quantitatively describe the concept of function, it was concluded that the most general function of organisms is reproduction, precisely in order to survive. This led to the biological concept of *reproductive success* as the ultimate measure of everything an organism is doing, including all its behavior and structure. The concept was formalized and quantified as *fitness* by J. B. S. Haldane and Ronald A. Fisher around 1930.

Here something very interesting happened with meaning. Introducing the relative rate of reproduction as a universal measure, any hint of purpose would not be necessary any more. This was understood as a mathematical solution of the teleology problem. However, what paradoxically and widely happened in biological discourse was that reproductive success became the ultimate sign of meaningfulness. As a result, "survival" became a dominant biological metaphor.

Also, organisms' features were said to have meaning as representing the fate of genes (or, more precisely, of alleles), and their reproductive success. This is the core of genocentric or neo-Darwinian view. Here, meaning is understood by its relation to fitness. During the period of hardening of the modern synthesis (as it was called by Stephen Jay Gould, 1983), some branches of neo-Darwinian theory were introduced: sociobiology (by Edward O. Wilson), evolutionary psychology (by Leda Cosmides and John Tooby), behavioral ecology (by Amotz Zahavi), and memetics (by Richard Dawkins). All these have derived meaning from fitness, or inclusive fitness.

In a model that relates or reduces all behavior to reproductive success, there appears an *illusory need to survive*. The need to survive became a dominant notion in biological discourse of the second half of the twentieth century, followed by the centrality of this concept (and metaphor) in popularization of biology via literature and films.

It's rather interesting to observe that such meaning in relation to survival is purely anthropomorphic. Looking closely, it is easy to see that there is no such general need to survive in nonhuman species. Organisms have many needs, but these are all more particular, such as the need for food, for water, for warmth, for a partner. As far as organisms do not have a concept of their own death, they cannot identify survivorship—although they can identify food, for instance. For nonhuman organisms, there is no integrative need to live, as different from particular needs related to their particular homeostatic systems, or habits. An organism cannot have any inherited information about nonliving for the simple reason that during all its genetic history back to the origin of life, it never died. Necessary conditions for capacity of worrying about one's own death obviously include episodic memory and chronesthesia, which are hominid capacities and not at all universal features of organisms.

Thus, with such neo-Darwinian models, other types of meaning besides survival became downplayed and marginalized, or were interpreted as illusory.

9.2.4   Meaning from the Evolutionary Past

A slightly different emphasis than that of survival, which refers more to the future, is exhibited by some other approaches that refer to the historical past, thus explaining organisms' features and their meaningfulness through earlier evolutionary experience. "Genomic information is 'meaningful' in that it generates an organism able to survive in the environment in which selection has acted" (Maynard Smith, 2000: 190). Here is another place for teleonomy. Accordingly, an organism's behavior is informed by its earlier evolutionary history which gives certain meaning to current life processes. Teleonomy in this sense can be understood rather broadly, so that it covers several other approaches in our list.

Evolutionary past may be understood as providing meaning to the life process of the organism as a whole. It is probably not possible to distinguish individual historical factors that correspond to the current characteristics of the organism. Therefore, some additional approaches will be required for distinguishing particular meanings.

9.2.5   Meaning from Coexistence

Meaning is always a part of something, in the sense that it requires context, and moreover, that there are always some additional components in the same living system. Meaning in this sense is a part–whole relation: an organ has a meaning in relation to the organism, an organelle is meaningful in relation to the cell, a species in relation to the ecological community, and so on.

There are various versions of such relations. Consider how Ernst Mayr (1969: 316) explains the meaning of species: "The division of the total genetic variability of nature into discrete packages, the so-called species, which are separated from each other by reproductive barriers, prevents the production of too great a number of disharmonious incompatible gene combinations. This is the basic biological meaning of species, and this is the reason why there are discontinuities between sympatric species."

9.2.6   Meaning from Code

In the midst of the neo-Darwinian era, a new biological home for the meaning-concept was created: meaning as a correspondence based on codes. *Code* itself is defined as a regular correspondence, which is arbitrary by its nature. Code concept was formalized in information theory, and it describes relations analogous to phoneme–letter correspondences.

Erwin Schrödinger (1944) was among the first to use the term *code* for the genetic process. George Gamow (1954) saw that the code concept could be applied to the inner working of biological cells as he started to use the term "translation" for protein synthesis. It took another decade until the genetic code (the particular correspondence between the primary structures of gene and protein) was described in the 1960s, and even more time until the process of protein synthesis based on ribosomes and sequential patterns of RNA was discovered. Eventually, it was accepted that the meaning of a gene (a sequence of DNA) is the protein produced on its basis. The meaning of the triplet or codon UGC is *cysteine*, the meaning of UGA and UAG is *stop*, and so on. An approach to the origin of meaning via genetic code has been called a mechanistic model of meaning (Barbieri, 2001; 2011).

A code is commonly defined as an arbitrary mapping. The arbitrariness is established by a mediating structure that creates the code. Accordingly, the genetic code is a mapping, or a translation. In order to persist, mapping should have a reproducible mediator that would function as the map-master, that is, building or constructing the mapping. The genetic code really has such a thing: its mediator is the set of transport-RNAs together with ribosomes.

A code was also seen as a *program*, the genetic program. This implies that a genotype has a meaning for coding (or representing, or mapping) the phenotype, the form of the organism.

Analogously, coding makes it possible to have *memory*, not only genetic, but also immunological and neural memory. Furthermore, specific codings work for instincts, and for habits. The process that creates all these various codes can be thought of as learning, in a broad sense.

The biological code-discourse, however, is susceptible to computer metaphors. In this case, what to call "meaning" looks rather arbitrary. One might say that there is a communication process that transforms digital patterns. Meaning may then be assumed through the involvement of an agency. However, the concept of agent or agency is often rather ambiguously defined (for instance, as a capacity to construct). The main problem with computer metaphors is related to the computer code as an algorithm, which is deterministic, whereas meaning requires some indeterminacy.

Manfred Eigen described such meaning-relation in biology: "Genetic information, apart from being 'syntactically' organised, is uniquely of a semantic nature. A gene encodes the amino acid sequence and thereby the precise spatial structure of a protein molecule, which in a 'suitable' chemical environment materialises as a certain enzymic function. This function is the 'meaning' of the genetic message, quite analogous to the meaning of a text communicated to us. The 'suitable chemical environment' for a genetic program corresponds—in McKay's formulation—to the 'conditional readiness' of the receiver of a message, the text of which is processed in his brain and responded to by 'goal-directed activity'" (Eigen, 2013: 445).

It is relevant to add here a view of John Maynard Smith (2000: 185): "in molecular biology, inducers and repressors are 'symbolic': in the terminology of semiotics, there is no necessary connection between their form (chemical composition) and meaning (genes switched on or off)." Brian Goodwin (2008: 145) uses the term *meaning* in a somewhat similar sense: "a reading of the genetic text by an organism is a process that makes meaning of the text through the self-construction of the organism."

9.2.7   Meaning from Signaling

The concept of code, of course, took its origin from the studies of communication. What is communicated obviously has some meaning. A signal is about something, thus it is meaningful (see Krohs, 2004: 209ff).

Seyfarth and Cheney provide a review on the signal concept as used in animal communication studies. They write: "Signals in a variety of social contexts are adaptive because they convey information. For recipients, meaning results from the integration of information from the signal and the social context. As a result, communication in animals—particularly in long-lived, social species where the same individuals interact repeatedly—constitutes a rich system of pragmatic inference in which the meaning of a communicative event depends on perception, memory and social knowledge" (Seyfarth & Cheney, 2017: 339).

Biological literature about animal communication, and also about intercellular and neural communication, is rather extensive. However, it is remarkable that the aspect of interpretation has received very little attention: biological communication studies are mainly about the carriers of chemical, electrical, or other nature, or about the communication channels—about the etic, and almost never about the emic, aspect of communication.

9.2.8 Meaning from Communication

Signaling and code are usually interpreted in the context of communication. Communication is distinguished from physical interaction by its meaningfulness, that is, by the transfer of a kind of message that is referencing something other than the carrier itself.

For instance, "insects cannot see the hidden nectar in flowers from a distance. It therefore became necessary for them to associate the nectar deposits and frequently also the covered pollen with some definite flower type or with some special characteristics, such as color, symmetry, smell, shape" (Leppik 1956: 452). This led to the introduction of new terms: "With the term *semaphyll* (sema, semeion; [. . .] in Greek means sign, mark) we wish to indicate [. . .] all colored leaves of plants, like petals, sepals, bracts, ligulate flowers, etc., which serve to attract pollinators. A 'trophosemeion' [. . .] or 'food mark' of insects is accordingly a coalescence of semaphylls, which is identical to but need not necessarily be a flower morphologically" (Leppik 1956: 452).

Communication studies were widely introduced into biology through the ethological studies undertaken since the 1930s (by Karl Frisch and others; e.g., Frisch, 1954), and received further attention after the introduction of information theory into biology (Tembrock, 1971). A popular example is alarm calls of vervet monkeys (Seyfarth et al., 1980). Through the usage of concepts like message, code, and interpretation, studies of biocommunication have often been directly related to semiotics (Sebeok, 1977), including in recent years (e.g., Witzany, 2012, 2014). A particular approach worth mentioning here was the one developed by Adolf Portmann, who studied the self-representation of organic form (see Kleisner, 2008).

Sociobiological studies of communication in behavioral ecology emphasize reproductive success and deal less with meaning-making per se. With communication and learning, though, another type of meaning arises which points to the anticipatory aspect of signaling.

9.2.9 Meaning from Anticipation and Need

A habit that is acquired by learning represents a certain regularity or rhythmicity in the world of a living system. As far as the particular environmental regularity persists, the habit has a predictive feature: its meaning is a correspondence to something that may come in the near future. Thus, a habit is also a basis for anticipation. When using the concept of feed-forward in behavioral studies, this refers to an expected or desired future. Anticipation is close to the phenomenon of need. In this case, the meaning is a relation to something absent, although the absent is real through its representation.

Anticipation is related to a rich variety of meanings, among them the proactive and the retroactive (abductive) meanings (Piaget, 1971; Rosen, 1985; Kurismaa, 2015; Nadin, 2016). Nevertheless, anticipation is too often described in the context of cybernetic models, which refer to meaning without always assuming a subjective world. Nevertheless, there is also a biological concept—*umwelt*—that relates meaning to the subjective space.

9.2.10 Meaning from Umwelt

Meaning requires interpretation. Interpretation presupposes possibilities, the options for choice. Therefore, interpretation is not a deterministic process, by definition. Options—at least two—should be available simultaneously; otherwise they are not options. These conditions define umwelt. Therefore, the existence of meaning is equivalent to the existence of umwelt.

That meaning is a feature of an organism's umwelt was Jakob von Uexküll's (1928; 1940) view. A fundamental feature of umwelt is co-presence or co-localization, which means simultaneity of something different. Meaning from umwelt is also meaning from perception.

The absolutely minimal umwelt consists of a difference. Difference, in order to be a difference, requires at least two functional circles to register the incompatibility that makes the difference. This cannot occur without simultaneity—which means that umwelt creates subjective time (Kull, 2020).

All this makes the concept of meaning in the framework of umwelt quite rich. That umwelt is related to simultaneity, gestalt, and choice was demonstrated in the works of Viktor von Weizsäcker (1940). If there is no umwelt, there is no sign or semiosis. If there is no semiosis, there is no meaning.

In case of umwelt, meaning appears as sensed. Meaning is here a relation as sensed—which means that in the case of umwelt, the relation—as well as the meaning—obtains reality. Existence of umwelt makes the relations real for an organism.

9.3 An Important Observation

In all the explications of meaning described above, meaning can be presented as a mediated relation between some relata. Indeed, mediatedness is an important general feature of meaning. Using a simple minimal account, some relata that may be linked for creating meaning are listed in table 9.1.

In the introduction we formulated two conditions for the phenomenon of meaning. Firstly, meaning is a relation in which relata are separated and linked by a mediating biological process. This condition was met in all listed approaches to meaning. However, this condition is too broad to guarantee the existence of meaning-relation. Now we turn to the second condition of meaning, according to which the relation should be real for the organism itself (not only for a human observer).

Assuming that meaning should be a relation, and presuming the minimal models of corresponding relations, we can see that in almost all the described relations (of purpose, of function, of reproductive success or survival, of competition, of coexistence, of code, of signal, of communication, of anticipation), the relata are separated in time: alfa appears before beta. An exception is the case of umwelt, because the space of distinction requires simultaneity. For a relation to be real for an organism, the relata should exist together in the given moment.

Minimal umwelt is just a difference, though the difference of something simultaneous. If the relata are separated in time, the relationship can have an algorithmic, reductionist

Table 9.1
Main types of meaning-relations as used in biology

meaning from . . .	Relatum alfa	is mediated by	relatum beta
. . . purpose	goal B	initiates	process A
. . . function	process A	works and results	condition B
. . . survival	feature A	permits reproduction for	staying alive as B
. . . competition	replicator A	in struggle for existence with	replicator B
. . . coexistence	agent A	is a part of	whole system B
. . . code	sequence A	is used for translating into	sequence B
. . . signal	recognized A	by transduction leads to	release of B
. . . communication	expression A	is interpreted as	referring to B
. . . anticipation	habit A	works to meet	future condition B
. . . umwelt (choice)	option A	is perceived as alternative to	option B

explanation. For umwelt, however, such explanation is impossible. For semiosis, the irreducibility, and in that sense noncausal, relation is necessary.

The condition for simultaneity is easy to overlook, since on the level of human description, every proposition is graspable due to its simultaneity in the human mind. The meaning of what is sequential (as a sentence that describes a relation) can be understood only if it is perceived as a whole (in the simultaneity of its components in the mind).

Thus, the concepts of meaning as these exist in traditional biology are poor: they do not include sufficient conditions for meaning-making (for semiosis) and cannot make the semiotic model necessary. Instead, the relationships can be explained by physicalist models which do not describe meaning-relations. An exception is the concept of umwelt. In umwelt, co-presence of relata is irreducible.

9.4   Discussion

If a relation is described as sequential, then it can be a meaning-relation only from the point of view of an external observer. Claus Emmeche (1997: 260–261) writes: "We could say that it is the meaning of the biological side of life to assure replication, or autopoiesis, or the ecological balance, or whatever, but (as stated by many neo-Darwinians as well as by the theory of autopoiesis) such ascriptions are due to the human observer, and should not be understood as a scientific teleology." Representation of relations by sequential means is a specific feature of human language. From the perspective of other organisms, relations for being real should be co-presented simultaneously (synchronously in their space), which is possible by the means of umwelt.

Meaning and purpose are rather close phenomena. Both are mediated relations and both require semiosis. To be clear with their relationship to teleology and teleonomy, the relevant distinctions should be specified.

(a) Teleology is meant as external intentionality. For life as a whole, teleology in this sense is scientifically no longer accepted. However, artifacts (like machines or tools) are designed by life on the basis of external intention.

(b) Teleonomy refers to directedness that self-reproducible systems obtain as a result of natural selection, while not due to subjectivity. If the concept of teleonomy is distinguished from internal teleology, then the next two (c and d) are namely this (cf. Corning 2014; 2019).

(c) Habit is a directedness as a product of semiotic (i.e., nondeterministic) learning. Habit may not have any goal. An essential feature of habit is that the organism can, at least potentially, deviate from it; that is, it can choose a not-yet-learned move or a habit "not appropriate" to the given situation. Accordingly, habit presupposes subjectivity, the latter defined as possession of umwelt.

(d) Purposefulness in a proper sense is directionality based on a represented goal. This is a kind of internal teleology that refers to consciousness.

For a relation to be a relation for a system itself, its relata have to exist simultaneously for this system. For a human as an external viewer, the relata of each of the relations described in table 9.1 can be perceived simultaneously, whereas for most other organisms only the last type of relations (relation in umwelt) is simultaneously viewable.

Once again: When is a relation real for an organism itself? When is behavior purposeful from an organism's own perspective? A simple homeostasis is obviously not such. In more complex cases, if all processes are either deterministic or stochastic, then behavior cannot depend on an organism's own decisions, despite its seeming as such for an external viewer who observes its behavior. To prove the existence of meaning-relations for an organism itself, we have to demonstrate the organism's capacity to make at least simple choices. This is equivalent to a demonstration that an organism has an umwelt: a synchronous space in which the perceived elements are not temporally preordered. As much as umwelt and semiosis are codependent, this is equivalent to the existence of semiosis (Kull, 2018). As humans, we know the existence of umwelt from our personal experience. For other organisms, we can know this if we demonstrate that those other organisms are equipped with the mechanism that is responsible for creating an umwelt. Despite the fact that Uexküll developed methods for experimental study of umwelt, the ambiguity of the concept of meaning together with underdeveloped methods of experimental semiotics have not allowed much progress in the study of umwelt and meaning in biology until recent decades. Nevertheless, research in many areas of biology—communication studies, cognitive biology, neurobiology, physiology, ethology, and others—have provided rich preliminary knowledge about organisms' species-specific capacities to make distinctions in their own world.

9.5   Conclusion

A major problem in contemporary biology is that meaning is usually described from a human perspective. What is commonly meant by meaningfulness is the relationship between an organism's features and reproductive success via use of the term *survival*. As we explained, for most organisms, meaning in this consecutive sense cannot be real for themselves.

Besides meaning as survival, we analyzed other approaches to the concept of meaning in biology. As we demonstrated, most of them appear semiotically insufficient. The only reliable biological model that is irreducible to causal (in the sense of sequential) modeling is umwelt. Therefore, umwelt is the best existing model for further building of semiotic biology, and for descriptions of organisms' behavior from their own perspective.

All the approaches to biological meaning discussed herein, however, have contributed importantly to a scientific study of interpretation processes and their role in living systems, including to the extended synthesis of evolutionary theory and the theory of biology as a whole, one of the aims of which is to include semiosis—the process of interpretation and meaning-making (Kull, 2022). The biological research program to study umwelten—the worlds of living systems from their own perspective—is incomparably rich.

Acknowledgments

I deeply thank Richard I. Vane-Wright and Peter A. Corning for their invitation and critical comments, and participants in biosemiotics seminars where the topic was discussed. I thank Elena Fimmel for her initiative to analyze foundations of theoretical biology. I also thank the Estonian Research Council grant PRG314 for support.

References

Alexander, V. N. (2011). *The biologist's mistress: Rethinking self-organization in literature, art, and nature.* Litchfield Peak, AZ: Emergent Publications.

Auletta, G. (2011). *Cognitive biology: Dealing with information from bacteria to minds.* Oxford: Oxford University Press.

Bailey, K. D. (2005). Typology construction, methods and issues. In K. Kempf-Leonard (Ed.), *Encyclopedia of social measurement*, vol. 3 (pp. 889–898). Amsterdam: Elsevier.

Barbieri, M. (2001). *The organic codes: The birth of semantic biology.* Ancona, Italy: PeQuod.

Barbieri, M. (2011). A mechanistic model of meaning. *Biosemiotics, 4*(1), 1–4.

Canguilhem, G. (2008/1965). *Knowledge of life.* New York: Fordham University Press.

Cordelli, A., & Galleni, L. (2003). Towards a theory of meaning in biology: A proposal for an operative definition. *Rivista di Biologia, 96*(1), 145–158.

Corning, P. A. (2014). Evolution 'on purpose': How behaviour has shaped the evolutionary process. *Biological Journal of the Linnean Society, 112*, 242–260.

Corning, P. A. (2019). Teleonomy and the proximate–ultimate distinction revisited. *Biological Journal of the Linnean Society, 127*, 912–916.

Deacon, T. W. (2010). What is missing from theories of information. In P. Davies & N. H. Gregersen (Eds.), *Information and the nature of reality: From physics to metaphysics* (pp. 146–169). Cambridge: Cambridge University Press.

Deacon, T. W. (2012). *Incomplete nature: How mind emerged from matter.* New York: W. W. Norton.

Donovan, C. (2019). Biological function. In T. K. Shackelford & V. A. Weekes-Shackelford (Eds.), *Encyclopedia of evolutionary psychological science* (pp. 1–4). Cham: Springer.

Eigen, M. (2013). *From strange simplicity to complex familiarity: A treatise on matter, information, life and thought.* Oxford: Oxford University Press.

Emmeche, C. (1997). Autopoietic systems, replicators, and the search for a meaningful biologic definition of life. *Ultimate Reality and Meaning, 20*(4), 244–264.

Emmeche, C. (2002). The chicken and the Orphean egg: On the function of meaning and the meaning of function. *Sign Systems Studies, 30*(1), 15–32.

Emmeche, C., & Kull, K. (Eds.). 2011. *Towards a semiotic biology: Life is the action of signs.* London: Imperial College Press.

Favareau, D. (Ed.). (2010). *Essential readings in biosemiotics: Anthology and commentary.* Berlin: Springer.

Frisch, K. von. (1954). *Symbolik im Reich der Tiere* [Symbolism in the animal kingdom]. (Münchener Universitätsreden, neue Folge 7.) Munich: Max Hueber Verlag.

Gamow, G. (1954). Possible relation between deoxyribonucleic acid and protein structures. *Nature, 173*, 318.

Goodwin, B. C. (1972). Biology and meaning. In C. H. Waddington (Ed.), *Towards a theoretical biology* 4 (pp. 259–275). Edinburgh: Edinburgh University Press.

Goodwin, B. C. (2008). Bateson: Biology with meaning. In J. Hoffmeyer (Ed.), *A legacy of living systems: Gregory Bateson as precursor to biosemiotics* (pp. 145–152). Berlin: Springer.

Gould, S. J. (1983). The hardening of the modern synthesis. In M. Grene (Ed.), *Dimensions of Darwinism: Themes and counterthemes in twentieth century evolutionary theory* (pp. 71–93). Cambridge: Cambridge University Press.

Halpern, O. (2005). Dreams for our perceptual present: Temporality, storage, and interactivity in cybernetics. *Configurations, 13*(2), 283–319.

Hennig, B. 2011. Teleonomy. In D. Perler & S. Schmid (Eds.), *Final causes and teleological explanation* (pp. 185–202). Paderborn, Germany: Mentis.

Hoffmeyer, J. (2008). *Biosemiotics: An examination into the signs of life and the life of signs*. Scranton, PA: Scranton University Press.

James, W. (1878). Remarks on Spencer's definition of mind as correspondence. *The Journal of Speculative Philosophy, 12*(1), 1–18.

Kauffman, S. (2012). From physics to semiotics. In S. Rattasepp & T. Bennett (Eds.), *Gatherings in biosemiotics* (pp. 30–46). Tartu, Estonia: Tartu University Press.

Kleisner, K. (2008). The semantic morphology of Adolf Portmann: A starting point for the biosemiotics of organic form? *Biosemiotics, 1*(2), 207–219.

Krohs, U. (2004). *Eine Theorie biologischer Theorien: Status und Gehalt von Funktionsaussagen und informationstheoretischen Modellen* [A theory of biological theories: Status and content of functional statements and information-theoretic models]. Berlin: Springer.

Krohs, U. (2009). Functions as based on a concept of general design. *Synthese, 166*, 69–89.

Kull, K. (1999). Biosemiotics in the twentieth century: A view from biology. *Semiotica, 127*(1/4), 385–414.

Kull, K. (2018). Choosing and learning: Semiosis means choice. *Sign Systems Studies, 46*(4), 452–466.

Kull, K. (2020). Jakob von Uexküll and the study of primary meaning-making. In F. Michelini & K. Köchy (Eds.), *Jakob von Uexküll and philosophy: Life, environments, anthropology* (pp. 220–237). London: Routledge.

Kull, K. (2022). The aim of extended synthesis is to include semiosis. *Theoretical Biology Forum, 115* (1/2), 119–132.

Kull, K., & Favareau, D. (2022). Semiotics in general biology. In J. Pelkey & S. Walsh Matthews (Eds.), *Bloomsbury semiotics*, vol. 2: *Semiotics in the natural and technical sciences* (pp. 35–56). London: Bloomsbury.

Kurismaa, A. (2015). On the origins of anticipation as an evolutionary framework: Functional systems perspective. *International Journal of General Systems, 44*(6), 705–721.

Leppik, E. (1956). The form and function of numerical patterns in flowers. *American Journal of Botany, 43*(7), 445–455.

Maran, T., Martinelli, D., & Turovski, A. (Eds.) (2011). *Readings in zoosemiotics*. Berlin: De Gruyter Mouton.

Maynard Smith, J. (2000). The concept of information in biology. *Philosophy of Science, 67*(2), 177–194.

Mayr, E. (1969). The biological meaning of species. *Biological Journal of the Linnean Society, 1*(3), 311–320.

Millikan, R. G. (2009). Biosemantics. In A. Beckermann, B. P. McLaughlin, & S. Walter (Eds.), *The Oxford handbook of philosophy of mind* (pp. 1–14). Oxford: Oxford University Press. https://doi.org/10.1093/oxfordhb/9780199262618.003.0024

Morris, C. (1938). *Foundations of the theory of signs*. (*International Encyclopedia of Unified Science* vol. 1, no. 2.) Chicago: University of Chicago Press.

Nadin, M. (Ed.). (2016). *Anticipation across disciplines*. Cham: Springer.

Ogden, C. K., & Richards, I. A. (1923). *The meaning of meaning: A study of the influence of language upon thought and of the science of symbolism*. New York: Harcourt, Brace & World.

Peterson, E. L. (2022). The third-way third wave and the enduring appeal of bioexceptionalism. *Theoretical Biology Forum, 115.*

Pharoah, M. (2020). Causation and information: Where is biological meaning to be found? *Biosemiotics, 13*(3), 309–326.

Piaget, J. (1971). *Biology and knowledge: An essay on the relations between organic regulations and cognitive processes*. Chicago: University of Chicago Press.

Pittendrigh, C. S. (1958). Adaptation, natural selection, and behavior. In A. Roe & G. G. Simpson (Eds.), *Behavior and evolution* (pp. 390–416). New Haven, CT: Yale University Press.

Prodi, G. (2021/1977). *The material bases of meaning*. Tartu, Estonia: University of Tartu Press.

Queiroz, J., & El-Hani, C. N. (2006). Towards a multi-level approach to the emergence of meaning in living systems. *Acta Biotheoretica, 54*, 179–206.

Reynolds, A. S. (2022). *Understanding metaphors in the life sciences*. Cambridge: Cambridge University Press.

Roli, A., & Kauffman, S. A. (2020). Emergence of organisms. *Entropy, 22*, 1163. https://doi.org/10.3390/e22101163

Rosen, R. (1985). *Anticipatory systems: Philosophical, mathematical, and methodological foundations*. New York: Springer.

Rosenblueth, A., & Wiener, N. (1950). Purposeful and non-purposeful behavior. *Philosophy of Science, 17*, 318–326.

Rosenblueth, A., Wiener, N., & Bigelow, J. (1943). Behavior, purpose and teleology. *Philosophy of Science, 10*(1), 18–24.

Schrödinger, E. (1944). *What is life?* Cambridge: Cambridge University Press.

Sebeok, T. A. (Ed.). (1977). *How animals communicate*. Bloomington: Indiana University Press.

Sebeok, T. A. (2001). *Signs: An introduction to semiotics* (2nd ed.). Toronto: University of Toronto Press.

Sebeok, T. A., & Umiker-Sebeok, J. (1981). You know my method: A juxtaposition of Charles S. Peirce and Sherlock Holmes. In T. Sebeok (Ed.), *The play of musement* (pp. 17–52). Bloomington: Indiana University Press.

Seyfarth, R. M., & Cheney, D. L. (2017). The origin of meaning in animal signals. *Animal Behaviour, 124*, 339–346.

Seyfarth, R. M., Cheney, D. L., & Marler, P. (1980). Vervet monkey alarm calls: Semantic communication in a free-ranging primate. *Animal Behaviour, 28*(4), 1070–1094.

Sinnott, E. W. (1950). *Cell and psyche: The biology of purpose*. New York: Harper.

Sonesson, G. (2006). The meaning of meaning in biology and cognitive science: A semiotic reconstruction. *Sign Systems Studies, 34*(1), 135–211.

Stillwaggon Swan, L., & Goldberg, L. J. (2010). How is meaning grounded in the organism? *Biosemiotics, 3*(1), 131–146.

Taylor, R. (1950). Comments on a mechanistic conception of purposefulness. *Philosophy of Science, 17*(4), 310–317.

Tembrock, G. (1971). *Biokommunikation: Informationsübertragung im biologischen Bereich* [Biocommunication: Transmission of information in the biological field]. Teil 1, 2. Berlin: Akademie-Verlag.

Uexküll, J. von. (1928). *Theoretische Biologie* [Theoretical biology] (2nd ed.). Berlin: Verlag von Julius Springer.

Uexküll, J. von. (1940). *Bedeutungslehre* [Theory of meaning]. (Bios, *Abhandlungen zur theoretischen Biologie und ihrer Geschichte sowie zur Philosophie der organischen Naturwissenschaften*, Bd. 10). Leipzig: Verlag von J. A. Barth.

Uexküll, J. von. (1982/1940). The theory of meaning. *Semiotica, 42*(1), 25–82.

Waddington, C. H. (1972). Epilogue. In C. H. Waddington (Ed.), *Towards a theoretical biology 4: Essays* (pp. 283–289). Edinburgh: Edinburgh University Press.

Weizsäcker, V. von. (1940). *Der Gestaltkreis: Theorie der Einheit von Wahrnehmen und Bewegen* [The gestalt circle: Theory of the unity of perception and movement]. Leipzig: Georg Thieme Verlag.

Winslow, R. (2014). Biological meaning. *Epoché: A Journal for the History of Philosophy, 19*(1), 65–85.

Witzany, G. (Ed.). (2012). *Biocommunication of fungi*. Dordrecht: Springer.

Witzany, G. (Ed.). (2014). *Biocommunication of animals*. Dordrecht: Springer.

10 Collective Intelligence of Morphogenesis as a Teleonomic Process

Michael Levin

Overview

Multiscale competency is a central phenomenon in biology: molecular networks, cells, tissues, and organisms all solve problems via behavior in various spaces (metabolic, physiological, anatomical, and the familiar 3D space of movement). These capabilities require being able to reach specific goal states despite perturbations and changes in their own parts and in the environment: effective teleonomy. Strong examples of the remarkable scaling of such goal states during teleonomic processes are seen across development, regeneration, and cancer suppression. In this paper I illustrate examples of regulative morphogenesis of multicellular bodies as the teleonomic behavior of a collective intelligence composed of cells. This perspective helps to unify many phenomena across multiscale biology, and suggests a framework for understanding how teleonomic capacity increased and diversified during evolution. Thus, teleonomy is a linchpin concept that helps address key open questions around evolvability, biological plasticity, and basal cognition, and is a powerful invariant that drives novel empirical research programs.

10.1 Introduction

To paraphrase a famous quote (Dobzhansky, 1973), nothing in biology makes sense except in light of teleonomy (Auletta, 2011; Ellis, Noble, & O'Connor, 2012; Noble, 2010, 2011). Most observers, including biologists, physicists, and engineers, have watched with wonder as biological systems expend energy to achieve a specific state of affairs different from the current one, despite changing circumstances. This phenomenon includes workhorse concepts such as stress (the system-level effects of the inability to reach desired states, and the driver of change), memory (the ability to represent specific states that are not present right now), intelligence (competency in navigating problem spaces toward desired goals), and preferences (inherent valence of specific states over others). The capacity to work toward goals (preferred future states) is ubiquitous across the biosphere and present at all scales of organization, from the planning capacities of primates to the abilities of cellular collectives to modify their activity to achieve a specific embryonic anatomy despite perturbations. It is a defining feature of life (Monod, 1972), of great importance to evolutionary biologists (in their quest to understand the origin of various functions); exobiologists (seeking ways to recognize unconventional life forms); researchers in artificial intelligence, robotics, and artificial life (trying to develop autonomous synthetic systems); and workers in regenerative biomedicine (whose goal requires the reprogramming of cellular and tissue functions toward desired goal states associated with health). How living systems establish, encode, and pursue goals is a fundamental question at the heart of numerous fields, including biology, philosophy, cognitive science, and the information technology sciences.

Teleology and related concepts have been the subject of much debate (Bertalanffy, 1951; Lander, 2004; Maturana & Varela, 1980; McShea, 2016; Rosenblueth, Wiener, & Bigelow, 1943; Turner, 2017; Varela & Maturana, 1972; Varela, Maturana, & Uribe, 1974; Varela, Thompson, & Rosch, 1991). Here, I focus on *teleonomy*: a naturalistic concept of purposeful functionality, in which systems expend energy to navigate toward preferred regions of some action space (Ashby, 1952; Corning, 2022). The key aspect of teleonomy as discussed herein is that it refers to models of complex systems' activity that are empirically testable based on the degree of explanation, prediction, and control it affords to an observer.

Teleonomy is a lens (akin to the pragmatic intentional stance [Dennett, 1987]) through which scientists see biological systems, creatures see each other, and parts of living systems model other parts and themselves to gain optimal causal control (Mar et al., 2007; Wood, 2019). Here, I focus on teleonomy as a profound way to understand morphogenesis as the teleonomic behavior of a multiscale collective agent (molecular networks, cells, etc.). A key aim is to show that goal-directed function is not just the province of advanced brains with self-aware agency, but rather is a primary principle scaled up from basal functions in the most primitive life forms. More than that, it is an essential invariant that pervades, and reveals actionable symmetries across, diverse aspects of biology.

The philosophical assumptions of this perspective (Levin, 2022) can be explicitly stated as follows. First, there is a primary goal to drive empirical research, not to preserve philosophical positions that make armchair decisions on questions of agency in the absence of specific experiment. Second, there is a commitment to evolutionary continuity of bodies and minds and to a search for minimal examples of key capacities, which will necessarily blur the boundaries between cognitive phenomena and "just physics." Proposals for sharp phase transitions in terms of agency carry the burden of having to show how discrete changes across one generation create a novel agential capacity in offspring that did not exist in the parents. Thus, I assume gradualism and continuous (not binary) metrics of all important parameters, such as agency, cognition, intelligence, memory, goal-directedness, and so on. Finally, empirical experiments are considered to be arbiters of truth value of positions in this field. Thus, teleonomic models (and agency claims in general) are taken fundamentally to be *engineering* claims about what a system can be relied upon to achieve on its own, as a module within some other functional adaptive system, without micromanagement and despite some degree of uncertain circumstances (from the perspective of human engineers, parasites, conspecifics, or evolution itself).

10.2   Teleonomy as a Lens on Collective Intelligence

All agents are made of parts, which work together to solve problems with various degrees of competency (intelligence). Goals belong to agents at various scales, and it is imperative to understand how novel agents and their novel goals emerge from the cooperation of active subunits and microstates (Hoel, 2018). The most obvious example is individual cognition arising from collections of neurons in a brain, but we must learn to recognize this phenomenon in unconventional guises as well. Cybernetics (Rosenblueth, Wiener, & Bigelow, 1943; Wiener, 1961) gives us a mature framework for understanding goal-directed behavior without resort to mysterianism, and dynamical systems and control theories offer rigorous formalisms in which attractor states are causes of system-level behavior (Manicka & Levin, 2019). Indeed, the engineering advances of the past few decades have shifted the burden of magical thinking to those who believe that humans possess some sort of unique ability to pursue goals that cannot exist in simpler life forms or bioengineered systems. *Anthropomorphism*, as a critique of agential models in biology, is a term often used to conceal a view of human capacity that is inconsistent with modern understanding of evolutionary origins of all capacities (Balazsi, van Oudenaarden, & Collins, 2011; Baluška & Levin, 2016; Keijzer, van Duijn, & Lyon, 2013; Lyon, 2006, 2015).

A most important aspect of cybernetic approaches is that they are substrate-invariant, and remind us that no specific materials (cytoplasm, neurons, etc.) or scale of organization are required for a capacity as fundamental as teleonomic action. This independence from specific implementation details removes traditional cataracts from the lens through which we view "agents" that exhibit teleonomic behavior (familiar creatures acting in three-dimensional space): self-imposed filters that have restricted research because our perceptual systems are tuned to recognize only some kinds of goal-directed behavior. Not only is there no unique material (brains) in which to find goal-directed behavior, but there is also no unique spatio-temporal scale (Noble, 2012). As occurred in physics (for quantum theory and relativity), we must go beyond the medium size, medium timeframe systems and be open to examining the evidence for agency in the very small (e.g., molecular networks) and the very large (e.g., whole lineages acting over evolutionary time scales) (Fields, Bischof, & Levin, 2020; Fields & Levin, 2020b; Friston, 2013; Ramstead et al., 2019).

Teleonomy is not the final step on a continuum of agency. Rather, it is a primary capacity (Monod, 1972), present in many unconventional substrates, that makes all others possible and catalyzes the climb from self-maintaining metabolic cycles all the way through human-level cognition and beyond (figure 10.1). Goal-directed behavior is, at the very least, uncontroversial in human animals. It is thought that this capacity is enabled by collectives of neurons (brains) exhibiting memory, error minimization capacity, and second-order metacognition that enables us to think about those goals (and perhaps reset them) in addition to executing them. However, brains evolved from much more ancient bioelectric networks that are formed by all cells in the body, and are as old as bacterial biofilms (Fields, Bischof, & Levin, 2020; Prindle et al., 2015; Yang et al., 2020). These networks readily form circuits with memory that enable basal homeostatic function (Cervera, Levin, & Mafe, 2020; Cervera et al., 2019; Cervera et al., 2018; Pietak & Levin, 2017). The remarkable capacities for both robustness and novelty in morphogenesis reveal the central role of the *scaling of goals* as an explanatory, facilitating concept for new basic research and biomedical applications (Levin, 2019), and the need to understand how evolution potentiates teleonomy.

Biological systems are not only structurally hierarchical, but also functionally hierarchical: each layer solves unique problems in its own relevant problem space, exhibiting teleonomy (*A*). The degree of competency and complexity that can be handled by a system in its pursuit

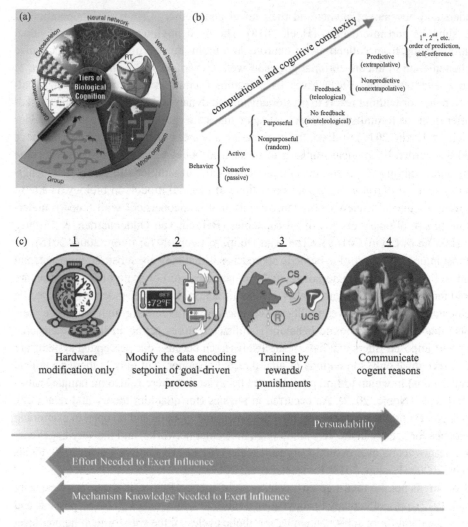

Figure 10.1
Goal-directedness is an invariant for a continuous spectrum of cognition. Images in panels A, C by Jeremy Guay of Peregrine Creative.

of goal states defines an order parameter for major transitions along a continuum of cognition ranging from passive matter to advanced self-reflective minds (*B*), which can be used to compare highly diverse intelligences (Rosenblueth, Wiener, & Bigelow, 1943). An empirically useful way to exploit teleonomy across systems is to use it as a guide to the most efficient prediction and control strategy: an "axis of persuadability" (*C*), in effect seeking to determine the optimal level of control (ranging from brute force micromanagement to persuasion by rational argument). Figure 10.1(C) shows only a few representative waypoints. On the far left (*C1*) are the simplest physical systems (e.g., mechanical clocks). These cannot be persuaded, argued with, or even rewarded/punished—only physical hardware-level "rewiring" is possible if one wants to change their behavior. On the far right (*C4*) are human beings (and others to be discovered; Bostrom, 2003; Kurzweil, 2005) whose behavior can be radically changed by a communication that encodes a rational argument that changes the motivation, planning, values, and commitment of the agent receiving it. Between these extremes lies a rich panoply of intermediate agents, such as simple homeostatic circuits (*C2*) which have setpoints encoding goal states, and more complex systems such as animals which can be controlled by training using stimuli that communicate to the system how it can achieve its goal of receiving a reward (*C3*). This continuum is not meant to be a linear *scala naturae* that aligns with any kind of "direction" of evolutionary progress; evolution is free to move in any direction in this option space of cognitive capacity. The goal of the scientist is to find the optimal position for a given system. Too far to the right, and one ends up attributing hopes and dreams to thermostats or simple artificial intelligences (AIs) in a way that does not advance prediction and control. Too far to the left, and one loses the benefits of top-down control in favor of intractable micro-management. Note also that this forms a continuum with respect to how much knowledge one has to have about the system's details in order to manipulate its function: for systems in class C1, one has to know a lot about their workings to modify them. For class C2, one has to know how to read/write the setpoint information, but need not know anything about how the system will implement those goals. For class C3, one doesn't have to know how the system modifies its goal encodings in light of experience, because the system does all of this on its own; one only has to provide suitable rewards and punishments. Ascertaining the optimal level of teleonomy in the objects around us is a key task for scientists interested in understanding and managing novel complex systems, and a built-in cognitive module for animals navigating complex environments, conspecifics, prey, and the like.

10.3   Evolution Scales Up Goal-Directed Activity: Anatomical Homeostasis

To recognize teleonomic behavior in unconventional contexts, it is helpful to start with the clear case of human goal-directed behavior and work backward. Nervous systems exhibit specific structure-function relationships that bind collections of neural cells into coherent selves with associative memories and goals that do not belong to any of the cells alone but only to the collective (i.e., all intelligences are collective intelligences). Complex brains enable memories of desired goal states and perceptual control loops which efficiently orchestrate behavior in 3D space in order to optimize specific parameters and satisfy drives (Allen & Friston, 2018; Pezzulo, Rigoli, & Friston, 2015; Powers, 1973). However, this same basic scheme can be applied to action in many spaces, including metabolic, transcriptional, and physiological ones. On this view, "environment" is extended to include the internal affordances (components and their capacities) that molecular pathways, cells, and tissues have access to, and "embodiment" is extended to other problem spaces, not just the familiar 3D space of motion. Indeed, William James' definition of *intelligence* as the capacity of an agent to achieve "the same goal via different means" (James, 1890) is suitably generic to encompass diverse intelligences of navigation of many different kinds of problem spaces. Here we consider one example, which likely served as the evolutionary origin for conventional goal-driven behaviors (Fields, Bischof, & Levin, 2020): bioelectric networks of non-neural cells that enable metazoan organisms to navigate morphospace (Levin, 2021a; Levin & Martyniuk, 2018).

Morphospace is the space of possible anatomical configurations that any group of cells can achieve (Stone, 1997). Multicellular organisms move through morphospace during embryogenesis, regeneration, and remodeling such as metamorphosis. Because genomes encode micro-level protein hardware, not directly specifying growth and form, it is essential to understand not only the molecular mechanisms *necessary* for morphogenesis, but also the information-processing dynamics that are *sufficient* for the swarm intelligence of cell groups to create, repair, and reconstruct large-scale anatomical features (Friston et al., 2015; Pezzulo & Levin, 2015, 2016). Examples abound of cellular collectives being able to reach the desired region of morphospace despite diverse starting positions and perturbations along the way—an activity which is strongly isomorphic to aspects of cognitive and behavioral science (Friston et al., 2015; Grossberg, 1978).

Embryogenesis itself is often thought about in terms of pure emergence: complex forms appear via the parallel action of large numbers of cells following local rules. However, it is not at all as brittle as this kind of emergent cellular automata paradigm would predict. Mammalian embryos cut in half produce monozygotic twins (not half-embryos), and embryos created with radically different numbers of cells still produce properly scaled bodies (Cooke, 1981). Perhaps the most instructive example from the perspective of teleonomy is that of the kidney tubule in the newt. Kidney tubules of the correct cross-sectional geometry and diameter typically arise from numerous cells working together. However, if the cells are made to be very large, just one cell will bend around itself to create the same structure (Fankhauser, 1945a, 1945b). This reveals that diverse underlying molecular mechanisms (cell:cell communication vs. cytoskeletal bending) can be called up as needed, diverging from the normal course of events in embryogenesis, in the service of a large-scale goal in anatomical morphospace. The ability to achieve the same outcome with highly altered components, requiring no retraining on ontogenetic or phylogenetic time scales, is something our engineering and machine-learning technologies cannot yet achieve. However, workers in regenerative medicine are already beginning to exploit this capacity, such as the physiological "need-of-function" that coordinates the growth and physiology of liver fragments transplanted to lymph nodes in the context of liver failure (DeWard, Komori, & Lagasse, 2014).

Development is incredibly reliable, producing bodies to very tight tolerances despite considerable deviations and noise at the level of gene expression and cellular activity (Eritano et al., 2020; Gonze et al., 2018; Simon, Hadjantonakis, & Schroter, 2018). This robustness, and its occasional failure in the case of birth defects, immediately suggests teleonomic perspectives, because only goal-directed agents can make mistakes. Biophysics alone cannot make mistakes; every micro-scale process proceeds according to the laws of physics and chemistry. Developmental defects are mistakes relative to the correct outcome toward which they strive (Matthewson & Griffiths, 2017). Embryonic bodies do a remarkable job of detecting and correcting such mistakes; for example, embryonic salamander tails grafted to the flank slowly remodel into a limb, altering the existing tissue structure to become correct with respect to the large-scale body plan (Farinella-Ferruzza, 1956). But this capacity is not just for the rare cases of teratogenic influences: it may drive all of development. From the perspective of each embryonic stage, the prior stage has incorrect anatomy: it is a "birth defect" that must be corrected by actuation of gene expression, physiology, and cell movement. One can view the progression of development as a series of repairs that drive the system toward the correct anatomical setpoint.

Regulative development is thus a special case of the more generic process of regeneration: moving an incorrect state closer to the target setpoint (figure 10.2). Many organisms can do this as adults, repairing drastic injury. Examples include salamanders (which can regenerate eyes, limbs, jaws, and other organs) and planarian flatworms (which regenerate every part of the body from even small fragments, while scaling the remaining tissue down so that perfect proportion results) (Beane et al., 2013; Oviedo, Newmark, & Sanchez Alvarado, 2003).

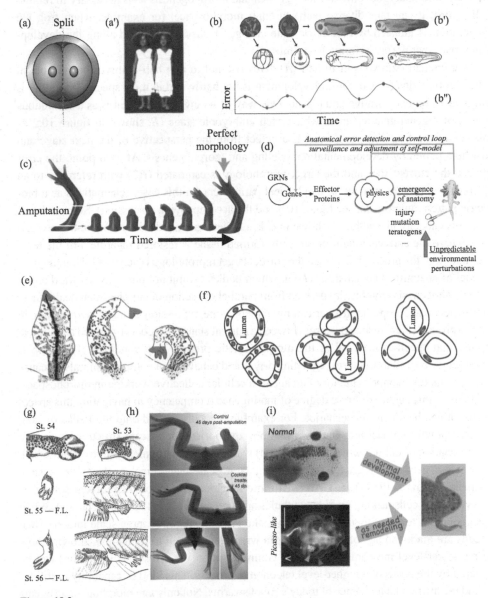

Figure 10.2
Robustness and plasticity: morphogenesis as a problem-solving agent.
Panel sources are as follows: A′ is reproduced with permission from Wikimedia Commons (Oudeschool; https://commons.wikimedia.org/wiki/File:Power20302.jpg; licensed under the Creative Commons Attribution 3.0 Unported license), B is courtesy of Brenda de Groot, C and F are by Jeremy Guay of Peregrine Creative Inc., E is from Farinella-Ferruzza (1956), and G is from Xenbase at http://www.xenbase.org.

Regeneration offers numerous examples of teleonomic activity (figure 10.2). First and most remarkably, it stops. The rapid growth and remodeling of regeneration (which can be as fast as any tumor) stops precisely when a correct organ shape has been achieved. This indicates that the collective can certainly detect when its goal has been achieved, which results in the cessation of numerous molecular-biological and biophysical processes. Second, it achieves its goal from diverse starting positions, as a limb can be cut at any point along the proximo-distal axis and undergoes only as much growth and morphogenesis as is necessary to rebuild itself. Third, it can take diverse paths through morphospace: for example, when frog leg regeneration is induced by bioelectric state change, it does not proceed along the develop-mental path that normally forms frog limbs.

A mammalian embryo split in half (A) gives rise not to two half-embryos but to normal monozygotic twins (A'), because development is not hardwired in most species but rather is remarkably context-sensitive and plastic. One way of seeing development is as a continuous process of regenerative repair, in which each embryonic stage (B, shown in figure 10.2 as embryos of the frog $Xenopus$ $laevis$) is a defect from the perspective of the next stage and must be repaired by developmental remodeling and morphogenesis. At each point, the error between the current state and the target morphology is estimated (B'') with reference to an information structure encoded in biophysical parameters (in this case, schematized as a bio-electric pattern memory, B', see figure 10.4 and the accompanying discussion). Some animals retain this capacity in adulthood; shown in C is a typical salamander limb amputation experi-ment, where the correct amount of perfectly formed tissue is restored regardless of the level of amputation; the process halts when the correct target morphology is achieved. This suggests a model of anatomical homeostasis (D) in which bodies exhibit not only feed-forward emer-gent morphogenesis (complexity derived from parallel execution of simpler microscale rules) but also feedback loops that trigger cell movement, gene expression changes, and the like in order to progressively reduce the error between a current state and a coarse-grained anatomical setpoint that specifies the goal of the morphogenetic process stop condition. This loop is homologous to similar structures regulating drives and behaviors of complex animals, because it reflects the teleonomic behavior of an agent: a cellular collective working in morphospace. Importantly, this agent exhibits a degree of intelligence (competency in navigating this space) because it can handle novel scenarios. For example (E), tails grafted onto the flanks of sala-manders slowly remodel into limbs—the more appropriate large-scale structure, including re-specification of tail-tip tissue (labeled in red) whose local environment is correct but which nevertheless gets remodeled by the emergent large-scale anatomical goals of the system (Farinella-Ferruzza, 1953, 1956). An even more remarkable example of problem solving is observed when cells making up kidney tubules are increased in size (F, cross-sections). When cells get larger, fewer of them cooperate to make the same required large-scale lumen; when the cells are made too big, one single cell can wrap around itself to do the job, showing how diverse lower-level mechanisms (cell:cell communication vs. cytoskeletal bending) can be triggered by the needs of a higher-level teleonomic process. This kind of capability is still far beyond the artificial intelligence of today's robot swarms. Not only can morphogenetic agents reach the correct region of morphospace despite significant perturbations of environment and self-structure, but they can also take different paths to reach those same goals. The normal stages of frog limb development (G) are not the same intermediate stages observed in induced frog leg regeneration (H), which creates a normal limb but does so in a central "stalk" with side branches for toes, instead of a paddle sculpted by programmed cell death. The paths through morphospace are sometimes associated with actual movements, such as the remodeling of tadpole to frog (*I*) which creates largely normal frog faces even when starting with scrambled tadpole faces with all the organs in the wrong position: the primordia move around in novel paths until a correct frog face is reached, showing that genetics specifies not a machine with hardwired motions in specific directions but rather a process that can minimize error from a target morphology and thus handle novelty. Two other cases are instructive because they emphasize knowledge gaps with respect to how teleonomy in anatomical space relates to genomes (figure 10.3). Planarian regeneration is extremely stable, invariably resulting in a correct flatworm after amputation along any plane. Because of their reproduction by fission and regeneration, some species of planaria do not use Weismann's barrier: every mutation that doesn't kill a stem cell is amplified in the soma in the subsequent generations. As a result

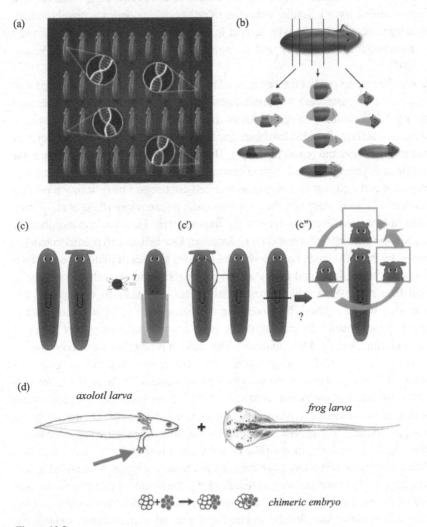

Figure 10.3
Knowledge gaps for predictions of morphogenesis: genetics and teleonomy.
Panels A–C″ by Jeremy Guay of Peregrine Creative. Panels in D courtesy of John Clare at https://www.axolotl.org/biology.htm

of this (reviewed in Levin, Pietak, & Bischof, 2018), their genomes are incredibly messy—the animals can be mixoploid (containing cells bearing different numbers of chromosomes). This illustrates the remarkable ability to reliably implement the same anatomy despite chaos within the underlying molecular components. A different kind of chaos is tamed by the remodeling actions at the beginning of frog metamorphosis. In order to build a frog face, the face of the tadpole must be strongly remodeled. However, it is now clear that this is not achieved by some sort of genetic hard coding of the amount and direction of movement for each component (Pinet et al., 2019; Pinet & McLaughlin, 2019; Vandenberg, Adams, & Levin, 2012). When "Picasso tadpoles" are created, with eyes, jaws, and nostrils in aberrant locations (scrambled), largely normal frogs result because all of these organs move through novel, unnatural paths, until a proper frog face results. This reveals that what the evolution of the frog genome discovered is not a machine that performs rote steps to emergently produce a frog face, but rather one that executes an error minimization scheme toward a specific setpoint (the basis of teleonomic activity). Similar examples of regenerative repair via novel (unnatural) paths through morphospace exist with respect to the left-right axial patterning pathway (McDowell, Rajadurai, & Levin, 2016), and are already seen in simple chordates (Voskoboynik et al., 2007).

A key goal is to be able to predict the behavior of morphogenetic agents: what shape will result under specific circumstances? Important knowledge gaps in this area exist because our knowledge of the genomically-specified hardware is much greater than our understanding of the teleonomic activity that this hardware implements. For example, some species of planaria reproduce by fission and regeneration (A). This implements somatic inheritance: for hundreds of millions of years, they accumulate mutations as each change to the genome that doesn't kill the stem cell ends up amplifying as that cell contributes to restoring a portion of the next generation's body. And yet, they are champion regenerators (Saló et al., 2009), restoring perfect little worms from any type of cut fragment (B). How can the morphology be so reliable when the genomic information is so fungible? Our inability to predict outcomes is clearly revealed by the following experiment. Consider two species of planaria, one which has cells that make a round head and then stop morphogenesis, and one whose cells make a flat head and then stop. What will happen if half of the neoblasts in one species are destroyed by irradiation and replaced by those from another species (C'): when the head is removed, what shape will result? Despite all the progress in molecular biology of stem cell differentiation in planaria, the field has no models that make a prediction about outcomes—one dominant shape, an intermediate shape, or a continuous remodeling that never ceases because neither set of cells is ever satisfied with the current shape of the head (C''). We lack computational models that link molecular details about the cellular hardware with large-scale decisions that cell collectives make in navigating morphospace. Similarly (D): despite being able to read both axolotl and frog genomes, we don't know how to predict whether chimeric larvae will have legs (like axolotls) or not (like tadpoles), and if so, whether those legs will be made of frog cells whose behaviors have been altered toward a novel anatomical task.

All of these examples illustrate, per James (James, 1890), the ability of this unconventional agent to achieve the same goals (a specific functional anatomy) by different means—taking novel paths through morphospace despite external and internal perturbations. Indeed, the remarkable robustness and plasticity of these teleonomic processes are the envy of workers in robotics and AI. The fundamental origins of these goals will be discussed later in this paper, but it is instructive to consider how these anatomical setpoints are physically encoded (being a precursor to representation of goals within advanced brain-mind systems). The computational medium in which the collective intelligence of cells operates to so competently navigate morphospace is the same as that of the brain: bioelectric networks. This design principle, which evolution discovered long before human engineers used it for reprogrammable computers (Levin, 2014; Sullivan, Emmons-Bell, & Levin, 2016), enables a software-hardware distinction that allows genomes to encode biophysical hardware, not final anatomical outcomes, while the software dynamics of this hardware hold the goal states and enable measurement and action of the anatomical homeostatic loop (Pezzulo & Levin, 2015, 2016).

10.4 Bioelectricity as a Medium for Teleonomic Control of Growth and Form

Evolution exploits three main modalities to coordinate morphogenesis: biochemical signals, biomechanical forces, and bioelectric communication (Levin, 2014; Newman, 2019). It is likely that all of these can be used to illustrate the ubiquity of teleonomy in anatomical control, but the bioelectric layer of the software of life makes the most direct connection to goal-directed behavior of brains. Importantly, control of morphogenesis and control of behavior are not only functionally isomorphic, but also share molecular mechanisms. This is not an accident, because nervous systems evolved by speed-optimizing ancient bioelectric circuits that evolved first to navigate morphospace and were then pivoted by evolution to navigate 3D space when nerves and muscles evolved. All of the key components of nervous systems—ion channels, electrical synapses (gap junctions), and neurotransmitter signaling—are much older than brains (Fields, Bischof, & Levin, 2020; Levin, Buznikov, & Lauder, 2006). Indeed, bioelectrics are already seen in the behavior at the single cell level (Eckert & Naitoh, 1970; Eckert, Naitoh, & Friedman, 1972; Naitoh & Eckert, 1969a, 1969b; Naitoh, Eckert, & Friedman, 1972). The ion channels and gap junctions in a plasma membrane together form a powerful interface provided by cells which enables their collective programming by ontogenic- and phylogenic-scale processes both within and outside of the nervous system (figure 10.4).

The most familiar goal-driven system, the brain, operates via a network of electrically active cells, whose resting potential is set by the activity of ion channels and can be propagated to their neighbors via gap junctions (A). Consistent with the fact that this architecture evolved from much more ancient cell types already using bioelectric signaling, all cells in the body (B) do the same thing (but on slower timescales than neural spiking). Patterns of resting potential thus arise in tissues (C), and are a complex, nonlinear property of large numbers of cells driving coupled electric circuits. Such patterns are often instructive scaffolds for gene expression and anatomy, such as the "electric face" observed in frog embryos (D) which guides the position of the eyes, mouth, and other organs (shown by the depolarization in light-colored cells, revealed by a voltage-sensitive fluorescent reporter dye). The functional role of these bioelectric patterns is revealed by experiments in which ion channels are introduced or opened in ways that alter the standing bioelectric patterns; for example, specific potassium channel misexpression can trigger a "build an eye here" pattern on the gut, resulting in the creation of an ectopic eye (E). Lineage marker labeling of such ectopic structures (e.g.,

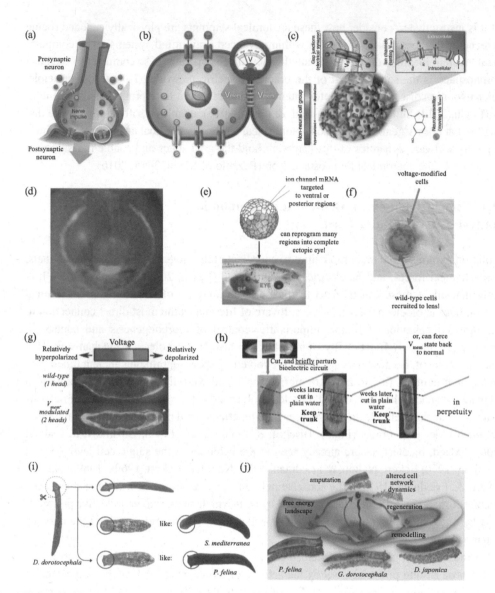

Figure 10.4
Bioelectric circuits encode teleonomic goals as pattern memories.
Panels A–C by Jeremy Guay of Peregrine Creative. Panels D, E, G, and I from Vandenberg, Morrie, & Adams, 2011; Pai et al., 2012; Pezzulo et al., 2021, and Emmons-Bell et al., 2015, respectively. Panel J by Alexis Pietak.

the lens induced in a tail, *F*) reveals that after some cells (labeled voltage-modified) are bio-electrically instructed, they further instruct neighboring cells (labeled wild-type cells forming the bottom half of the lens) which were not themselves altered in any way, showing that the patterning goals encoded by bioelectric states are not single-cell–level properties but can trigger a teleonomic process of instruction toward a new organ-level goal. In planaria, the number and location of heads are indicated by an endogenous bioelectric pattern which can, with drugs targeting ion channels and gap junctions, be reset to a new pattern (*G*). As befits a kind of memory, the circuit not only leads to the creation of two-headed animals, but keeps the new pattern permanently, as these two-headed animals continue to generate two-headed regenerates in further rounds of cutting with no new manipulation (*H*). Teleonomic models of planarian regeneration as a goal-directed process that builds to a specific, directly represented pattern memory have led to the ability to produce permanent lines with a different anatomical body plan despite their wild-type genetics (no genomic editing or transgenes need be used in this process). Remarkably, not only head number, but also head shape can be altered by disruption of bioelectric communication after head amputation, resulting in the formation of head (and brain) shapes belonging to other species of planaria (*I*), as the system is pushed out of the normal region of morphospace by injury, and is confused by a general anesthetic on its way to find the correct attractor in the space of possible planarian heads (*J*, discussed in Sullivan, Emmons-Bell, & Levin, 2016).

Recent work suggests a unification of neural and non-neural physiology because all of the techniques of neuroscience are now being used outside the brain to understand development, regeneration, and cancer (Adams et al., 2014). The extreme portability of tools, concepts, and reagents from neuroscience (ion channel constructs, optogenetics, and computational models) suggests that the distinction between neurons and other somatic cell types is artificial. These techniques do not distinguish neural from non-neural tissues, revealing the opportunity to expand neuroscience well beyond neurons (Pezzulo & Levin, 2015). Modulation of native bioelectric signaling (by targeting ion channels, gap junctions, and downstream neurotransmitter machinery) has enabled the modular induction of organs such as eyes (Pai et al., 2012), the rational repair of birth defects of complex organs such as the brain induced by mutation or teratogenesis (Pai et al., 2020; Pai et al., 2018), the induction of regeneration of appendages in nonregenerative contexts (Adams, Masi, & Levin, 2007; Tseng et al., 2010), and the reversal or duplication of major body axes (Durant et al., 2019; Levin et al., 2002).

A brief experience of a particular voltage state can change cellular decision making from "tail" to "head," from "gut" to "eye," and from "scar" to "limb" (McLaughlin & Levin, 2018); this is not micromanagement but large-scale setting of goals. Indeed, the target morphology—the shape to which cells regenerate after damage—can be permanently modified by transient changes of global bioelectric patterns. Genetically wild-type planaria can be induced to form two heads instead of a head and tail, and this pattern is then permanently propagated in the animals regenerating from subsequent cuts in plain water with no further manipulation (Oviedo et al., 2010). Planarian fragments can also be induced to form heads appropriate to other species, with no genomic editing (Emmons-Bell et al., 2015). Voltage-sensitive fluorescent dyes now allow the visualization of these pattern memories; for example, showing a two-headed bioelectric prepattern induced in a transcriptionally and anatomically normal one-head worm. The memory is latent until injury causes it to be recalled by the cellular collective (Durant et al., 2017).

The parallels with cognitive neuroscience are strong, including the abilities to do "neural decoding" to extract the semantics (in this case, in morphospace) of the electric states (Beane et al., 2011; Durant et al., 2019; Durant et al., 2017; Vandenberg, Morrie, & Adams, 2011), incept false pattern memories (Levin, 2021a) without having to edit the genome, and detect and manipulate perceptual bistability (create planaria that randomly regenerate as one- or two-headed animals because the circuit cannot quite decide between two memories; Durant et al., 2017; Pezzulo et al., 2021), all by using the same tools and conceptual framework as used in manipulation of goal-directed agents with brains. A key concept emphasized by this work is the storage and manipulation of rewritable information in bioelectric state. This control of modular decision making in software via experiences, rather than by hardware rewiring, offers precisely the same enormous advantages that evolution exploited in nervous systems (learning) and that we exploit in our computers (reprogrammability). These attempts to view morphogenesis as not merely an emergent physical process but a goal-directed control loop have led to many new discoveries and novel capabilities in the prediction and control of anatomical outcomes that had not been discovered from prior bottom-up approaches, and which offer numerous advantages for regenerative medicine (Levin, 2021a; Mathews & Levin, 2018).

Importantly, bioelectric signaling is not just another piece of biophysics. First, it is a medium for representing morphogenetic goals: the memories of the collective intelligence of morphogenesis (Pezzulo & Levin, 2015, 2016). Stable distributions of resting potential in tissues encode the target morphology—the setpoint for anatomical homeostasis—toward which cells work to repair and maintain. For example, the number of heads in a "correct" planarian body (defined as that number of heads which, once complete, causes further regeneration to cease) is not set genetically but rather is determined by the memory of a bioelectric circuit, which can be reset externally (Durant et al., 2019; Oviedo et al., 2010). By manipulating the ion channels and gap junctions to induce states encoding "2 heads" instead of the default one-head state, planaria were produced that continue to regenerate as two-headed *permanently*, across future rounds of regeneration in plain water with no more manipulation. A different state of the bioelectric circuit, enabling counterfactual memories that do not (yet) correspond to the current anatomy, and exhibiting the kind of perceptual bistability found in visual processing, can also be induced (Pezzulo et al., 2021). This reveals not only the stable yet rewritable memory of the morphogenetic process, but also the fact that techniques of developmental bioelectricity now allow us to directly *read and write the teleonomic goals* of a complex system. These goal states are ontologically real in the most important sense of all: they serve as the target of powerful experimental perturbations (Durant et al., 2016) and enable novel capabilities, results, and research progress.

Memory (implemented by bioelectric networks or other mechanisms) is central to teleonomy as a mechanism for encoding future goal states. More generally, however, bioelectric states are a medium that binds individual cells toward large-scale goals; it underlies scale-up (figure 10.5) and emergence of higher-level teleonomic individuals (Levin, 2019), much as it does to create brains with emergent unified mental content out of a collection of individual neuronal cells. This is why disruptions of bioelectric communication, in the absence of genetic alterations or carcinogens, can initiate cancer in vivo—a shrinking of the size of goals from morphogenetic activity of normal maintenance to unicellular goals of maximum proliferation and migration (metastasis) (Levin, 2021b). Conversely, forcing appropriate bioelectric communication can normalize cells despite strong expression of oncogenes that otherwise induce tumors (Chernet & Levin, 2013, 2014). The framework focused on inflating or shrinking the scale of the teleonomic activity leads directly to novel capabilities, in this case in the context of the cancer problem (Levin, 2021b; Moore, Walker, & Levin, 2017).

Tools from dynamical systems theory and connectionist machine learning (A) are examples of how to rigorously conceptualize the goal states needed for teleonomy: attractor states in specific spaces which serve as memories for processes such as morphogenesis that direct lower-level systems (cells and pathways) toward higher-level goals. This facilitates hypotheses about the scaling of goals from those of single cells (B) such as metabolic states, which

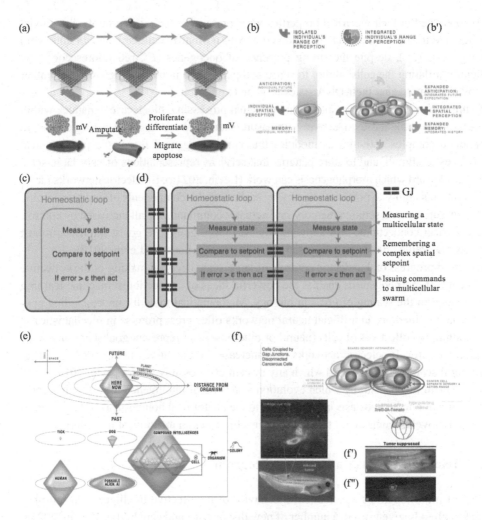

Figure 10.5
Conceptual tools for understanding scaling of goals in morphogenesis.

require very little memory, spatial measurements, and forward anticipation, into those of tissues and organs, which have larger goals (*B'*) because the collective is able to measure and act in a larger spatio-temporal sphere because of size and computational power. The homeostatic loops of single cells (*C*) are readily scalable as cells join into networks via gap junctions, forcing the measured states, actions, and instructive pattern memories to necessarily be larger and more complex (*D*). A model of cognition for truly diverse intelligences, focused on the scale of the goals they are able to work toward (*E*; Levin, 2019), shows how teleonomy can serve as a central invariant—a symmetry that enables comparison and synthesis despite huge variance in diverse agents' construction and origins. In biology, the scaling of the "cognitive light cone" (as a measure of the spatio-temporal size of the goals of a given system) shows how cells can electrically detach from the network, scaling their goals to those of an amoeba, which leads to treating the rest of the animal as external environment. This is exemplified by the transition to metastatic cancer (Levin, 2021b), which can be observed in their electrical properties via reporter dyes (F). Importantly, this model of cancer led to specific research that showed how to normalize cellular behavior and avoid tumorigenesis (F'), despite the strong presence of oncogenes (F'', red fluorescence), by artificially inflating the cells' ability to sense and participate in morphological goals, rather than by micromanaging their DNA states or gene expression.

By implementing long-range integration of signal processing, bioelectric dynamics within cell networks enable these collectives to measure states that are larger than single cells, to encapsulate complex activities as modules that can be triggered by simple physiological experiences or stimuli, and to store patterns that serve as representations of very large-scale goal states toward which morphogenesis can work (Levin, 2021a). Bioelectric networks facilitate evolution's ability to potentiate agency by scaling up the components of tiny homeostatic loops: measured states, setpoint memory, and actuator commands are all increased by controllable electrical connections, thus allowing for ever more grandiose goals, improved robust plasticity (Di Paolo, 2000), and expansion of the cognitive horizon (Levin, 2019).

Increased progress on bioelectric controls of large-scale decision making of the collective intelligence of morphogenesis enables the powerful ideas of connectionist machine learning to be applied to the scaling of goals in biology. Mathematical tools for understanding generalization and memory in artificial neural networks offer great promise in mechanistically explaining how collectives of cells (neural or otherwise) can represent goal states and work to minimize error. Bioelectric networks help increase the cognitive "light cone"—the spatiotemporal scale of goals toward which any system can possibly work (Levin, 2019)—and are a powerful mechanism by which evolution scales basal intelligence from the tiny, local loops of metabolic homeostasis in single cells to the anatomical homeostasis of large bodies navigating novel circumstances to achieve their objectives in anatomical morphospace.

10.5   Teleonomy Drives a Research Program

A view of morphogenesis as teleonomic behavior of a collective intelligence in morphospace has already given rise to a number of new discoveries and capabilities (Levin, 2021a; Mathews & Levin, 2018; Pezzulo & Levin, 2016). The emerging field at the intersection of synthetic developmental biology, computer science, and cognitive science generates numerous opportunities for next steps and further progress driven by a focus on recognizing, quantifying, and learning to exploit goal-directedness of diverse biological levels. From the perspective of theory/conceptual advances and specific research directions, the following questions should be developed:

• What is an effective Eigenspace for modeling agency? What would be the minimal axes for the space of all possible teleonomic agents? And how do we recognize, quantify, and compare teleonomic agents in radically diverse embodiments? Even gene regulatory networks, a paradigmatic case of deterministic genetic hardware, appear to have learning capacity (Biswas et al., 2021; Gabalda-Sagarra, Carey, & Garcia-Ojalvo, 2018; Herrera-Delgado et al., 2018; Szabó, Vattay, & Kondor, 2012; Watson et al., 2010); it is imperative that we abandon the tendency for armchair pronouncements of what can and cannot be seen as cognitive, and develop toolkits for generating and testing teleonomic models of arbitrary systems.

• If evolution is blind and always prefers immediate fitness payoffs, how is it that it not only gives rise to creatures highly adapted for specific environments, but also evolves hardware that can problem-solve in numerous novel configurations never seen before? How does evolution capitalize on the laws of physics and computation to generalize so well from specific examples to highly diverse possible instantiations? Work on the impacts of teleonomic behavior on evolutionary processes (Vane-Wright, 2014) can now be extended toward behavior in unconventional spaces such as morphological space, transcriptional space, and physiological space (Fields & Levin, 2022).

• How do we formulate and test specific teleonomic models of scaling from metabolic homeostatic loops (Turner, 2019) to large-scale morphogenetic goals via a balance of local dynamics and global stress loops? More generally, can we develop a science of prediction of collective systems' goals from a knowledge of their parts? Where else do goals come from, in cases where evolutionary history and genomic information is not a plausible origin, such as Xenobots made from wild-type *Xenopus laevis* skin cells, or wild-type planaria bioelectrically induced to create head shapes of other species (Emmons-Bell et al., 2015; Kriegman et al., 2021; Sullivan, Emmons-Bell, & Levin, 2016)?

• Can the same models be used to understand the role of the changing environment in plasticity and adaptability and the contribution of changing internal structure and function? Can the notion of external environment be extended to a multiscale concept in which adjacent cells, tissues, and the like are each other's environment? Can molecular pathways and biophysical dynamics be thought of as affordances for systems to compete and cooperate within and across levels in the organism (Gawne, McKenna, & Levin, 2020; Queller & Strassmann, 2009)?

• What is the relationship or overlap between the sets demarcated by "life" and "cognition"? If all (most?) components in living things are teleonomic agents and are thus somewhere on the continuum of cognition (figure 10.1), are all living things cognitive? What is a useful definition of *life*, given that teleonomic agents can be produced by engineering with organic or inorganic parts? Although modern life is necessarily teleonomic (in order to survive in the biosphere), could there have been very early life forms that were not teleonomic? Could current efforts at truly minimal synthetic life (Cejkova et al., 2017; Hanczyc, Caschera, & Rasmussen, 2011) clarify the relationship between teleonomy and physics?

• Can we find new ways to control system-level goals via the tools of behavioral neuroscience (Pezzulo & Levin, 2015), rather than by trying to solve the inverse problem? Reading genomes is now easy, and clean genomic editing will surely be solved in a few years, but this technology reveals the hard problem of Lamarckism facing workers in regenerative medicine: knowing what to edit at the genetic level to achieve desired morphology and behavior outcomes in cellular collectives. We now have the opportunity to adopt tools of neuroscience (Pezzulo & Levin, 2015) to learn to distort the perception and action spaces for cells and tissues via modulation of bioelectric and neurotransmitter dynamics, using computer models to help guide their behaviors by approaches focused on their teleonomic control loops. Existing data (Levin & Martyniuk, 2018) have begun to show how pattern memories are encoded, but the other parts of the loop—how tissues monitor current anatomical state and perform computations to measure error relative to the remembered setpoint—are almost entirely unknown and remain to be probed. Discoveries of ways to control morphogenesis top-down, and avoid the complexity barrier of engineering at the protein hardware level, could be applied in such areas as repairing birth defects, normalizing cancer, and inducing growth of healthy organs after injury or degenerative disease.

10.6   Conclusion: The Future of Teleonomy

The increasingly reductive (single-cell, single-molecule focus) advances of big-data biology risk significantly delaying deep insight because they focus on contingent details of specific mechanisms. How long would we have had to wait for the discovery of thermodynamics principles, if the physicists of the time had had the possibility of actually tracking every single molecule in a gas (as modern biologists now have), and didn't feel the need to develop meso-scale laws? Teleonomy is an excellent candidate for scale-free fundamental principles driving the unique capabilities of life, enabling a study of the software of life, as a complement to the molecular biology and medicine which dive ever deeper into the hardware. A key aspect is to recognize the need for pragmatic, observer-centered formalisms (Fields, 2018; Fields, Glazebrook, & Levin, 2021; Torday & Miller, 2016), thus avoiding the pseudo-question of whether systems "really" have teleologic goals in favor of empirical research focused on understanding and exploiting the robust control provided by a teleonomic perspective through which systems can model the environment and themselves. At stake are transformative advances in regenerative medicine (to get beyond the low-hanging fruit reachable by conventional stem cell biology and genomic editing approaches), robotics, and general AI.

Teleonomy is also central to developing deeper definitions of intelligence, selves, organisms, stress, robustness, and so on that can survive the coming advances in biological and software engineering, which will produce novel living forms that bear little relationship to any touchstone within the tree of life on Earth—biobots, cyborgs, hybrots, and the like (Kamm & Bashir, 2014). What are the classic "model systems" (from yeast to mouse) used in biological research models *of*? Teleonomy is a conceptual tool that allows us to move beyond the history of frozen accidents of evolutionary lineages and explore the truly general laws of biology instantiated by existing and novel beings (Maturana & Varela, 1980; Rosen, 1985). The science of cybernetics, and the deep lessons of neuroscience that extend well beyond neurons (Fields, Bischof, & Levin, 2020; Fields & Levin, 2020a, 2020b; Friston, Sengupta, & Auletta, 2014; Ramstead et al., 2019) to address the scaling of goals in biological collectives, will be key components of this future.

Acknowledgments

I thank Joshua Bongard, Avery Caulfield, Anna Ciaunica, Peter A. Corning, Pranab Das, Daniel Dennett, Thomas Doctor, Bill Duane, Christopher Fields, Jacob Foster, Karl Friston, James F. Glazebrook, Paul Griffiths, Erik Hoel, Eva Jablonka, Santosh Manicka, Patrick McGivern, Noam Mizrahi, Aniruddh Patel, Giovanni Pezzulo, Julia Poirier, Andrew Reynolds, Matthew Simms, Elizaveta Solomonova, Richard Vane-Wright, Richard Watson, Olaf

Witkowski, and numerous others from the Levin Lab and the Diverse Intelligences community for helpful conversations and discussions, and Julia Poirier for assistance with the manuscript. I gratefully acknowledge support by the Templeton World Charity Foundation (TWCF0606), the John Templeton Foundation (62212), and The Elisabeth Giauque Trust.

References

Adams, D. S., Lemire, J. M., Kramer, R. H., & Levin, M. (2014). Optogenetics in developmental biology: Using light to control ion flux-dependent signals in xenopus embryos. *The International Journal of Developmental Biology 58*(10–12), 851–861.

Adams, D. S., Masi, A., & Levin, M. (2007). H+ pump-dependent changes in membrane voltage are an early mechanism necessary and sufficient to induce xenopus tail regeneration. *Development 134*(7), 1323–1335.

Allen, M., & Friston, K. J. (2018). From cognitivism to autopoiesis: Towards a computational framework for the embodied mind. *Synthese 195*(6), 2459–2482.

Ashby, W. R. (1952). *Design for a brain*. London: Chapman & Hall.

Auletta, G. (2011). Teleonomy: The feedback circuit involving information and thermodynamic processes. *Journal of Modern Physics 2*(3), 136–145.

Balazsi, G., van Oudenaarden, A., & Collins, J. J. (2011). Cellular decision making and biological noise: From microbes to mammals. *Cell 144*(6), 910–925.

Baluška, F., & Levin, M. (2016). On having no head: Cognition throughout biological systems. *Frontiers in Psychology 7*, 902.

Beane, W. S., Morokuma, J., Adams, D. S., & Levin, M. (2011). A chemical genetics approach reveals h,k-atpase-mediated membrane voltage is required for planarian head regeneration. *Chemistry & Biology 18*(1), 77–89.

Beane, W. S., Morokuma, J., Lemire, J. M., & Levin, M. (2013). Bioelectric signaling regulates head and organ size during planarian regeneration. *Development 140*(2), 313–322.

Bertalanffy, L. (1951). Towards a physical theory of organic teleology: Feedback and dynamics. *Human Biology 23*(4), 346–361.

Biswas, S., Manicka, S., Hoel, E., & Levin, M. (2021). Gene regulatory networks exhibit several kinds of memory: Quantification of memory in biological and random transcriptional networks. *iScience 24*(3), 102131.

Bostrom, N. (2003). Human genetic enhancements: A transhumanist perspective. *Journal of Value Inquiry 37*(4), 493–506.

Cejkova, J., Banno, T., Hanczyc, M. M., & Stepanek, F. (2017). Droplets as liquid robots. *Artificial Life 23*(4), 528–549.

Cervera, J., Levin, M., & Mafe, S. (2020). Bioelectrical coupling of single-cell states in multicellular systems. *The Journal of Physical Chemistry Letters 11*(9), 3234–3241.

Cervera, J., Pai, V. P., Levin, M., & Mafe, S. (2019). From non-excitable single-cell to multicellular bioelectrical states supported by ion channels and gap junction proteins: Electrical potentials as distributed controllers. *Progress in Biophysics & Molecular Biology 149*, 39–53.

Cervera, J., Pietak, A., Levin, M., & Mafe, S. (2018). Bioelectrical coupling in multicellular domains regulated by gap junctions: A conceptual approach. *Bioelectrochemistry 123*, 45–61.

Chernet, B. T., & Levin, M. (2013). Transmembrane voltage potential is an essential cellular parameter for the detection and control of tumor development in a xenopus model. *Disease Models & Mechanisms 6*(3), 595–607.

Chernet, B. T., & Levin, M. (2014). Transmembrane voltage potential of somatic cells controls oncogene-mediated tumorigenesis at long-range. *Oncotarget 5*(10), 3287–3306.

Cooke, J. (1981). Scale of body pattern adjusts to available cell number in amphibian embryos. *Nature 290*(5809), 775–778.

Corning, P. A. (2022). A systems theory of biological evolution. *Biosystems 214*, 104630.

Dennett, D. C. (1987). *The intentional stance*. Cambridge, MA: MIT Press.

DeWard, A. D., Komori, J., & Lagasse, E. (2014). Ectopic transplantation sites for cell-based therapy. *Current Opinion in Organ Transplantation 19*(2), 169–174.

Di Paolo, E. A. (2000). Homeostatic adaptation to inversion of the visual field and other sensorimotor disruptions. SAB2000 Sixth International Conference on Simulation of Adaptive Behavior: From Animals to Animats, Paris.

Dobzhansky, T. (1973). Nothing in biology makes sense except in the light of evolution. *American Biology Teacher 35*(3), 125–129.

Durant, F., Bischof, J., Fields, C., Morokuma, J., LaPalme, J., Hoi, A., & Levin, M. (2019). The role of early bioelectric signals in the regeneration of planarian anterior/posterior polarity. *Biophysical Journal 116*(5), 948–961.

Durant, F., Lobo, D., Hammelman, J., & Levin, M. (2016). Physiological controls of large-scale patterning in planarian regeneration: A molecular and computational perspective on growth and form. *Regeneration (Oxford) 3*(2), 78–102.

Durant, F., Morokuma, J., Fields, C., Williams, K., Adams, D. S., & Levin, M. (2017). Long-term, stochastic editing of regenerative anatomy via targeting endogenous bioelectric gradients. *Biophysical Journal 112*(10), 2231–2243.

Eckert, R., & Naitoh, Y. (1970). Passive electrical properties of paramecium and problems of ciliary coordination. *Journal of General Physiology 55*(4), 467–483.

Eckert, R., Naitoh, Y., & Friedman, K. (1972). Sensory mechanisms in paramecium. I. Two components of the electric response to mechanical stimulation of the anterior surface. *Journal of Experimental Biology 56*(3), 683–694.

Ellis, G. F. R., Noble, D., & O'Connor, T. (2012). Top-down causation: An integrating theme within and across the sciences? *Interface Focus 2*(1), 1–3.

Emmons-Bell, M., Durant, F., Hammelman, J., Bessonov, N., Volpert, V., Morokuma, J., Pinet, K., Adams, D. S., Pietak, A., Lobo, D., & Levin, M. (2015). Gap junctional blockade stochastically induces different species-specific head anatomies in genetically wild-type *Girardia dorotocephala* flatworms. *International Journal of Molecular Sciences 16*(11), 27865–27896.

Eritano, A. S., Bromley, C. L., Bolea Albero, A., Schutz, L., Wen, F. L., Takeda, M., Fukaya, T., Sami, M. M., Shibata, T., Lemke, S., & Wang, Y. C. (2020). Tissue-scale mechanical coupling reduces morphogenetic noise to ensure precision during epithelial folding. *Developmental Cell 53*(2), 212–228 e12.

Fankhauser, G. (1945a). The effects of changes in chromosome number on amphibian development. *The Quarterly Review of Biology 20*(1), 20–78.

Fankhauser, G. (1945b). Maintenance of normal structure in heteroploid salamander larvae, through compensation of changes in cell size by adjustment of cell number and cell shape. *Journal of Experimental Zoology 100*(3), 445–455.

Farinella-Ferruzza, N. (1953). Risultati di trapianti di bottone codale di urodeli su anuri e vice versa [Results of tail button transplants of urodeles on anurans and vice versa]. *Rivista di Biologia 45*, 523–527.

Farinella-Ferruzza, N. (1956). The transformation of a tail into a limb after xenoplastic transformation. *Experientia 15*, 304–305.

Fields, C. (2018). Sciences of observation. *Philosophies 3*(4), 29.

Fields, C., Bischof, J., & Levin, M. (2020). Morphological coordination: A common ancestral function unifying neural and non-neural signaling. *Physiology (Bethesda) 35*(1), 16–30.

Fields, C., Glazebrook, J. F., & Levin, M. (2021). Minimal physicalism as a scale-free substrate for cognition and consciousness. *Neuroscience of Consciousness 2021*(2), niab013.

Fields, C., & Levin, M. (2020a). How do living systems create meaning? *Philosophies 5*(4), 36.

Fields, C., & Levin, M. (2020b). Scale-free biology: Integrating evolutionary and developmental thinking. *BioEssays 42*(8), e1900228.

Fields, C., & Levin, M. (2022). Competency in navigating arbitrary spaces as an invariant for analyzing cognition in diverse embodiments. *Entropy* (Basel) *24*(6), 819.

Friston, K. (2013). Life as we know it. *Journal of the Royal Society Interface 10*(86), 20130475.

Friston, K., Levin, M., Sengupta, B., & Pezzulo, G. (2015). Knowing one's place: A free-energy approach to pattern regulation. *Journal of the Royal Society Interface 12*(105), 20141383.

Friston, K., Sengupta, B., & Auletta, G. (2014). Cognitive dynamics: From attractors to active inference. *Proceedings of the IEEE 102*(4), 427–445.

Gabalda-Sagarra, M., Carey, L. B., & Garcia-Ojalvo, J. (2018). Recurrence-based information processing in gene regulatory networks. *Chaos 28*(10), 106313.

Gawne, R., McKenna, K. Z., & Levin, M. (2020). Competitive and coordinative interactions between body parts produce adaptive developmental outcomes. *BioEssays 42*(8), e1900245.

Gonze, D., Gerard, C., Wacquier, B., Woller, A., Tosenberger, A., Goldbeter, A., & Dupont, G. (2018). Modeling-based investigation of the effect of noise in cellular systems. *Frontiers in Molecular Biosciences 5*, 34.

Grossberg, S. (1978). Communication, memory, and development. In R. Rosen & F. Snell (Eds.), *Progress in theoretical biology, Vol. 5* (pp. 183–232). New York: Academic Press.

Hanczyc, M. M., Caschera, F., & Rasmussen, S. (2011). Models of minimal physical intelligence. *Procedia Computer Science 7*, 275–277.

Herrera-Delgado, E., Perez-Carrasco, R., Briscoe, J., & Sollich, P. (2018). Memory functions reveal structural properties of gene regulatory networks. *PLoS Computational Biology 14*(2), e1006003.

Hoel, E. P. (2018). Agent above, atom below: How agents causally emerge from their underlying microphysics. In A. Aguirre, B. Foster, & Z. Merali (Eds.), *Wandering towards a goal: How can mindless mathematical laws give rise to aims and intention?* (pp. 63–76). Cham: Springer International Publishing.

James, W. (1890). *Principles of psychology*. New York: Henry Holt.

Kamm, R. D., & Bashir, R. (2014). Creating living cellular machines. *Annals of Biomedical Engineering 42*(2), 445–459.

Keijzer, F., van Duijn, M., & Lyon, P. (2013). What nervous systems do: Early evolution, input-output, and the skin brain thesis. *Adaptive Behavior 21*(2), 67–85.

Kriegman, S., Blackiston, D., Levin, M., & Bongard, J. (2021). Kinematic self-replication in reconfigurable organisms. *Proceedings of the National Academy of Sciences of the U S A 118*(49), e2112672118.

Kurzweil, R. (2005). *The singularity is near: When humans transcend biology*. New York: Viking.

Lander, A. D. (2004). A calculus of purpose. *PLoS Biology 2*(6), e164.

Levin, M. (2014). Endogenous bioelectrical networks store non-genetic patterning information during development and regeneration. *The Journal of Physiology 592*(11), 2295–2305.

Levin, M. (2019). The computational boundary of a "self": Developmental bioelectricity drives multicellularity and scale-free cognition. *Frontiers in Psychology 10*, 2688.

Levin, M. (2021a). Bioelectric signaling: Reprogrammable circuits underlying embryogenesis, regeneration, and cancer. *Cell 184*(4), 1971–1989.

Levin, M. (2021b). Bioelectrical approaches to cancer as a problem of the scaling of the cellular self. *Progress in Biophysics & Molecular Biology 165*, 102–113.

Levin, M. (2022). Technological approach to mind everywhere: An experimentally-grounded framework for understanding diverse bodies and minds. *Frontiers in Systems Neuroscience 16*, 768201.

Levin, M., Buznikov, G. A., & Lauder, J. M. (2006). Of minds and embryos: Left-right asymmetry and the serotonergic controls of pre-neural morphogenesis. *Developmental Neuroscience 28*(3), 171–185.

Levin, M., & Martyniuk, C. J. (2018). The bioelectric code: An ancient computational medium for dynamic control of growth and form. *Biosystems 164*, 76–93.

Levin, M., Pietak, A. M., & Bischof, J. (2018). Planarian regeneration as a model of anatomical homeostasis: Recent progress in biophysical and computational approaches. *Seminars in Cell & Developmental Biology 87*, 125–144.

Levin, M., Thorlin, T., Robinson, K. R., Nogi, T., & Mercola, M. (2002). Asymmetries in h+/k+-atpase and cell membrane potentials comprise a very early step in left-right patterning. *Cell 111*(1), 77–89.

Lyon, P. (2006). The biogenic approach to cognition. *Cognitive Processing 7*(1), 11–29.

Lyon, P. (2015). The cognitive cell: Bacterial behavior reconsidered. *Frontiers in Microbiology 6*, 264.

Manicka, S., & Levin, M. (2019). The cognitive lens: A primer on conceptual tools for analysing information processing in developmental and regenerative morphogenesis. *Philosophical Transactions of the Royal Society of London B-Biological Sciences 374*(1774), 20180369.

Mar, R. A., Kelley, W. M., Heatherton, T. F., & Macrae, C. N. (2007). Detecting agency from the biological motion of veridical vs animated agents. *Social Cognitive and Affective Neuroscience 2*(3), 199–205.

Mathews, J., & Levin, M. (2018). The body electric 2.0: Recent advances in developmental bioelectricity for regenerative and synthetic bioengineering. *Current Opinion in Biotechnology 52*, 134–144.

Matthewson, J., & Griffiths, P. E. (2017). Biological criteria of disease: Four ways of going wrong. *The Journal of Medicine and Philosophy 42*(4), 447–466.

Maturana, H. R., & Varela, F. J. (1980). *Autopoiesis and cognition: The realization of the living. Boston Studies in the Philosophy of Science* vol. 42. Dordrecht: D. Reidel Publishing.

McDowell, G., Rajadurai, S., & Levin, M. (2016). From cytoskeletal dynamics to organ asymmetry: A nonlinear, regulative pathway underlies left-right patterning. *Philosophical Transactions of the Royal Society of London B-Biological Sciences 371*(1710), 20150409.

McLaughlin, K. A., & Levin, M. (2018). Bioelectric signaling in regeneration: Mechanisms of ionic controls of growth and form. *Developmental Biology 433*(2), 177–189.

McShea, D. W. (2016). Freedom and purpose in biology. *Studies in History and Philosophy of Biological and Biomedical Sciences 58*, 64–72.

Monod, J. (1972). *Chance and necessity: An essay on the natural philosophy of modern biology.* New York: Vintage Books.

Moore, D., Walker, S. I., & Levin, M. (2017). Cancer as a disorder of patterning information: Computational and biophysical perspectives on the cancer problem. *Convergent Science Physical Oncology 3*(4), 043001.

Naitoh, Y., & Eckert, R. (1969a). Ciliary orientation: Controlled by cell membrane or by intracellular fibrils? *Science 166*(3913), 1633–1635.

Naitoh, Y., & Eckert, R. (1969b). Ionic mechanisms controlling behavioral responses of paramecium to mechanical stimulation. *Science 164*(3882), 963–965.

Naitoh, Y., Eckert, R., & Friedman, K. (1972). A regenerative calcium response in paramecium. *Journal of Experimental Biology 56*(3), 667–681.

Newman, S. A. (2019). Inherency of form and function in animal development and evolution. *Frontiers in Physiology 10*, 702.

Noble, D. (2010). Biophysics and systems biology. *Philosophical transactions. Series A, Mathematical, physical, and engineering sciences 368*(1914), 1125–1139.

Noble, D. (2011). The aims of systems biology: Between molecules and organisms. *Pharmacopsychiatry 44*, Suppl. 1, S9–S14.

Noble, D. (2012). A theory of biological relativity: No privileged level of causation. *Interface Focus 2*(1), 55–64.

Oviedo, N. J., Morokuma, J., Walentek, P., Kema, I. P., Gu, M. B., Ahn, J. M., Hwang, J. S., Gojobori, T., & Levin, M. (2010). Long-range neural and gap junction protein-mediated cues control polarity during planarian regeneration. *Developmental Biology 339*(1), 188–199.

Oviedo, N. J., Newmark, P. A., & Sanchez Alvarado, A. (2003). Allometric scaling and proportion regulation in the freshwater planarian *Schmidtea mediterranea. Developmental Dynamics 226*(2), 326–333.

Pai, V. P., Aw, S., Shomrat, T., Lemire, J. M., & Levin, M. (2012). Transmembrane voltage potential controls embryonic eye patterning in xenopus laevis. *Development 139*(2), 313–323.

Pai, V. P., Cervera, J., Mafe, S., Willocq, V., Lederer, E. K., & Levin, M. (2020). Hcn2 channel-induced rescue of brain teratogenesis via local and long-range bioelectric repair. *Frontiers in Cellular Neuroscience 14*, 136.

Pai, V. P., Pietak, A., Willocq, V., Ye, B., Shi, N. Q., & Levin, M. (2018). Hcn2 rescues brain defects by enforcing endogenous voltage pre-patterns. *Nature Communications 9*(1), 998.

Pezzulo, G., Lapalme, J., Durant, F., & Levin, M. (2021). Bistability of somatic pattern memories: Stochastic outcomes in bioelectric circuits underlying regeneration. *Philosophical Transactions of the Royal Society of London B-Biological Sciences 376*(1821), 20190765.

Pezzulo, G., & Levin, M. (2015). Re-membering the body: Applications of computational neuroscience to the top-down control of regeneration of limbs and other complex organs. *Integrative Biology 7*(12), 1487–1517.

Pezzulo, G., & Levin, M. (2016). Top-down models in biology: Explanation and control of complex living systems above the molecular level. *Journal of the Royal Society Interface 13*(124), 20160555.

Pezzulo, G., Rigoli, F., & Friston, K. (2015). Active inference, homeostatic regulation and adaptive behavioural control. *Progress in Neurobiology 134*, 17–35.

Pietak, A., & Levin, M. (2017). Bioelectric gene and reaction networks: Computational modelling of genetic, biochemical and bioelectrical dynamics in pattern regulation. *Journal of the Royal Society Interface 14*(134), 20170425.

Pinet, K., Deolankar, M., Leung, B., & McLaughlin, K. A. (2019). Adaptive correction of craniofacial defects in pre-metamorphic xenopus laevis tadpoles involves thyroid hormone-independent tissue remodeling. *Development 146*(14), dev175893.

Pinet, K., & McLaughlin, K. A. (2019). Mechanisms of physiological tissue remodeling in animals: Manipulating tissue, organ, and organism morphology. *Developmental Biology 451*(2), 134–145.

Powers, W. T. (1973). *Behavior: The control of perception.* Chicago: Aldine.

Prindle, A., Liu, J., Asally, M., Ly, S., Garcia-Ojalvo, J., & Suel, G. M. (2015). Ion channels enable electrical communication in bacterial communities. *Nature 527*(7576), 59–63.

Queller, D. C., & Strassmann, J. E. (2009). Beyond society: The evolution of organismality. *Philosophical Transactions of the Royal Society of London B-Biological Sciences 364*(1533), 3143–3155.

Ramstead, M. J. D., Constant, A., Badcock, P. B., & Friston, K. J. (2019). Variational ecology and the physics of sentient systems. *Physics of Life Reviews 31*, 188–205.

Rosen, R. (1985). *Anticipatory systems: Philosophical, mathematical, and methodological foundations.* 1st ed. Vol. 1, *IFSR international series on systems science and engineering.* Oxford: Pergamon Press.

Rosenblueth, A., Wiener, N., & Bigelow, J. (1943). Behavior, purpose, and teleology. *Philosophy of Science 10*, 18–24.

Saló, E., Abril, J. F., Adell, T., Cebrià, F., Eckelt, K., Fernandez-Taboada, E., Handberg-Thorsager, M., Iglesias, M., Molina, M. D., & Rodriguez-Esteban, G. (2009). Planarian regeneration: Achievements and future directions after 20 years of research. *The International Journal of Developmental Biology 53*(8–10), 1317–1327.

Simon, C. S., Hadjantonakis, A. K., & Schroter, C. (2018). Making lineage decisions with biological noise: Lessons from the early mouse embryo. *Wiley Interdisciplinary Reviews: Developmental Biology 7*(4), e319.

Stone, J. R. (1997). The spirit of D'Arcy Thompson dwells in empirical morphospace. *Mathematical Biosciences 142*(1), 13–30.

Sullivan, K. G., Emmons-Bell, M., & Levin, M. (2016). Physiological inputs regulate species-specific anatomy during embryogenesis and regeneration. *Communicative & Integrative Biology 9*(4), e1192733.

Szabó, Á., Vattay, G., & Kondor, D. (2012). A cell signaling model as a trainable neural nanonetwork. *Nano Communication Networks 3*(1), 57–64.

Torday, J. S., & Miller, W. B., Jr. (2016). Biologic relativity: Who is the observer and what is observed? *Progress in Biophysics & Molecular Biology 121*(1), 29–34.

Tseng, A. S., Beane, W. S., Lemire, J. M., Masi, A., & Levin, M. (2010). Induction of vertebrate regeneration by a transient sodium current. *Journal of Neuroscience 30*(39), 13192–13200.

Turner, J. S. (2017). *Purpose & desire: What makes something "alive" and why modern Darwinism has failed to explain it.* 1st ed. New York: HarperOne.

Turner, J. S. (2019). Homeostasis as a fundamental principle for a coherent theory of brains. *Philosophical Transactions of the Royal Society of London B-Biological Sciences 374*(1774), 20180373.

Vandenberg, L. N., Adams, D. S., & Levin, M. (2012). Normalized shape and location of perturbed craniofacial structures in the xenopus tadpole reveal an innate ability to achieve correct morphology. *Developmental Dynamics 241*(5), 863–878.

Vandenberg, L. N., Morrie, R. D., & Adams, D. S. (2011). V-ATPase-dependent ectodermal voltage and pH regionalization are required for craniofacial morphogenesis. *Developmental Dynamics 240*(8), 1889–1904.

Vane-Wright, R. I. (2014). What is life? And what might be said of the role of behaviour in its evolution? *Biological Journal of the Linnean Society 112*(2), 219–241.

Varela, F. J., & Maturana, H. (1972). Mechanism and biological explanation. *Philosophy of Science 39*(3), 378–382.

Varela, F. J., Maturana, H. R., & Uribe, R. (1974). Autopoiesis: The organization of living systems, its characterization and a model. *Currents in Modern Biology 5*(4), 187–196.

Varela, F. J., Thompson, E., & Rosch, E. (1991). *The embodied mind: Cognitive science and human experience.* Cambridge, MA: MIT Press.

Voskoboynik, A., Simon-Blecher, N., Soen, Y., Rinkevich, B., De Tomaso, A. W., Ishizuka, K. J., & Weissman, I. L. (2007). Striving for normality: Whole body regeneration through a series of abnormal generations. *The FASEB Journal 21*(7), 1335–1344.

Watson, R. A., Buckley, C. L., Mills, R., & Davies, A. (2010). Associative memory in gene regulation networks. Artificial Life Conference XII, Odense, Denmark.

Wiener, N. (1961). *Cybernetics; or, control and communication in the animal and the machine.* 2nd ed. New York: MIT Press.

Wood, C. C. (2019). The computational stance in biology. *Philosophical Transactions of the Royal Society of London B-Biological Sciences 374*(1774), 20180380.

Yang, C. Y., Bialecka-Fornal, M., Weatherwax, C., Larkin, J. W., Prindle, A., Liu, J., Garcia-Ojalvo, J., & Suel, G. M. (2020). Encoding membrane-potential-based memory within a microbial community. *Cell Systems 10*(5), 417–423 e3.

11 Form, Function, Agency: Sources of Natural Purpose in Animal Evolution

Stuart A. Newman

Overview

Attempts to reconstruct the evolutionary origins of the metazoan animals most often assume that the multicellularity of this group began with clusters that formed when cells failed to detach after division. This would have ensured that the multicellular entities were clonal, genetically uniform individuals, and thus appropriate targets for natural selection. The genotypes of variant individuals (their defining property in this picture) would become more prominent in each generation if they left more offspring than their competitors, and this would occur if they did better at gaining access to scarce resources or more favorable habitats (Futuyma & Kirkpatrick, 2017).

In this view, any morphological attributes (e.g., segments, appendages) or functional specializations (e.g., muscle or nerve cells) would be built up gradually over many generations to meet adaptive requirements, necessitating the concomitant evolution of finely honed developmental mechanisms that caused cells to take up assigned positions at appropriate stages. Additional mechanisms must also have evolved to keep individual organisms clonal, thus suppressing evolution that could lead to the emergence of cellular "cheaters" that undermined the cell societies (Strassmann & Queller, 2011). Most of these changes are presumed to have occurred by selection for genetic variants with small effects on the phenotype. After many such cycles, this would produce a panoply of morphologically diverse but phenotypically stable organisms, well adapted to their environments, with refined capacities to inhabit the ecological niches in which they arose.

In this paper I question all the basic assumptions of this framework and present evidence that instead favors a view in which evolution of multicellular organisms is more directional and purposeful than the opportunistic, random-search-based scenario just described suggests. Concerning the specific elements of the standard picture, I show that (i) morphological motifs and patterns that arise during animal development are based on inherent material properties of cell clusters and are therefore readily accessible to these systems without repeated cycles of selection for marginally distinct variants; (ii) development does not depend on genetic uniformity of the embryo's cells and therefore the beginnings of metazoan evolution were unlikely to have required it; (iii) functionalities that provided the physiological bases for specializations of differentiated cell types and organs were inherent to the single-celled organisms directly ancestral to the metazoans; (iv) a unique gene regulatory system whose origin accompanied animal evolution was capable of readily appropriating and parceling out cell functionalities to novel differentiated cell types; (v) the capacity of organisms to behave as autonomous agents, able to define their own boundaries and sustain themselves according to internal motives, was already present in unicellular antecedents; and (vi) novel organismal characters, drawing on intrinsic cellular or material properties, often appeared abruptly, and in preferred, or partly predictable directions, serving as enablements for new ways of life, rather than as adaptations to existing or emerging challenges.

Among other heterodox implications, these proposed departures from the standard evolutionary narrative indicate a less deterministic relationship between genotype and phenotype than generally believed. They will be taken up in order.

11.1   Inherencies of Multicellular Matter Imply Nonadaptive Origins of Morphology

The metazoan animals emerged from the eukaryotic cellular lineage termed holozoans around 800 million years ago (Sperling & Stockey, 2018). These animals are not the only extant holozoans: present-day choanoflagellates and icthyosporeans are also in this group, which in turn is a sister clade (within the opisthokonts) of the fungi (Ros-Rocher et al., 2021). In all metazoan phyla, cell-cell attachment during development is mediated by members of the cadherin family of cell adhesion molecules. These cell surface proteins were based on earlier evolving holozoan ones but in addition contained a novel transmembrane domain that linked the extracellular portion to the internal motility-generating cytoskeleton. This had the novel effect of permitting cells to remain tightly bound to their neighbors while they move past one another in random directions, giving the resulting cell clusters the properties of liquid-state matter (reviewed in Newman, 2016a).

This defining condition of animal life, the "liquid-tissue" state, gave the ancestral colonial holozoans that embodied it strikingly novel material properties. Like the spherical shape of nonliving liquid droplets, the shapes of early-stage embryos and newly formed organ primordia are spherical by default (Steinberg & Poole, 1982). In general, the behaviors and morphological outcomes of biological materials that have nonliving counterparts are relatively predicable, or "generic" (Newman & Comper, 1990). Liquids with two different kinds of subunits can undergo phase separation (like shaken salad dressing), and this is seen in embryonic tissues containing cells with different cadherins on their surfaces (reviewed in Forgacs & Newman, 2005). Gastrulation, the reorganization of a mass of cells into adjacent nonmixing layers, is the first step in the development of animal bodies. Experiments with some species' embryos show that phase separation-like effects drive gastrulation (Krieg et al., 2008), suggesting that these effects may have been the basis for such rearrangements in the first metazoans (Newman, 2016b; Newman, 2022a).

Along with their unique cadherins, all animals, including the ones with the simplest body plans (the labyrinthine sponges and sandwich-like placozoans), produce the protein Wnt, which (like the cadherins themselves) is not present outside the animals. Wnt is secreted and diffuses between cells (thus serving as a morphogen), inducing them to selectively express certain genes and thus differentiate and to rearrange their cadherins and to change their shapes.

Due to altered distributions of adhesive proteins, cell clusters exposed to Wnt can form interior spaces. Further, by causing cells to elongate, Wnt can promote their intercalation (analogous to the formation of liquid crystals by long polymers) and thereby reshape an embryo or organ primordium (reviewed in Forgacs & Newman [2005] and Newman [2019a]).

Bodies more complex than those of sponges and placozoans, such as the two-layered (diploblastic), tentacled ctenophores (comb jellies) and cnidarians (hydra, jellyfish), appeared when metazoans became capable of producing a basal lamina, a stiff but flexible planar substratum to which cells could adhere. This structure, a type of extracellular matrix (ECM), permitted the formation of the stable, sharply defined tissue layers of these organisms, and introduced mechanical effects along with merely liquid-like effects. The diploblasts, and the even more complex and abundant three-layered (triploblastic) forms derived from them (which include arthropods, mollusks, and our own group, chordates), depended on an enzyme called peroxidasin, still another protein unknown outside the animals. Although the ECM protein type IV collagen is present in all metazoans, only in diploblasts and triploblasts (and one type of sponge) is it converted (by peroxidasin) into a basal lamina (Fidler et al., 2017). The capacity to form elongated bodies, appendages, tissue ridges, folds, and clefts is only possible with the presence of a basal lamina, and this in turn depends on the action of peroxidasin (Fidler et al., 2017).

The interpolation of additional ECM molecules between the two epithelial layers of some diploblastic lineages led to a third tissue layer, the mesoderm, which typically took on a loosely packed "mesenchymal" physical state. The resulting tissue-tissue interactions led to novel morphological elements involving folding, clefting, invagination, and evagination of the epithelium, as well as condensation (compaction of dispersed cells) and solidification (skeletogenesis) of the mesenchyme, a process that eventually gave rise to about 32 triploblastic body plans (reviewed in Newman, 2016b). Triploblastic (or bilaterian) bodies are therefore more complex than those of the diploblasts and could develop true organs, a feature absent in earlier evolving forms.

Several of the most important genes and gene products accompanying the original constitution of animal life and the major transitions within this group have not been detected in nonmetazoan organisms and have no known evolutionary intermediates. Their origins are therefore obscure. Once they were in place, the physical processes they mobilized could be employed optionally or in different temporal sequences in primitive cell aggregates, leading to extensive variations on structural themes, even within genetically related types. This brought vastly expanded developmentally accessible morphospaces into existence, and because the operative physical effects were often nonlinear and condition dependent, novel forms could be produced by small genetic steps or changes of externalities (temperature, nutrients).

After this combination of mobilizing gene products and physical effects produced a new range of anatomies, evolution did not stand still. Innovation remained within a restricted phyletic envelope—a "natural kind" (Newman, 2020a)—because, although development could navigate each physico-genetically defined morphospace, it was essentially impossible to move outside it. Changes in genetic circuitry can occur even as forms become increasingly autonomized and persistent (Müller & Newman, 1999). Even the physical processes by which structures originated could be supplanted by more reliable ones without changing the forms themselves (Newman, 2019b). The molecular and physical bases of somitogenesis in fishes, birds, and mammals, for example, differ in significant ways even though these groups had a common vertebrate ancestor (Stern & Piatkowska, 2015). More generally, when physics-based development of morphological motifs became overlain and reinforced by new regulatory processes during eons of evolution, it caused organismal subgroups and individual identities to become increasingly unique and internally defined. The philosopher Lenny Moss and I have compared these new forms of biological matter to the "natural purposes" (i.e., causes and effects of themselves) posited by Immanuel Kant as the definition of living organisms as teleological beings (Moss & Newman, 2015; Kant, 1790/1966).

In the present-day descendants of the necessarily pleiomorphic early metazoans, each phylum is characterized by developmental toolkit genes which are either (i) shared with all the animal groups, (ii) characteristic of the superphylum it belongs to (e.g., protostomes, deuterostomes, ecdysozoans), or (iii) specific to that phylum. Among the toolkit genes and gene products that participated in these early-trial stages of morphological evolution were morphogens additional to Wnt, such as Shh, BMPs, and FGFs, each acting via spatial gradients and thus making the morphological outcomes less uniform than they would have been in the absence of the morphogens. The interaction of these diffusible factors with the biosynthetic responses they induce in their target cells could lead, under certain effect-ratios (parameters), to physical effects termed reaction-diffusion or Turing-type instabilities (Turing, 1952; Kondo & Miura, 2010), which produce spatial periodicities in cell type distributions (e.g., skeletal primordia) (Newman & Frisch, 1979). Added to this were cell surface proteins (e.g., the Notch-Delta juxtacrine system) that exerted, in some contexts, lateral inhibition, ensuring that cell groups followed different developmental pathways from ones immediately adjacent to them. In other contexts, Notch activation could also induce feedback oscillations in the expression of transcription factors, another source of periodicity (temporal in these cases) which was eventually employed in the generation of tandem tissue blocks such as the segments of insects or the somites of vertebrate embryos. Collectively, the modulation by diffusible factors and lateral inhibition is termed *pattern formation* (Newman, 2019a; Salazar-Ciudad et al., 2003).

Two implications concerning development of form emerge from this analysis, both of which are at odds with generally accepted views. The first is that the morphological features of animals (whatever their roles may be in present-day species) likely originated by expression and elaboration of organizational propensities inherent to multicellular masses. There is no reason to believe that they were built up gradually by the greater reproductive success of marginally better adapted morphological variants. While this is difficult to prove unequivocally (like the Darwinian view), recent experiments suggest that multicellular entities of animal origin can survive and reproduce, at least in laboratory settings, with no discernible morphological elaborations. Levin and coworkers isolated fragments of embryonic epithelium from the top surface ("animal cap") of frog embryos and permitted them to reassociate into aggregates of several thousand cells, and found that these cellular masses behaved as autonomous organisms despite having no tissue-level morphological characteristics (Blackiston et al., 2021). Like the morphologically simplest animals, the placozoans, they could navigate by the action of cilia (specializations intrinsic to the source cells) on their surfaces. Notwithstanding the authors' use of terms like "synthetic," "machines," and "bots" for these entities, they are purely biological. They can socially engage with one another and navigate diverse environments. Placed in a suitable culture medium, they can survive up to 90 days.

Unexpectedly, when their environment contains dissociated cells of the same type, they can actively collect and form them into replicas of themselves, thereby "reproducing" by construction (Kriegman et al., 2021).

If tissue masses composed of a uniform population of cells but lacking salient morphological features can perform so many organism-level activities (i.e., exhibit agency; see "Alignment and Export of Cell Agency in the Evolution of Multicellularity" following), it follows that organisms with the capacity to generate more elaborate attributes—interior spaces, segments, appendages, and so forth—could explore and establish themselves in more complex environments which, by making use of novel affordances (a term introduced by Gibson, 1979), they can even participate in constructing. In such scenarios (see West-Eberhard, 2003), rather than new phenotypes resulting from the incrementalism of adaptive natural selection in defined niches, developmental shifts would produce new organisms capable of new ways of life. While morphological complexity can increase versatility and promote exploration and ecological inhabitance, it is not necessary for existence, as the case of Placozoa exemplifies.

The idea that structural attributes can be generated as indirect consequences of other processes and recruited to uses for which they were not naturally selected is familiar from the conceptual work of Gould, with Lewontin and Vrba, respectively, on "spandrels" (Gould & Lewontin, 1979) and "exaptations" (Gould & Vrba, 1982) and from Müller's "side-effect" hypothesis inspired by his experiments on the avian fibular crest (Müller, 1990). The view advanced here generalizes these phenomena, making them not exceptions, but rather (in consideration of inherencies) the primary modes of morphological innovation.

A second implication of this picture, anticipated by nineteenth-century laws-of-form theorists such as J. W. von Goethe and Etienne Geoffroy Saint-Hilaire (discussed in Webster & Goodwin, 1996), and in the twentieth century, D'Arcy W. Thompson (1942), derives from the conclusion that morphological attributes are spontaneous or elicited expressions of the inherent physical properties of tissue masses and not the outcomes of successive generations of natural selection on genetically based minor variants. Thus, body plan and organ development of an animal should not depend on the precise details of its genotype. So long as its embryonic cells contain a relevant set of toolkit genes under the regulatory control typical of its phylum, it is not important for all the cells to have the same genomes or even a shared recent evolutionary history. This inference is confirmed by the production of healthy full-term animals from experimentally constructed embryo chimeras (Fehilly, Willadsen, & Tucker, 1984). These are formed from the blastomeres of mammalian species as evolutionary distant as sheep and goats which are estimated to have diverged 14–16 million years ago (Mya; resulting in "geeps"), or from the fishes medaka and zebrafish, teleost lineages that separated on the order of 320 Mya. The cells of chimeras (unlike those of hybrids such as the mule) retain their species-specific genomes and gene regulatory molecules but are nonetheless capable of interpreting each other's patterning signals in constructing an evolutionarily unprecedented animal (discussed in Newman, 2014). Specifically, the morphological phenotypes of chimeric bodies and organs are intermediates, or compromises, between those of the originating species, thus representing structures that were not fine-tuned by natural selection.

The only distinctive components that appear necessary to produce the stereotypical anatomical elements of the encompassing group (genus, order) of each of these chimeric forms are the respective species' homologous developmental-genetic toolkits, not any species-specific developmental programs (Newman, 2011). Here, the key patterning processes are the meso-scale physical effects, termed "dynamical patterning modules" (Newman & Bhat, 2009) that come into existence during the stage of early (normal or chimeric) development (variously, the blastula or inner cell mass) when enough cells are present to constitute the embryo as a liquid tissue.

The patterning processes discussed above as modulators of morphogenesis also have the role of specifying locations within the tissue mass at which differentiated cells appear. Representing discrete expression states of the organism's genome, these cell types contributed to organismal physiological versatility during evolution, leading to present-day forms. Whereas morphological motifs may have emerged (as argued above) as optional attributes stemming from physical inherencies of metazoan cell masses, functions are based on intricate biochemical pathways and have specific life-sustaining roles. They are poor candidates for spandrels, side effects, or exaptations and are never treated as such in the evolutionary biology or philosophy literature.

In the next section, however, I provide evidence that the functions of differentiated cells and organs of animals are in fact based on inherent properties of holozoan cells that were appropriated and partitioned to perform specialized roles in the multicellular aggregates that became metazoans (see also Newman, 2022b). Like the morphological motifs discussed above, they are attributes that facilitate niche exploration and, at least at their inception, were nonobligatory. The basis of this appropriation is a unique metazoan system of selective gene regulation.

11.2   Origination of Cell Differentiation by Appropriation of Unicellular Functions

The notion of biological function has a long history in the philosophy of biology (Wouters, 2005). It has broken into the empirical discourse when, for example, physiological roles have been ascribed to system components of debated significance, such as non-protein-coding DNA (Germain, Ratti, & Boem, 2014). In the animals, organs, in performing life-sustaining functions, will often do other things as well. The heart, for example, pumps blood but also makes sounds. It is only the first that is considered a "proper function" (Thomas, 2017), and because natural selection is usually considered the only possible explanation for how organs and their functions arise, proper functions have also been synonymized by many philosophers, following Neander, to "selected effects" (Neander, 1991).

But other activities of organs and body parts can also contribute importantly to an organism's life even when they were not specifically selected to do so. Staying with the example of the heart, its ventricular wall can become thickened, or hypertrophic, when arterial blood flow is impaired, a change associated with enhanced ejection force. Ventricular hypertrophy thus performs a health-sparing compensatory function, but one that is unlikely to have been selected for, as the need for it typically appears after reproductive age. Further, it can lead to exacerbation of the blood flow problem by narrowing the ventricle (Nagueh, 2021).

Some commentators argue for excluding nonselected activities as functions at all. But adaptationist scenarios for some components with unambiguous causal roles can be elusive. This is particularly true when it comes to functions performed by individual cells. The origins of cytoplasmic contractility and vesicular packaging of proteins are unknown, but both appear to have roots in archaea (Akıl et al., 2021; Gould, Garg, & Martin, 2016). They have essential causal roles, but are they selected effects? There is no evidence one way or the other, and as we noted in the previous section, features can arise by means other than natural selection. Once having appeared, they can be propagated simply by not undermining survival, constituting what John Tyler Bonner termed "neutral phenotypes" (Bonner, 2013).

Without speculating on the origin of functions in the holozoan cellular ancestors of the animals, I will present recent findings on the coincident emergence of metazoans with a gene regulatory apparatus capable of appropriating ancestral unicellular functions into specialized cell types. An implication is that animal tissue and organ functions of increasingly complex phyla and species were pre-existing in, and inherent to, holozoan cellular life and that they originally were optional embellishments of novel metazoans which only later became indispensable in certain lineages.

Against the standard dichotomy of the function discourse, this view proposes that specialized functions in multicellular organisms such as animals need not be selected effects or, in their original instances, causally essential to their bearers' survival. (Their intracellular versions are, of course, causally essential.) The plausibility of this add-on notion of organismal function is supported by further work of Levin and coworkers in which they added functionalities to their multicellular constructs such as light-responsive genes, and found that the biobots acquired unforeseen novel agent-like behaviors and "invented" new ways of life (Blackiston et al., 2021).

Like their present-day counterparts, premetazoan opisthokonts and holozoans almost certainly exhibited alternative cell phenotypes and states. While some of these may have been antecedents of metazoan cell types (Brunet & King, 2017; Sogabe et al., 2019), complex animals have a hundred or more cell types (vertebrates over 250), and most of these emerged as the metazoans diversified. Gene regulatory networks (GRNs), the interacting sets of transcription factors (TFs) and their cognate genes that control the expression of one or more function-related genes of specific cell types, are often conserved across animal phyla (Arendt et al., 2016). The GRNs that mediate differentiated functions of animal cell types are organized into "gene coexpression networks" (Singh et al., 2018; Cao & Cheng, 2015; Laarman et al., 2019; Stodola et al., 2018). These networks are conserved over broad taxonomic groups (Tanizawa et al., 2010; Diament & Tuller, 2017) where they mediate alternative physiological states (Sebé-Pedrós, Degnan, & Ruiz-Trillo, 2017; Sebé-Pedrós et al., 2016). In early emerging metazoans such as sponges, they provide the bases of stem cells and of a few differentiated cell types (Sogabe et al., 2019).

Cell types of later-diverging animals with apparent roots in coexpression networks in ancestral single-cell functions include myoblasts (motility), fibroblasts, osteocytes, and chondrocytes (extracellular matrix production), neurons (electrical excitability), hepatocytes (detoxification), adipocytes (lipid storage), retinal rods (light responsivity), and erythrocytes (oxygen capture). The hypothesis that differentiated cell functions of metazoans were recruited from cellular functionalities of direct ancestors implies that functions only found in distant lineages would not appear in animals except, rarely, through endosymbiosis or lateral transfer of genes. In confirmation, while cell types that capture light or store starch are seen in the vascular plants, they are not present in animals. The cellulosic tunic of urochordates appears to represent an exceptional laterally transferred skeletogenic functionality (Nakashima et al., 2004).

The means by which animal embryos co-opt pre-existing gene coexpression networks into specialized cell types is a gene regulatory system different from those in any other form of life. This system is based on chromatin, the complex of DNA and proteins found in all eukaryotic (nucleated) cells, but holozoan chromatin has components in addition to those found in other eukaryotes, and the chromatin of metazoans is still more elaborate. Briefly (because this system has been described in detail elsewhere), the nuclei of animal cells contain an architectonic complex of "function-amplifying centers" (FACs) (Newman, 2020b). The main components of the FACs are a "write-read-rewrite" transcription regulatory system (Prohaska, Stadler, & Krakauer, 2010), based on reversible chemical modification of the histone proteins that organize the cell's DNA into nucleosomes and have the capacity to restrict or facilitate access to it. This is a feature shared with all eukaryotic cells, as is the scaffolding of the transcriptionally active stretches of DNA by Mediator, a large protein complex of 21–26 subunits that brings relevant TFs and cofactors to the genes' promoters (Verger, Monte, & Villeret, 2019).

However, with the nonmetazoan holozoans the composition and function of chromatin began to diverge from those of other eukaryotes. Holozoans contain two closely related proteins, p300 and CBP (with a homolog in nonmetazoan holozoan genomes), which, by chemically modifying (acetylating) nucleosomal histones, make the associated DNA available for transcription. In metazoan species they also help recruit linearly distant DNA segments termed *enhancers* to liquid-like biomolecular condensates termed *topologically associating domains* (TADs). These are molecular hubs containing sets of co-expressed genes contained in chromatin loops drawn by mechanical forces from far-flung chromosomal sites (Furlong & Levine, 2018).

During development and terminal cell differentiation, an intensification of the role of p300 occurs. Now it binds "pioneer" TFs (Iwafuchi-Doi & Zaret, 2016) which physically open up regions of chromatin within TADs which contain genes destined to be regulated at certain times and places during embryogenesis. In addition, "lineage-determining" TFs (Obier & Bonifer, 2016) are recruited to these sites. These promote the expression of genes associated with broad developmental pathways and families of related cell types. Concomitantly, superenhancers, containing several hundred or more enhancer sequences as well as noncoding RNAs transcribed from them, transform the respective TADs into FACs.

The FACs, which are unique to metazoans, are essentially engines for appropriation of cellular functions. They amplify life-sustaining activities based on pathways consisting of coordinated sets of genes, which were already present in the unicellular and transiently colonial holozoan cell populations they arose from, partitioning them into specialized cell types. Which cells in a developing embryo acquire which specific functions is determined by the morphogen-based and lateral-inhibition–based pattern-forming systems described in the previous section.

While much remains to be discovered about how a particular pathway is plucked from the intracellular repertoire and elevated, often depending on gene duplication, to a novel cell type, these findings underscore the fact that cell and organ functions are based on inherencies of metazoan existence. Unlike morphological motifs, which are based on physical propensities of multicellular aggregates, the functionalities of cell types are intrinsic to individual cells. Together, however, they imply an evolutionary scenario based on the mobilization of pre-existing properties and capabilities. This is a departure from one in which these characteristics are acquired gradually, based on external challenges, with no preferred direction.

The picture presented here, considered along with the comparative phylogenomics of metazoans and their sister clades, provides a logic for the evolutionary emergence of the main tissues of the animal body, something unaddressed in the Darwin-influenced theory of undirected evolution. A starting point is that the morphologically simplest animals, marine sponges and placozoans (inferred to have resembled the earliest emerging metazoans), have the fewest cell types. *Trichoplax adhaerens*, a well-studied placozoan, has 6-9 somatic cell types (Smith et al. 2014; Schierwater et al., 2021), and *Spongilla lacustris*, a demosponge, has 18 (Sogabe et al., 2019). Some form of epithelium, mediated by the homophilic binding of cadherins, is a defining characteristic of metazoans, and cadherin expression is induced by Grainyhead, one of a small number of simultaneously pioneer and lineage-determining transcription factors (Boivin & Schmidt-Ott, 2018). An early step of metazoan evolution was apparently the coordinated production (likely at an early-appearing FAC) of a metazoan-type cadherin, and Grainyhead, which already existed in opisthokontal lineages where it was also involved in regulating cell attachment (Paré et al., 2012).

Once constituted in this fashion, primitive animals could have made use of their nuclear FAC engines in various ways, first generating muscle and true connective tissue cells using additional simultaneously pioneer and lineage-determining transcription factors (MyoD and Twist). This would have been followed by neurons (hints of appropriation of the respective ancestral circuitry already existing in sponges; Musser et al., 2021) and other cell types that are often refinements and embellishments (again, by co-optation of existing cell circuitry) of these canonical tissue categories. In such cases, the broad-stroke effects of the rare TFs that are both pioneer and lineage-determining types in bringing about macroevolutionary transformations may have been exhausted. Generating new cell types then came to depend on the synergistic effects of generically acting pioneer factors (e.g., Pax7) and lineage-determining TFs without pioneer function (e.g., Nkx-3.1 for heart muscle, Sox9 for cartilage).

These specialized cell types were nonessential add-ons, however. Placozoans and poriferans, which lack muscle and nerves, are perfectly good organisms despite their disabilities, and not all diploblasts and triploblasts have all the capabilities of their respective groups. All of them nonetheless occupy ecological niches suited to their biology. Further, each of these forms exhibits its own manner of exploratory, self-preserving, and cooperative agency, a universal characteristic different from either form or function. The next section takes up this enigmatic set of properties.

11.3   Alignment and Export of Cell Agency in the Evolution of Multicellularity

We have seen that major morphological motifs and physiological functions of animals can, respectively, be manifestations of the inherent forms of the materials constituted by cell aggregates and appropriations of life-sustaining activities of their originating and constituent cells. But organisms also exhibit *agency*, the capacity to define and regulate their own boundaries and sustain their integrity according to a set of internal motives and rules

(Wilson, 2005, Moreno & Mossio, 2015). This property is not reducible to either form or function. Is organismal agency finally only explicable by natural selection?

Addressing this question requires recognizing that agency was in fact inherent to animal life from its start, a property of its founding cells (Arnellos & Moreno, 2015). Owing to their intracellular chemical dynamics and capacity to generate mechanical forces and electrical fields, cells actively modify their behavior in response to their environment and modify their surroundings in ways that can further affect cell–environment interactions (Wan & Jékely, 2021). Although the origins of cellular life from nonliving chemistry are obscure, we know of no cells, from prokaryotes to archaea and eukaryotes, that do not exhibit a form of agency (Lyon et al., 2021; Baluška & Levin, 2016). Whatever the evolutionary origins of individual cell agency may have been, its role in the more recent and better documented evolution of agentive multicellular organisms is more amenable to investigation.

Agentive processes of cells include exploratory behaviors, migration to sources of nutrients or social cues (chemotaxis), cessation of movement, and quiescence. Entering a biochemically oscillatory mode or acquiring surface or shape polarity can also be elements of agent-type activities (Arias Del Angel et al., 2020). Cells also engage in active processes that are not agentive ones, and it is important for our purposes not to conflate them. The randomly directed cytoskeletally driven perambulation of cells, discussed above in relation to the liquid-like properties of animal tissues, is evidence of energy-consuming intracellular mechanical activity, unique to living systems. But it is not indicative of agency, which implies goal-directedness. A similar case is the random motile effects of ciliary beating on animal cells or flagellar motion of bacteria. There is nothing inherently purposeful about these activities. Like the differentiated cell functions that were proposed to have been co-opted from them (see § 11.2, Origination of Cell Differentiation by Appropriation of Unicellular Functions), the functionalities of unicellular organisms facilitate ways of life but were not necessarily selected for their ability to increase the numbers of offspring of the cells that bear them.

Individual cell agency is largely obscured in animals and plants by virtue of being marshaled into collective modes of behavior in the tissues of developing organisms (see § 11.1, Inherencies of Multicellular Matter Imply Nonadaptive Origins of Morphology). In animal embryos, cell motility is mainly of the randomly oriented, nonagentive type that gives the tissues their liquid-like properties. This enables morphogenetic effects such as layering, lumen formation, and segmentation. The associated patterning of cell arrangements is typically based on reaction-diffusion or other self-organizing effects that depend on biochemical activities and therefore positive and negative feedback effects, but not, in any obvious way, on the autonomous agency of cells.

Only rarely does cell agency come to the fore in animal development: in sperm activity during fertilization, in the neural crest cells of vertebrates as they detach from the embryo's central axis and migrate to distant sites where they differentiate into disparate cell types, and pathologically, in cancer metastasis. The atypicality of these cases, and the persistent but unfounded notion of the development-orchestrating "genetic program" (critically discussed in Sarkar, 1998), led to the idea of cell agency being written out of the theoretical paradigms not only of developmental but also of evolutionary biology. However, animal cells in culture, derived from either tumors and normal tissue, can navigate mazes (Tweedy et al., 2020) and make decisions that are both nonrandom and productive for their survival (SenGupta,

Parent, & Bear 2021). Is there a relationship between these latent agent-like, apparently foresightful properties of these isolated cells and the agency that is uncontroversially exhibited by mature animals?

The idea that classes of individuals in the populations of single-celled organisms that eventually gave rise to multicellular entities have a property termed "fitness"—an index of their relative adaptation to their ecological setting (Futuyma & Kirkpatrick, 2017)—has given rise to a widely accepted model for the evolution of multicellular organisms: multilevel selection (MLS) theory (Folse & Roughgarden, 2010). When applied to the emergence of multicellularity, this model implies that the cellular ancestors of multicellular organisms were originally well adapted to their circumstances and only began to form multicellular structures with selection for a genetic propensity to bring their fitnesses into *alignment* with one another. This would have allowed the collective to explore new ecological niches and form new lineages having a multicellular stage in their life cycle. Once entering the multicellular state, however, the possibility of cells individually mutating into a noncooperative condition and nonetheless reaping benefits of the collective required a second phase of selection to ensure that such "cheating" was suppressed (Strassmann & Queller, 2011). The multicellular entity then became the new evolutionary unit of selection as fitness was *exported* to the supracellular level.

The concept of fitness in MLS theory is problematic, however. That specific cells or multicellular organisms are able to establish themselves and survive is readily demonstrable.[1] It is not the same as attributing to them a relative fitness, a quantitative measure based on the comparative potential (often difficult to assess) of leaving different numbers of offspring. Appealing to fitness to account for the emergence of multicellular organisms assumes the validity of the Darwinian hypothesis that organisms and the cells that constitute them arose by a struggle for survival based on fortuitous, undirected variations.

While it is difficult to disentangle individual cellular agency from integrated whole-organism agency in animals, studies of species in which aggregation-based morphological development is part of their life cycle can provide insight into the unicellular-multicellular transition in agentive behavior. Dictyostelia and Myxobacteria, respectively eukaryotic and prokaryotic multicellular lineages, have life cycles comprising a vegetative and a developmental stage. The social amoebae Dictyostelia behave as solitary cells during their vegetative stage, whereas the social bacteria Myxobacteria exist as cell consortiums through their entire life cycles in the wild. Both lineages are commonly found in soils where they feed on bacteria (Myxobacteria on other species). Once nutrients have been depleted, they enter a developmental stage characterized by a substratum- and intercellular-signal–dependent cellular aggregation, formation of fluid cell streams, and, in the case of *Dictyostelium*, a motile pseudoplasmodium 2–4 mm long that resembles a metazoan slug. In both organisms, development culminates in the formation of multicellular structures (up to 10^5–10^6 cells) called fruiting bodies, where cell differentiation takes place (Whitworth, 2008). In each case, individual agents reappear when the fruiting body releases spores that mature and recreate a population like the original one.

There is a remarkable resemblance of developmental processes and resulting morphologies between eukaryotic fruiting-body–forming amoebae such as *Dictyostelium* and the prokaryotic Myxobacteria. The fruiting bodies in both lineages, for example, display a similar range of diversity ranging from simple mound-like to highly branched tree-like structures. This is despite the enormous evolutionary distance between them, reflected in entirely different cellular structures and lack of homology in the signaling and cell–cell interaction modalities mechanisms employed in these groups.

These evolutionary-developmental convergences were explored systematically by Arias Del Angel et al. (2020). The aggregative nature of these organisms, and the need for cells individually (in the case of *Dictyostelium*) or in groups (in the case of Myxobacteria) to detect and consume bacterial prey, and then, when these are no longer available, to switch to different collective modes of behavior, require that cell agency plays a prominent role in their development. This includes, for *Dictyostelium*, directed migration, signaling and response to other cells, joining fluid streams, and participation in the slug. During this period the cells become subject to mesoscopic physical effects like those that drive animal development, and these lead to liquid-liquid- and solidification-type phase transformations in which the agency of individual cells is curtailed. But even during collective activities like slug translocation and fruiting body construction, some cells behave autonomously, moving independently to new locations within the aggregates.

For the rod-shaped cells of Myxobacteria, protein complexes that promote motility define a lagging and a leading pole. Reversals in polarity, which flip the direction of movement, are a major agent-type behavior in Myxobacteria motility (Wu et al., 2009). At the molecular level, reversals are controlled by intracellular oscillators, and cell-cell contacts, by affecting the rate of reversal, convert individual agential behavior to collective motion. As in *Dictyostelium*, the cell mass become liquid-like, sustaining viscous streams. Directed migration in Myxobacteria is also facilitated by *stigmergy*, in which individual cellular movement is biased by cues left behind by other cells (Gloag et al., 2016).

Although most amoebae and bacteria of species related to the social ones do not exhibit their dramatic collective behaviors, they use corresponding modalities to inhabit and explore their respective *umwelts* (i.e., species-characteristic worlds) (Ginn, 2014). The evolution of cellular agents to participate in multicellular life cycles occurred with small changes in their agential toolkits. Though mechanistically and molecularly very different in the amoeboid and bacterial lineages, cell agency came to serve the analogous functions of mediating cell aggregation, the effect of which was to constitute mesoscopic materials. In many, if not most cases, the inherent morphogenetic properties of such materials, rather than the genetic disciplining of individual cells not to cheat, mobilized them to the collective purpose of building complex structures.

From this consideration of the aggregative microorganisms, it can be inferred that the first steps toward metazoan autonomy began in holozoan cell populations as hybrid quasi-tissue associations in which aggregates acquired their own agency. The example of the *Dictyostelium* slug shows that an association of cells can have its own, different form of agency, albeit transient. While it might be argued that this stage of the *Dictyostelium* life cycle is a product of natural selection, even single-cell-type tissue fragments, with no evolutionary history behind their independent existence (e.g., the biobots described above), spontaneously exhibit modes of agency different from the cells that constitute them or from the organisms from which they were derived.

The potential forms and functions of these proto-metazoans may have been (as Darwin suggested) "endless," but they were also constrained and limited by the material inherencies of their developing tissues and the co-optable functionalities of their constituent cells.

These processes account for innovation within bounds, but the fossil record suggests that evolutionarily stasis most often follows novelty (Eldredge & Gould, 1972). In the following section I describe some of the inherent forces that normalize the generation and regeneration of phenotypes, adding generative entrenchment and stability to what is otherwise, in our discussion up to now, a highly plastic process.

11.4 Field Effects in Developmental Precision, Templating, and Regeneration

The standard evolutionary narrative based on differential fitness treats innovation and conservation of phenotypes as two sides of one process. The variant individuals in a population are stringently tested against previously evolved physiological systems and ecological relationships, and the ones with improved fitness will in general be only marginally different from their progenitors. Evolution is held to produce genetic programs, and programs function with precision. In contrast, in the scenario described here the innovational processes are more unbridled and anarchic: forms can be elicited in cell aggregates based on physical inherencies that are manifested, or not, depending on externalities or slight shifts in the tuning of genetic circuitry. Functions can be appropriated, or not, from a repertoire intrinsic to constituent cells. Survival or extinction of deviant phenotypes does not depend on the capacity to leave more offspring than competitors in the niche from which they originated, as in the Darwinian framework. Rather, because they will be agents with new capabilities, they can potentially set out on their own and, through physiological and behavioral plasticity, fashion new affordances (Carneiro & Lyko, 2020; Little, Chapman, & Hillier, 2020).

But the biosphere is filled with stable types of organisms that can often regenerate their canonical states when damaged, and the evolutionary record is marked by anatomical stasis. So there must be strong forces of stabilization and normalization to balance those of innovation. One set of phenomena that occurs during embryonic development (again in metazoans) is the establishment of developmental or morphogenetic fields—poorly understood effects that coordinate tissue organization over distances greater than known modes of cell-cell communication can achieve (Goodwin & Trainor, 1980; Brandts & Trainor, 1990; Levin, 2012). The processes that arrange cell types in stereotypical patterns are molecular ones involving cell-cell contact, and diffusion and related transport processes. The effective spatial range of these during the relevant periods would typically be on a scale of less than 10 cell diameters. But the scale of precise developmental patterning is typically over 100 or more cell diameters. In certain cases this discordance is resolved by the cells in a prospectively patterned field undergoing periodic changes in one or more key transcriptional regulators, often Hes1, with the oscillations then coming into synchrony (Özbudak & Lewis, 2008; Bhat et al., 2019). This effect can occur through weak, nonspecific interactions, with no exchange or transport of molecules (Strogatz, 2003; Garcia-Ojalvo, Elowitz, & Strogatz, 2004). When the developing primordium is then induced to undergo patterning, the constituent cells will now all be "on the same page" and respond in a coordinated fashion despite not being in communication across the domain. This smoothing-out mechanism makes development (a dynamical, condition-dependent process, not a programmed one) a less chaotic and more reproducible phenomenon than it would otherwise be. Synchronization is also used elsewhere in the organism: in the brain, for example, where the resulting coordination of neuronal firing is thought to be a basis for cognitive integration (Mamashli et al., 2021; Palva et al., 2010).

The physics underlying synchronization of oscillators is understood, at least in principle. A less familiar, but apparently more general globally acting effect is the representation, recording, and templating of morphological structures by spatial patterns of voltage gradients (Levin, 2021). Levin and his colleagues call this phenomenon "anatomical homeostasis." The phenomenon is based on V_{mem}, or resting membrane potential, mediated by transport of ions and other small molecules through gap junctions. The group found that in the developing face of the frog, patterns of hyperpolarization map out the prospective locations of the eyes and other structures (Vandenberg, Morrie, & Adams, 2011). When these voltage patterns are perturbed by chemicals or manipulation of expression of connexin gap junction proteins, the boundaries of expression of face patterning genes are altered, with corresponding deleterious effects on craniofacial anatomy (Levin, Pezzulo, & Finkelstein, 2017). Spatial differences of resting potential can thus scaffold and guide subsequent morphogenesis. Importantly, they are instructive, capable of placing normal structures in ectopic positions.

Similar studies on planaria showed that development of the body axis, regeneration, and number of heads, and even species identity of head morphology are guided (even when the tissues normally containing portions of the field are missing) by spatially nonuniform electric fields (Sullivan, Emmons-Bell, & Levin, 2016). The influences of V_{mem} are on cell proliferation, migration, and differentiation and the proximate bases of these effects. The experimental production of interspecies chimeras such as geeps (see preceding discussion), with physiologically healthy intermediate morphological phenotypes that develop notwithstanding the genetically mosaic nature of the embryos and mature tissues, suggests that the homeostatic setpoints of the electrical field-based pattern-memory effects is something other than species- or genome-level characters.

For the present discussion, it is notable that the efficacious representation of tissue morphology in endogenous patterned bioelectric fields, like the animal functions and agential properties discussed above, has deep roots in the tree of life. Organisms as phylogenetically distant from the metazoans as bacteria (in patterned communities and biofilms) (Yang et al., 2020), and *Physarum*, a single-celled amoebozoan with exploratory properties (Murugan et al., 2021), show evidence of the same guidance by self-generated electrical fields (Zheng et al., 2015). This suggests that morphological scaffolding by bioelectric fields was present from the start of metazoan evolution, with its locus of activity being transferred to the multicellular collective as soon as one existed. How anatomical memory is implemented at the level of cell physiology is not understood, but once in place it would have had an inertial or conservative effect on evolution, most likely contributing to phyletic stasis.

11.5   Conclusion: What We Know and What We Don't

I have presented an account of the diversification of animals using several recent bodies of investigation not usually included in theories of multicellular origins and evolution. These are:

• Phylogenomics of the nonmetazoan holozoans, identifying key innovations in the developmental genetic toolkit accompanying the origination of animals and their phylum-level radiation.

• Experimental tissue physics of animal morphogenesis and pattern formation, including findings that, although liquid-tissue effects are universal in this group, liquid-crystal, liquid-substratum, and tissue condensation and solidification effects are added successively during phylogenesis.

• Comparative analysis of chromatin structure and gene regulation, which has disclosed unique features of metazoans, including super-enhancers, and pioneer and lineage-determining transcription factors.

• Recognition that all cells, cell aggregates, and tissue fragments exhibit autonomous and agentive behavior.

• Studies of tissue-level field effects that normalize, coordinate, template, and restore morphology and pattern, including synchronization of transcriptional oscillators and bio-electrical pattern-memory.

These matters, none of them controversial or even contested (though all with incompletely understood aspects) have not been considered together, but when they are, they gel into a coherent alternative to the Darwinian narrative of adaptive evolution by gradual natural selection. This new view accounts for several phenomena that have eluded satisfactory explanation in the standard framework. These include: (i) the abruptness of the appearance of animal forms in the fossil record (the Ediacaran and Cambrian radiations; Sperling & Stockey, 2018); (ii) the tempo and mode of subsequent evolution (saltation, stasis, punctuated equilibria; Gould and Eldredge, 1977); (iii) discordances between phenotype and genotype (developmental system drift; True & Haag, 2001, and homomorphy; Newman, 2019b); (iv) the recurrence of morphological motifs across the animal kingdom (inherency) (Newman, 2019a), homoplasy (Wake, 1991); (v) the use of a conserved developmental-genetic toolkit to generate analogous structures in lineages in which they were not present in common ancestors (Shubin, Tabin, & Carroll, 2009); aka the "developmental analogy-homology paradox" (Newman, 2006); (vi) the delay in appearance of phylum-characteristic body plans until mid-embryogenesis (the evolutionary-developmental hourglass) (Newman, 2011); (vii) the origin of animal-characteristic differentiated cell types, and tissue and organ functions (Arendt et al., 2016); (viii) the ability of interspecies embryo chimeras to develop into viable organisms with class-characteristic intermediate phenotypes (Fehilly, Willadsen, & Tucker, 1984); (ix) the nonadaptive origins of morphological characters (spandrels; Gould & Lewontin, 1979), developmental side-effects (Müller, 1990), and developmentally based evolutionary innovation (Müller, 2021); (x) adaptive appropriation of forms arising by unrelated processes (exaptation; Gould & Vrba, 1982); (xi) the ability of organisms to prevail in ecological settings in which they had no prior evolutionary history (invasive species; Carthey & Banks, 2014), and transgressive hybrids (Dittrich-Reed & Fitzpatrick, 2013).

The increasingly acknowledged phenomenon of niche construction (Odling-Smee, Laland & Feldman, 2003), a process in which organisms influence their abiotic and biotic environments to suit their own attributes, has been discussed as part of an "extended evolutionary synthesis," where niches provide behaviorally initiated venues for adaptive natural selection (Laland et al., 2015). In the perspective presented here (earlier versions of which also contributed to the initiative toward a new synthesis [Newman & Müller, 2000; see also Müller, 2021]), niche construction follows from the more fundamental concept of organismal agency (including that of organisms with novel, originally nonadaptive features; see points ix–xi) but is only incidentally a target or locus of selection. If organisms, even the simplest of the animals, like placozoans, and artificially constructed ones like biobots, exhibit agency, the identification of suitable affordances becomes a default activity of life rather than a set of evolved coadaptations.

This justifies dispensing with the concept of adaptation, as well as relative reproductive fitness (i.e., the propensity to leave different numbers of offspring) as the key driving processes of evolution (as asserted in the population genetics framework). Metazoan organisms can acquire complex morphological features by the manifestation of inherencies of multicellular matter, and functional attributes by the appropriation for novel cell types of pre-existing cellular life-sustaining pathways. Such add-ons are not products of adaptive natural selection. They may become useful after the fact but are more properly considered enablements than Darwinian adaptations.

Darwin's mechanism for the gradual building of complex traits ("Darwin's dangerous idea"; Dennett, 1995) requires differential fitness and competition for resources in a common niche. But if elaborate characters are produced by other means, the relative numbers of progeny any variant leaves, while potentially significant ecologically and biogeographically, are irrelevant to the origination of phenotypes. Agential activities and adjustments of lifestyle can spare even relatively low-producing variants, and the fecundity of organisms that occupy different niches are rarely compared in evolutionary scenarios.

Another element of the proposed extended synthesis with a relationship to the view presented here is facilitated variation, which looks to the reorganization of the established components of developmental systems to generate novelty, often by reconfiguration of developmental systems (Gerhart & Kirschner, 2007). However, the variants contemplated by facilitated variation, while more loosely associated with genetic change than in the standard perspective, do not draw on the inherencies of form and function discussed here, and do not incorporate a role for organismal agency. The facilitated variation model remains fully within the Darwinian paradigm, as exemplified by its formulators' statement: "By such reductions and increases, the conserved core processes facilitate the generation of phenotypic variation, which selection thereafter converts to evolutionary and genetic change in the population" (Gerhart & Kirschner, 2007, p. 8582).

This finally brings us to the unknowns of the items in the bulleted list above:

• The phylogenomics of the transition from nonmetazoan holozoans to animals contains many surprises, but none more mysterious than the appearance of genes and gene modules specifying the cytoskeletal-binding domain of the metazoan cadherins, the morphogen Wnt, the enzyme peroxidasin, and the cell shape polarity modulator Stbm/Vang, all of which thus far have no protein counterparts in other life forms (reviewed in Newman, 2019a). These are not just any proteins, but ones that mobilize new physical effects underlying new body plans. Where did they come from? There are various models for, and speculations on, the origination of new genes (Neme & Tautz, 2013; Bornberg-Bauer, Schmitz, & Heberlein, 2015; Schmitz, Ullrich, & Bornberg-Bauer, 2018; Werner et al., 2018), but none of these scenarios has yet been brought to bear on this unique set of metazoa-constituting and transition-mediating genes.

• The assembly of topologically associating domains and of super-enhancers in metazoan nuclei have been the subject of scores of research papers in the past decade. But little is known about how the function-amplifying centers that they form (apparently amorphous liquid-like droplets rich in intrinsically disordered protein domains and noncoding RNAs) mediate extreme differences in the regulation of differentiation-specific genes.

• Cell agency is incontrovertible, but very little is understood about its origin and physical bases. This is also the case with the different forms of agency of multicellular aggregates.

• The processes by which protein concentrations are caused to oscillate in cells, and how these oscillations come into synchrony to establish developmental fields, are relatively well understood. The means by which nonuniform patterns of membrane resting potential in developing tissues anticipate, guide, remember, and restore morphological arrangements lack explanation at present.

To answer these questions, some of which are new to science, we can echo the call of the early twentieth-century embryologist E. E. Just (in contemplating the strange, nonmechanistic behaviors of the developing marine organisms he had been studying) for "a physics and chemistry in a new dimension . . . superimposed upon the now known physics and chemistry" (Just, 1939: 3). Concerning the theory of evolution, however, relevant progress has been made ever since Darwin, more than 160 years ago, advanced his theory of replacement of organisms by better-adapted ones. The organismal subjects of his theory of natural selection were conceived as relatively passive entities. Darwin and his contemporaries, specifically, knew nothing about physically determined inherent forms, of the multicellular means for co-opting unicellular functions, of cell-based agency, or of oscillation and electrical field-based developmental homeostasis. Given these advances, it is difficult not to conclude that Theodosius Dobzhansky's famous tenet that "nothing in biology makes sense except in the light of evolution" (Dobzhansky, 1973) must also include the evolution of the theory of evolution itself.

Note

1. Ecological persistence is included in some versions of fitness (Bouchard, 2008). In that case, variants of any degree of departure from the originating phenotype can establish themselves, so it is difficult to see what Darwinian gradualism would add to explaining phyletic transformation. The processes described here would plausibly serve this role.

References

Akıl, C., Y. Kitaoku, L. T. Tran, D. Liebl, H. Choe, D. Muengsaen, W. Suginta, & R. C. Schulte-Robinson. 2021. "Mythical origins of the actin cytoskeleton." *Curr Opin Cell Biol* 68:55–63. https://doi.org/10.1016/j.ceb.2020.08.011

Arendt, D., J. M. Musser, C. V. H. Baker, A. Bergman, C. Cepko, D. H. Erwin, M. Pavlicev, G. Schlosser, S. Widder, M. D. Laubichler, and G. P. Wagner. 2016. "The origin and evolution of cell types." *Nat Rev Genet* 17 (12):744–757. https://doi.org/10.1038/nrg.2016.127

Arias Del Angel, J. A., V. Nanjundiah, M. Benítez, and S. A. Newman. 2020. "Interplay of mesoscale physics and agent-like behaviors in the parallel evolution of aggregative multicellularity." *EvoDevo* 11 (1):21. https://doi.org/10.1186/s13227-020-00165-8

Arnellos, A., and A. Moreno. 2015. "Multicellular agency: an organizational view." *Biology & Philosophy* 30 (3):333–357. https://doi.org/10.1007/s10539-015-9484-0

Baluška, F., and M. Levin. 2016. "On having no head: cognition throughout biological systems." *Frontiers in Psychology* 7 (902). https://doi.org/10.3389/fpsyg.2016.00902

Bhat, R., T. Glimm, M. Linde-Medina, C. Cui, and S. A. Newman. 2019. "Synchronization of Hes1 oscillations coordinates and refines condensation formation and patterning of the avian limb skeleton." *Mech Dev* 156:41–54. https://doi.org/10.1016/j.mod.2019.03.001

Blackiston, D., E. Lederer, S. Kriegman, S. Garnier, J. Bongard, and M. Levin. 2021. "A cellular platform for the development of synthetic living machines." *Sci Robot* 6 (52). https://doi.org/10.1126/scirobotics.abf1571

Boivin, F. J., and K. M. Schmidt-Ott. 2018. "Functional roles of Grainyhead-like transcription factors in renal development and disease." *Pediatr Nephrol* 35: 181–190. https://doi.org/10.1007/s00467-018-4171-4

Bonner, J. T. 2013. *Randomness in evolution*. Princeton, NJ: Princeton University Press.

Bornberg-Bauer, E., J. Schmitz, and M. Heberlein. 2015. "Emergence of de novo proteins from 'dark genomic matter' by 'grow slow and moult.'" *Biochem Soc Trans* 43 (5):867–873. https://doi.org/10.1042/BST20150089

Bouchard, F. 2008. "Causal processes, fitness, and the differential persistence of lineages." *Philosophy of Science* 75 (5):560–570. https://doi.org/10.1086/594507

Brandts, W. A. M., and L. E. H. Trainor. 1990. "A non-linear field model of pattern formation: Intercalation in morphalactic regulation." *J Theoretical Biol* 146 (1):37–56. https://doi.org/10.1016/S0022-5193(05)80043-8

Brunet, T., and N. King. 2017. "The origin of animal multicellularity and cell differentiation." *Dev Cell* 43 (2):124–140. https://doi.org/10.1016/j.devcel.2017.09.016

Cao, R., and J. Cheng. 2015. "Deciphering the association between gene function and spatial gene-gene interactions in 3D human genome conformation." *BMC Genomics* 16:880. https://doi.org/10.1186/s12864-015-2093-0

Carneiro, V. C., and F. Lyko. 2020. "Rapid epigenetic adaptation in animals and its role in invasiveness." *Integr Comp Biol* 60 (2):267–274. https://doi.org/10.1093/icb/icaa023

Carthey, A. J. R., and P. B. Banks. 2014. "Naïveté in novel ecological interactions: lessons from theory and experimental evidence." *Biological Reviews* 89 (4):932–949. https://doi.org/10.1111/brv.12087

Dennett, D. C. 1995. *Darwin's dangerous idea: evolution and the meanings of life*. New York: Simon & Schuster.

Diament, A., and T. Tuller. 2017. "Tracking the evolution of 3D gene organization demonstrates its connection to phenotypic divergence." *Nucleic Acids Res* 45 (8):4330–4343. https://doi.org/10.1093/nar/gkx205

Dittrich-Reed, D. R., and B. M. Fitzpatrick. 2013. "Transgressive hybrids as hopeful monsters." *Evol Biol* 40 (2):310–315. https://doi.org/10.1007/s11692-012-9209-0

Dobzhansky, T. 1973. "Nothing in biology makes sense except in the light of evolution." *The American Biology Teacher* 35:125–129.

Eldredge, N., and S. J. Gould. 1972. "Punctuated equilibria: an alternative to phyletic gradualism." In *Models in paleobiology*, edited by T. J. M. Schopf, 82–115. San Francisco: Freeman Cooper.

Fehilly, C. B., S. M. Willadsen, and E. M. Tucker. 1984. "Interspecific chimaerism between sheep and goat." *Nature* 307 (5952):634–636.

Fidler, A. L., C. E. Darris, S. V. Chetyrkin, V. K. Pedchenko, S. P. Boudko, K. L. Brown, W. Gray Jerome, J. K. Hudson, A. Rokas, and B. G. Hudson. 2017. "Collagen IV and basement membrane at the evolutionary dawn of metazoan tissues." *Elife* 6. https://doi.org/10.7554/eLife.24176

Folse, H. J., 3rd, and J. Roughgarden. 2010. "What is an individual organism? A multilevel selection perspective." *Q Rev Biol* 85 (4):447–472.

Forgacs, G., and S. A. Newman. 2005. *Biological physics of the developing embryo*. Cambridge: Cambridge University Press.

Furlong, E. E. M., and M. Levine. 2018. "Developmental enhancers and chromosome topology." *Science* 361 (6409):1341–1345. https://doi.org/10.1126/science.aau0320

Futuyma, D. J., and M. Kirkpatrick. 2017. *Evolution* (4th ed.) Sunderland, MA: Sinauer Associates.

Garcia-Ojalvo, J., M. B. Elowitz, and S. H. Strogatz. 2004. "Modeling a synthetic multicellular clock: Repressilators coupled by quorum sensing." *Proc Natl Acad Sci USA* 101 (30):10955–10960.

Gerhart, J., and M. Kirschner. 2007. "The theory of facilitated variation." *Proc Natl Acad Sci USA* 104, Suppl. 1:8582–8589. https://doi.org/10.1073/pnas.0701035104

Germain, P.-L., E. Ratti, and F. Boem. 2014. "Junk or functional DNA? ENCODE and the function controversy." *Biol & Philosophy* 29 (6):807–831. https://doi.org/10.1007/s10539-014-9441-3

Gibson, J. J. 1979. *The ecological approach to visual perception*. Boston: Houghton Mifflin.

Ginn, F. 2014. "Jakob von Uexküll beyond bubbles: on umwelt and biophilosophy." *Science as Culture* 23 (1):129–134. https://doi.org/10.1080/09505431.2013.871245

Gloag, E. S., L. Turnbull, M. A. Javed, H. Wang, M. L. Gee, S. A. Wade, and C. B. Whitchurch. 2016. "Stigmergy co-ordinates multicellular collective behaviours during *Myxococcus xanthus* surface migration." *Sci Rep* 6:26005. https://doi.org/10.1038/srep26005

Goodwin, B. C., and L. E. Trainor. 1980. "A field description of the cleavage process in embryogenesis." *J Theor Biol* 85 (4):757–770. https://doi.org/10.1016/0022-5193(80)90270-2

Gould, S. B., S. G. Garg, and W. F. Martin. 2016. "Bacterial vesicle secretion and the evolutionary origin of the eukaryotic endomembrane system." *Trends Microbiol* 24 (7):525–534. https://doi.org/10.1016/j.tim.2016.03.005

Gould, S. J., and N. Eldredge. 1977. "Punctuated equilibria: The tempo and mode of evolution reconsidered." *Paleobiology* 3:115–151.

Gould, S. J., and R. C. Lewontin. 1979. "The spandrels of San Marco and the panglossian paradigm." *Proc. Roy. Soc. London B* 205:581–598.

Gould, S. J., and E. Vrba. 1982. "Exaptation—a missing term in the science of form." *Paleobiology* 8:4–15.

Iwafuchi-Doi, M., and K. S. Zaret. 2016. "Cell fate control by pioneer transcription factors." *Development* 143 (11):1833–1837. https://doi.org/10.1242/dev.133900

Just, E. E. 1939. *The biology of the cell surface*. Philadelphia: P. Blakiston's Son & Co.

Kant, I. 1790/1966. *Critique of judgment*. J. H. Bernard, trans. New York: Hafner.

Kondo, S., and T. Miura. 2010. "Reaction-diffusion model as a framework for understanding biological pattern formation." *Science* 329 (5999):1616–1620. https://doi.org/10.1126/science.1179047

Krieg, M., Y. Arboleda-Estudillo, P. H. Puech, J. Kafer, F. Graner, D. J. Muller, and C. P. Heisenberg. 2008. "Tensile forces govern germ-layer organization in zebrafish." *Nat Cell Biol* 10 (4):429–436. https://doi.org /10.1038/ncb1705

Kriegman, S., D. Blackiston, M. Levin, and J. Bongard. 2021. "Kinematic self-replication in reconfigurable organisms." *Proc Natl Acad Sci USA* 118 (49). https://doi.org/10.1073/pnas.2112672118

Laarman, M. D., G. Geeven, P. Barnett, Bank Netherlands Brain, G. J. E. Rinkel, W. de Laat, Y. M. Ruigrok, and J. Bakkers. 2019. "Chromatin conformation links putative enhancers in intracranial aneurysm-associated regions to potential candidate genes." *J Am Heart Assoc* 8 (9):e011201. https://doi.org/10.1161/JAHA.118.011201

Laland, K. N., T. Uller, M. W. Feldman, K. Sterelny, G. B. Muller, A. Moczek, E. Jablonka, and J. Odling-Smee. 2015. "The extended evolutionary synthesis: Its structure, assumptions and predictions." *Phil. Trans. Roy. Soc. B: Biological Sciences* 282 (1813). https://doi.org/10.1098/rspb.2015.1019

Levin, M. 2012. "Morphogenetic fields in embryogenesis, regeneration, and cancer: Non-local control of complex patterning." *Biosystems* 109 (3):243–261. https://doi.org/10.1016/j.biosystems.2012.04.005

Levin, M. 2021. "Bioelectric signaling: Reprogrammable circuits underlying embryogenesis, regeneration, and cancer." *Cell* 184 (8):1971–1989. https://doi.org/10.1016/j.cell.2021.02.034

Levin, M., G. Pezzulo, and J. M. Finkelstein. 2017. "Endogenous bioelectric signaling networks: Exploiting voltage gradients for control of growth and form." *Annual Review of Biomedical Engineering* 19 (1):353–387. https://doi.org/10.1146/annurev-bioeng-071114-040647

Little, C. M., T. W. Chapman, and N. K. Hillier. 2020. "Plasticity is key to success of *Drosophila suzukii* (Diptera: Drosophilidae) invasion." *J Insect Sci* 20 (3). https://doi.org/10.1093/jisesa/ieaa034

Lyon, P., F. Keijzer, D. Arendt, and M. Levin. 2021. "Reframing cognition: Getting down to biological basics." *Phil Trans Roy Soc B: Biological Sciences* 376 (1820):20190750. https://doi.org/10.1098/rstb.2019.0750

Mamashli, F., S. Khan, M. Hämäläinen, M. Jas, T. Raij, S. M. Stufflebeam, A. Nummenmaa, and J. Ahveninen. 2021. "Synchronization patterns reveal neuronal coding of working memory content." *Cell Rep* 36 (8):109566. https://doi.org/10.1016/j.celrep.2021.109566

Moreno, A., and M. Mossio. 2015. "Biological autonomy: A philosophical and theoretical enquiry." In *History, Philosophy and Theory of the Life Sciences*. Dordrecht: Springer Netherlands.

Moss, L., and S. A. Newman. 2015. "The grassblade beyond Newton: The pragmatizing of Kant for evolutionary-developmental biology." *Lebenswelt* 7:94–111.

Müller, G. B. 1990. "Developmental mechanisms at the origin of morphological novelty: A side-effect hypothesis." In *Evolutionary innovations*, edited by M. H. Nitecki, 99–130. Chicago: University of Chicago Press.

Müller, G. B. 2021. "Evo-Devo's contributions to the extended evolutionary synthesis." In *Evolutionary developmental biology: A reference guide*, edited by Laura Nuño de la Rosa and Gerd B. Müller, 1127–1138. Cham: Springer.

Müller, G. B., and S. A. Newman. 1999. "Generation, integration, autonomy: Three steps in the evolution of homology." In *Homology (Novartis Foundation Symposium 222)*, edited by G. K. Bock and G. Cardew, 65–73. Chichester, UK: Wiley.

Murugan, N. J., D. H. Kaltman, P. H. Jin, M. Chien, R. Martinez, C. Q. Nguyen, A. Kane, R. Novak, D. E. Ingber, and M. Levin. 2021. "Mechanosensation mediates long-range spatial decision-making in an aneural organism." *Advanced Materials* 33 (34):2008161. https://doi.org/10.1002/adma.202008161

Musser, J. M., K. J. Schippers, M. Nickel, G. Mizzon, A. B. Kohn, C. Pape, P. Ronchi, N. Papadopoulos, A. J. Tarashansky, J. U. Hammel, F. Wolf, C. Liang, A. Hernandez-Plaza, C. P. Cantalapiedra, K. Achim, N. L. Schieber, L. Pan, F. Ruperti, W. R. Francis, S. Vargas, S. Kling, M. Renkert, M. Polikarpov, G. Bourenkov,

R. Feuda, I. Gaspar, P. Burkhardt, B. Wang, P. Bork, M. Beck, T. R. Schneider, A. Kreshuk, G. Worheide, J. Huerta-Cepas, Y. Schwab, L. L. Moroz, and D. Arendt. 2021. "Profiling cellular diversity in sponges informs animal cell type and nervous system evolution." *Science* 374 (6568):717–723. https://doi.org/10.1126/science.abj2949

Nagueh, S. F. 2021. "Heart failure with preserved ejection fraction: Insights into diagnosis and pathophysiology." *Cardiovasc Res* 117 (4):999–1014. https://doi.org/10.1093/cvr/cvaa228

Nakashima, K., L. Yamada, Y. Satou, J. Azuma, and N. Satoh. 2004. "The evolutionary origin of animal cellulose synthase." *Dev Genes Evol* 214 (2):81–88. https://doi.org/10.1007/s00427-003-0379-8

Neander, K. 1991. "Functions as selected effects: The conceptual analyst's defense." *Philosophy of Science* 58 (2):168–184.

Neme, R., and D. Tautz. 2013. "Phylogenetic patterns of emergence of new genes support a model of frequent de novo evolution." *BMC Genomics* 14:117. https://doi.org/10.1186/1471-2164-14-117

Newman, S. A. 2006. "The developmental-genetic toolkit and the molecular homology-analogy paradox." *Biological Theory* 1 (1):12–16.

Newman, S. A. 2011. "Animal egg as evolutionary innovation: A solution to the 'embryonic hourglass' puzzle." *J Exp Zool B Mol Dev Evol* 316 (7):467–483. https://doi.org/10.1002/jez.b.21417

Newman, S. A. 2014. "Why are there eggs?" *Biochem Biophys Res Commun* 450 (3):1225–1230. https://doi.org/10.1016/j.bbrc.2014.03.132

Newman, S. A. 2016a. "'Biogeneric' developmental processes: drivers of major transitions in animal evolution." *Philos Trans R Soc Lond B Biol Sci* 371 (1701). https://doi.org/10.1098/rstb.2015.0443

Newman, S. A. 2016b. "Origination, variation, and conservation of animal body plan development." *Reviews in Cell Biology and Molecular Medicine* 2 (3):130–162. https://onlinelibrary.wiley.com/doi/abs/10.1002/3527600906.mcb.200400164.pub2

Newman, S. A. 2019a. "Inherency of form and function in animal development and evolution." *Front Physiol* 10:702. https://doi.org/10.3389/fphys.2019.00702

Newman, S. A. 2019b. "Inherency and homomorphy in the evolution of development." *Curr Opin Genet Dev* 57:1–8. https://doi.org/10.1016/j.gde.2019.05.006

Newman, S. A. 2020a. "The origins and evolution of animal identity." In *Biological identity: perspectives from metaphysics and the philosophy of biology*, edited by Anne Sophie Meincke and John Dupré, 128–148. New York: Routledge.

Newman, S. A. 2020b. "Cell differentiation: What have we learned in 50 years?" *J Theor Biol* 485:110031. https://doi.org/10.1016/j.jtbi.2019.110031

Newman, S. A. 2022a. "Self-organization in embryonic development: myth and reality." In *Self-Organization as a New Paradigm in Evolutionary Biology: From Theory to Applied Cases in the Tree of Life*, edited by A. D. Malassé, 195–222. Cham: Springer.

Newman, S. A. 2022b. "Inherency and agency in the origin and evolution of biological functions." *Biological Journal of the Linnean Society*, blac109. https://doi.org/10.1093/biolinnean/blac109

Newman, S. A., and R. Bhat. 2009. "Dynamical patterning modules: A 'pattern language' for development and evolution of multicellular form." *Int J Dev Biol* 53 (5–6):693–705. https://doi.org/10.1387/ijdb.072481sn

Newman, S. A., and W. D. Comper. 1990. "'Generic' physical mechanisms of morphogenesis and pattern formation." *Development* 110 (1):1–18.

Newman, S. A., and H. L. Frisch. 1979. "Dynamics of skeletal pattern formation in developing chick limb." *Science* 205 (4407):662–668.

Newman, S. A., and G. B. Müller. 2000. "Epigenetic mechanisms of character origination." *J. Exp. Zool. B (Mol Dev Evol)* 288 (4):304–317.

Obier, N., and C. Bonifer. 2016. "Chromatin programming by developmentally regulated transcription factors: Lessons from the study of haematopoietic stem cell specification and differentiation." *FEBS Lett* 590 (22):4105–4115. https://doi.org/10.1002/1873-3468.12343

Odling-Smee, F. J., K. N. Laland, and M. W. Feldman. 2003. *Niche construction: The neglected process in evolution* (Monographs in Population Biology; 37). Princeton, NJ: Princeton University Press.

Özbudak, E. M., and J. Lewis. 2008. "Notch signalling synchronizes the zebrafish segmentation clock but is not needed to create somite boundaries." *PLoS Genet* 4 (2):e15.

Palva, J. M., S. Monto, S. Kulashekhar, and S. Palva. 2010. "Neuronal synchrony reveals working memory networks and predicts individual memory capacity." *Proceedings of the National Academy of Sciences* 107 (16):7580–7585. https://doi.org/10.1073/pnas.0913113107

Paré, A., M. Kim, M. T. Juarez, S. Brody, and W. McGinnis. 2012. "The functions of grainy head-like proteins in animals and fungi and the evolution of apical extracellular barriers." *PLoS One* 7 (5):e36254. https://doi.org/10.1371/journal.pone.0036254

Prohaska, S. J., P. F. Stadler, and D. C. Krakauer. 2010. "Innovation in gene regulation: The case of chromatin computation." *J Theor Biol* 265 (1):27–44. https://doi.org/10.1016/j.jtbi.2010.03.011

Ros-Rocher, N., A. Perez-Posada, M. M. Leger, and I. Ruiz-Trillo. 2021. "The origin of animals: An ancestral reconstruction of the unicellular-to-multicellular transition." *Open Biol* 11 (2):200359. https://doi.org/10.1098/rsob.200359

Salazar-Ciudad, I., J. Jernvall, and S. A. Newman. 2003. "Mechanisms of pattern formation in development and evolution." *Development* 130 (10):2027–2037.

Sarkar, S. 1998. *Genetics and reductionism* (Cambridge Studies in Philosophy and Biology). Cambridge: Cambridge University Press.

Schierwater, B., H. J. Osigus, T. Bergmann, N. W. Blackstone, H. Hadrys, J. Hauslage, P. O. Humbert, K. Kamm, M. Kvansakul, K. Wysocki, and R. DeSalle. 2021. "The enigmatic Placozoa part 1: Exploring evolutionary controversies and poor ecological knowledge." *Bioessays* 43 (10):e2100080. https://doi.org/10.1002/bies.202100080

Schmitz, J. F., K. K. Ullrich, and E. Bornberg-Bauer. 2018. "Incipient de novo genes can evolve from frozen accidents that escaped rapid transcript turnover." *Nat Ecol Evol* 2 (10):1626–1632. https://doi.org/10.1038/s41559-018-0639-7

Sebé-Pedrós, A., C. Ballare, H. Parra-Acero, C. Chiva, J. J. Tena, E. Sabido, J. L. Gomez-Skarmeta, L. Di Croce, and I. Ruiz-Trillo. 2016. "The dynamic regulatory genome of *Capsaspora* and the origin of animal multicellularity." *Cell* 165 (5):1224–1237. https://doi.org/10.1016/j.cell.2016.03.034

Sebé-Pedrós, A., B. M. Degnan, and I. Ruiz-Trillo. 2017. "The origin of Metazoa: A unicellular perspective." *Nat Rev Genet* 18 (8):498–512. https://doi.org/10.1038/nrg.2017.21

SenGupta, S., C. A. Parent, and J. E. Bear. 2021. "The principles of directed cell migration." *Nature Reviews Molecular Cell Biology* 22 (8):529–547. https://doi.org/10.1038/s41580-021-00366-6

Shubin, N., C. Tabin, and S. Carroll. 2009. "Deep homology and the origins of evolutionary novelty." *Nature* 457 (7231):818–23. https://doi.org/10.1038/nature07891

Singh, A. J., S. A. Ramsey, T. M. Filtz, and C. Kioussi. 2018. "Differential gene regulatory networks in development and disease." *Cell Mol Life Sci* 75 (6):1013–1025. https://doi.org/10.1007/s00018-017-2679-6

Smith, C. L., F. Varoqueaux, M. Kittelmann, R. N. Azzam, B. Cooper, C. A. Winters, M. Eitel, D. Fasshauer, and T. S. Reese. 2014. "Novel cell types, neurosecretory cells, and body plan of the early-diverging metazoan *Trichoplax adhaerens*." *Curr Biol* 24 (14):1565–1572. https://doi.org/10.1016/j.cub.2014.05.046

Sogabe, S., W. L. Hatleberg, K. M. Kocot, T. E. Say, D. Stoupin, K. E. Roper, S. L. Fernandez-Valverde, S. M. Degnan, and B. M. Degnan. 2019. "Pluripotency and the origin of animal multicellularity." *Nature* 570 (7762):519–522. https://doi.org/10.1038/s41586-019-1290-4

Sperling, E. A., and R. G. Stockey. 2018. "The temporal and environmental context of early animal evolution: Considering all the ingredients of an 'explosion.'" *Integr Comp Biol* 58 (4):605–622. https://doi.org/10.1093/icb/icy088

Steinberg, M. S., and T. J. Poole. 1982. "Liquid behavior of embryonic tissues." In *Cell Behavior*, edited by R. Bellairs and A. S. G. Curtis, 583–607. Cambridge: Cambridge University Press.

Stern, C. D., and A. M. Piatkowska. 2015. "Multiple roles of timing in somite formation." *Semin Cell Dev Biol* 42:134–139. https://doi.org/10.1016/j.semcdb.2015.06.002

Stodola, T. J., P. Liu, Y. Liu, A. K. Vallejos, A. M. Geurts, A. S. Greene, and M. Liang. 2018. "Genome-wide map of proximity linkage to renin proximal promoter in rat." *Physiol Genomics* 50 (5):323–331. https://doi.org/10.1152/physiolgenomics.00132.2017

Strassmann, J. E., and D. C. Queller. 2011. "Evolution of cooperation and control of cheating in a social microbe." *Proc Natl Acad Sci USA* 108, Suppl 2:10855–10862. https://doi.org/10.1073/pnas.1102451108

Strogatz, S. H. 2003. *Sync: The emerging science of spontaneous order* (1st ed.). New York: Theia.

Sullivan, K. G., M. Emmons-Bell, and M. Levin. 2016. "Physiological inputs regulate species-specific anatomy during embryogenesis and regeneration." *Communicative & Integrative Biology* 9 (4):e1192733. https://doi.org/10.1080/19420889.2016.1192733

Tanizawa, H., O. Iwasaki, A. Tanaka, J. R. Capizzi, P. Wickramasinghe, M. Lee, Z. Fu, and K. Noma. 2010. "Mapping of long-range associations throughout the fission yeast genome reveals global genome organization linked to transcriptional regulation." *Nucleic Acids Res* 38 (22):8164–8177. https://doi.org/10.1093/nar/gkq955

Thomas, P. D. 2017. "The gene ontology and the meaning of biological function." In *The gene ontology handbook*, edited by Christophe Dessimoz and Nives Škunca, 15–24. New York: Springer New York.

Thompson, D'Arcy W. 1942. *On growth and form* (2nd ed.). Cambridge: Cambridge University Press.

True, J. R., and E. S. Haag. 2001. "Developmental system drift and flexibility in evolutionary trajectories." *Evol Dev* 3 (2):109–119.

Turing, A. M. 1952. "The chemical basis of morphogenesis." *Phil Trans Roy Soc London B* 237:37–72.

Tweedy, L., P. A. Thomason, P. I. Paschke, K. Martin, L. M. Machesky, M. Zagnoni, and R. H. Insall. 2020. "Seeing around corners: Cells solve mazes and respond at a distance using attractant breakdown." *Science* 369 (6507). https://doi.org/10.1126/science.aay9792

Vandenberg, L. N., R. D. Morrie, and D. S. Adams. 2011. "V-ATPase-dependent ectodermal voltage and ph regionalization are required for craniofacial morphogenesis." *Developmental Dynamics* 240 (8):1889–1904. https://doi.org/10.1002/dvdy.22685

Verger, A., D. Monte, and V. Villeret. 2019. "Twenty years of Mediator complex structural studies." *Biochem Soc Trans* 47 (1):399–410. https://doi.org/10.1042/BST20180608

Wake, D. B. 1991. "Homoplasy: The result of natural selection or evidence of design limitations?" *American Naturalist* 138:543–567.

Wan, K. Y., and G. Jékely. 2021. "Origins of eukaryotic excitability." *Phil Trans Roy Soc B: Biological Sciences* 376 (1820):20190758. https://doi.org/10.1098/rstb.2019.0758

Webster, G., and B. C. Goodwin. 1996. *Form and transformation: Generative and relational principles in biology.* Cambridge: Cambridge University Press.

Werner, M. S., B. Sieriebriennikov, N. Prabh, T. Loschko, C. Lanz, and R. J. Sommer. 2018. "Young genes have distinct gene structure, epigenetic profiles, and transcriptional regulation." *Genome Res* 28 (11):1675–1687. https://doi.org/10.1101/gr.234872.118

West-Eberhard, M. J. 2003. *Developmental plasticity and evolution.* Oxford: Oxford University Press.

Whitworth, D. E. 2008. *Myxobacteria: Multicellularity and differentiation.* Washington, DC: ASM Press.

Wilson, Robert A. 2005. *Genes and the agents of life: The individual in the fragile sciences, biology.* New York: Cambridge University Press.

Wouters, A. 2005. "The function debate in philosophy." *Acta Biotheoretica* 53 (2):123–151. https://doi.org/10.1007/s10441-005-5353-6

Wu, Y., A. D. Kaiser, Y. Jiang, and M. S. Alber. 2009. "Periodic reversal of direction allows Myxobacteria to swarm." *Proc Natl Acad Sci USA* 106 (4):1222–1227. https://doi.org/10.1073/pnas.0811662106

Yang, C. Y., M. Bialecka-Fornal, C. Weatherwax, J. W. Larkin, A. Prindle, J. Liu, J. Garcia-Ojalvo, and G. M. Suel. 2020. "Encoding membrane-potential-based memory within a microbial community." *Cell Syst* 10 (5):417–423 e3. https://doi.org/10.1016/j.cels.2020.04.002

Zheng, Y., R. Jia, Y. Qian, Y. Ye, and C. Liu. 2015. "Correlation between electric potential and peristaltic behavior in *Physarum polycephalum.*" *Biosystems* 132–133:13–19. https://doi.org/10.1016/j.biosystems.2015.04.005

12 How Purposive Agency Became Banned from Evolutionary Biology

Denis Noble and Raymond Noble

Overview

Purposive "agency" featured strongly in the nineteenth-century ideas on evolution of Charles Darwin. It was removed by the neo-Darwinians, Alfred Russel Wallace and August Weismann. The modern synthesis of evolutionary biology, as it was originally formulated by Julian Huxley in 1942, incorporated the ideas of Wallace and Weismann, but nevertheless included a wide range of evolutionary processes. The scope of the modern synthesis became vastly reduced during its simplification in the 1960s and 1970s. In this paper we identify the main historical developments leading to these outcomes. We also summarize the utility and predictive ability as reasons for which purposive agency should to be restored in the study of living organisms.

12.1 Introduction

This chapter documents the historical process by which purposive agency was removed from theories of biology. We then summarize the conceptual and practical reasons for which it ought to be restored.

12.2 The Seventeenth Century: Newton, Galileo, and the Opposing Philosophical Views of Descartes and Spinoza

The seventeenth century saw the triumph of mechanism. Isaac Newton formulated his equations of motion, Galileo observed the moons of Jupiter and concluded that the Earth circulates the sun, just as the moons circulate Jupiter. The physical world was yielding to the mechanics of inert objects. The question that exercised the greatest thinkers of the period was whether living organisms could be explained in a similar way. Could the development and movements of animals also be accounted for with purely physical principles?

That question became important because people were impressed by the behavior of mechanical toys installed in the gardens of the rich, and even in puppet theaters (Chapuis, 1952), the movements of which were powered by hydraulic pressure communicated through tubes so that they could be made to move their limbs and perform movements and make sounds similar to those of living creatures. The philosopher and mathematician René Descartes was so impressed that he based his mechanical explanation of the movements of muscles on the same hydraulic principle. He wrote:

Now according as these spirits enter thus into the concavities of the brain, they pass thence into the pores of the substance, and from these pores into the nerves; where according as they enter or even only tend to enter more or less into one rather than the others, they have the power to change the shape of the muscles into which these nerves are inserted, and by this means to make the limbs move. Just as you can have seen in the grottoes and foundations of our kings, that the same single force by which the water moves, coming out from its source, is enough to move diverse machines and even to make them play instruments or to pronounce words according to the different disposition of the tubes which conduct it. (Descartes, *Oeuvres* (ed. Cousin), cited by Needham, 1971: 14–15)

In this text, "spirits" means a fluid flowing through various parts of the body. William Croone clarified this idea in his 1664 *De ratione motus musculorum*:

As often therefore as I say the Animal Spirits, I mean the most subtle, active and highly volatile liquor of the nerves, in the same way as we speak of the spirit of wine or salt. . . . those spirits in the nerve called Animal, are nothing else than a rectified and enriched juice of this kind. (Trans. Wilson, cited by Needham, 1971: p. 16)

Descartes' and Croone's account of muscle contraction was disproved in the same century by the experiments of Jan Swammerdam in 1663 and Francis Glisson in 1667, which showed that a muscle contracting in a fluid bath did not raise the level of the fluid in the bath, and so did not expand. Details of this work can be found in Needham (1971: 18). Croone later calculated ways in which muscle contraction could occur with no change in volume (Needham, 1971: 20).

Descartes went further with his mathematical and mechanical vision of animals. He imagined that all the properties necessary for the development of the embryo leading to the infant could be used mathematically to predict its development. He wrote:

If one had a proper knowledge of all the parts of the semen of some species of animal in particular, for example of man, one might be able to deduce the whole form and configuration of each of its members from this alone, by means of entirely mathematical and certain arguments, the complete figure and the conformation of its members. (Descartes, "Treatise on the fetus," in *Treatise on man* [1665])

This text anticipates the nineteenth-century Weismann barrier by two centuries and the twentieth-century central dogma of molecular biology by three centuries.

The mechanical fluid idea led Descartes to regard animals as pure automata, but he famously thought that humans are exceptions through possessing an immaterial soul which could interact with the body. This idea was the beginning of treating the human species as an exception possessing purpose, capable therefore of actively influencing their behavior and development.

12.3 The Eighteenth-Century Discovery of Animal Electricity

What was missing, of course, from Descartes' and others' seventeenth-century ideas was the replacement of the fluids idea with electrical changes in excitable tissues like nerves and muscles. It was the achievement of Galvani in the eighteenth century (1780) to show that nerves and muscles could be excited by electric currents. The flow of electricity has negligible volume effects. So, the fluids of Descartes and Croone became the movement of charged atoms (ions) through minute holes (channels in the membranes of cells and their organelles), eventually leading to the nineteenth- and twentieth-century developments of electrophysiology, culminating in our present understanding of a large range of ionic channels formed by proteins using the DNA templates we now call genes, but functioning in membranes for which there are no DNA templates.

12.4 Nineteenth-Century Developments on Active and Passive Darwinism

Charles Darwin fully espoused the ability of organisms to actively influence the direction of their evolution. He did so in at least two ways.

In the first chapter of *The Origin of Species* (Darwin, 1859) he compared two forms of selection: artificial and natural. His model for artificial selection was the consciously active selection by plant and animal breeders to develop new varieties. He introduced his idea of natural selection as a metaphorical contrast with human breeders' selection. Later, in his 1871 book *The Descent of Man, and Selection in Relation to Sex*, he extended artificial selection to animals that "consciously exert their mental and bodily power": "Just as man can give beauty [in his breeding of animals]. . . . so it appears that female birds in a state of nature, have by a long process of selection of the more attractive males, added to their beauty." If Darwin was right, then he thought that purposive action must have evolved many hundreds of millions of years ago. We will draw the remarkable contrast with the modern synthesis view later.

In his 1868 book *The Variation of Animals and Plants Under Domestication*, he formulated his theory of gemmules to explain how changes in the soma during an organism's lifetime might be transmitted to the germ cells, thus permitting directed evolution through the inheritance of acquired characteristics. Those acquired characteristics could include those due to life-time/style choices. Darwin accepted Lamarck's use-disuse idea.

The co-discoverer of natural selection, Alfred Russel Wallace, disagreed with Darwin on sexual selection and on the agency (ability to choose) of organisms. In his 1878 book *Tropical Nature and Other Essays*, he proposed alternative explanations for a number of cases Darwin had attributed to sexual selection (Slotten, 2004:353–356). He revisited the issue in his 1889 book *Darwinism* and in 1890, he wrote a critical review in *Nature* (Wallace, 1890) specifically attacking claims on the "aesthetic preferences of the insect world."

This was the first move to exclude Darwin's explanation of purposive behavior in animals from the theory of evolution.

The next move was made by August Weismann, who introduced his idea of the Weismann barrier in 1890 to claim that communication between the soma and the germ cells must be impossible. The most complete account of his views was published in 1892 (Weismann, 1892). Both of Weismann's works appeared after Darwin's death in 1882, so we cannot know for certain how Darwin would have reacted. However, it is unlikely that he would have agreed with Weismann, as he had great hopes that his pangenesis idea could be proved through physiological experimentation with George Romanes, as described in Noble (2022).

Wallace and Weismann were therefore jointly responsible for the development of what became known as neo-Darwinism. Romanes was strongly critical of the idea; his physiological training led him to be suspicious of germ-line theory and the "all-sufficiency" (Weismann's term: *allmacht*) of natural selection.

By denying agency to organisms, Wallace was faced with the same dilemma as Descartes before him. Just as Descartes did, Wallace exempted humans from the "all-sufficiency" of natural selection. He even became a spiritualist (Wallace, 1875/2009). Far from making things simpler and clearer, the nineteenth-century founder of neo-Darwinism ended up with an even worse philosophical muddle than Descartes.

Meanwhile, Weismann anticipated the twentieth-century hardening of the modern synthesis through his insistence that the "all-sufficiency" of natural selection was a *necessary* requirement, not open to experimental proof (see his quotation "not because we can demonstrate it" in section 12.12 of this chapter).

It is hard to credit how deeply these two nineteenth-century founders of neo-Darwinism misunderstood basic philosophy. It is incoherent for someone to deny agency while, as a human being, exercising precisely that agency in an attempt to convince others of the correctness of his ideas. To deny agency to animals then automatically leads to the idea of exceptionalism for human beings. There is no wonder that Darwin and Romanes opposed Wallace, and in the case of Romanes, also opposed Weismann. The very nature of physiology as an integrative functional discipline was at stake, since the ultimate aim of the discipline in its contribution to evolutionary biology must be to account for *agency*, the active ability of organisms to choose and influence their development and evolution. There are physiological processes that can explain how organisms are able to do this (Noble & Noble, 2020).

12.5   Twentieth-Century Developments on Active and Passive Darwinism

The further developments to remove Darwin's concepts of active evolution and replace them with the simpler idea that all evolution could be attributed to passive natural selection occurred with the formulation of the modern synthesis, through the synthesis between Mendelism and neo-Darwinism during the first half of the twentieth century. This development culminated in Julian Huxley's seminal book *Evolution: The Modern Synthesis* (Huxley, 1942), which was reprinted in 2010 with a foreword by Massimo Pigliucci and Gerd Müller. They are also authors and editors of the book titled *Evolution: The Extended Synthesis* (Pigliucci & Müller, 2010), which documents many of the developments outside the modern synthesis up to that time.

Julian Huxley, like his grandfather, Thomas Henry Huxley, was a popularizer of evolution. But his 1942 work is much more than that. It is an impressive work of scholarship. It was also written well before what Stephen J. Gould described as the hardening of the modern synthesis (Gould, 2002: 541). It is therefore important to document both the original formulation of the modern synthesis and its later conversion into a very dogmatic view of evolution.

It is refreshing to read Huxley's book. It gives a view of the modern synthesis theory of evolutionary biology as containing many ideas that seem to have either disappeared or become downplayed during the later hardening (Shapiro & Noble, 2021).

Before we document Huxley's more nuanced position, it is important to explain why he was nevertheless one of the founders of the modern synthesis. This section of our chapter is not an attempt to recruit Julian Huxley to our viewpoint!

12.6 Three Main Reasons Why Julian Huxley Was the Founding Exponent of the Modern Synthesis

12.6.1 The Weismann Barrier and the Exclusion of Lamarckism

Huxley notes (1942: 17):

Darwin . . . was always inclined to allow some weight to Lamarckian principles, according to which the effects of use and disuse of environmental influences were supposed to be in some degree inherited.

He then fully accepts Weismann's idea with the categorical statement that "modifications. . . . are not inheritable." No experimental basis is given for this statement. On the contrary, Huxley notes, in a very significant passage:

the distinction between soma and germplasm is *not always so sharp as Weismann supposed*, and although the principle of Baldwin and Lloyd Morgan, usually called organic selection, shows how Lamarckism may be simulated by the later replacement of adaptive modifications by adaptive mutations, Weismann's conceptions resulted in a great clarification of the position. (Huxley, 1942:17–18 of the 2010 reprint; emphasis ours)

Uncharacteristic of his usual careful logical analysis, Huxley sees no conflict here. It is hard to understand how "great clarification" could be obtained without any relevant experimental information to justify it. What in fact happened was a gratuitous great simplification.

What Huxley calls "organic selection" is effectively Waddington's (1957) idea of genetic assimilation. This was a missed opportunity. If Waddington had been brought into the modern synthesis founding group, this muddle could have been tidied up. Instead, he was deliberately excluded (Peterson, 2011). One of us (Noble, 2016) has explained elsewhere why genetic assimilation should be viewed as a form of the inheritance of acquired characteristics. The relevant passage is this:

Orthodox Neo-Darwinists dismissed Waddington's findings as merely an example of the evolution of phenotype plasticity. That is what you will find in many of the biology textbooks. I think that is to misrepresent what Waddington showed. Of course, plasticity can evolve, and that itself could be by a Neo-Darwinist or Darwinist or any other mechanism. But Waddington was not simply showing the evolution of plasticity *in general*; he was showing how it could be exploited to enable a *particular* acquired characteristic *in response to an environmental change* to be inherited and become assimilated into the genome. To repeat: the characteristic was acquired as a result of an environmental change, and it was inherited. That is the definition of Lamarckian inheritance that I and many others now use. But the designation doesn't really matter. What does is that he discovered a protocol by which an acquired characteristic could be inherited. Evolution could have used the same kind of protocol. (Noble, 2016:217–218)

In the 1950s, Waddington clearly distanced himself from what he called neo-Darwinism, but did not at that time describe himself as a Lamarckian. He probably wished to distance himself from Lysenko, who was opposing genetics in the Soviet Union. Waddington was not opposing genetics, but rather explaining how genes are controlled epigenetically by the organism. Indeed, he invented the term *epigenetics* precisely to account for what he called "the strategy of the genes" (Waddington, 1957).

Huxley's more specific dismissal of Waddington is found in the introduction to the second edition of his book, published in 1963:

Meanwhile in Britain, Waddington (1957, 1960) has made a notable contribution to evolutionary theory by his discovery that Lamarckian inheritance may be simulated by a purely neo-Darwinian mechanism. This he called genetic assimilation. It operates through the natural selection of genes which dispose the developing organism to become modified in reaction to some environmental stimulus. (page 580 of the 2010 reprint)

This is incorrect. Waddington specifically said in his 1957 book that he was a Darwinian, *not a neo-Darwinian*. The process he described in 1957 certainly does not operate through natural selection alone; Waddington himself was doing the selection in each of the generations of his experiment, thus mimicking what a change in environment might do to guide evolution toward an acquired characteristic. That was the whole point of his experiment. It completely misrepresents Waddington to classify him as a neo-Darwinian. Waddington himself provides the best and most direct evidence in his contribution to the 1971–1973 Gifford Lectures in Edinburgh, in which he wrote:

Say we have a population of animals which has to meet some new challenge offered to it by an altered environment in a way which makes them better able to deal with the challenge—they show a capacity for adaptive modification. They will, therefore, be favoured by natural selection. After this selection has gone on for some considerable number of generations the new pathways of development will have gradually acquired a more pronounced chreodic character which can itself be difficult to modify. In fact should the environment now revert to what it was before the whole process started, the organism may well go on developing in the way in which it adapted to the changed environment. This process, which I have called genetic assimilation, gives exactly the same end result as the theory proposed by Lamarck, at one time espoused by Darwin, but rejected by modern biology, of the direct inheritance of acquired characteristics. (Kenny et al., 1972)

Clearly, the hesitation Waddington understandably had at being described as a Lamarckian in the 1950s had disappeared completely 20 years later, when it was no longer likely that an espousal of Lamarck could be confused with agreement with Lysenko's dismissal of genetics.

12.6.2   Agreement with Wallace in Dismissing Darwin on Sexual Selection

The second neo-Darwinian aspect of Huxley's work is that he was quite clear in dismissing sexual selection as interpreted by Darwin.

Under the impetus of Darwin's great work, *The Descent of Man*, what may be called the orthodox Darwinian view came to be generally held, namely that all bright colours of higher animals which are restricted to the male sex, are, in the absence of definite evidence to the contrary, to be interpreted as owing their origin to sexual selection; the same was assumed for the songs of birds. . . . [But] the Darwinian presumption in its sweeping form was erroneous. Only when the bright colour or other performance is solely or mainly used in display before the female can it hold. If so, however, the presumption is very strong that its origin is to be sought in sexual selection in the modern sense, which differs considerably from that which was originally employed by Darwin, and the burden of proof is on the other side. (Huxley, 1942: 35 of 2010 reprint)

Note that Huxley says that, both ways, it is a *presumption* whether one uses the Darwinian or what he calls the modern sense. So, what may have convinced the modern synthesis founders that Darwin was wrong? We speculate that they dismissed the idea of purposive action, whether conscious or unconscious, in animals other than humans. The evidence for that conclusion is found on page 572 of the 2010 reprint of Huxley's book, where he writes:

Conscious and conceptual thought is the latest step in life's progress. It is, in the perspective of evolution, a very recent one, having been taken perhaps only one or two and certainly less than ten million years ago.

It seems that Huxley thought that conscious (and, perhaps, also purposive?) action may be restricted to what were called the higher animals, including humans. As physiologists, we find this view strange. Consciousness and the neural structures that support it and the capacity to make intentional choices are great consumers of energy. Their evolution required payback in terms of selective advantages, which the nervous system has given us. Consciousness, like life itself, must be functional to justify its cost. We therefore prefer the thesis of Ginsburg and Jablonka (2019) in their magisterial *Evolution of the Sensitive Soul: Learning and the Origins of Consciousness*. Conscious processes evolved because they serve a purpose. As that book and one of their prior articles (Ginsburg & Jablonka, 2010) argue, this development may even have been one of the drivers of the Cambrian explosion. On this interpretation, consciousness may have evolved vastly more than 2–10 million years ago. It would be more like 300–500 million years ago.

12.6.3 Support for Eugenics

It is also sad to see Huxley writing in support of eugenics even as late as 1963, two decades after the Holocaust: "eugenic improvement will become an increasingly important goal of evolving man" (Introduction to the second edition of his book, page 587).

No further comment is needed.

12.7 Huxley's Extensively Nuanced View of the Modern Synthesis

Having clarified the main reasons why Huxley supported the modern synthesis, became a major spokesperson for it, and wrote its founding text, we now detail the extent to which his position was nevertheless far more nuanced than the later hardening.

12.7.1 Permeability of the So-Called Weismann Barrier

We have already noted that Huxley was not completely wedded to the Weismann barrier, as he wrote that "the distinction between soma and germplasm is not always so sharp as Weismann supposed."[1] He disagreed with Weismann on more than the barrier idea:

Contrary to the views of the Weismann School, selection alone has been shown to be incapable of extending the upper limit of variation. . . . [M]utation alone has been shown to be incapable of producing directional change. . . . the third process of recombination is almost equally essential, not only for conferring plasticity on the species and allowing for a sufficient speed of evolutionary change, but also for *adjusting the effects of mutations to the needs of the organism.* (Huxley, 1942, p. 29 of 2010 reprint; emphasis ours)

This quotation of course refers to recombination in sexual reproduction, but it contains the seeds of a realization that other recombination processes could also ensure that evolution proceeds much faster and in a directed sense to meet "the needs of the organism."

12.7.2   The Speed of Evolution

The reference to "allowing for a sufficient speed of evolutionary change" is further supported on a later page:

The main lines of evolution in the more abundant forms of sea urchins and horses may depend upon gradual change: but this is no reason for assuming that this holds for all organisms. And as a matter of fact . . . abrupt changes of large extent do play a part in certain kinds of evolution in certain kinds of plants." (Huxley, 1942: 3 in 2010 reprint)

One is tempted to think that Huxley, like Gould and Eldredge (1977), was a saltationist, and was even tempted to classify Darwin also as a saltationist, based on this quote from *The Origin of Species*:

Species of different genera and classes have not changed at the same rate, or in the same degree. (Darwin, 1859:340)

Huxley was referring to the speed with which hybridization may trigger speciation in plants. We now know that such processes occur in animals too; witness the rapid speciation of finches on the Galapagos Islands via hybridization (Bell, 2015). On the next page (1942: 39) Huxley refers to "the impossibility of ascribing all kinds of evolutionary change to a single mechanism."

And again, on page 41:

We have already mentioned that species *may be formed abruptly*, and other large variations are known to occur and to serve, in some cases, as building blocks of evolutionary change.

And yet further on pages 45–46:

It would seem clear that *we cannot expect to find a single cause of evolution*: rather we must look for several agencies which alone or in combination will account for the very various processes lumped together under that comprehensive term. . . . [D]ifferent groups may be expected to show different kinds of evolution . . . Just as there is no one method of the origin of species, so there is no one type of variation (emphasis ours).

On page 147, discussing the influence of human cultivation, Huxley refers to "a degree of species-hybridization which must be unprecedented in evolutionary history."

Huxley's (1942) concept of the modern synthesis was clearly very different from the hardened version.

12.8   The Role of Mutation Rates

Huxley also has the idea of variable mutation rates and that rates may change in response to the environment:

Zuitin (1941: 137) in *Drosophila* finds that the mutation rate is increased by sudden environmental changes. (Huxley, 1942: 55 in 2010 reprint)

This is remarkable. We are almost in the field of the control of hypermutation—except that Huxley always denies that any such processes can be directed. He even refers to what must be the concept of hypermutation:

Either mutation rates *many times higher* than any as yet detected must sometimes be operative, or else the observed results can be better accounted for by selection. (Huxley, 1942: 56, emphasis ours)

Of course, as a neo-Darwinian, he opts for the latter. But, surely, if he had known of hypermutation rates many thousands, even hundreds of thousands, of times above normal in response to environmental stress he would have had little difficulty in accepting the former option in his sentence. He knew about smaller, but still significant, mutation rate changes in response to environmental stress:

Timakov (1941) in wild *Drosophila* has detected a gene increasing mutation rate at least 40 times, and possibly to a level higher than that induced by X-rays. (Huxley, 1942: 51 in 2010 reprint)[2]

On page 137 he even specifically states that the rate varies in response to evolutionary needs:

We may thus expect with reasonable assurance that mutation-rate also will be *in some degree adjusted to evolutionary needs*. (Huxley, 1942: 137 in 2010 reprint, emphasis ours)

So close, and yet so far!

12.9   Natural Genetic Engineering

Is it too far-fetched to see shades of what Barbara McClintock discovered in maize when Huxley wrote:

X-rays also induce the rearrangement of chromosome-sections by translocation, inversion etc. (Huxley, 1942: 50)

[T]here are *genome-mutations*, involving one or more whole sets of chromosomes. (Huxley, 1942: 87)

The rearrangement of chromosome sections is precisely McClintock's great discovery. McClintock realized, of course, that the organism must be sensing the environmental stress and reorganizing its genetic material. She described the genome as an "organ of the cell." Huxley does not seem to have had that physiological feedback process in his repertoire, but it couldn't have required much persuading if he had known what we know today. All he had to do was to add feedback control from the organism to its chromosomes and he would have had a directional mechanism in his theory. Control and constraint by higher levels of organization in living systems is precisely what function and purpose are about.

12.10   Buffering Against Genetic Change

One of the key concepts we have developed in our recent articles is that of the buffering provided by functional physiological networks which ensure that organisms are relatively shielded from genetic variations (Noble & Noble, 2017, 2018; Noble et al., 2019). Among other advantages in relation to evolutionary biology, this idea readily explains the very low association scores found in Genome Wide Association studies, leading many to the polygenic or even omnigenic hypothesis (Boyle et al., 2017).

Huxley had precisely the same idea. He wrote:

The detailed analysis of the last ten or fifteen years, however, has revealed large numbers of gene differences with extremely small effects, down almost to the limit of detectability. . . . Specific differences in *Drosophila* depend on many single genes, *often grouped in polygenic systems*. (Huxley, 1942:115 in 2010 reprint, emphasis ours)

12.11   Why Did Huxley Not See the Physiological Control He Missed?

We have shown that Huxley's interpretation of the modern synthesis was much more nuanced than when it became a hardened, dogmatic theory. Why did he not draw the obvious physiological conclusion: that many of the processes he admitted could be controlled physiologically? The reason is that if he had included feedback control, he would also have had to admit that the organism, within which such active feedback occurs, can influence its own evolution.

As we have shown, on this key point, he can only appeal to one presumption over another. Like all of us, Huxley was a man of his time. The influence of eschewing any form of teleology was simply too strong. Otherwise all he needed to do was: (a) to add feedback control from the organism to the chromosomes, which we now know exists; (b) take Darwin more seriously on sexual selection, which should never have been removed from Darwin's theories; and (c) to take Darwin seriously on his gemmule theory, which we can more easily do today after the discovery of the transmission of nucleotides and proteins to the germline from the soma via extracellular vesicles, and even as naked molecules.

12.12   Why Did the Modern Synthesis Become Hardened?

In his historical and philosophical magnum opus, *The Structure of Evolutionary Theory*, Stephen J. Gould (2002) devotes a complete chapter (7) of 80 pages to what he called the hardening of the modern synthesis. Just as we have found in analyzing the breadth of evolutionary processes revealed in Julian Huxley's 1942 book, Gould notes that "the first phase also included a vigorously pluralistic range of permissible mechanisms within the primary restriction" (2002:505).

That restriction was, of course, the elimination of Lamarckian processes and of active sexual selection from Darwin's range of theory. Gould must also have known that his and Eldredge's punctuated equilibrium ideas, first expressed as different speeds of evolution by Darwin himself in *The Origin of Species*, were also to be found in Huxley's presentation of the range of processes regarded as acceptable and part of the modern synthesis. Yet, by the time Eldredge and Gould published their work 30 years later in 1972, their theory was regarded as heretical. The date is significant, since by then the hardening was well under way.

We don't need to fully review Gould's chapter here. Instead, we will focus on the aspect of the hardening process that is most relevant to our chapter. It consists in the progressive, and unjustified, adulation of the ideas of August Weismann, praised by Ernst Mayr in his 1982 book *The Growth of Biological Thought* as "one of the great biologists of all time." Weismann's ideas, about which Huxley was quite critical, progressively became the severely restricted, acceptable theory. Those ideas completely infuse the popularizations and text-books with dogmatic insistence on (a) complete isolation of the germ line and (b) all evo-lutionary change being attributed solely to natural selection. Yet the only experimental test of Weismann's ideas on inheritance was his work on amputating the tails of mice and finding no tail-less mice even after several generations. As Weismann himself recognized, this exper-iment only disproves the inheritance of surgical or accidental mutilations. It establishes nothing relevant to testing Lamarck's theory that acquired characteristics that were *functional* could be inherited. Tail amputations are not functional. Animals need to *use* their tails. Amputation removes that use!

Furthermore, Weismann was guilty of the cardinal sin in science of denying the need for any empirical test of his theory. In his debate with Herbert Spencer, Weismann wrote that because Lamarckism is wrong, no evidence is needed one way or the other for his own theory since it is the only remaining possible theory. Therefore:

We accept it, not because we are able to demonstrate the process [of natural selection] in detail, not even because we can with more or less ease imagine it but simply *because we must, because it is the only possible explanation* that we can conceive. For there are only two a priori explanations of adaptation for the naturalist—namely the transmission of functional adaptations [i.e., Lamarckism] and natural selection; but as the first of these can be excluded, only the second remains. . . . We are thus able to prove by exclusion the reality of natural selection, and once that is done, the general objections which are based on our inability to demonstrate selection-value in individual cases, must collapse, as being of no weight. . . . It does not matter whether I am able to do so or not, or whether I could do it well or ill; once it is established that natural selection is the only principle which has to be considered, it necessarily follows that the facts can be correctly explained by natural selection. (Weismann, 1893:336–337).

To fully savor this text, recall that Weismann's criticism of Spencer was that Spencer was a philosopher, not a scientist. Yet, as we have already shown, the mistakes of Weismann and Wallace are themselves *philosophical* mistakes, independent of any experimental evidence.

How could the modern synthesis have been reduced to relying on this kind of circular logic independent of any experimental test? The important question, which Gould could not answer in 2002, is whether Weismann's barrier really exists. That is the central issue. The plain fact is that it doesn't. It is better described as a physiological boundary, con-trolled actively by the organisms to determine what may and may not cross. The boundary between the soma and the germ line is permeable to precisely the molecules Weismann wished to exclude.

As physiologists, we are used to thinking about the role of boundaries in organisms (Noble et al., 2019). Since living organisms are open systems, there are no absolute barriers. No cells, tissues, organs, or systems, or indeed the organism as a whole, can be alive if they cannot exchange matter, energy, and information with the rest of the organism and with the environment. Boundaries are therefore important *functional* entities in living organisms. The organism necessarily has to control what can pass and what cannot pass.

Also, as physiologists we know that our colleagues in reproductive biology have been observing transgenerational maternal and paternal effects for many years (Hanson & Gluck-man, 2014; Hanson & Skinner, 2016; Ben Maamar et al., 2018). We have therefore been in precisely the same position as Darwin was when he formed his theory of gemmules to explain how small particles could communicate material from the soma to the germline. The logic here is similar to that of one of the greatest physiological discoveries of all time: William Harvey's seventeenth-century demonstration of the circulation of blood in the body. After demonstrating the role of the valves in blood vessels, and the unidirectional flow, he deduced that there must be what he called anastomoses, though he could not see them. Thirty years later, capillaries were seen for the first time by Malpighi using the light microscope.

Darwin's problem was similar, but his idea has had to wait for much longer. Even the nineteenth-century light microscope was not up to proving what we now know to exist: small lipid membrane vesicles pouring out from cells of the body and being transmitted all round the body. It has taken a ten-fold increase in the resolution of the light microscope using fluorescent markers of specific RNAs, DNAs, and proteins to make that higher resolution possible. Those vesicles (also called exosomes) fit the requirements of Darwin's gemmules since they have been shown to transfer their cargo across the so-called barrier (Noble, 2019). Of course, it is not a *barrier*, it is a purposeful *boundary* which the organism controls to enable it to function and maintain itself.

May the Weismann barrier rest in peace! Barriers, as functional boundaries, turn out to be one of the most goal-directed processes in living organisms.

12.13   The Restoration of Purpose in Biology

The banning of purpose in biology is both a philosophical category mistake and an empirically testable hypothesis. As we have shown, the category mistake leads to the dualist muddles encountered by Descartes and his neo-Darwinian successors. As an empirically testable idea, it is incorrect in that it has resulted in expensive consequences for healthcare,[3] but with very poor outcomes for predicting disease states for the multifactorial diseases of aging populations. If a theory is to be judged by its scientific and practical utility, then the theory that purposive agency does not exist has failed spectacularly.

12.14   A Category Mistake

The category mistake is that of assuming that, if purpose cannot be found at a molecular (e.g., DNA or protein) level, then it cannot exist at all. The absurdity of this assumption is easily proven. It is equivalent to arguing that since the particles of ink in a printed message are not purposive, the message cannot be. Similarly, the particles of blood know nothing of the function and purpose of the heart and circulation. Functions and purposes can logically be applied only at the levels of organization to which it makes sense to ascribe them. We have illustrated this point in detail with examples from physiological research in a recent article for the *Biological Journal of the Linnean Society* (Noble & Noble, 2021).

12.15   Incorrect Hypotheses and Unfulfilled Promises

Our analysis of those examples revealed a general rule: higher-level functionality can be used to make testable predictions about events and processes at lower, including molecular, levels, but the reverse is only rarely true. We also showed that this rule is vindicated by the genome-wide association studies showing only low association scores for nearly all genes for most diseases, particularly multifactorial diseases. The predictive outcome for healthcare therefore falls far short of what was promised when the Human Genome Project was launched in the later part of the twentieth century. We also explained how quantitative integrative physiological modeling could be used to reveal where functional processes are sufficiently robust (i.e., relatively independent of particular genes). Those functions can resist the effects of gene-knockouts or the equivalent block of protein function by pharmaceutical agents, thus producing negligible associations between DNA sequences and disease states.

The prediction that genome sequencing would lead to "previously unimaginable insights, and from there to the common good [including] a new understanding of genetic contributions to human disease and the development of rational strategies for minimizing or preventing disease phenotypes altogether" (Collins, 1999) is therefore very far from having been confirmed even after 20 years of intensive genome sequencing of ever-larger cohorts of humans. The elimination of purpose in biology has thus been an expensive mistake. The mistake is compounded even further by the idea that gene association studies do not even need hypotheses; according to that idea, the data alone is supposed to be sufficient, since it is claimed that "a hypothesis is a liability" (Yanai & Lercher, 2020, 2021; see the responses by Felin et al., 2021a, b). To be sure, data is critical to science, but:

Data itself is passive and inert. Data is not meaningful until it encounters an active, problem-solving observer. And in science, data gains relevance and becomes data in response to human questions, hypotheses, and theories. (Felin et al., 2021b)

Furthermore, the idea that data alone is sufficient is itself a hypothesis. Like all hypotheses, it is subject to verification or refutation. In the case of genome-wide association studies, the hypothesis has clearly failed after two decades of intensive research. How much longer do we need to wait before admitting that such research is looking for explanations at the wrong level of biological organization? Do we conclude that "forty years of *The Selfish Gene* [Dawkins 1989/1976] are not enough" (Yanai & Lercher, 2016a, b), or do we finally wake up to how that idea has seriously misled biology for nearly half a century (Noble & Noble, 2022)? How long do we continue to invest huge funding into a project that fails to deliver what it promised?

Physiology must now come to the rescue of genomics and of a gene-centric evolutionary biology. Thanks to the Physiome Project, we now have quantitatively precise tools to do so (Noble & Hunter, 2020). Being able to estimate the difference between association and causality is itself an empirical vindication of the concepts of function and purpose in living organisms. The existence of purposive "agency" turns out to be one of the most productive and predictive theories of science.

Acknowledgments

No funding sources supported this work. There are no conflicts of interest. The chapter is partly based on an oral presentation to the Linnean Society under the title "Physiology and Telos: Is Teleology a Sin?": https://vimeo.com/562981272

We thank Perry Marshall and Anthony Kenny for comments on an early draft of this article.

Notes

1. It is noteworthy that a later supporter of the modern synthesis, John Maynard Smith, made a similar point in his book *Evolutionary Genetics* (Maynard Smith, 1998: 8) where he wrote: "It [Lamarckism] is not so obviously false as is sometimes made out."

2. Unusually for Huxley's careful documentation, this reference is missing from his book. We suspect it can be found in Timakov and Kagan (1973). Timakov was a leading USSR microbiologist, head of the N. F. Gamaleia Institute of Epidemiology and Microbiology of the Academy of Medical Sciences of the USSR in Moscow, and president of the Academy of Medical Sciences of the USSR. https://encyclopedia2.thefreedictionary.com/Tima kov%2c+Vladimir

3. Hundreds of billions of dollars have been spent worldwide on ever-larger cohorts of genome sequencing.

References

Bell, G. (2015). Every inch a finch: A commentary on Grant (1993) 'Hybridization of Darwin's finches on Isla Daphne Major, Galapagos.' *Philosophical Transactions of the Royal Society* B, 3702014028720140287. http://doi .org/10.1098/rstb.2014.0287

Ben Maamar, M., Sadler-Riggleman, I., Beck, D., & Skinner, M. (2018). Epigenetic transgenerational inheritance of altered sperm histone retention sites. *Scientific Reports*, 8, 5308. https://doi.org/10.1038/s41598-018-23612-y

Boyle, E. A., Li, Yang, I., & Pritchard, J. K. (2017). *An expanded view of complex traits: From polygenic to omnigenic. Cell, 169(7), 1177–1186.* https://doi.org/10.1016/j.cell.2017.05.038

Chapuis, A. (1952). Les jeux d'eaux et les automates hydraulique du parc d'Hellbrunn pres Salzburg. *La Suisse Horlogere*, 67, 89. (Cited in Needham 1971:15.)

Collins, F. S. (1999). Shattuck Lecture: Medical and societal consequences of the Human Genome Project. *New England Journal of Medicine*, 341, 28–37. https://doi.org/10.1056/NEJM199907013410106

Darwin, C. R. (1859). *On the origin of species by means of natural selection, or the preservation of favoured races in the struggle for life.* London: John Murray.

Darwin, C. R. (1868). *The variation of animals and plants under domestication.* London: John Murray.

Darwin, C. R. (1871). *The descent of man, and selection in relation to sex.* London: John Murray.

Dawkins, R. (1989/1976). *The selfish gene.* New York: Oxford University Press.

Descartes, R. (1665). *"Treatise on the fetus."* In *Treatise on man.*

Eldredge, N., & Gould, S. J. (1972). Punctuated equilibria: An alternative to phyletic gradualism. In: T. J. M. Schopf (Ed.). *Models in paleobiology* (pp. 82–115). San Francisco: Freeman Cooper.

Felin, T., Koenderink, J., Krueger, J. I., Noble, D., & Ellis, G. F. R. (2021a). The data-hypothesis relationship. *Genome Biology*, 22, art. 57. https://doi.org/10.1186/s13059-021-02276-4

Felin, T., Koenderink, J., Krueger, J. I. Noble, D., & Ellis, G. F. R. (2021b). Data bias. *Genome Biology* 22, art. 59. https://doi.org/10.1186/s13059-021-02278-2

Ginsburg, S., & Jablonka, E. (2010). The evolution of associative learning: A factor in the Cambrian explosion. *Journal of Theoretical Biology*, 266, 11–20. https://doi.org/10.1016/j.jtbi.2010.06.017

Ginsburg, S., & Jablonka, E. (2019). *The evolution of the sensitive soul: Learning and the origins of conscious-ness.* Cambridge, MA: MIT Press.

Gould, S. J. (2002). *The structure of evolutionary theory.* Cambridge, MA: Harvard University Press.

Gould, S. J., & Eldredge, N. (1977). Punctuated equilibria: The tempo and mode of evolution reconsidered. *Paleobiology*, 3(2), 115–151. https://www.jstor.org/stable/2400177

Hanson, M. A., & Gluckman, P. (2014). Early developmental conditioning of later health and disease: Physiology or pathophysiology? *Physiological Reviews*, 94(4), 1027–1076. https://doi.org/10.1152/physrev.00029.2013

Hanson, M. A., & Skinner, M. (2016). Developmental origins of epigenetic transgenerational inheritance. *Environmental Epigenetics*, 2(1), dvw002. https://doi.org/10.1093/eep/dvw002

Huxley, J. (1942). *Evolution: The modern synthesis.* New York: Harper & Row. Reprinted in 2010 by MIT Press with foreword by Massimo Pigliucci & Gerd Müller.

Kenny, A. J. P., Longuet-Higgins, H. C., Lucas, J. R., & Waddington, C. H. (1972). *The nature of mind* (Gifford Lectures). Edinburgh: Edinburgh University Press.

Mayr, E. (1982). *The growth of biological thought.* Cambridge, MA: Belknap Press.

Needham, D. (1971). *Machina carnis.* Cambridge: Cambridge University Press.

Noble, D. (2016). *Dance to the tune of life*. Cambridge: Cambridge University Press.

Noble, D. (2019). Exosomes, gemmules, pangenesis and Darwin. In: L. Edelstein, J. Smythies, P. Quesenberry, & D. Noble (Eds.). *Exosomes: A clinical compendium* (pp. 487–501). Amsterdam: Elsevier. https://doi.org/10.1016/B978-0-12-816053-4.00021-3

Noble, D. (2022). Modern physiology vindicates Darwin's dream. *Experimental Physiology*, 107, 1015–1028. https://doi.org/10.1113/EP090

Noble, D., & Hunter, P. (2020). How to link genomics to physiology through epigenomics. *Epigenomics, 12*(4), 285–287. https://doi.org/10.2217/epi-2020-0012

Noble, R., & Noble, D. (2017). Was the watchmaker blind? Or was she one-eyed? *Biology*, 6(4), 47. https://doi.org/10.3390/biology6040047

Noble, R., & Noble, D. (2018). Harnessing stochasticity: How do organisms make choices? *Chaos*, 28, 106309. https://doi.org/10.1063/1.5039668

Noble, R., & Noble, D. (2020). Can reasons and values influence action: How might intentional agency work physiologically? *Journal for General Philosophy of Science*, 52, 277–295. https://doi.org/10.1007/s10838-020-09525-3

Noble, R., & Noble, D. (2021). Physiology restores purpose to evolutionary biology. *Biological Journal of the Linnean Society*, blac049. https://doi.org/10.1093/biolinnean/blac049

Noble, R., & Noble, D. (2022). Origins and demise of selfish gene theory. *Theoretical Biology Forum*, 115 (1–2), 29–43.

Noble, R., Tasaki, K., Noble, P. J., & Noble, D. (2019). Biological relativity requires circular causality but not symmetry of causation: So, where, what and when are the boundaries? *Frontiers in Physiology*, 10, 827. https://doi.org/10.3389/fphys.2019.00827

Peterson, E. L. (2011). The excluded philosophy of evo-devo? Revisiting C. H. Waddington's failed attempt to embed Alfred North Whitehead's "organicism" in evolutionary biology. *History & Philosophy of the Life Sciences*, 33(3), 301–20. https://pubmed.ncbi.nlm.nih.gov/22696826

Pigliucci, M., & Müller, G. (Eds.). (2010). *Evolution: The extended synthesis*. Cambridge, MA: MIT Press.

Shapiro, J., & Noble, D. (2021). What prevents mainstream evolutionists teaching the whole truth about how genomes evolve? *Progress in Biophysics and Molecular Biology*, 165, 140–152.

Slotten, R. A. (2004). *The heretic in Darwin's court: The life of Alfred Russel Wallace*. New York: Columbia University Press.

Timakov, V. D. (1941), in Timakov, V. D., & Kagan, G. Y. (1973). *Bacterial L-forms and mycoplasmataceae in pathology*. Moscow. (Cited by Huxley, 1942:51 in 2010 reprint)

Waddington, C. H. (1957). *The strategy of the genes*. London: Allen and Unwin.

Wallace, A. R. (1878). *Tropical nature and other essays* (Scholar's Choice ed.). London: Macmillan.

Wallace, A. R. (1889). *Darwinism: An exposition of the theory of natural selection, with some of its applications*. London: Macmillan.

Wallace, A. R. (1890). The colours of animals. *Nature, 42(1082), 289–291*. https://doi.org/10.1038/042289a0

Wallace, A. R. (1875/2009). *On miracles and modern spiritualism. Three essays*. Cambridge: Cambridge University Press.

Weismann, A. (1892). *Das keimplasma: Eine theorie der vererbung*. Jena, Germany: Fischer.

Weismann, A. (1893). *Die allmacht der Naturzüchtung: eine Erwiderung an Herbert Spencer*. (The all-sufficiency of natural selection. A reply to Herbert Spencer.) *Contemporary Review*, 64, 309–338.

Yanai, I., & Lercher, M. J. (2016a). Forty years of *The Selfish Gene* are not enough. *Genome Biology*, 17, 39. https://doi.org/10.1186/s13059-016-0910-7

Yanai, I., & Lercher, M. J. (2016b). *The society of genes*. Cambridge, MA: Harvard University Press.

Yanai, I., & Lercher, M. J. (2020). A hypothesis is a liability. *Genome Biology*, 21, 231. https://doi.org/10.1186/s13059-020-02133-w

Yanai, I., & Lercher, M. (2021). The data-hypothesis conversation. Genome Biology, 22, 58. https://doi.org/10.1186/s13059-021-02277-3

Zuitin, A. I. (1941). The changes in the environment as the principle external factor in natural mutations. *Drosophila Information Service*, 15, 41. (Cited by Huxley, 1942: 55 in 2010 reprint).

13 Goal Attributions in Biology: Objective Fact, Anthropomorphic Bias, or Valuable Heuristic?

Samir Okasha

Overview

Goal-attributing statements—which attribute a goal or endpoint to an organismic activity or process—arise in three different biological contexts. The first context is the mid-twentieth-century debate among biologists and philosophers over how to understand what was widely considered to be the "goal-directedness" in the living world. The second context is the debate in cognitive ethology over whether nonhuman animals are capable of goal-directed behavior—that is, behavior that results from having a mental representation of a goal state. The third is the practice (common in evolutionary biology) of treating evolved traits, including behaviors, as means by which an organism furthers its overall goal of survival and reproduction (or maximization of fitness). In each of these contexts, a similar philosophical issue arises: Are the goal-attributing statements literally true? And if not, do they represent an anthropomorphic bias that should be expunged from science, or a valuable heuristic?

13.1 Introduction

The label "teleonomy" was introduced into biology by Pittendrigh (1958) and subsequently taken up by Williams (1966), Mayr (1974), and others—though they did not all define it in the same way. The original point of the teleonomy concept was to demarcate a notion of goal-directedness that was both objective and scientifically important, and to sharply distinguish it from discredited forms of teleology, such as the idea that the evolutionary process itself unfolds in accordance with a plan. As Corning (2022) explains, the teleonomy/teleology distinction was meant to capture the idea that living organisms exhibit a kind of "internal purposiveness," in both their ontogenetic development and their daily activities, that has evolved by natural selection. So, while Darwinism forces us to reject the idea that the process of evolution is in any sense goal-directed, this is quite compatible with recognizing that the products of that process—well-adapted organisms and their parts—exhibit goal-directed behavior. In short, we should not throw out the valid forms of teleology with the invalid.

The notion that organisms exhibit an internal purposiveness was endorsed by many mid-twentieth-century biologists. Thus, for example, Monod (1973: 9) wrote: "one of the most fundamental characteristics common to all living thing [is] that of being endowed with a project or purpose." In a similar vein, Waddington (1957: 2) wrote that "most of the activities of a living organism are of such a kind that they tend to produce a certain characteristic end result," a phenomenon he referred to as "directiveness." A more explicit statement came from Mayr (1988: 45): "goal-directed behavior . . . is extremely widespread in the organic world; for instance, most activity connected with migration, food-getting, courtship, ontogeny, and all phases of reproduction is characterized by such goal orientation." These quotations convey a reasonably clear sense of what was meant by teleonomy.

The concept of goal-directedness receives relatively little attention in contemporary philosophy of biology, in sharp contrast with the concept of function, which continues to be a focal point.[1] It was not always thus. In the philosophical literature of the mid- to late-twentieth century, function and goal-directedness enjoyed equal airtime and were often discussed together (Sommerhoff, 1950; Braithwaite, 1953; Beckner, 1969; Nagel, 1977a, 1977b; Woodfield, 1976; Wright, 1976). What explains this change? To some extent it may reflect the changing scientific climate, in particular the decline in prominence of cybernetics. Garson (2008: 539) suggests that goal-directedness fell from favor in philosophy because the prominent attempts to analyze it suffered from "conceptual shortcomings." In addition, the philosophical literature's reorientation away from goal-directedness and toward function likely stemmed in part from the conviction that while "function talk" is ubiquitous in day-to-day biological practice, "goal talk" is neither. So naturalistically-inclined philosophers, who often conceive their task as (in part) to analyze the meaning of scientific terms, were led to focus on function rather than goal-directedness.

My aim in this paper is to reconsider goals and goal-directedness from (what I hope is) a new angle. I examine three biological contexts in which goal attributions arise, all of which are of philosophical interest. (By a "goal attribution" I mean a statement of the form "the goal of x is y," where x stands for an organismic behavior or process and y stands for an endpoint. In some cases, x may stand for a whole organism.) The first context is the classic mid-twentieth-century debate, alluded to above, about how to understand the apparent purposiveness of organismic activities and processes (including development). The second context is the ongoing debate about whether nonhuman animals are capable of goal-directed behavior, meaning (roughly) behavior that results from having a mental representation of a goal. This is part of the broader question of animal intentionality, long a source of controversy among ethologists and philosophers. The third context is the practice found in evolutionary biology of treating evolved organisms *as if* they were rational agents pursuing a goal—as when we explain why female rats kill their malformed offspring by saying that they know the offspring won't survive and don't want to waste resources on them. This mode of explanation is commonly used by evolutionists studying the adaptive significance of organismic traits.

These three contexts are separate, involving goal attributions of different sorts, and raise distinct philosophical issues. However, a common theme runs through them all, namely: Do the goal attributions in question state objective facts about the world, or not? Scholars sympathetic to the notion of teleonomy, including many of the contributors to this book, have typically assumed that the answer is yes, at least in certain cases, but this assumption is not inevitable. An alternative is that goals are "projected" onto the biological world by humans (as proposed by Kant, 2000/1790). One version of this projectivist view holds that goal attributions reflect an anthropomorphic tendency and so have no place in science. Another version holds that goal attributions are of heuristic value even if they are not literally true. The main aim of this paper is to try to adjudicate between these views, in each of three contexts described above.

The structure of the paper is as follows. Section 13.2 offers a re-examination of the traditional debate on teleonomy and goal-directedness, focusing on the work of Mayr and E. Nagel. Section 13.3 examines the ethological debate over whether animal behavior is ever goal-directed in the sense of stemming from belief-like and desire-like mental representations. Section 13.4 discusses the practice of treating evolved organisms as if they were rational agents pursuing a goal, in the context of evolutionary explanation. Section 13.5 draws together the pieces and concludes.

13.2   The Traditional Debate over Goal-Directedness: A Retrospective

I start by briefly outlining Mayr's views on goal-directedness, juxtaposing them with the views of Nagel. Though divergent, Mayr's and Nagel's positions share certain presuppositions, and their respective analyses touch on most of the important points in the traditional debate.

13.2.1 Mayr's Views

Mayr (1992) set out his mature views in an essay that builds on ideas he first developed in the 1970s. His aim is to distinguish between various categories of biological phenomena to which the label "teleological" has been applied. One such category is what Mayr calls "teleonomic processes," which he describes as "goal-directed" and "goal-oriented" (Mayr, 1992: 52). In addition to processes, Mayr also applies the adjective *teleonomic* to activities and behaviors. Though Mayr does not try to define "process," "activity," or "behavior," his examples show that the category of the teleonomic as he conceives it is broad. It includes ontogenetic processes such as gastrulation, physiological processes such as thermoregulation, and whole-organism behaviors such as migration and mating displays. Quite often, Mayr uses *process* as a catch-all term that subsumes activities and behaviors, a policy I will follow here in the interest of brevity.

Teleonomic processes have two key features, according to Mayr. Firstly, they have an "endpoint, goal, or terminus," the attainment of which leads the process to stop (Mayr, 1992: 52). This endpoint might be a developmental structure, physiological state, or behavioral outcome, Mayr tells us. Thus, the goal of meiosis is to produce haploid germ cells; the goal of a bacterium's movement is to reach an area of higher oxygen; and the goal of the salmon's homing behavior is to arrive at its natal stream. Secondly, the process is guided by an evolved *program* that encodes a set of instructions for the unfolding of the process, and which thus encodes the goal. The significance of this for Mayr is that it dispels any specter of backwards causality (which haunts other forms of teleology), since the program exists before the teleonomic process begins to unfold. The attribute of being program-guided thus shows how to reconcile goal-directedness with ordinary causality.

When referring to an "evolved" program, Mayr is primarily thinking of a genetic program, in the sense of a set of instructions written in DNA. He alludes to the well-known idea that natural selection, acting over phylogenetic time, has led to the accumulation of information in a species' genome, which then guides ontogenetic development and gives rise to species-specific activities and behaviors. However, Mayr emphasizes that adaptive behavior in higher animals is influenced by learning as well as genes; he thus distinguishes between a "closed program," which contains genetically hard-wired instructions, and an "open program" which can be modified by acquired information. Most complex animal behaviors are the result of an open program, he claims.[2]

Mayr contrasts his account of goal-directedness with a rival analysis, inspired by cybernetics, that defines it in terms of negative feedback, as for example in Rosenbleuth et al. (1943). Mayr allows that negative feedback often plays a role in explaining how the endpoint of a teleonomic process is reached despite perturbations, but denies that this is the essence of such a process; rather, it is something that improves its precision. The essential feature, he argues, is rather that a mechanism exists that initiates the goal-directed process, and this occurs because of the program-directed nature of the process.

Although most organismic behavior counts as teleonomic for Mayr, he reserves the label "purposive behavior" for behavior that involves a deliberate attempt to attain a goal—that is, where the organism has a mental representation of the goal and performs the action in order to attain it. Following the lead of cognitive ethologists, Mayr suggests that such purposive behavior is not confined to humans, but is also common among mammals and birds. Though Mayr puts purposive behavior in a separate category from teleonomic behavior, he describes them both as goal-directed; thus, they are species of a single genus.

Mayr puts "adapted features" in a separate category again. These are structural and morphological attributes of organisms, including whole organs, that have evolved by natural selection. Mayr emphasizes that such features "do not involve movements," and thus in his view are not appropriately described as teleological. However, Mayr does see a link between adapted features and teleonomy, for such features *perform* teleonomic processes and are thus "executive organs" of such processes (Mayr, 1992: 58). (So, circulation of the blood is a teleonomic process of which the heart is an executive organ.) Though he does not put it this way, in effect Mayr's point is that "static" adaptive features are not in the right ontological category to count as goal-directed, for they are not activities, behaviors, or processes. This seems right: it makes good sense to enquire what a stag's antlers are *for* but no sense to ask what the antlers' *goal* is.

Mayr's analysis of goal-directedness is interesting, but from a modern perspective it contains questionable aspects, in particular its reliance on the notion of genetic program; this is discussed below.

13.2.2   Interlude: Function and Goal

Mayr's point about adapted features shows that a biological item's having a goal, in the sense that he is trying to capture, cannot be equated with having a function (in any of the usual senses of that term). A biological trait of any sort is a candidate for having a function, but only a subclass of traits are candidates for having goals. Indeed, this observation was a commonplace among mid-twentieth century writers on teleology, who emphasized that goal-directedness and functionality were distinct aspects of teleology that should not be confused (Beckner, 1969; Wright, 1973; Woodfield, 1976; Nagel, 1977a). But what exactly is their relation?

One natural suggestion is that function and goal will coincide for any biological item that has both. That is, if we pick something that *is* in the right ontological category to have a goal, such as an organismic behavior, then the behavior's function, if it has one, will in general be its goal, and vice versa. This sounds plausible, for observe that the pair of statements "the function of the bird's dance is to attract mates" and "the goal of the bird's dance is to attract mates" seem practically synonymous.

However, an argument by Wright (1973) suggests that this coincidence does not always hold. Wright gives the example of freshwater plankton that diurnally vary their distance below the surface. He writes that "the goal of this behavior is to keep light intensity in their environment relatively constant. This can be determined by experimenting with artificial light sources. The function of this behavior, on the other hand, is keeping constant the oxygen supply, which normally varies with sunlight intensity" (Wright, 1973: 140). The moral, Wright says, is that the function of an organismic behavior need not be the behavior's goal but can instead be "some natural concomitant or consequence of the . . . goal" (1973: 140).

Wright is partly correct here (though the biological details are not quite as he says). Light intensity is the (main) cue to which the plankton's daily vertical movement responds, but the selective advantage that the behavior confers—hence the function *sensu* the standard "selected effect" theory of function—is likely some correlate of light intensity, such as predator avoidance.[3] So, it is indeed possible that a behavior's function may be a "natural concomitant" (i.e., correlate) of its goal. However, the other possibility that Wright mentions—that the function may be a downstream consequence of the goal—is less clear-cut. For instance, consider a causal sequence of the form $b \rightarrow g \rightarrow c$, where b is the behavior, g is the goal, and c is the consequence. If c confers a selective advantage (i.e., enhances present and/or past fitness), and so is a candidate for being b's function, then the same must be true of g as well. It is a familiar point that singling out an effect as "the" function of a trait involves a partly conventional decision about how to chop up a complex causal chain. (The same is likely true of "the" goal.) Thus, in causal sequences of this sort, it will always be possible to identify goal and function so long as the behavior in question does have a goal.

One final difference between function and goal deserves mention. Where a goal is attributed to an organismic behavior (though not to an internal process), it is generally possible to re-express this in terms of the organism's goal. Thus, instead of saying that the goal of a swallow's migration is to reach warmer climes, we can equivalently say that the swallow's goal (in flying south) is to reach warmer climes. Functional attributions, by contrast, cannot always be so re-expressed.

13.2.3 Nagel's Tripartite Taxonomy

It is useful to contrast Mayr's view of goal-directedness with alternative approaches that were prevalent at the time. In a well-known paper, Nagel (1977a) laid out three views of goal-directedness that he called "the *intentional* view, the *program* view and the *system-property* view." Nagel criticizes the first two views and defends the third. The intentional view posits a close link between goal-directedness and intentionality. Proponents of this view argue that the paradigm goal-directed phenomenon is the conscious pursuit of a goal by an intelligent agent who has a mental representation of the goal the agent wishes to achieve. Nagel recognizes that the intentional view is attractive because it avoids the threat of reverse causality:

the cause of a goal-directed action is not the goal itself but rather the agent's intention to bring the goal about. However, he regards the intentional view as being ill-suited as a general analysis of goal-directedness, for it cannot make sense of goal attributions "in connection with organisms . . . which are incapable of having intentions and beliefs, in connection with subsystems of organisms; or in connection with inanimate systems" (Nagel, 1977a: 265).

Nagel briefly considers the position of Woodfield (1976), an adherent of the intentional view, who argues that while our "core concept" of goal-directedness involves wanting to achieve a goal, we also have a "broader concept" that applies to any process produced by an inner state that is suitably analogous to wanting. Nagel argues that Woodfield's "analogical extension" story cannot make sense of what he regards as *bona fide* goal-directed processes in biology, such as a tadpole developing into a frog. He argues that there is no useful sense in which "the inner state representing the goal . . . which is perhaps a complex subsystem of genetic materials in the tadpole . . . resemble[s] (or is analogous to) a desire or a belief" (Nagel, 1977a: 266). Nagel seems on strong ground here: morphogenesis is not much like the deliberate pursuit of a goal. However, the broader implication of Woodfield's view—that goal attributions in biology involve an element of anthropomorphism—may still be defensible.

The program view is essentially the position of Mayr. Nagel opposes Mayr in part because he regards the notion of an "open program" as obscure, in part because he thinks Mayr's analysis misclassifies clear-cut cases, and in part because he doubts that it reflects how we actually make judgments of goal-directedness. Nagel argues that some programmed organismic processes, such as the knee reflex, intuitively do not count as goal-directed, so the program view is too liberal. He also argues that Mayr's attempt to exclude "automatic" processes (which Mayr calls "teleomatic") from the domain of the goal-directed does not work, because if applied consistently it would exclude phenomena that Mayr wishes to include, such as the workings of a clock. But Nagel's main objection to the program view is that we cannot observe the program that controls an organismic process, so our judgment about whether it is goal-directed, and if so what the goal is, must derive from another source. Since Nagel regards such judgments as reliable, he is led to seek a purely behavioral definition of goal-directedness.

From a modern perspective, this last criticism is uncompelling because it conflates the metaphysical question of what *makes* something an x with the epistemological question of how we can *tell* whether something is an x. The latter is at best at a defeasible guide to the former. Nagel may well be right that the usual way for us humans, given our cognitive limitations, to determine whether a process is goal-directed is to observe the process unfold, rather than to inspect the program (if any) that controls it. However, this is compatible with Mayr's claim that *what it is to be* a goal-directed process is for that process to be controlled by an internal program. Mayr's view may be untenable for other reasons, but Nagel's criticism should not convince.

The systems view, Nagel tells us, says that "being goal-directed is a property of a system, in virtue of the organization of its parts" (Nagel, 1977a: 273). The systems view is essentially the cybernetical view of Rosenbleuth et al. (1943) and Sommerhoff (1950). The latter, Nagel says, sought a general account of what makes a system goal-directed "irrespective of whether the goal is pursued by purposive human agents, by living systems incapable of having intentions, or by inanimate systems," an aim of which he approves (Nagel 1977a: 271–272).

Nagel says that the systems view correctly identifies two of the essential properties of goal-directedness: plasticity and permanence. *Plasticity* means the system can reach the goal via different pathways, and from multiple starting points. *Persistence* means that that the system maintains its goal-directed behavior despite external perturbations, thanks to internal compensatory adjustments that keep it on course; this is closely related to the negative feedback idea. Nagel illustrates plasticity and permanence with the example of the homeostatic regulation of the water content of blood in humans, a textbook example of a goal-directed process (the goal being to keep the water content at 90%). Following Sommerhoff (1950), Nagel adds to permanence and plasticity a somewhat obscure requirement that he calls "orthogonality." The point of this requirement is to exclude processes that are an upshot of simple physical law, such as a marble rolling down the side of a bowl coming to rest at its base, which Nagel thinks should not count as goal-directed.

13.2.4 A Natural Kind?

Looking back at Mayr's and Nagel's discussions, one point stands out. Both authors take for granted that, aside from a few problem cases, they know which phenomena count as goal-directed and which do not—these are the "data" against which their analyses are to be tested. That is, they assume that goal-directedness is an objective scientific property that defines a natural kind, not something projected onto nature by humans. These commitments are not merely idiosyncrasies of Mayr and Nagel; they are widely shared by participants in the traditional debate. But are they justified?

There are three reasons for doubt. Firstly, the terms *goal-directed* and *goal-directedness* are not technical terms in biology nor widely used in the general biological literature (though they are common in neuroscience and psychology).[4] There is a contrast here with the term *function*, which *is* widely used in biology. Now this does not prove that goal-directedness is not a scientifically important phenomenon; as Nagel himself notes, not all goal-attributing statements necessarily use the word "goal" (Nagel, 1977a: 263). However, it does lead one to wonder whether Mayr and Nagel were right to take the reality and ubiquity of goal-directedness as their starting point.

Secondly and relatedly, the class of things that are supposed to count as goal-directed is rather heterogenous. The class includes the entire process of development from embryo to adult; development subprocesses such as cleavage and gastrulation; physiological processes such as thermoregulation and tissue repair; organismic activities such as foraging and migration; conscious human behaviors such as writing an article; and the operation of man-made artifacts such as engines and thermometers. (Perhaps it is unsurprising that science has no single term covering all these.) The supposedly unifying feature is that in each case there is an "endpoint" which is reliably reached despite perturbations. No doubt there is some truth to this. Certainly, these processes involve striking regularities that cry out for explanation. However, the explanation of how a tadpole develops into a frog, how an injured salamander regenerates its limbs, and how a salmon reaches its natal stream are all quite different. Understanding how one of these processes works tells us nothing about how the others work. It is not obvious that we have a natural kind here, rather than a class of processes at most superficially alike.

Thirdly, the idea that goal-directedness is something we project onto nature is consonant with findings in experimental psychology. Many studies suggest that humans have an inbuilt cognitive bias that leads us, from a young age, to see intention and purpose where there is none, to anthropomorphize, and to favor teleological over mechanistic descriptions of the world (Barrett, 2011). (This bias may have evolved because it conferred a survival advantage.) If this bias is real and universal, it is conceivable that it may influence scientists and philosophers too.

This suggests that there may be a purely psychological explanation for why the processes that theorists regard as goal-directed have been co-classified. Perhaps all these processes elicit a certain psychological reaction in us. They strike us as akin to what we would see if the system were consciously aiming at a goal (in the case of whole-organism activities and behaviors) or had been designed by an agent with a goal in mind (in the case of internal processes). According to this view, what unites a firefly's mating display, a tadpole's metamorphosis, and an immune system's production of T cells is not any similarity of actual behavior or underlying causal mechanism; rather, it is that in each case the system behaves *as if* it were goal-directed in the sense of having, or stemming from, a conscious mental representation of a goal. And if that is so, then goal-directedness is something that we project into nature rather than discover in it.

This is somewhat similar to what Nagel calls the intentional view, but there is a difference. The intentional view as Nagel characterizes it is based on the idea that "real" goal-directedness requires intentional causation, so other cases belong only if their etiology is analogous or isomorphic. What I am suggesting is something different (though compatible): namely, that the disparate class of processes that theorists have treated as goal-directed may be unified only in that they all provoke a certain psychological reaction in us, that leads us to anthropomorphically assimilate the endpoint of the process to the intentional object of an agent. This is not a claim about isomorphism between goal-directedness processes and the conscious pursuit of goals, but rather about the unity (or lack of it) among the processes that have been pretheoretically classed as goal-directed.

13.2.5 Upshot

Where does this leave us? There seem to me to be two options. Either something *like* Mayr's program view can be made to work, or we should jettison the idea that goal-directed processes form a natural kind in favor of the projectivist alternative sketched above.

I argue this for two reasons. Firstly, attempts to make the systems view work have arguably failed (Garson, 2018). Despite much ingenuity, no version of the systems view has provided a definition of goal-directedness that can simultaneously rule out pseudo-cases such as the marble rolling down the bowl; can uniquely identify "the" goal toward which a process is directed; can explain how a process can be goal-directed and yet the goal not be achieved or the goal object not exist (as when a salmon's natal stream has dried up). (This last problem undermines the attempt to define goal-directedness in terms of feedback alone.) The basic problem is simply that the goal toward which a process is directed, if any, is heavily underdetermined by the process's actual behavior. Appeal to hypothetical behavior may perhaps help; but philosophical experience with similar situations shows that it is preferable to simply abandon the behaviorist pretense and appeal directly to internal factors, as Mayr does.

Secondly, seeking unity among goal-directed processes at the level of observable behavior, rather than internal factors, sits uneasily with the inclusion of purposive human behaviors in the class of the goal-directed, which most theorists have agreed with. Features such as persistence and plasticity, on which systems theorists focus, hardly characterize all human behaviors. Some humans behave in a rather erratic way, fleetingly pursuing a goal and then doing something else (so not persisting); when a given action fails to achieve a goal, humans sometimes choose an alternative means to achieve the goal (thus exhibiting plasticity), but sometimes they change goals or give up entirely. It seems that the real reason for including purposive behaviors in the class of the goal-directed is that they derive from a conscious mental representation of the goal, not that the behaviors themselves are particularly similar to other standard examples of goal-directed processes (such as thermoregulation).

If Mayr's program view were right, then goal-directed processes would indeed form a natural kind, despite their apparent heterogeneity. However, the program view sits uneasily with modern biological knowledge and is at best a metaphor. Certainly, genes play a crucial causal role in all of the processes—ontogenetic, physiological, and behavioral—that Mayr regards as goal-directed, but the idea that the genome is a program controlling these processes is highly doubtful (Newman, 2022). Developmental genetics teaches us that that the genome is a highly *reactive* entity, not a fixed repository of instructions, and that gene expression often depends crucially on environmental triggers and conditions (Gilbert, 2003; Fox-Keller, 2014). Nor does Mayr's attempt to liberalize his view by allowing that programs can be "open" help much, for the program is still supposed to initiate the goal-directed process, which is hard to square with the context-sensitivity of gene action. In short, the program view rests on an undermotivated computer science analogy and an a priori privileging of genetic over environmental causes.

However, something in the spirit of Mayr's view may be salvageable even if we jettison the notion of an evolved program. We may be able to retain the idea that the common feature of goal-directed processes is that they arise (in part) from a system having an inner state that *represents* the goal-state (or endpoint); and that this plays an essential role in explaining how such processes work. Now, what exactly this means is a difficult question. But we can say the following. Representation does not mean conscious mental representation (though it includes this as a special case). It is a commonplace of contemporary cognitive science that organisms and their subsystems contain "sub-personal representations" of both internal and external states, which could in principle include the goal-states of teleonomic processes. Philosophers have made considerable progress with articulating this notion of representation and with showing how representing-involving explanations work (Shea, 2018).

There is some reason to think that inner representations can help understand goal-directedness. One way for a system to produce goal-directed behavior is to have an inner state representing the goal, to compare this inner state with its actual state, then to suitably alter the latter. This is how some (though not all) goal-directed animal behaviors work. Whether this inner representation story covers all cases of putative goal-directedness in biology, including developmental and physiological processes, is not clear. For the story to work in full generality, the inner states doing the representing would presumably have to include genomic states (per Nagel's suggestion that an inner state of the tadpole's genome represents the adult frog's form). The idea that genes "represent" phenotypic outcomes has been defended before, but it is more controversial than the idea that the state of a cognitive system represents the world.[5] So, if the defender of the reality of goal-directedness wishes to go the inner representation route, some work would be needed to show that it picks out the desired class of phenomena.

Be that as it may, the inner representation story seems preferable to Mayr's program view for two reasons. Firstly, the notion of representation is arguably in better standing than that of internal program and is needed in some of areas of biology anyway. Secondly, it more easily explains why purposive human behaviors count as goal-directed, given that conscious mental representations are a subtype of representation. By contrast, Mayr cannot easily explain this.

To conclude: the assumption that goal-directedness is an objective feature of the world is not obviously true. The assumption seems defensible only if goal-directed processes share a commonality at the level of internal mechanism rather than observable behavior. The most promising such candidate is the existence of an inner representation of the goal-state that is causally implicated in producing the process. Should such a commonality turn out not to exist, or to pick out the "wrong" class of processes, we should conclude that the disparate phenomena traditionally treated as goal-directed do not form a natural kind.

13.3 Goal Attributions in Cognitive Ethology

A quite different debate about goal-directedness occurs in the fields of ethology and comparative cognition. At issue is the correct explanation of certain complex animal behaviors. The starting point is the apparent contrast between the stimulus-response behavior that is common throughout the living world and the more sophisticated behaviors found in some vertebrate taxa. For example, contrast a bacterium moving toward oxygen with a rat navigating its way out of an intricate maze. Whereas the former is simply a direct response to an environmental stimulus, the latter is more of a cognitive achievement, requiring memory, learning, and inference. Impressed by this intuitive contrast, some researchers propose that certain nonhuman animals are capable of goal-directed actions that stem from internal mental representations, such as belief-like and desire-like states, pointing to intriguing experimental findings that seem to show this (Ristau, 1991; Dickinson, 2011; Clayton, Emery, & Dickinson, 2006). These researchers thus posit continuity between the intentional behavior of humans and nonhuman animals, rejecting the suggestion that this is anthropomorphic.

Importantly, the concept of goal-directed behavior at work in this debate is not the same as that at work in the debate over teleonomy examined above. Many if not all of the "simple" stimulus-response behaviors (such as bacterial chemotaxis) that are supposed to *contrast* with goal-directed behavior in the ethological debate would be classified as goal-directed by Mayr, Nagel, and others in the first debate. Rather, goal-directed behavior in the ethological discussion corresponds to Mayr's category of purposive behavior (that is, behavior that requires explanation in intentional-psychological terms) or equivalently that stems from belief-like and desire-like mental representations. Recall that Mayr regards purposive behavior as a subcategory of teleonomy.

Whether nonhumans are capable of goal-directed behavior in this intentional sense, and if so how widespread the capability is, is controversial. This reflects disagreement both about how to interpret the empirical data and how to define the relevant concepts. At one extreme is the view that only humans exhibit true goal-directed actions, because nonhumans lack the necessary cognitive requirements. For example, Davidson (1982) has argued, on essentially a priori grounds, that language is a prerequisite for having beliefs, desires, and other intentional mental states; and thus that the behavior of nonhuman animals cannot stem from their being in such states. This view is not popular among scientists of animal behavior, though the position of Kennedy (1992), who regards all attribution of conscious mental states to nonhumans as rooted in anthropomorphism and unwarranted by the data, comes close.

At the other extreme are those researchers who regard goal-directed behavior, or its cognitive preconditions, as pervasive in the living world, even extending beyond the animal kingdom. Thus, Trewavas (2014) argues that plants are intelligent, insisting that he is speaking literally; Bray (2009) suggests that even single cells can be credited with "knowledge" of their environment. What underpins such arguments is the idea that wherever organisms exhibit adaptive plasticity and/or learning, and hence can vary their behavior in response to the environment, it is legitimate to attribute to them (a rudimentary form of) cognition. Such arguments should not be dismissed out of hand, though whether the cognitive states in question should really be thought of as mental representations is debatable. We should also note that much organismic behavior is only plastic within narrow bounds; ingenious experimental interventions can often make apparently intelligent behavior seem rather dumb.

In between these extremes, one finds a spectrum of positions allowing that some complex behavior of animals with nervous systems is goal-directed in the intentional sense. Thus, for example, Ristau (1991) has studied the behavior of piping plovers that feign a wing injury when a predator approaches in order to lead the predator away from its young. The plover's broken-wing display is highly sensitive to the predator's position, location, and movement. Ristau argues that the best explanation of the plover's behavior is that it wants to lead the intruder away from its young; only this accounts for the precision and timing of its actions. Similarly, Clayton, Emery, and Dickinson (2006) study the food-caching behavior of scrub jays. The jays not only store and retrieve food, but also use strategies to reduce the chance that their food will be pilfered, such as delaying caching if other birds are watching, and choosing locations that are concealed from others' view. Clayton et al. insist that the jays' behavior should be explained by attributing to them beliefs, desires, and memories, arguing that alternative nonintentional explanations fail. Other behaviors that have been thought to require intentional explanation include navigation, tool use, and future planning.

These middle positions prompt the question of whether precise behavioral criteria for goal-directedness can be laid down. This question is addressed by Dickinson (2011), who specifies two criteria for an action to count as goal-directed as opposed to "habitual." His *goal criterion* says that a goal-directed action must be "sensitive to whether or not [its] outcome is currently a goal for the animal" (Dickinson, 2011: 80). Dickinson illustrates this with the example of a rat pressing a lever to receive an outcome which has previously been devalued by conditioning. If the rat presses the lever anyway, despite the devaluing of the reward, its behavior is purely habitual, Dickinson argues. His "instrumental criterion" says that the animal's action should be sensitive to the causal relationship between action and reward. Thus, if a food reward that is usually contingent on one action is suddenly made contingent on a different action by the experimenter, the animal must be capable of learning this for its action to count as goal-directed. In effect, this is to say that goal-directed action requires that an animal have "causal knowledge," or more precisely a mental representation of the causal dependence of outcome on action.

Dickinson's criteria are quite strict (though he argues that rats meet them), making goal-directed action fairly uncommon. His instrumental criterion, in particular, seems overly demanding, as a number of authors have argued (Allen & Bekoff, 1995; Carruthers, 2004). Though it is quite plausible that learning of some sort is a requirement for having intentional states, and thus for goal-directed action, requiring that an animal be able to learn the pattern of causal contingency of reward on action seems too strong. The capacity for such causal learning is plausibly needed for complex means-end reasoning of the form "if I were to do x, I would get y," but this goes beyond merely acting from belief-like and desire-like mental representations; thus, the instrumental criterion is too restrictive. A more plausible, though admittedly more vague, criterion is that animal behavior counts as goal-directed when it is sufficiently flexible, intelligent-seeming, and complex that no nonintentional explanation is feasible. Experimental intervention is generally necessary to probe this.

Two further positions on the issue merit brief mention. The first is Dennett's idea that there is no sharp distinction between genuine and "as if" intentionality anyway. In Dennett's view, it is a mistake to ask whether a particular behavior (whether human or nonhuman) really stems from belief-like and desire-like states; the only question is whether it is heuristically useful to study the behavior from the "intentional stance." Dennett (1983) argues, perhaps somewhat optimistically, that adopting the intentional stance is often helpful for cognitive ethologists, as it leads to interesting hypotheses that can be tested. The second is the position of the prominent twentieth-century ethologist D. McFarland (1989a, 1989b) who argues, radically, that not even human behavior is genuinely goal-directed; our belief in the goal-directedness of our own and others' actions is a delusion that has been programmed into us by natural selection. McFarland's reason for thinking this appears to be that he does not believe in mental representation at all, as he does not see a way of squaring our folk-psychological talk of beliefs and desires with the underlying facts of neuroscience. In philosophy, a similar position has been advocated by P. Churchland (1981) under the label "eliminative materialism."

This debate raises complex issues, both philosophical and scientific, that I cannot hope to resolve here, so I will confine myself to a number of points. Firstly, *if* we accept that much human behavior is caused by inner mental representations, there seems every reason to allow that the same may be true of some nonhumans, both on grounds of evolutionary continuity and known neurophysiological similarity (Glimcher, 2003). A priori arguments to the contrary should carry little weight. So, there is no reason to believe that all attributions of goal-directed behavior to nonhumans are the result of anthropomorphic projection.

Secondly, the accusation of anthropomorphism is probably justified in some cases. It is well-established that complex adaptive behavior may arise from mechanisms that bear no resemblance to belief-like and desire-like inner representations. Barrett (2011) argues that simple internal mechanisms can often produce complex behavior by taking advantage of environmental regularities. Barrett gives the example of predatory *Portia* spiders, which show a remarkable ability to detour around obstacles while hunting for prey. The spiders' behavior conveys the impression of advance planning, for they need to let the prey out of their sight to get around an obstacle, and they appear to carefully scan the terrain before starting a detour. However, experiments reveal that no planning is going on; the spider is using a simple rule of thumb, based on the presence or absence of horizontal lines in its field of vision, to determine which direction to move in at each moment; this leads it to trace out the most direct route toward its prey. So, if someone were to explain the spider's behavior in belief-desire terms, this *would* be anthropomorphic; moreover, it would detract from rather than conduce toward a correct scientific understanding of the behavior.

Thirdly, it seems unrealistic to hope for fully explicit behavioral or experimental criteria for when attributions of goal-directedness, in the intentional sense, are justified and/or of heuristic value. Some key variables are clear, including the degree of behavioral flexibility, the capacity for novelty, and the ability to learn. But the link between mental representations and observed behavior is too indirect to expect fully explicit criteria; an element of judgment will always be needed.

Fourthly, Dennett's idea that no hard-and-fast line separates "real" from "as if" intentionality, though out of fashion in contemporary philosophy, may well be true, and fits with the fact that most biological attributes come in degrees. But this is still compatible with some explanations of behavior being clearly anthropomorphic. That a distinction is not hard-and-fast, and thus admit borderline cases, is compatible with the existence of clear-cut cases on either side. Finally, even if Dennett is wrong and there is an objective distinction between behavior that is genuinely intentional and behavior that to all intents and purposes appears as if it were, this distinction is (by construction) impossible to operationalize. Empirical work on goal-directedness in animals is thus insensitive to the distinction and cannot resolve the question of whether it exists. This latter question is inherently a philosophical one.

13.4   Goal Attributions in Evolutionary Biology

The final context in which goal attributions arise is evolutionary biology. A well-known project in evolutionary biology seeks to explain an organism's evolved traits in terms of the traits' adaptive significance (or function, in one sense of that term). Such explanations quite often treat the organism as if it were an agent with a goal. The organism's overall goal is often said to be survival and reproduction (or maximizing its fitness); to achieve this goal it needs to pursue intermediate goals such as finding food, attracting mates, and raising its young, to which its various evolved traits, including its behaviors, make distinct contributions. Thus, for example, Roff (1992) writes: "the primary goal of any organism is to reproduce . . . the first 'decision' it must make . . . is when to start reproducing." West and Gardner (2013) describe maximizing its inclusive fitness as the "objective" (goal) of an organism's social behavior. Grafen (2007) describes an evolved organism's phenotype as an "instrument" which it wields "in pursuit of a maximand" (i.e., the quantity that the organism is trying to maximize).

The idea of an organism as pursuing a goal toward which its evolved traits conduce may seem innocuous, at least on the assumption that those traits are broadly adaptive. But it becomes philosophically interesting when the traits in question are evolved behaviors. For then, the idea often assumes a particular form, in which the behavior's function is treated as if it were the organism's goal, and the evolutionary explanation is recast in an intentional idiom. Why do swallows migrate? Because they want to escape the cold. Why do female rats kill their offspring? Because they know that the offspring will not survive and don't want to waste resources on them. Why do worker honeybees eat the eggs laid by fellow workers? Because they want the offspring of the queen to be reared instead. In this way, the language of instrumental rationality (as philosophers call it) is used to describe and theorize about evolved behavior: the organism is treated as an agent who acts for reasons, makes decisions, and pursues goals.

Note that this evolutionary use of intentional-psychological language arises in the context of giving ultimate or evolutionary explanations. As such, it raises quite different issues from those discussed in the ethological literature examined earlier, where the focus was on proximate explanation. A honeybee's nervous system is too simple, and its behavior too rigid, for a proximate explanation of its egg-eating behavior in intentional terms to be plausible. The bee does not really *want* anything; it is simply obeying its hormonal impulses. But this fact does not prevent us from construing the evolutionary explanation of its behavior, metaphorically, in terms of what the bee wants: that is, the goal that it is trying to achieve. We just need to be clear that "goal" in this context refers to the behavior's evolutionary function, not its proximate cause.

Expressing evolutionary explanations in this psychological fashion is fairly common and has been explicitly defended by Dennett (1987) and Dawkins (1976). But one might reasonably wonder what its point is, given that, unless used carefully, it invites a confusion of ultimate with proximate explanation (Scott-Phillips et al., 2011). Moreover, given that evolutionary biology in any case needs the notion of function (in the sense of adaptive significance), what if anything is to be gained by treating a trait's function as if were the organism's goal and then introducing psychological descriptors? Does this add anything important?

I think that it does, for the following key reason. Functional talk applies to traits, but intentional-psychological talk applies to the whole organism. A particular trait has a function; but it is the whole organism, not its traits, that pursues a goal, or prefers one thing to another, or adjusts its behavior to achieve a certain end. Thus, when intentional language is used in an evolutionary context, the subject of the intentional attribution is the whole organism, not one of its traits. The meerkat's warning behavior has a function, but it is the meerkat that sees the danger and wants to warn its companions.

Why does this matter? Because it highlights an implicit theoretical commitment of the "organism as agent with goal" idea: namely, that the organism exhibits what may be called a *unity of purpose*. This means that its different traits have evolved because of their contributions to a *single* overall goal: enhancing the organism's fitness (or perhaps its inclusive fitness). Where this unity does not obtain, the organism cannot be regarded as agent-like, and treating its behavior as a means by which it furthers its goal will impede, not facilitate, evolutionary understanding of its behavior. I develop this theme below.[6]

Consider unity of purpose in the human context first, where it is a fundamental aspect of human agency.[7] This unity has two components. Firstly, a person's goals should cohere with each other in the sense of being mutually reinforcing, or at least not clearly inconsistent; secondly, their actions should tend to further their goals (that is, they should be instrumentally rational). Minor deviations from this unitary ideal are common, but if they are too many, or too great, it becomes impossible to treat the person as a unified agent, and to rationalize their actions in terms of their goals. Indeed, if a person is sufficiently disunified, psychological descriptors lose their grip entirely: we cannot say what they believe or are trying to achieve. In the biological case, an analogous unity of purpose is necessary in order to sensibly treat an evolved organism as akin to an agent with a goal and is presupposed when intentional-psychological idioms ("wants," "tries") are applied to the organism in an evolutionary context, in the manner described above. By contrast, the functional idiom, because it applies on a trait-by-trait basis, involves no such presupposition.

To illustrate this point, let us consider three cases in which the required unity of purpose partly breaks down, two actual and one hypothetical. All three involve within-organism conflict. In *Drosophila pseudoobscura*, males that carry a particular X-chromosome variant produce no Y-bearing sperm at all, as a result of "sperm killer" genes on the X chromosome which disrupt spermatogenesis. As a result, far fewer viable sperm are produced than in normal males. This trait—failure to produce Y-bearing sperm—evolved not because it benefits the organism, which it does not, but rather because it benefits the X-chromosome itself (and the genes on it). So, the fly exhibits a partial disunity of purpose. Most of its traits (e.g., its mating behavior) pull in the direction of maximizing its reproductive success, but the trait of producing no Y-bearing sperm pulls in a completely different direction. If biologists studying fly spermatogenesis treat the fly as a fully unified agent, they will not understand what they see.

Examples of this sort could easily be multiplied, because a certain amount of intragenomic conflict is found in many species (Burt & Trivers, 2006). For the most part, though, organisms have evolved mechanisms to suppress such internal conflict, and thus to ensure that all their constituent genes work for the common good. It is precisely because this suppression is so effective that we are usually able to treat organisms as agents pursuing goals, and to describe their evolved behavior using the language of instrumental rationality.

Our second example involves parasitic manipulation. Consider an ant infected by the liver fluke parasite *Dicrocoelium dendriticum*. This parasite induces a change in the ant's behavior, causing it to climb to the top of a blade of grass every evening and stay there, clamped to the tip with its mandible. This increases the chance that the ant will be ingested by a sheep, which is what the parasite needs to complete its life cycle. An infected ant thus exhibits a partial disunity of purpose. Most of its traits, such as its foraging behavior, further its goal of survival and reproduction, but its nightly ascent of a blade of grass detracts from that goal. It is as if the ant is simultaneously pursuing two incompatible goals. If biologists try to treat the ant as a fully unified agent, they will not understand what they see.

These two failures of biological unity of purpose are the analogue of a human agent whose goals conflict. Our third, hypothetical example is the analogue of a human agent performing an action that detracts from, rather than furthers, one of its goals. Imagine a mouse gene, expressed in females, that causes a female mouse to kill any pups in its litter which do not contain a copy of that gene. (This is not as far-fetched as it sounds.[8]) Such a gene could easily spread by natural selection (though genes at unlinked loci in the mouse genome would be selected to suppress it). If the gene does spread, then although an evolutionary explanation of the mouse's infanticidal behavior could be given, it could not be couched in terms of what the mouse "wants" or is "trying" to achieve, for the behavior detracts from, rather than furthers, the mouse's goal of leaving surviving offspring. The mouse thus lacks (the second component of) unity of purpose.

The general moral is this. To treat an evolved organism as agent-like requires that we can treat the organism's various traits as instruments for achieving subgoals—finding food, keeping warm, producing gametes, mating—which contribute to a single overarching goal, namely enhancing the organism's fitness. Empirically, this requires that the genes coding for the traits have evolutionary interests that are aligned, so that the traits evolve functions which are complementary rather than antagonistic. This in turn requires that intragenomic conflict and parasitic manipulation are largely absent or suppressed. For otherwise, then although each trait considered individually can be given a functional explanation, the traits cannot be treated as contributions to a single overarching goal.

The contrast between intentional attributions to organisms and functional attributions to traits is reminiscent of a traditional contrast in the philosophy of psychology, between personal and subpersonal attributions. Folk-psychological notions, such as believing and desiring, are personal-level: it is whole persons that occupy these states. By contrast, the computational processes described by cognitive psychology are subpersonal; they are carried out not by persons but by parts of their brains. Thus, the cerebral cortex processes visual information, but the person *sees* the approaching car and moves out of the way. Essentially, we have here a biological analogue of this distinction. It is the parts of an organism—that is, its traits—that have Darwinian functions, but it is the whole organism that has aims, goals, and preferences. Moving from the former sort of attribution to the latter is only possible insofar as the organism exhibits a biological unity of purpose.

This point is a corollary of a widely accepted evolutionary principle, namely that internal conflict tends to undermine the integrity of a larger unit (be it an organism or a social group). Thus, multicelled organisms have evolved numerous mechanisms for suppressing conflict among their constituent genes and cells, including fair meiosis, uniparental inheritance of organelles, and programmed cell death (Frank, 2003; Bourke, 2011). It is because these mechanisms usually work well that organisms are as cohesive and integrated as they are. This biological principle is an empirical one, but it has a conceptual counterpart. It is only because an organism's constituent traits typically cohere with each other in this way that the organism can be treated as akin to an agent pursuing a goal. For the same reason, groups of multicelled organisms can only rarely be treated as agent-like.

Because the unity-of-purpose requirement is not always satisfied, this might be regarded as a limitation of expressing evolutionary explanations in terms of organisms pursuing goals. In a way this is so, but it also shows that this practice has a genuine rationale and is not idle metaphor. Most of the time, the requisite organismic unity does obtain, at least to a high degree of approximation. There is thus a real pattern in nature that is captured by treating the organism as if it were an agent pursuing a goal, which the functional idiom alone does not capture. Therein lies the heuristic value of psychologizing evolutionary explanations in this way, a practice that at first blush may seem unmotivated.

This point should be sharply distinguished from a suggestion of Trivers (2009), which postulates a causal link between intragenomic conflict and internal conflict in the human psyche. Trivers focuses on genomic imprinting, in which a gene has different phenotypic effects depending on whether it is paternally or maternally inherited. This leads to intragenomic conflict, for if a gene in an organism is paternally inherited, then it has no genetic interest in the future reproduction of the organism's mother, while genes at other loci do. Trivers suggests that this will have psychological consequences: "we literally have a paternal self and a maternal self, and they are often in conflict" (Trivers, 2009: 163). Haig (2006) argues similarly.

The Trivers/Haig hypothesis is interesting though speculative. I take no stand on the matter here. I do not claim a direct connection between biological unity of purpose and unity of purpose in human agents (or their absence). Rather, my point is that the attribution of goals and intentions to a subject only makes sense if the subject exhibits sufficient unity, that is, the subject is an undivided self, or close enough. So, to treat an evolved organism as akin to an agent trying to achieve a goal, for the purposes of evolutionary theorizing, requires that the organism's traits do not have mutually antagonistic functions; and empirically, this requires the absence (or near-absence) of intragenomic conflict. This is a claim about the presuppositions of a particular psychologically derived idiom that we apply, usually meta-phorically, to evolved organisms, not a claim about the evolutionary roots of psychological unity or disunity in humans.

13.5   Conclusion

We have examined three different biological contexts in which goal attributions arise. The first is the mid-twentieth-century debate over goal-directedness among both biologists and philosophers, where the driving concern was to better understand teleology (or the appear-ance of it). It is here that the concept of teleonomy has its natural home. The second is the ongoing debate among cognitive ethologists about the presence or otherwise of goal-directed behavior, in the intentional sense, in nonhuman animals. The third is the evolu-tionary biologist's penchant for treating organisms as if they were rational agents pursuing a goal, in the context of seeking evolutionary explanations for their behavior.

The three contexts are distinct, but there are interrelations among them. The first and second are related because the sort of goal-directed behavior that the cognitive ethologists are concerned with (Mayr's "purposive behavior") was usually treated as a special case of goal-directedness by participants in the first debate. The second and third are connected because the intentional-psychological descriptors that are used in an as-if sense in the third, evolutionary context, are exactly those whose literal applicability is at stake in the second, ethological context. Finally, the first and third are related because one theme in the first debate was the need to distinguish something's having a goal from its having a function; whereas in the third context, the mode of evolutionary explanation in question precisely involves treating a behavior's function as if it were the organism's goal.

Moreover, there is a thematic question that runs through all three contexts, captured in the subtitle of this paper: objective fact, anthropomorphic bias, or valuable heuristic? In the first context, the underlying issue is whether there really is an objective property of goal-directedness in the first place. Does "goal-directed process" pick out a natural kind? We expressed skepticism on this score, because of the heterogeneity of the class and the lack of clarity regarding its membership; and we suggested that something like Mayr's "program view" would have to be defensible if the objectivity of goal-directedness were to be sus-tained. In the second context, the issue is whether nonhuman animals are really capable of (intentional) goal-directed behavior, or whether this is just anthropomorphism. We tentatively suggested that goal-directed behavior is likely a reality in some nonhumans, but that the anthropomorphism accusation is not without basis in other cases, as complex adaptive behav-ior can be produced by mechanisms that have no resemblance to belief-like and desire-like inner states. In the third context, the issue is whether psychologizing evolutionary explana-tions by treating behavior-functions as organismic goals has any real point to it. We argued that it does, because the unity of purpose that this practice presupposes reflects a real and biologically important fact about organisms: namely, that their evolved traits (for the most part) are designed to achieve the single overall goal of maximization of fitness.

These answers are provisional; they are not intended as the final word on how we should understand goal attributions in the three contexts. But I hope that the foregoing discussion has shown that goals and goal-directedness in biology—a central aspect of the concept of teleonomy—are still topics worthy of philosophers' and biologists' attention.

Acknowledgments

This paper is part of a project that has received funding from the European Research Council (ERC) under the European Union's Horizon 2020 research and innovation program (grant agreement number 101018533). Thanks to Dick Vane-Wright and Peter Corning for comments on an earlier draft.

Notes

1. A notable exception is McShea (2012).

2. Mayr also introduces the category of a "somatic program," but defines it, confusingly, in essentially the same way as an open program.

3. There are various hypotheses about why vertical migration is advantageous, of which predator avoidance is the leading contender. Maintenance of the oxygen supply is unlikely to be the reason.

4. A Web of Science search for articles published between 1900 and 2021 with the terms *goal-directed* or *goal-directedness* in the title, abstract, or keywords produces 14,105 results, of which only 198 are in biology journals. Neuroscience and psychology journals account for the largest share.

5. Shea (2013) defends the idea that genes contain "representational content." Maynard Smith (2000) defends the related idea that genes contain "semantic information" about phenotypes.

6. A fuller exposition of this line of argument is given in Okasha (2018, ch. 1).

7. Unity of purpose is discussed by Kennett & Matthews (1993) and Okasha (2018, ch. 1). It is closely related to the "rational unity" discussed by Rovane (1998) and the "unity of agency" described by Korgsaard (1989). Bratman (1987) emphasizes that agents are rationally required to have consistent intentions, and to exhibit means-end coherence.

8. David Haig's kinship theory of genomic imprinting has uncovered phenomena of exactly this sort (Haig, 2002).

References

Allen, C., & Bekoff, M. (1995). Cognitive ethology and the intentionality of animal behavior. *Mind and Language*, 10(4), 313–328.

Barrett. L. (2011). *Beyond the brain*. Princeton NJ: Princeton University Press.

Beckner, M. (1969). Function and teleology. *Journal of the History of Biology*, 2(1), 151–64.

Bourke, A. F. G. (2011). *Principles of social evolution*. Oxford, UK: Oxford University Press.

Braithwaite, R. B. (1953). *Scientific explanation*. Cambridge: Cambridge University Press.

Bratman, M. (1987). *Intention, plans, and practical reason*. Cambridge, MA: Harvard University Press.

Bray, D. (2009). *Wetware: A computer in every living cell*. New Haven, CT: Yale University Press.

Burt, A., & Trivers, R. (2006). *Genes in conflict*. Cambridge, MA: Harvard University Press.

Carruthers, P. (2004). On being simple minded. *American Philosophical Quarterly*, 41(3), 205–220.

Churchland, P. (1981). Eliminative materialism and the propositional attitudes. *Journal of Philosophy*, 78(2), 67–90.

Clayton, N., Emery, E., & Dickinson, A. (2006). "The rationality of animal memory." In S. Hurley, S., & Nudds, M. (Eds.). *Rational animals*, (pp. 197–216). Oxford, UK: Oxford University Press.

Corning, P. A. (2022). "Teleonomy in evolution: The ghost in the machine." In P. A. Corning, S. A. Kauffman, D. Noble, J. A. Shapiro, & R. Vane-Wright (Eds.). *Evolution 'On Purpose': Teleonomy in Living Systems*, (pp. forthcoming) Cambridge, MA: MIT Press.

Davidson, D. (1982). Rational animals. *Dialectica*, 36, 318–27.

Dawkins, R. (1976). *The selfish gene*. Oxford, UK: Oxford University Press.

Dennett, D. C. (1983). Intentional systems in cognitive ethology. *Behavioral and Brain Sciences*, 6, 343–355.

Dennett, D. C. (1987). *The intentional stance*. Cambridge MA: MIT Press.

Dickinson, A. (2011). "Goal-directed behavior and future planning in animals." In R. Menzel, & J. Fischer. (Eds.), *Animal thinking: Contemporary issues* (pp. 79–91). Cambridge, MA: MIT Press.

Fox-Keller, E. (2014). From gene action to reactive genomes. *Journal of Physiology*, 592(11), 2423–2429.

Frank, S. A. (2003). Repression of competition and the evolution of cooperation. *Evolution*, 57, 693–705.

Garson, J. (2008). "Function and teleology." In A. Plutynski & S., Sarkar (Eds.). *A companion to the philosophy of biology*, (pp. 525–549). Malden, MA: Blackwell.

Garson, J. (2018). "Nature and normativity: Biology, teleology and meaning". *Notre Dame Philosophical Reviews*. https://ndpr.nd.edu/reviews/ature-and-normativity-biology-teleology-and-meaning/

Gilbert, S. F. (2003). "The reactive genome." In G. B. Müller & S. A. Newman (Eds.). *Origination of organismal form: Beyond the gene in developmental and evolutionary biology*, (pp. 87–101). Cambridge, MA: MIT Press.

Glimcher, P. (1983). *Decisions, uncertainty and the brain*. Cambridge, MA: MIT Press.

Grafen, A. (2007). The formal Darwinism project: a mid-term report. *Journal of Evolutionary Biology*, 20(4), 1243–1254.

Haig, D. (2002). *Genomic imprinting and kinship*. New Brunswick, NJ: Rutgers University Press.

Haig, D. (2006). Intrapersonal conflict. In M. K. Jones, & A. C. Fabian (Eds.). *Conflict*, (pp. 8–22). Cambridge: Cambridge University Press.

Kant, I. (2000/1790). *Critique of the power of judgment*. (P. Guyer, Ed.; P. Guyer & E. Matthews, Trans.) Cambridge: Cambridge University Press.

Kennedy, J. S. (1992). *The new anthropomorphism*. Cambridge: Cambridge University Press.

Kennett, J., & Matthews, S. (2003). The unity and disunity of agency. *Philosophy, Psychiatry and Psychology*, 10, 308–312.

Korsgaard, C. M. (1989). Personal identity and the unity of agency. *Philosophy and Public Affairs*, 18, 103–131.

Maynard Smith, J. (2000). The concept of information in biology. *Philosophy of Science*, 67 (2), 177–194.

Mayr, E. (1974). Teleological and teleonomic. a new analysis. *Boston Studies in the Philosophy of Science*, 14, 91–117.

Mayr, E. (1988). *Toward a new philosophy of biology*. Cambridge, MA: Harvard University Press.

Mayr, E. (1992). The idea of teleology. *Journal of Historical Ideas*, 53, 117–535.

McFarland, D. J. (1989a). *Problems of animal behavior*. London: Longmans.

McFarland, D. J. (1989b.) "Goals, no-goals and own goals." In A. Montefiore & D. Noble (Eds.) *Goals, no-goals and own Goals*, (pp. 39–57). London: Unwin Hyman.

McShea, D. W. (2012). Upper-directed systems: A new approach to teleology in biology. *Biology and Philosophy*, 27, 663–684.

Monod, J. (1973). *Chance and necessity*. New York: Vintage Books.

Nagel, E. (1977a). Teleology revisited: Goal-directed process in biology. *Journal of Philosophy*, 74(5), 261–279.

Nagel, E. (1977b.) Functional explanations in biology. *Journal of Philosophy*, 74(5), 280–301.

Newman, S. (2022). "Self-organization in embryonic development: Myth and reality." In A. D. Malassé (Ed.) *Self-Organization as a new paradigm in evolutionary biology: From theory to applied cases in the tree of life*. New York: Springer.

Okasha, S. (2018). *Agents and goals in evolution*. Oxford, UK: Oxford University Press.

Pittendrigh, C. S. (1958). "Adaptation, natural selection and behavior." In A. Roe & G. G. Simpson (Eds.) *Behavior and evolution*, (pp. 390–416). New Haven, CT: Yale University Press.

Ristau, C. (1991). Aspects of the cognitive ethology of an injury-feigning bird, the piping plover. In C. Ristau (Ed.), *Cognitive Ethology* (pp. 93–124). Hillsdale, NJ: Lawrence Erlbaum Associates.

Roff, D. A. (1992). *Evolution of life histories*. New York: Chapman and Hall.

Rosenblueth, A., Wiener, N., & Bigelow, J. (1943). Behavior, purpose and teleology. *Philosophy of Science*, 10, 18–24.

Rovane, C. (1998). *The bounds of agency*. Princeton, NJ: Princeton University Press.

Scott-Phillips, T. C., Dickins, T. E., & West, S. A. 2011. Evolutionary theory and the ultimate/proximate distinction in the human behavioral sciences. *Perspectives on Psychological Science*, 6, 38–47.

Shea, N. (2013). Inherited representations are read in development. *British Journal for the Philosophy of Science*, 64 (1), 1–31.

Shea, N. (2018). *Representation in Cognitive Science*. Oxford: Oxford University Press.

Sommerhoff, G. (1950). *Analytical Biology*. Oxford: Oxford University Press.

Trewavas, A. (2014). *Plant Behaviour and Intelligence*. Oxford: Oxford University Press.

Trivers, R. (2009). Genetic conflict within the individual. *Sonderdruck der Berliner-Brandenburgische Akademie der Wissenschschaften*, 14, 149–199, Berlin: Akademie Verlag.

Waddington, C. H. (1957). *The strategy of the genes*. London: Ruskin House.

West, S. A., & Gardner, A. (2013). Adaptation and inclusive fitness. *Current Biology*, 23, R577-R584.

Williams, G. C. (1966). *Adaptation and natural selection*. Princeton, NJ: Princeton University Press.

Woodfield, A. (1976). *Teleology*. Cambridge: Cambridge University Press.

Wright, L. (1973). Functions. *Philosophical Review*, 82, 139–68.

Wright, L. (1976). *Teleological explanations: An etiological analysis of goals and functions*. Berkeley, CA: University of California Press.

14 Toward the Physicalization of Biology: Seeking the Chemical Origin of Cognition

Robert Pascal and Addy Pross

Overview

The seeming incompatibility between central life attributes, such as cognition on the one hand, and the objective laws of physics and chemistry on the other, continues to torment modern science. In this chapter we investigate whether the existence of the recently proposed chemical domain of dynamic kinetic chemistry, and its underlying stability kind, dynamic kinetic stability (DKS), may provide a physico-chemical basis for understanding the different behavioral characteristics of living and nonliving things, including that of cognition. It is suggested that all living systems, together with many of their subsystems, are in the DKS state. In contrast to the thermodynamic state, the DKS state is defined as one that is open to both energy and material exchange, and critically dependent on its environment for that energy and material support. That dependence allows for the emergence of "self" and "nonself" since both manifest through an ongoing dynamic interaction between the DKS system and its supporting environment. Once a chemical system in the DKS state acquires a replicative capability, it would then express rudimentary elements of cognitive function: perception, memory, learning, decision making, information transfer. A subsequent process of kinetic selection would then direct the system toward greater DKS stability, leading to improvements in both physical and cognitive capabilities. Darwin's radical suggestion, that life's mental dimension is as much a part of the evolutionary process as its physical counterpart, 160 years on, appears to have been magnificently prescient. The physical basis for life's mental dimension, with its associated teleonomic character, may be beginning to emerge from the chemical shadows.

14.1 Introduction

Despite the enormous advances in our understanding of the natural world that have taken place during the past century, there lies a black hole at the center of that understanding, one that continues to cast a long shadow on science's three central disciplines of physics, chemistry, and biology: the nature of the life phenomenon. The issue has a variety of puzzling physical facets, but life's nonmaterial characteristics—cognition, agency, mind, consciousness—remain awkwardly incompatible with a physical material view of nature.

Though molecular biology's string of extraordinarily successful revelations over the past 70 years, beginning with Watson and Crick's landmark DNA structure paper (Watson & Crick, 1953), is clear testament to life's molecular basis, life's unique emergent properties remain inexplicable in existing physico-chemical terms. How can life's "purposiveness" and unambiguous teleonomic character (Monod, 1972) be understood in material terms? As Stuart Kauffman put it: "despite the fine work . . . in the past three decades of molecular biology, the core of life itself remains shrouded from view. We know chunks of molecular machinery, metabolic pathways, means of membrane biosynthesis—we know many of the parts and many of the processes. But what makes a cell alive is still not clear to us" (Kauffman, 2000). Life's central question—what is it?—continues to confound.

The mystery is of course a long-standing one. During the seventeenth century, in a flash of inspiration, the Flemish physicist Jan van Helmont postulated that "all life is chemistry" (van Helmont, 1648). That insight was born from a growing belief in a reductionist approach to understanding nature, within what is commonly termed the seventeenth-century modern scientific revolution (Butterfield, 1997). After all, it was that revolution in scientific thinking that led to the dramatic scientific and technological advances of the past several centuries. Yet, in life, matter has seemingly managed to transcend its material constraints. Erwin Schrödinger, the renowned twentieth-century physicist, stated it pointedly: "Consciousness cannot be accounted for in physical terms. For consciousness is absolutely fundamental. It cannot be accounted for in terms of anything else" (Moore, 1994, 181). Indeed, to rationalize the existence of mind and consciousness, Schrödinger and other leading physicists of the twentieth century were increasingly drawn toward an anti-mechanistic, anti-reductionist view, or, at the very least, to the possibility that unknown laws of physics must be involved (Schrödinger, 1944; Delbrück, 1949; Wigner, 1970). Well into the twenty-first century, the source of life's teleonomic character remains chemically enigmatic.

Of course, that anti-reductionist sentiment was not universal. Several of biology's leading twentieth-century scientific figures, such as Francis Crick, Linus Pauling, and Jacques Monod, remained faithful to reductionism and continued to insist that it was the only scientific game in town (Fuerst, 1982). More recently, Evelyn Fox Keller, though continuing to argue for a reductionist approach to biology, pointed out that the concept of function, so indispensable to biology, is absent from the vocabulary of physics and chemistry (Keller, 2010). How can one expect physics and chemistry to explain a phenomenon whose existence is not even acknowledged in physical terminology? Indeed, biology's seeming inability to adequately connect with physics and chemistry has ended up leading the biological sciences toward a separate philosophy of biology, one whose central theme has been an "autonomy of biology" approach to the subject (Mayr, 1988, 24–25). Given life's seemingly irreducible complexity, the philosopher Thomas Nagel went as far as to question reductive materialism as a legitimate philosophic basis for scientific understanding (Nagel, 2012). Could there be a world beyond the physical (Kauffman, 2019)? The scientific stakes could not be higher.

But there is one significant piece of knowledge that argues against biology's irreducibility to physics and chemistry. The current overwhelming scientific consensus is that life emerged from nonlife through some coherent physico-chemical process (for recent reviews, see: Ruiz-Mirazo et al., 2014; Ashkenasy et al., 2017). However, if that scientific presumption is valid, then that fact, on its own, suggests that biology *is* reducible to physics and chemistry. Yes, a clear description of that emergence process is still tantalizingly out of reach. But, as for all physico-chemical processes, it would presumably be constituted from coherent steps, each of which would be both logically possible and physically consistent with established physical/chemical theory. The point is that once the process becomes known—revealing how and why matter of *any* kind would be directed toward life—life's essence would have begun to be exposed. Like in a magician's conjuring trick, once the trick's workings have been uncovered, the mystery would dissipate. Importantly, it will not only be the mystery of life's emergence that will have been resolved. Understanding the process of emergence would likely also throw new light on the biological evolutionary process which followed that emergence (so-called Darwinian evolution) given that the two processes increasingly appear to have been two phases of one single continuous process (Pross, 2011, 2016). More importantly, however, life's unique characteristics—its inordinate complexity, its functional character, its mental dimension—would be there for inspection and analysis. The challenge is there before us.

14.2 Seeking to Physicalize Biology

So, where to begin? Though the process by which life emerged continues to mystify, there is one facet of the process that is beyond doubt: the extraordinary degree of complexification that took place during that extended physico-chemical transformation. The conversion of some prebiotic chemical system, though of unknown identity, into simplest life would have involved an increase in size/mass of some nine orders of magnitude. (Molecular systems are typically of mass of *ca.* 10^{-21} g while that of a bacterium is *ca.* 10^{-12} g). That's a staggering change in both size and structural complexity—the equivalent of something the size of a coin growing into something the size of our planet. In fact, it seems reasonable to conclude that life's unique nonmaterial characteristics must have emerged as a direct consequence of that extraordinary degree of complexification (Pross, 2013). So how can that process of complexification be explained?

Complexity is a multifaceted problem in both the physical and biological sciences (Gell-Mann, 1995; Whitesides & Ismagilov, 1999; Adami, 2002), so different approaches, reflecting the different perspectives, have been taken. Thus, Corning and Szathmáry (2015) have taken a Darwinian approach to the problem by proposing the explanatory concept of "synergistic selection." A physico-chemical perspective suggests that at least one source of the complexity derives from the process of self-organization (Karsenti, 2008). Such self-organization is observed at the molecular level within cellular processes, as well as in the collective behavior of organisms, as observed, for example, in bacterial communication (Lyon, 2015). A physical-mathematical approach to the problem, broadly termed *complexity theory*, was in vogue in the late twentieth century (Gell-Mann, 1995; Holland, 1998; Heylighen et al., 1999; Kauffman, 2000), while Goldenfeld and Woese recently proposed that the problem of biological complexity lies within nonequilibrium statistical physics (Goldenfeld & Woese, 2011). What is clear is that the search for simplicity within the complexity continues, as it must. For buried within that complexity is where the essence of the life phenomenon presumably lies hidden.

One hint as to where a missing element may lie comes about from a simple observation. A living thing is not just a particular physico-chemical assembly, an aggregate of various molecular components synchronously functioning together like the cogs and wheels within a mechanical watch. A living system is better understood as an *energized* physico-chemical assembly (Ragazzon & Prins, 2018; Das et al., 2021), one which both stores and depends on a continuous supply of material and energy for its maintenance. If the material and energy supplies are cut off, the living system ceases to exist—it dies. Only then does that chemical aggregate behave as chemical aggregates normally do. That explains why life is so difficult to create. Simply, we still do not know how to generate dynamic energized systems of that kind. Or, to put it in less scientific terminology, we still lack the physico-chemical capability of being able to breathe life into a dead chemical system. We have yet to identify the physico-chemical basis of life's so-called "vital force." However, as will now be described, a possible way in which inanimate chemical systems may be activated in a biologically relevant way has come to light in recent years (Boekhoven et al., 2010, 2015; Ragazzon & Prins, 2018; Das et al., 2021). That landmark discovery has led to a broader understanding of the kinetic stability concept and its different manifestations within physico-chemical systems (Pascal et al., 2013; Pascal & Pross, 2019; Pross 2011, 2016; Pross & Khodorkovsky, 2004; Pross & Pascal, 2013; Yang et al., 2021; Liu et al., 2022). In the life context, that greater kinetic understanding has provided new insights into the life phenomenon: its physical basis as well as its physically intractable biological characteristics.

14.3 Dynamic Kinetic Stability (DKS) as a Physico-Chemical Characterization of Life

14.3.1 DKS as an Energized State

Change characterizes the world around us, with matter undergoing continual transformation from less stable to more stable forms. Such change in nature is traditionally understood in thermodynamic terms, in accordance with the second law of thermodynamics (Lineweaver & Egan, 2008). More recent developments in nonequilibrium thermodynamics, however, as formulated by Prigogine and co-workers, led to the concept of dissipative structures, thereby enabling thermodynamic understanding to be extended to nonequilibrium systems (Kondepudi & Prigogine, 1998). Thus, given the nonequilibrium state that characterizes all living things, it was suggested that living things can be viewed as dissipative structures, that life was but a further means by which nature is able to dissipate energy (England, 2013).

In contrast to a thermodynamic approach to the understanding of life processes, the possible role of kinetics in governing such processes has become more apparent in recent years. Beginning with the early work of Lotka (1922), the discovery of oscillatory chemical reactions (Belousov, 1959; Zhabotinsky, 1964), and subsequent analyses of how kinetics and thermodynamics relate to one another (Pross, 2003; Ross et al., 2012; Pascal et al., 2013; Pross & Pascal, 2017), the view of life as a *kinetic* phenomenon is increasingly acknowledged. It is that kinetic perspective that lies at the heart of the following discussion.

Central to the understanding of change in the world is the concept of *stability*. Broadly speaking, within the physico-chemical domain, one can specify two general stability kinds: thermodynamic and kinetic (see figure 14.1). A system is considered thermodynamically stable when it exists in its lowest possible energy state on a potential energy surface (figure 14.1, state a). In contrast, a system is considered kinetically stable if it exists in a local energy minimum that is separated from the lowest energy state by a significant energy barrier (figure 14.1, state b). That kinetic stability is of a *static* kind (as exemplified by the stability of an unreacted mixture of hydrogen and oxygen gases). Such a stability kind is well-understood and needs no further discussion. However, there exists an alternative kinetic stability kind that is *dynamic*, termed *dynamic kinetic stability* (DKS) (figure 14.1, state c). As will now be discussed, that stability kind bears directly on both chemical and biological reactivity patterns (Pross & Khodorkovsky, 2004; Pross, 2011, 2016; Pascal & Pross, 2019; Pross & Pascal, 2013; Pascal et al., 2013), and is a central focus of this chapter. It turns out that *all* living things exist in a dynamic kinetically stable state.

As helpful background for a description of the chemical DKS state, let us begin by describing a simple physical DKS system, a water fountain. An operational water fountain is a thermodynamically unstable system, and its existence only comes about through the action of a pump. Notice, however, that though the water in the fountain at any given moment is in an unstable energy state, the fountain itself is stable in a *time* sense; provided that the water supply and energy operating the pump are maintained, the fountain continues to exist. That stability kind can be termed *dynamic kinetic*. The term "dynamic" is applied because the water that constitutes the fountain is continually turning over—same fountain, continually different water. The "kinetic" term signifies that both the fountain's existence and its nature depend on *rates*— the rate at which water is ejected from the fountain nozzle, and the rate at which it then falls away. Surprisingly but significantly, thinking about the stability/persistence of a water fountain in this fashion can offer useful insights into the nature of living things: as *transient, yet persistent*. Metaphorically speaking, living things can be thought of as "chemical fountains."

The water-fountain analogy helps us understand the significance of the 2010 discovery by Jan van Esch, Rienk Eelkema, and co-workers that chemical fountains (chemical systems in

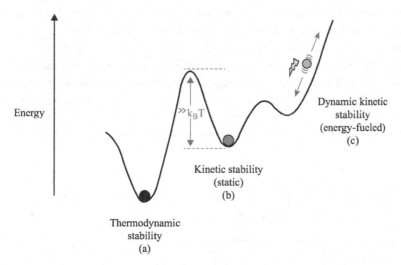

Figure 14.1
Possible stability kinds of a chemical system depicted on a potential energy profile. (a) Thermodynamic stability, where a system is in its lowest energy state. (b) Kinetic stability (static), where a system is in a local energy minimum, but separated from the lowest energy minimum by a kinetic barrier. (c) Dynamic kinetic stability, which describes the stability of a far-from-equilibrium chemical system, whose stable state reflects the dynamic energy-fueled continuous interconversion of material through a balance of synthesis and degradation processes.

the DKS state) actually exist (Boekhoven et al., 2010, 2015). That discovery was a chemical game-changer, as prior to that work dynamically stable chemical systems were not generally recognized (Whitesides & Grzybowski, 2002). A chemical DKS system is defined as one that is in an energized, nonequilibrium, dynamic, cyclic state, in analogy to that physical DKS water-fountain system (Pascal & Pross, 2015). By taking one of the most basic reactions in organic chemistry, esterification, and activating it into the DKS state through the coupling of the reaction with material and energy resources, the chemical system was able to enter a kinetically stable state, one whose self-assembly characteristic (hydrogel formation) could be tuned through kinetic modification (Boekhoven et al., 2010, 2015). Accordingly, as illustrated schematically in figure 14.2, one can think of chemical space as being composed of two distinct domains: the traditional thermodynamic domain, encompassing most chemical processes, and the recently revealed dynamic kinetic domain. In the thermodynamic domain, two interconverting materials, X and Y, can exist in an equilibrium state, or in a nonequilibrium state that tends toward equilibrium, and that situation is the chemical norm. However, if a chemical system is appropriately activated into a DKS state, then X and Y could interconvert via an energy-fueled irreversible cyclic process (depicted in figure 14.2) potentially resulting in a different kind of system with different properties. Of course, if a DKS system's material and energy support are withdrawn, the system immediately collapses. Both physical and chemical DKS systems respond in that way. Recall the simple water fountain analogy: if either the water or the energy supply to the fountain is cut off, the fountain collapses.

The significance of the existence of the DKS chemical domain lies in the fact that chemical DKS systems have been found to express biological-like properties, such as self-healing, adaptivity, and to communicate in a limited sense (Maiti et al., 2016; Zhao et al., 2016; Leira-Iglesias et al., 2018; Te Brinke et al., 2018; Merindol & Walther, 2017; Tena-Solsona et al., 2018). In fact, through the van Esch–Eelkema discovery, a new chemical world of opportunity appears to have opened up. Indeed, a rash of follow-up work on chemical DKS systems has led to the opening of an exciting new door in material science (Merindol & Walther, 2017).

Now back to the biological connection. First, though chemists discovered chemical DKS systems relatively recently, we now realize that biological entities have been exploiting that kind of chemistry—one might call it *dynamic kinetic chemistry*—for millions, if not billions, of years. The dynamic functional character of the cell cytoskeleton is a prime example of such chemistry. The cytoskeleton is composed of continually polymerizing and depolymerizing microtubules and actin filaments mediated by GTP/ATP in the energy-fueled polymerization step, followed

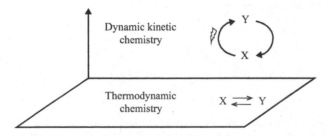

Figure 14.2
Schematic diagram illustrating the existence of two reactivity domains in chemical space: the traditional thermodynamic domain and the newly discovered dynamic kinetic domain.

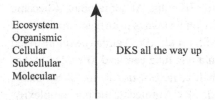

Ecosystem
Organismic
Cellular   DKS all the way up
Subcellular
Molecular

Figure 14.3
Life as a hierarchy of nested biological DKS systems, from molecular through to ecosystem.

by GTP/ATP hydrolysis to GDP/ADP, which induces the depolymerization step. That dynamic process is crucial for key cell functions, in particular transport of cellular components and sorting of chromosomes during cell division (Desai & Mitchison, 1997; Karsenti, 2008). More generally, however, cellular metabolism, based on cyclic, energy-fueled, irreversible, dynamic processes, also exemplifies DKS subcellular structures and associated processes.

However, it is more than the cell's various subsystems that are in an effective DKS state. The cell in toto, being an energized, nonequilibrium, dynamic system, can also be characterized as being in a DKS state. Indeed, cells within a multicelled organism undergo dynamic exchange at various rates, thereby enabling life to be viewed as a *set of nested DKS systems*—molecular, subcellular, cellular, organismic, ecological, as depicted in figure 14.3. The DKS state is central to operation and functionality at all of life's hierarchical levels (Pross, 2011, 2016, 2019; Pascal et al., 2013; Pross & Pascal, 2013; Pascal & Pross, 2019).

Interestingly, this DKS perspective lends some unexpected physico-chemical support to the Gaia hypothesis, the idea first proposed by Lovelock and Margulis half a century ago. They contended that life on Earth was more than just about individual organisms struggling to survive, and that it should instead be considered as an evolving global homeostatic system (Lovelock & Margulis, 1974). Biologists who, as a group, had in the past largely scorned the Gaia concept, are now reassessing. Ford Doolittle, the noted evolutionary biologist, offered a Darwinian perspective in support of Gaia (Doolittle, 2017). The DKS concept lends added support to the Gaia hypothesis because it allows life to be understood as an everextending dynamic replicative network. Network complexification takes place at multiple levels, from molecular through to ecological (figure 14.3). Thus, we see how the DKS concept that characterizes living systems in physico-chemical terms appears able to provide a more fundamental understanding of the life phenomenon.

Let us now turn to one of life's central characteristic, its replicative ability, to see how that fits into the DKS view of life.

14.3.2   Replicative DKS Dynamics and the Persistence Principle

The importance of molecular replication and its role in life processes was first hypothesized by Troland in the early twentieth century (Troland, 1914). However, it was the landmark discovery of DNA's molecular structure and the template mechanism by which replicating molecules replicate (Watson & Crick, 1953), that led to a replication-based hypothesis for the origin of life on Earth: the RNA-world hypothesis (Woese, 1965; Crick, 1968; Orgel, 1968; Gilbert, 1986). Its essence was that the emergence of life was initiated by the fortuitous appearance of some molecular replicating entity, possibly RNA or RNA-like, which then undertook an evolutionary process toward greater complexity and life.

It was not too long before that hypothesis ran into difficulty. At some point it became increasingly clear that a molecular replicative capability on its own is insufficient to facilitate an evolutionary process toward life. As Spiegelman (Mills et al., 1967) already demonstrated some 50 years ago, a molecular replicative process in a test tube can lead to evolution, but the evolutionary process is then directed *away* from life, not *toward* it. In Spiegelman's molecular evolutionary experiment, the initially extended RNA molecule did not complexify, but rather simplified—it shortened drastically. Evidently, a piece of the life puzzle was missing, and the identity of that missing puzzle piece has plagued the subject for more than half a century. How can an evolutionary process of a replicating entity toward complexity and life be induced?

The existence of the DKS state offers a feasible solution to that evolutionary conundrum. If in some manner a simple replicative system is induced into the DKS state, a new world of chemical possibilities opens up. Kinetic analyses by Szathmáry, Lifson, and others (Szathmáry & Gladkih, 1989; Lifson, 1997), though not dealing explicitly with DKS systems, indicated that the evolutionary dynamics of a replicative system governed by DKS kinetics would be quite different from that of a simple replicative chemical system following traditional kinetics. The evolutionary directive would be toward dynamic kinetic more stable replicating entities, rather than toward faster replicators or thermodynamically more stable ones (Pascal & Pross, 2015). Thus, the direction of change in nature, normally addressed in second-law terms, requires an alternative directing principle in such cases. The second law is less useful for predicting the direction of change for energy-fueled systems where change may be governed by *kinetic* rather than *thermodynamic* factors.

In fact, the DKS concept allows the formulation of what could be considered an extension of the second law, one that could be applied, at least in principle, to both kinetic and thermodynamic systems. That extended principle has been termed the *persistence principle* (Pascal & Pross, 2015). The principle may be stated most simply as follows: *All material entities are driven from less persistent to more persistent forms.* The origin of the principle may be traced back to Grand's provocative tautological pronouncement that he termed "the most important law of nature": *"things that persist, persist and things that don't, don't"* (Grand, 2001). Indeed, for replicative systems in the DKS state, the persistence principle enables the direction of change to be specified; that is, from *less* DKS stable to *more* DKS stable. Thus, whereas the second law is more useful for predicting the direction of change for thermodynamically-directed systems, the persistence principle can, in some cases at least, extend that predictive capability to cover kinetically-directed systems. That proves useful in the biological context, as an immediate evolutionary insight then follows. In contrast to a widespread biological viewpoint, the evolutionary process is not just a "blind algorithm" (Dennett, 1995) with no associated direction. A DKS perspective suggests that the evolutionary process for replicative systems does have a direction: toward more persistent forms, toward replicative systems of greater DKS. Indeed, preliminary experimental work by Otto and co-workers has lent support to that expectation (Yang et al., 2021; Liu et al., 2022). In recent chemical evolutionary experiments, it was found that the formation of the more complex (and DKS more stable) replicator was favored over replicators that replicated faster or were thermodynamically more stable. In this context it is also important to note that the concept of natural selection, at the heart of Darwin's epoch-making scientific contribution, can now be understood as a particular biological manifestation of a wider physical/chemical phenomenon: kinetic selection (Pross, 2011, 2016).

Interestingly, biologists, who are of course also concerned with survival issues, have reached some complementary conclusions, though couched in biological terminology. Thus, the central biological concept of homeostasis, termed by Turner as biology's second law (second to its first law, natural selection) (Turner, 2017), can be thought of as the biological equivalent of DKS (though, strictly speaking, *homeostasis* refers to individual biological entities, whereas DKS, in the context of replicative systems, is associated with replicating populations). The point is that both homeostasis, formally described as the living organism's state of internal constancy, and DKS are about *dynamic stability*, the former in biological jargon, the latter in physico-chemical jargon.

Some further comment on the persistence principle and how it manifests in living systems is now in order. A moment's consideration shows us that in seeking persistent forms, nature has demonstrated remarkable creativity through effectively having "discovered" its own laws (Pross, 2021). Just by watching life in action, we see that nature exploits those laws, and does so far more effectively than we humans. Take the common house fly, for example. Its ability to land on a ceiling upside down is quite staggering. It greatly exceeds the technological know-how of flight engineers. Engineers happily study high-speed videography of the fly's inverted landing to learn how to better program robot flight. The reality is that in living organisms, from lowly bacteria on up, nature has proven itself to be the ultimate technologist, exceeding human technological capabilities by far. The 2020 Nobel Prize in chemistry was awarded to Jennifer Doudna and Emmanuelle Charpentier for discovering the CRISPR-Cas9 procedure for DNA editing (Jinek et al., 2012). The procedure was proposed following their discovery that bacteria carry out that gene-editing function on their own genome as part of their immune system, and have been doing so for eons. Copy and paste! No doubt about it: nature, the technologist, can be seen to be an expert information theorist, polymer scientist, electrical engineer, systems chemist, photochemist, synthetic chemist, and molecular biologist, to mention just a few of the scientific skills that nature exhibits in carrying out life processes (Pross, 2021). And it does all of this through the remarkable directing power of kinetic selection or, in biological terminology, natural selection. Indeed, the emergence of mind and consciousness could also be understood in that light. Nature discovered the possible existence of a mental dimension just as it discovered the possibility of flight, photosynthesis, signal transduction, and gene editing. However, in contrast to flight, photosynthesis, signal transduction, and gene editing, science is still struggling to understand the nonmaterial world of mind and consciousness. We are still far from understanding the basis for a nonmaterial dimension and, in particular, the manner in which the material and nonmaterial dimensions interrelate.

Let us now take the most tentative of steps to begin to probe that nonmaterial dimension through the DKS prism. One of life's most striking characteristics is cognition, its study being a central focus of what is now termed cognitive science, the interdisciplinary study of the mind. It is hoped that by tracing life's origins back to its presumed chemical beginnings, the beginnings of life's mental dimension might be discerned and a physical basis for cognition possibly uncovered. A way of addressing Chalmer's so-called "hard question" (Chalmers, 1996)—why *any* set of physical and chemical process would be accompanied by mind (as we humans are all too aware)—might open up through a study of origins.

14.4   The Physical Basis for Cognition

14.4.1   What Is Cognition?

It is hard to come up with a topic in the life sciences that has proven more controversial than that of cognition (Akagi, 2018; Lyon, 2020; Keijzer, 2021). Notwithstanding those many uncertainties, however, there are some broad features of cognition that are widely agreed upon, and it is these that we will attempt to address. Our goal is to explore the possible origins of that biological phenomenon in physico-chemical terms. The methodological approach taken is that in tackling a complex problem, it is often useful to "go simple." If you want to understand the principles of flight, don't start by examining a Boeing 747. Start with kites, gliders, and the Wright brothers' 1903 model. In that context the cognitive capabilities of so-called simplest life, bacteria, are revealing, and constitute a first step toward a deeper understanding of the cognition phenomenon.

So, what *is* cognition? As noted, there is no generally accepted definition for the term, despite it being a central concept in neuroscience (Bechtel, 1998; Lyon, 2015, 2020; Akagi, 2018; Keijzer, 2021; Lyon et al., 2021; Shapiro, 2021). A widely cited definition offered by Shettleworth (1998) proposes it to be "the mechanisms by which living things acquire, process, store, and act on information from the environment. These include perception, learning, memory, and decision-making." That definition, which was formulated to cover animal cognition, was based on the widespread view that cognitive behavior derived from the existence of a brain and nervous system (Miller, 2003).

But that neural viewpoint/assumption is increasingly questioned (Maturana & Varela, 1980; Lyon, 2015; Shapiro, 2007, 2021; Beer, 2021; Lyon et al., 2021). With time, it has become evident that the Shettleworth definition could well apply to *aneural* life forms, from multicelled plant life (Gagliano, 2017) down to the simplest life form, bacteria. Thus, Maturana and Varela (1980), despite Maturana's background in neurobiology, took that broader view by hypothesizing that cognition is at the very core of all biological activity, that "living as a process is a process of cognition." More recently, Shapiro (2021), in his study of bacterial function, describes in detail the various means by which bacteria sense their internal condition, and the elaborate response systems that are routinely activated. As he puts it: "bacteria are small, but not stupid" (Shapiro, 2007). Lyon (2015) has gone as far as to claim that cognitive terms such as *decide, talk, listen, cheat, eavesdrop, lure,* and *vote* are appropriate for describing remarkably complex bacterial social behavior. Provocative, for sure. Beer paraphrases Maturana by pointing out that cognition is evident in all living things, and that "in the end, there is just an organism engaging its environment . . . in the service of its existence" (Beer, 2021). What is increasingly clear, therefore, is that simplest life—single-cell prokaryotes—already displays remarkable cognitive capabilities. That realization, on its own, necessitates a reassessment of the cognition concept. What is unexpected, however, is that through that realization, the interminable body–mind problem may be starting to emerge out of the shadows. Let us describe how.

An intriguing observation regarding that more basic, aneural, view of cognition is that Darwin already alluded to it in his *Origin of Species* (Darwin, 1860). As pointed out by Lyon (2021), Darwin proposed that the evolutionary process took place along both physical/material *and* mental axes, and that life's mental attributes were therefore present from life's beginnings. If that is true—if mental life was indeed present at life's beginnings, as Darwin speculated, and as more recent research into bacterial cognition is reaffirming (Shapiro, 2021)—this has profound implications, not only for what "mental" means, but also, crucially, for how it could have originated. Indeed, life's chemical origins might offer new insights into one of science's most challenging frontiers, the nature of mind.

The growing view that a mental axis was present from the outset of the evolutionary process strongly suggests that such mental evolution did not begin with earliest life, but would have existed in an earlier evolutionary phase, in some prelife chemical system; in other words, that cognition would have been initiated in chemistry. Chemical cognition? Initially the phrase might sound quite unreasonable. Surely molecules, or even molecular assemblies, cannot be cognizant. However, if bacteria can be cognizant, then the logic of evolution almost requires that some prelife chemical entity would also have exhibited at least rudimentary cognitive behavior. After all, that is the essence of Darwinism—continuity. Life did not just pop out of nowhere. But the following questions then arises: If cognition did in fact begin within chemistry, how would that manifest at the chemical level? What would be the simplest chemical system that could be considered cognitive, at least in a rudimentary way? Was there a chemical Rubicon that had to be crossed? As will now be discussed, the nature of the energized replicative DKS state, being conceptually relevant to both physico-chemical and biological systems, may well offer insight into the physical basis of cognition, and possibly other nonmaterial characteristics as well.

14.4.2   Cognition and the DKS State

Let us begin by stating the obvious: All material systems, living or not, respond to external perturbation, though, of course, the response of living and nonliving things to those perturbations is very different. Let us consider each in turn. Inanimate material forms respond to perturbations through the directing effects of the second law of thermodynamics. Thus, the heating of a physical body might cause it to expand, undergo a phase transition, or undergo some other physical (or chemical) process. Similarly, the addition of some material to a chemical system at equilibrium would drive the system out of chemical equilibrium, and the system would then respond by seeking to re-establish an equilibrium state. In neither the physical nor the chemical case do we consider the system's response as reflecting a cognitive capability. Also, in neither case is the system's response classified as adaptation, certainly not in the biological sense of the word. In both cases the system's response is explicable in thermodynamic terms.

In contrast, the response of living things to some perturbation or stimulus is quite different. Living systems respond through *adaptation*, defined as the biological mechanisms by which organisms adjust to modifications in their environment in a way that advances the living thing's agenda. Consider, for example, chemotaxis, the directed motion of an organism in response to some chemical stimulus, for example, the presence in the organism's vicinity of nutrition or a toxin (Wadham & Armitage, 2004). Living systems are agents, meaning they act on their own behalf. Accordingly, the response of biological systems to perturbations is associated with the system's agenda, ultimately being one of survival. This explains why bacteria propel themselves toward nutrition but away from toxins. A biological system's response to any perturbation, though *consistent* with the second law, is in no way *explained* by the second law. So, given the undeniable physico-chemical basis of all living things, how can we understand the response of living systems to perturbations in physico-chemical terms, rather than in biological terms? As we will now discuss, the DKS concept when applied to replicative systems may assist in this endeavor. Recall our declared strategy for attempting to unravel biological complexity: go simple. The relationship of a DKS system to its environment, whether that DKS system is physical or biological, is distinctly different from the relationship of a thermodynamic system to its environment. Let us consider the biological and physical DKS cases in turn, beginning with the biological one.

A living system is crucially dependent for its sustainability—indeed, its very existence—on its environment. Thus, a photosynthetic life form is critically dependent on a continual supply of solar energy, carbon dioxide, and water, while an aerobic respiring life form is critically dependent on a continual supply of organic resources and oxygen. Sunlight, carbon dioxide, water, organic resources, and oxygen are not merely perturbations on those living systems. They are a crucial support system that allows those living thing to live.

That description is, of course, consistent with the DKS view of life. All DKS systems, whether simple or complex, physical or biological, depend on environmental resources for them to be maintained. Indeed, a comparison with the physical DKS example of a water fountain may help clarify the point. In the fountain example, the fountain's stability (in fact, its very being) also depends on external factors—on a continued supply of water, on pump action, and on an energy source that enables pump action. Accordingly, perturbations in a DKS system, whether physical or biological, which affect the energy and material support for the DKS system are not perturbations in the thermodynamic sense. Such perturbations may well be existential. To exist or not to exist will depend on the nature and extent of the perturbation.

Clearly then, the relationship of a DKS system to its environment is of a totally different category than that of an inanimate system to its environment. The DKS system and its environmental support system are umbilically linked. The DKS system's existence is based on an *ongoing dynamic interaction* between the system and its supporting environment. Severing that umbilical connection between the DKS system and its environment is for the DKS state the physico-chemical equivalent of death. Taking the DKS system with its supporting environment as a super-system whole, the DKS system can be thought of as the super-system's "inside"—the dynamic, energy-fueled, energized physical or chemical part—while the external support system of material resources and energy (and to a lesser extent, the environment as a whole) on which the system depends is the super-system's "outside." Indeed, the chemical origin of cognition can be seen to lie, at least in part, in the nature of the DKS state. Dependence is a nonmaterial state; in fact, one might think of it as a first step toward a rudimentary state of mind. It is inherent in the DKS system's very existence, as without its environment, it ceases to exist. Samuel Johnson's famous hanging aphorism, that knowing one is to be hanged concentrates the mind wonderfully, may assist in driving that point home. Through that intimate dependence, the basis for rudimentary cognition can be understood to begin—a nonmaterial dimension, though one clearly derived from material circumstances. One might also view the beginnings of that rudimentary cognitive process as the beginning of information transfer in a biological sense. The existence of a continual flow of resources to the DKS system could be viewed as the transfer of information to the DKS system from its environment.

This DKS view of cognition can be taken a step further, as through that perspective the concept of self might take form. All living things, operationally at least, manifest this sense of self, a sense we do not associate with inanimate systems. But for this sense of self to exist, there would have to be a reference non-self. It is an "outside," intimately linked to the "inside" through continual information transfer, which provides that reference, though the full implication of what "self" can mean only comes about through an evolutionary process in which the sense of self is enhanced. Indeed, with the evolutionary process in mind, the crucial role of replication in the emergence of cognition can now be considered.

The rudimentary sense of self described above is not truly meaningful for a *nonreplicative* DKS system. A nonreplicative physical DKS system such as a water fountain, or a nonreplicative chemical DKS system such as the DKS ester system, would not be classified as cognitive. Only through a replicative capability in a chemical DKS system—one able to undergo an evolutionary process—could the emergence of cognitive function take place. In fact, central elements of cognition (perception, memory, learned response)—can already be discerned within a replicative DKS system. Perception is manifested through the dynamic interaction between the system and its environment, as described previously. Memory is manifested through the replicative capability that enables favorable heritable structural characteristics to be maintained over successive generations. The kinetic process by which less persistent DKS replicators are continually replaced by more persistent ones can be viewed as a manifestation of learned response. Through the kinetic selection process the system "learns" which structural forms are beneficial, and which are not. The replicative DKS system both "remembers" and "learns" through interaction with its environment. Thus, what might be considered rudimentary mental aspects associated with living processes can be understood in established physico-chemical terms, with both physical and mental capabilities being enhanced in line with Darwin's early supposition. Such an evolutionary process would also lead to the DKS system undertaking material modification of its environment, not just self-modification, through what biologists have termed "niche construction" (Odling-Smee et al., 1996)—again, all induced through kinetic selection. The bottom line: Both physical *and* mental evolution, working in tandem, would, through a process of kinetic selection, lead toward increasingly stable DKS systems.

The preceding discussion helps clarify the difference between a living organism and a mechanical entity such as a watch, a comparison that has been widely employed since Descartes. As noted earlier, a DKS system is not a mechanical, machine-like entity like that watch. It is a dynamic energized system in an obligatory dependence on its surroundings with inherent cognitive capabilities. Interestingly, that view is consistent with the relatively recent enactive approach to cognition, which is based on the idea that an autonomous system is one that generates and sustains its own activity, thereby enacting its own cognitive domain (Varela & Bourgine, 1991; Thompson & Stapleton, 2009). Accordingly, we suggest that the emergence of a replicative chemical DKS system, able to evolve in an open-ended manner, would have been the watershed moment which led to the subsequent emergence of physical life together with its nonmaterial accompaniment of cognition.

Some aspects of this way of thinking about cognition have already appeared in the biological literature. Buzsáki, in a text entitled *The Brain from Inside Out*, argues that the traditional outside-in perspective to understanding cognitive behavior—one based on the view of the brain as an organ to perceive and represent the world—is inadequate (Buzsáki, 2019). He argues that the brain's fundamental function is to induce actions in support of its host's survival. Accordingly, he argues for an inside-out approach to understanding cognition, thereby turning the traditional view on its head. However, further reflection on the DKS system–environment relationship suggests there is actually a two-way dynamic interaction between the system and its supporting environment. Maturana's comment that "living as a process is a process of cognition" now takes on greater physical significance (Maturana & Varela, 1980). Living is effectively the dynamic interplay between a DKS system and its supportive environment, between self and non-self, and leads directly to Maturana's view that cognition is the essence of all biological activity. Life's physical and "mental" facets are intimately linked.

Finally, it should be emphasized that the DKS term characterizes a population of entities, not individual entities, whose existence is necessarily transient. In the life context, it is a population of replicators in the DKS state that gives rise to the structural characteristics that determine the system's stability—characteristics that its forerunners would have been selected for and found to be advantageous. Indeed, in that population context, death plays a crucial role. Death is a central element of the life cycle, one that facilitates the dynamic process of population renewal, whereby new variants can be continually scoured for improved function, thereby leading to increased DKS. Ultimately DKS enhancement, by all possible means, is the evolutionary name of the game.

To sum up, though cognition has intrinsically nonmaterial aspects, it derives from material causes, and is able to induce material consequences in both the system and its external environment. Indeed, that is why cognition emerged. From this viewpoint, neither cognition nor its associated nonmaterial counterparts—agency, self, mind—can exist independent of their material base. Nature "discovered" and "learned" to exploit the nonmaterial dimension in the same way that it learned to discover and exploit the physical dimension: to facilitate nature's ultimate drive, the drive toward increasingly persistent forms. Cognition is an emergent property of a particular kind of replicative chemistry, one whose beginnings can be discerned at the chemical/molecular level.

14.5   Concluding Remarks

Darwin was a biologist, but he intuitively understood that to understand biology in a fundamental way, the subject had to be physicalized, with his principle of natural selection being a first step in that physicalization process. He also understood that despite the limitations of nineteenth-century physics and chemistry, the process of physicalization had to be based on a reductionist materialist methodology. Already at that time, materialism was a widely (though not universally) accepted scientific and philosophical position, built on the understanding that all living things are but physical and chemical stuff, that life was just a particular kind of physics and chemistry. Indeed, as described in his 1871 letter to J. D. Hooker (Darwin, 1871), Darwin's proposal of a chemical origin of life "in some warm little pond with all sorts of ammonia & phosphoric salts—light, heat, electricity" can be viewed as further evidence for his physicalizing way of thinking. In fact, Darwin's subsequent comment in that letter that "a protein compound was chemically formed ready to undergo still more complex changes" makes it eminently clear that Darwin intuitively understood that the requirement of continuity in evolution meant that biology's roots must ultimately lie in chemistry. Not surprisingly, however, given the limited chemical knowledge of the time, he was unable to proceed further. And, indeed, here we are, 160 years on, still struggling to piece together life's physical and chemical puzzle.

This brings us back to the reductionist methodology and its relevance to biological understanding. Though that methodology over the centuries has proven to be extremely powerful, it can only yield useful insights if the reductionist elements on which the methodology is based are known. Just as the properties of matter cannot be reduced to those involving molecular interactions if the existence of molecules is unknown, so biological entities and biological processes cannot be reduced to physics and chemistry if the physical and chemical elements on which that reduction are based are not firmly established. Concepts can only develop in tandem with empirical findings that underpin those concepts. The existence of a *stable energized state of matter*, the DKS state—an energized, dynamic, nonequilibrium, cyclic, kinetic state—has, until recently, been a missing piece of the puzzle. However, once such a state acquires a replicative capability, a new dimension of chemical possibility opens up.

But now to the chapter's central issue of cognition and mind. Life's nonmaterial capabilities, like its material ones, also have to be acknowledged and incorporated within a physical perspective. Darwin already hinted as much in his *Origins*: that life's mental dimension was as much a part of the evolutionary process as its physical counterpart, that natural selection operated on both "corporeal and mental endowments" (Darwin, 1860). As Lyon (2021) put it, "Just as we have come to think of our bodies as evolved from simpler forms of body, it is time to embrace Darwin's radical idea that our minds, too, are evolved from much simpler minds. Body and mind evolved together and will continue to do so." Indeed, it is on this key issue that reductionist thinking offers new understanding. Given the current view that simplest life emerged from chemical beginnings in an early phase of the evolutionary process, the strong implication is that the evolution of body and mind would have been initiated in *prelife*, within chemistry, extending the conclusions of recent studies that have identified advanced cognitive capabilities in simplest life. Through the DKS concept, the mind–body problem may well have become tractable.

The underlying message is clear: nonmaterial (pseudo-mental) relationships can emerge from a material world, at least from certain segments of it, and such relationships can and will affect the material forms that make up the world. Mind has its roots in matter. Life, together with its mental dimension, is an explicable physico-chemical form whose characteristics have to be fully consistent with nature's objective physical and chemical laws. As van Helmont already foresaw some five centuries ago, all life is chemistry. The physicalization of biology, Darwin's momentous scientific crusade initiated 160 years ago, continues apace. We may finally have begun to internalize Darwin's deepest insight: that to understand life, both physically and mentally, we must understand its origins.

Acknowledgments

The authors thank Sijbren Otto and Pamela Lyon for helpful comments on an earlier draft of this chapter.

References

Adami C. 2002. What is complexity? *BioEssays* 24, 1085–1094.

Akagi M. 2018. Rethinking the problem of cognition. *Synthese* 195, 3547–3570. https://doi.org/10.1007/s11229-017-1383-2

Ashkenasy G, Hermans TM, Otto S, and Taylor AF. 2017. Systems chemistry. *Chem. Soc. Rev.* 46, 2543–2554.

Bechtel W. 1998. Representations and cognitive explanations: assessing the dynamicist's challenge in cognitive science. *Cogn. Sci.* 22, 295–317.

Beer RD. 2021. Some historical context for minimal cognition. *Adapt. Behav.* 29(1), 89–92. https://doi.org/10.1177/1059712320931595

Belousov B. 1959. A periodic reaction and its mechanism. *Compilation of Abstracts on Radiation Medicine* 147, 145 (in Russian).

Boekhoven J, Brizard AM, Kowlgi KNK, Koper GJM, Eelkema R, et al. 2010. Dissipative self-assembly of a molecular gelator by using a chemical fuel. *Angew. Chem. Int. Ed.* 49(28), 4825–4828. https://doi.org/10.1002/anie.201001511

Boekhoven J, Hendriksen WE, Koper GJM, Eelkema R, and van Esch JH. 2015. Transient assembly of active materials fueled by a chemical reaction. *Science* 349(6252), 1075–1079. https://doi.org/10.1126/science.aac6103

Butterfield H. 1997. *The origins of modern science.* New York: Free Press.

Buzsáki G. 2019. *The brain from inside out.* New York: Oxford University Press.

Chalmers DJ. 1996. *The conscious mind: In search of a fundamental theory.* New York: Oxford University Press.

Corning PA and Szathmáry E. 2015. Synergistic selection: a Darwinian frame for the evolution of complexity. *J. Theoret. Biol.* 371, 45–58.

Crick FH. 1968. The origin of the genetic code. *J. Mol. Biol.* 38, 367–379.

Darwin CR. 1860. *On the origin of species by means of natural selection,* 2nd ed. London: Murray.

Darwin CR. 1871. Darwin Correspondence Project, University of Cambridge, Letter 7471. https://www.darwinproject.ac.uk/letter/DCP-LETT-7471.xml

Das K, Gabrielli L, and Prins LJ. 2021. Chemically-fueled self-assembly in biology and chemistry. *Angew. Chem. Int. Ed.* 60(37), 20120–20143. https://doi.org/10.1002/anie.202100274

Delbrück M. 1949. A physicist looks at biology. *Transactions Connecticut Academy of Arts and Sciences,* 38, 173–190.

Dennett, D. C. 1995. *Darwin's dangerous idea.* London: Penguin.

Desai A and Mitchison TJ. 1997. Microtubule polymerization dynamics. *Annu. Rev. Cell Dev.* Biol. 13, 83–117. https://doi.org/10.1146/annurev.cellbio.13.1.83

Doolittle WF. 2017. Darwinizing Gaia. *J. Theoret. Biol.* 434, 11–19.

England JL. 2013. Statistical physics of self-replication. *J. Chem. Phys.* 139, 121923. https://doi.org/10.1063/1.4818538

Fuerst JA. 1982. The role of reductionism in the development of molecular biology: peripheral or central? *Social Stud. Sci.* 12(2). https://doi.org/10.1177/030631282012002003

Gagliano M. 2017. The mind of plants: Thinking the unthinkable. *Communicative & Integrative Biol.* 10(2), e1288333. https://doi.org/10.1080/19420889.2017.1288333

Gell-Mann M. 1995. What is complexity? *Complexity* 1(1), 16–19. https://doi.org/10.1002/cplx.6130010105

Gilbert W. 1986. The RNA world. *Nature* 319, 618.

Goldenfeld N and Woese C. 2011. *Annu. Rev. Condens. Matter Phys.* 2, 375–399.

Grand S. 2001. *Creation.* Cambridge, MA: Harvard University Press.

Hershko A and Ciechanover A. 1982. Mechanisms of intracellular protein breakdown. *Annu. Rev. Biochem.* 51, 335–364. https://doi.org/10.1146/annurev.bi.51.070182.002003

Heylighen F, Bollen J, and Riegler A. (Eds.). 1999. *The evolution of complexity: The violet book of 'Einstein meets Magritte'.* Dordrecht, Germany: Kluwer Academic Press.

Holland JH. 1998. *Emergence: From chaos to order.* Reading, MA: Addison-Wesley.

Jinek M, Chylinski K, Fonfara I, Hauer M, Doudna JA, and Charpentier E. 2012. A programmable dual-RNA-guided DNA endonuclease in adaptive bacterial immunity. *Science* 337(6096), 816–821. https://doi.org/10.1126/science.1225829

Karsenti E. 2008. Self-organization in cell biology: a brief history. *Nat. Rev. Mol. Cell. Biol.* 9, 255–262. https://doi.org/10.1038/nrm2357

Kauffman SA. 2019. *A world beyond physics: The emergence and evolution of life.* Oxford: Oxford University Press.

Kauffman SA. 2000. *Investigations.* Oxford: Oxford University Press.

Keijzer F. 2021. Demarcating cognition. *Synthese* 198 (Suppl 1), 137–157. https://doi.org/10.1007/s11229-020-02797-8

Keller EF. 2010. It is possible to reduce biological explanations to explanations in chemistry and/or physics. In *Contemporary debates in philosophy of biology*, 19–31. (Eds.). Ayala FJ and Arp R. Hoboken, NJ: Wiley-Blackwell.

Kondepudi D and Prigogine I. 1998. *Modern thermodynamics: From heat engines to dissipative structures.* Chichester, UK: Wiley.

Leira-Iglesias J, Tassoni A, Adachi T, Stich M, and Hermans TM. 2018. Oscillations, travelling fronts and patterns in a supramolecular system. *Nat. Nanotechnol.* 13, 1021–1027. https://doi.org/10.1038/s41565-018-0270-4

Lifson S. 1997. On the crucial stages in the origin of animate matter. *J. Mol. Evol.* 44, 1–8. https://doi.org/10.1007/PL00006115

Lineweaver CH and Egan CA. 2008. Life, gravity and the second law of thermodynamics. *Phys. Life Rev.* 5(4), 225–242. https://doi.org/10.1016/j.plrev.2008.08.002

Liu B, Wu J, Geerts M, Markovitch O, et al. 2022. Out-of-equilibrium self-replication allows selection for dynamic kinetic stability in a system of competing replicators. *Angew. Chem. Int. Ed.* 61(18), e202117605. https://doi.org/10.1002/anie.202117605

Lotka A. 1922. Natural selection as a physical principle. *Proc. Natl. Acad. Sci. USA* 8(6), 151–154. https://doi.org/10.1073/pnas.8.6.151

Lovelock JE and Margulis L. 1974. Atmospheric homeostasis by and for the biosphere: the Gaia hypothesis. *Tellus* 26, 2–9.

Lyon P. 2015. The cognitive cell: bacterial behaviour reconsidered. *Front. Microbiol.* 6, 264. https://doi.org/10.3389/fmicb.2015.00264

Lyon P. 2020. Of what is 'minimal cognition' the half-baked version? *Adapt. Behav.* 28(6), 407–424. https://doi.org/10.1177/1059712319871360

Lyon P. 2021. On the origin of minds. *Aeon.* https://aeon.co/essays/the-study-of-the-mind-needs-a-copernican-shift-in-perspective

Lyon P, Keijzer F, Arendt D, and Levin M. 2021. Reframing cognition: getting down to biological basics. *Phil. Trans. R. Soc. B* 376: 20190750. https://doi.org/10.1098/rstb.2019.0750

Maiti S, Fortunati I, Ferrante C, Scrimin P, and Prins LJ. 2016. Dissipative self-assembly of vesicular nanoreactors. *Nat. Chem.* 8, 725–731. https://doi.org/10.1038/nchem.2511

Maturana HR and Varela FJ. 1980. *Autopoiesis and cognition: The realization of the living.* Dordrecht, Germany: Reidel Publishing.

Mayr E. 1988. *Toward a new philosophy of biology.* Cambridge, MA: Harvard University Press.

Merindol R and Walther A. 2017. Materials learning from life: concepts for active, adaptive and autonomous molecular systems. *Chem. Soc. Rev.* 46, 5588. https://doi.org/10.1039/c6cs00738d

Miller G. 2003. The cognitive revolution: a historical perspective. *Trends in Cognitive Sci.* 7, 141–144.

Mills DR, Peterson RL, and Spiegelman S. 1967. An extracellular Darwinian experiment with a self-duplicating nucleic acid molecule. *Proc. Natl. Acad. Sci. USA* 58(1), 217–224. https://doi.org/10.1073/pnas.58.1.217

Monod J. 1972. *Chance and necessity.* New York: Vintage Books.

Moore W. 1994. *A life of Erwin Schrödinger.* Canto ed. Cambridge: Cambridge University Press.

Nagel T. 2012. *Mind and cosmos.* Oxford: Oxford University Press.

Odling-Smee FJ, Laland KN, and Feldman MW. 1996. Niche construction. *The Am. Naturalist* 147(4), 641–648. https://doi.org/10.1086/285870

Orgel LE. 1968. Evolution of the genetic apparatus. *J. Mol. Biol.* 38, 381–393.

Pascal R and Pross A. 2015. Stability and its manifestation in the chemical and biological worlds. *Chem. Commun.* 51, 16160–16165. https://doi.org/10.1039/c5cc06260h

Pascal R and Pross A. 2019. Chemistry's kinetic dimension and the physical basis for life. *J. Syst. Chem.* 7, 1.

Pascal R, Pross A, and Sutherland JD. 2013. Towards an evolutionary theory of the origin of life based on kinetics and thermodynamics. *Open Biol.* 3(11), 130156. https://doi.org/10.1098/rsob.130156

Pross A. 2003. The driving force for life's emergence: kinetic and thermodynamic considerations. *J. Theoret. Biol.* 220(3), 393–406. https://doi.org/10.1006/jtbi.2003.3178

Pross A. 2011. Toward a general theory of evolution: extending Darwinian theory to inanimate matter. *J. Syst. Chem.* 2, 1. https://doi.org/10.1186/1759-2208-2-1

Pross A. 2013. The evolutionary origin of biological function and complexity. *J. Mol. Evol.* 76, 185–191. https://doi.org/10.1007/s00239-013-9556-1

Pross A. 2016. *What is life? How chemistry becomes biology* (2nd ed.). Oxford: Oxford University Press.

Pross A. 2019. Seeking to uncover biology's chemical roots. *Emerg. Top. Life Sci.* 3(5), 435–443. https://doi .org/10.1042/ETLS20190012

Pross A. 2021. How was nature able to discover its own laws—twice? *Life* 11(7), 679. https://doi.org/10.3390 /life11070679

Pross A and Khodorkovsky V. 2004. Extending the concept of kinetic stability: toward a paradigm for life. *J. Phys. Org. Chem.* 17(4), 312–316. https://doi.org/10.1002/poc.729

Pross A and Pascal R. 2013. The origin of life: what we know, what we can know and what we will never know. *Open Biol.* 3(3), 120190. https://doi.org/10.1098/rsob.120190

Pross A and Pascal R. 2017. How and why kinetics, thermodynamics, and chemistry induce the logic of biological evolution. *Beilstein J. Org. Chem.* 13, 665–674. https://doi.org/10.3762/bjoc.13.66

Ragazzon G and Prins LJ. 2018. Energy consumption in chemical fuel-driven self-assembly. *Nat. Nanotech.* 13, 882–889. https://doi.org/10.1038/s41565-018-0250-8

Ross J, Corlan AD, and Muller SC. 2012. Proposed principles of maximum local entropy production. *J. Phys. Chem. B* 116, 7858–7865. https://doi.org/10.1021/jp302088y

Ruiz-Mirazo K, Briones C, and de la Escosura A. 2014. Prebiotic systems chemistry: new perspectives for the origins of life. *Chem. Rev.* 114, 285–366.

Schrödinger E. 1944. *What is life?* Cambridge: Cambridge University Press.

Shapiro JA. 2007. Bacteria are small but not stupid: cognition, natural genetic engineering and socio-bacteriology. *Stud. Hist. Phil. Biol. & Biomed. Sci.* 38, 807–819.

Shapiro JA. 2021. All living cells are cognitive. *Biochem. Biophys. Res. Commun.* 564, 134–149. https://doi .org/10.1016/j.bbrc.2020.08.120

Shettleworth SJ. 1998. *Cognition, evolution and behavior.* Oxford: Oxford University Press.

Szathmáry E and Gladkih I. 1989. Sub-exponential growth and coexistence of non-enzymatically replicating templates. *J. Theor. Biol.* 138(1), 55–58. https://doi.org/10.1016/S0022-5193(89)80177-8

Te Brinke E, Groen J, Herrmann A, Heus HA, Rivas G, Spruijt E, et al. 2018. Dissipative adaptation in driven self-assembly leading to self-dividing fibrils. *Nat. Nanotechnol.* 13, 849–855. https://doi.org/10.1038/s41565 -018-0192-1

Tena-Solsona M, Wanzke C, Riess B, Bausch AR, and Boekhoven J. 2018. Self-selection of dissipative assemblies driven by primitive chemical reaction networks. *Nat. Commun.* 9, 2044. https://doi.org/10.1038/s41467 -018-04488-y

Thompson E and Stapleton M. 2009. Making sense of sense-making: reflections on enactive and extended mind theories. *Topoi* 28, 23–30. https://doi.org/10.1007/s11245-008-9043-2

Troland LT. 1914. The chemical origin and regulation of life. *Monist* 24, 92–133.

Turner JS. 2017. *Purpose and desire.* New York: HarperOne.

van Helmont JB. 1648/1961. *Ortus medicinae.* Amsterdam: Ludovicum Elzevirium. In Gabriel ML, Fogel S (eds. & trans.), *Great experiments in biology.* Englewood Cliffs, NJ: Prentice Hall (from Dutch).

Varela FJ and Bourgine P. (Eds.). 1991. Toward a practice of autonomous systems. In: *Proceedings of the first European conference on artificial life.* Cambridge, MA: MIT Press.

Wadham GH and Armitage JP. 2004. Making sense of it all: bacterial chemotaxis. *Nat. Rev. Mol. Cell Biol.* 5, 1024–1037.

Watson JD & Crick FHC. 1953. Genetical implications of the structure of deoxyribonucleic acid. *Nature* 171, 964–966.

Whitesides GM and Grzybowski B. 2002. Self-assembly at all scales. *Science* 295(5564), 2418–2421. https://doi .org/10.1126/science.1070821

Whitesides G M and Ismagilov R F. 1999. Complexity in chemistry. *Science* 284, 89–92.

Wigner E. 1970. Physics and the explanation of life. *Found. Phys.* 1, 35–45.

Woese C. R. 1965. On the evolution of the genetic code. *Proc. Natl. Acad. Sci. USA* 54, 1546–1552.

Yang S, Schaeffer G, Mattia E, Markovitch O, Liu K, Hussain AS, Ottele J, Sood A, and Otto S. 2021. Chemical fueling enables molecular complexification of self-replicators. *Angew. Chem. Int. Ed.* 60(20), 11344–11349. https://doi.org/10.1002/anie.202016196

Zhabotinsky A. 1964. Periodic processes of malonic acid oxidation in a liquid phase. *Biofizika* 9, 306–311 (in Russian).

Zhao H, Sen S, Udayabhaskararao T, Sawczyk M, Kučanda K, Manna D, et al. 2016. Reversible trapping and reaction acceleration within dynamically self-assembling nanoflasks. *Nat. Nanotechnol.* 11, 82–88. https://doi .org/10.1038/nnano.2015.256

15 Evolutionary Change Is Naturally Biological and Purposeful

James A. Shapiro

Overview

In trying to remove any innate biological action so that natural selection could be the only determinant directing evolutionary change, Darwin's followers in the twentieth century obliged themselves to assume that gradual selected accumulation of character changes due to random mutations could produce major differences in morphology, metabolism, and behavior over long periods of time. This process, phyletic gradualism, was proposed to account for all evolutionary diversity without any biological agency. However, the gradualist view of evolutionary change could not accommodate cytogenetic and genomic evidence that living organisms possess internal capacities for making large-scale modifications to their heredity and phenotype. In particular, McClintock's discovery in the late 1940s of transposable controlling elements that could move to new genomic positions and alter the expression of nearby coding sequences solved an evolutionary problem that was virtually impossible to explain by phyletic gradualism: the formation of coordinately regulated multilocus systems connected by shared transcription factor binding sites. Molecular genomic data in this century has confirmed the roles of transposable elements in formatting expression of new genomic networks at key stages in evolutionary change. Moreover, the movement of the various types of transposable elements to new genome locations is triggered by ecological challenge and organismal stress. Thus, mobile DNA cassettes serve as dedicated change operators when evolutionary transformation functions capable of causing major genome rewriting under stress are essential purposive tools needed for life to survive in dynamic ecologies.

15.1 Introduction

As students, we are taught that the idea of goal-directed activities—teleology or teleonomy (Pittendrigh, 1958)—is unscientific. We learn that the natural world has no inherent drives or purposeful actions, and we are further taught that to believe such nonmaterial impulses exist is to indulge in anthropomorphizing nature. Nowhere was the word *teleology* such a forbidden term as among mainstream neo-Darwinian evolutionary biologists. Their attempts to deny biological purposefulness or agency in any form were complete mysteries to me. I could see plenty of functionality and intentionality in all fields of the life sciences,[1] and informatics/ cybernetics has taught us that not all natural phenomena are material (Corning, 2007; Wiener, 1965). It seemed to me that many neo-Darwinists turned themselves into intellectual and linguistic contortionists when they discussed biological action. The following quotation from a 2014 blog post by my well-known emeritus colleague at the University of Chicago, Jerry Coyne, is an illustration of what I mean.

> Among virtually all scientists, dualism is dead. Our thoughts and actions are the outputs of a computer made of meat—our brain—a computer that must obey the laws of physics. Our choices, therefore, must also obey those laws. This puts paid to the traditional idea of dualistic or "libertarian" free will: that our lives comprise a series of decisions in which **we could have chosen otherwise**. We know now that we can never do otherwise, and we know it in two ways.
>
> The first is from scientific experience, which shows no evidence for a mind separate from the physical brain. This means that "I"—whatever "I" means—may have the *illusion* of choosing, but my choices are in principle predictable by the laws of physics, excepting any quantum indeterminacy that acts in my neurons. In short, the traditional notion of free will—defined by Anthony Cashmore as "a belief that there is a component to biological behavior that is something more than the unavoidable consequences of the genetic and environmental history of the individual and the possible stochastic laws of nature"—is dead on arrival.
>
> The illusion of agency is so powerful that even strong incompatibilists like myself will always act as if we had choices, even though we know that we don't. We have no choice in this matter. But we can at least ponder why evolution might have bequeathed us such a powerful illusion.
>
> —Coyne, J. A. (2014). On free will, in answer to the question, What scientific idea is ready for retirement? (https://www.edge.org/response-detail/25381)

15.2   The Delusion of Physical Bottoms-Up Reductionism: A Commitment to Natural Selection as the Unique Guiding Force in Evolution

Coyne's denial of human agency or choice, naming both as illusory artifacts of "libertarian" dualism, is a stunning example of physical reductionism run amok. What kind of thinking lies behind such an extreme declaration? I only realized the answer to that question when I was asked to write an extended definition of "evolution" for the SAGE *Encyclopedia of Theory in Science, Technology, Engineering, and Mathematics*. The exercise obliged me to examine the basic tenets of the neo-Darwinian modern synthesis and explain why it failed to incorporate much of what we have learned from genome sequence data and other aspects of molecular biology and genetics (Shapiro & Noble, 2021). The appeal to natural selection as a guiding principle was meant to replace Larmarck's internal and immaterial "le pouvoir de la vie" that was supposed to drive evolution toward more complex life forms (Lamarck, 1994). Among twentieth-century neo-Darwinists, natural selection and random mutation were also invoked to replace Lamarck's "L'influence des circonstances," which guided hereditary variation by use and disuse of particular traits, an idea that Darwin had fully accepted in *The Origin of Species* (Darwin, 1859). I came to see that the determination of Darwin's twentieth-century followers to eliminate any internal (i.e., biological) drives and make natural selection the sole determinant of direction in evolutionary change imposed stringent limits on their ability to explain the origins of biological processes.

Assigning to natural selection the sole responsibility for determining the course of hereditary variation meant that mutations had to be random events, devoid of organic causation, and of small effect on the character of the organism. This was the founding principle of phyletic gradualism. As Darwin wrote in chapter 6 of the first edition of *The Origin of Species*, "if it could be demonstrated that any complex organ existed, which could not possibly have been formed by numerous, successive, slight modifications, my theory would absolutely break down. But I can find out no such case" (Darwin, 1859). It was this overly simplified view of evolutionary change adopted for theoretical reasons and devoid of any biological input[2] that forced the neo-Darwinists like Coyne to invoke a purely physicalist model of how organisms evolve. There was consequently no place in the conceptual framework of the modern synthesis for biologically mediated punctuated (saltatory) evolutionary change, as proposed by non-Darwinian evolutionists like Hugo de Vries or Richard Goldschmidt (de Vries, 1905; Goldschmidt, 1940), or for responses by evolving organisms to inputs from ecological fluctuations. In addition, the modern synthesis was based on the mid-twentieth-century idea that "genes are the basic units of all living things" (Beadle, 1948). Accordingly, an organism's genome was conceptualized as a collection of more or less independent gene units subject to their individual random gene mutations.

15.3 How Did Discoveries After the Modern Synthesis Reveal the Neo-Darwinian Emphasis on Natural Selection to Be a Problem in Practical and Theoretical Evolution Science?

The fundamental epistemological error of the neo-Darwinists was to believe that they could abstract evolution away from any causative biological component. The idea that forbidding biological goals, agency, and intentionality to make their theoretical constructs more rigorous was to confuse purely mechanical accounts with scientific explanations. In their desire to eliminate "libertarian dualism," evolutionists such as Coyne have had to ignore mounting empirical evidence demonstrating the activity of numerous cellular and biomolecular processes capable of changing the structure and content of genomes (Shapiro, 2013; Shapiro, 2017; Shapiro, 2019; Shapiro & Noble, 2021; Shapiro, 2022).

To select one of the most important genome-change processes for detailed discussion, we will focus on the work of Barbara McClintock and genomic validation of her ideas. In the late 1940s, McClintock unexpectedly discovered that extreme genome stress could induce her maize plants to activate normally dormant "controlling elements," which can transpose from one position in the genome to other positions and alter the regulation and expression of adjacent genetic loci (McClintock, 1950; McClintock, 1952; McClintock, 1987). Transpositions and chromosome rearrangements associated with these elements were a far cry from the random gene mutations envisaged by the modern synthesis. Consequently, in the 1950s and 1960s, McClintock's work received a hostile and incredulous reception. Nonetheless, molecular and genomic studies beginning in the 1960s confirmed the presence of comparable transposable DNA elements playing important roles in evolution in all types of living organisms, from the simplest bacteria and archaea to higher plants and animals, and she received the Nobel Prize for her discovery in 1983 (Bukhari et al., 1977; McClintock, 1984; Shapiro, 1983). Mobile DNA elements (sometimes called "jumping genes") analogous to McClintock's come in a variety of forms: (1) DNA transposons that move as DNA and have defined ends, (2) retroviruses and structurally related retrotransposons with long terminal repeats (LTRs) integrated into the genome which move via RNA intermediates with defined ends, and (3) non-LTR retrotransposons which move via RNA intermediates that do not require defined ends (Bourque et al., 2018). To buttress his 1859 view of phyletic gradualism, Darwin had quoted Linnaeus' dictum *Natura non facit saltus* ("Nature does not make jumps") in *The Origin of Species* (Darwin, 1859). However, the prestige and authority of Linnaeus and Darwin notwithstanding, McClintock and her followers demonstrated that when it comes to genomic DNA, nature does indeed literally make jumps, and life has evolved several different molecular mechanisms to make those jumps take place.

Their dedication to natural selection led neo-Darwinists like Coyne to deny well-documented sensory, cognitive, computational, and decision-making abilities in living organisms. In doing so, they discounted the discoveries about intricately organized biological soft matter that molecular genetics revealed to exist and to exercise sensory, regulatory, and computational processes in even the simplest living organisms (Bray, 2009; Regolin & Vallortigara, 2021). These biological systems often display a sensitivity and robustness that far exceed the products of modern technology (Bray, 2012; Bray, 2015). In the interval between the mid-twentieth century and the third decade of the twenty-first century, we have experienced a theoretical revolution in our understanding of inherited organismal traits. In particular, we have abandoned one-gene, one-trait notions based on individual genes as "the basic units of life" (Beadle, 1948), instead embracing what has come to be called *complex systems biology* (Bornholdt, 2001; Pattee, 1973; Toussaint & von Seelen, 2007). Biological traits are currently viewed from the perspective of the genomic networks that encode and regulate the formation and operation of multicomponent functional systems and structures in all organisms (Bray, 2003; Heng, 2008; Maslowska, Makiela-Dzbenska, & Fijalkowska, 2019; Pal et al., 2017; Pezzulo & Levin, 2016). This network perspective is particularly relevant, for example, to thinking about the evolution of multicellular development (evo-devo) (Bowman, Briginshaw, & Florent, 2019; Rebeiz, Patel, & Hinman, 2015).

15.3.1 Genetic Regulatory Studies Indicate How Mobile DNA Elements Contribute to Hereditary Network Evolution

McClintock's genetic studies were far from the only research revolutionizing our understanding of genetics. At the same time as McClintock was working out the details of controlling-element action, Jacques Monod and his colleagues at the Institut Pasteur in Paris began studying the regulation of protein synthesis in bacteria (Jacob & Monod, 1961). They and their successors discovered novel noncoding genetic elements that were central to the control of transcription and genome expression (Reznikoff, 1992). These noncoding elements included DNA sequences recognized by DNA-binding proteins now generically known as transcription factors (TFs), which control and coordinate messenger mRNA synthesis from selected coding sequences as organisms elaborate complex traits and structures (Bondos & Tan, 2001; Sprague, 1991; Zhang et al., 2005). A key feature of the underlying genomic expression networks is the presence of shared TF binding-site DNA sequences controlling transcription at coding regions located at multiple places within the genome. These TF binding sites are now designated as *cis*-regulatory modules or CRMs (Pauls et al., 2015; Schwarzer & Spitz, 2014; Suryamohan & Halfon, 2015). It is very difficult to imagine how random mutations and phyletic gradualism could generate the functionally significant numbers of shared CRMs that appear at unlinked genetic loci encoding different proteins. The collective probability of the same sequence motifs independently evolving *ab initio* at multiple genomic locations is the product of their individual probabilities, and this product rapidly becomes impossibly small as the number of shared distal CRMs increases. However, the ability of controlling elements to transpose to multiple genome sites provides a feasible solution to the problem of organizing or wiring transcriptional networks across genomes (Rebollo, Romanish, & Mager, 2012).

In the twenty-first century, evolution research is chiefly based upon genome sequencing. As empirical scientists, we are obliged to ask: What does the genomic evidence have to say about saltatory network evolution by mobile DNA action? The answer is quite a bit, in fact, and the number and documentation of the roles that mobile DNA plays in plant and animal evolution increase every year. The presence of TF binding sequences in mobile DNA elements had been noted early in the twenty-first century (Bolotin et al., 2011; Bourque et al., 2008; Jordan et al., 2003; Polak & Domany, 2006; Polavarapu et al., 2008; van de Lagemaat et al., 2003). Genome researchers also noted early on that control sites derived from mobile elements are phylogenetically distinct. For example, one early study noted that "human-mouse genome wide sequence comparisons reveal that the regulatory sequences that are contributed by TEs are exceptionally lineage specific" (Marino-Ramirez et al., 2005). A decade after the publication of the initial draft of the human genome sequence (Lander et al., 2001), comparison of 29 mammalian genomes identified more than 280,000 examples of CRMs derived from all classes of transposable DNA elements in our genome (Lindblad-Toh et al., 2011; Lowe et al., 2011). A 2014 paper analyzing CRMs in the mouse and human genomes confirmed the species-specificity of transposable DNA-derived elements and found that anywhere from 2% to 40% of all genomic binding sites could be traced to mobile DNA insertions, depending on the particular TF bound (Sundaram et al., 2014).

Table 15.1 summarizes the range and diversity of biological characteristics that have been attributed to mobile DNA-derived networks. The cases are notable for the range of organisms (a single-celled diatom to complex multicellular plants and vertebrates) and the biological importance of the systems affected, including gamete formation (germline development, flowering), early embryonic development (gastrulation, stem cell pluripotency), embryo development and nourishment (placenta, endosperm), body plan development, and nervous system formation. The phylogenetic distribution of a particular family of mobile elements indicates when it appeared in evolutionary history (Jurka et al., 2012). For example, "at least 16% of eutherian specific CNEs (conserved non-coding elements) overlap currently recognized transposable elements in human" (Mikkelsen et al., 2007). That kind of phylogenetic "dating" is how many of the examples in table 15.1 were assigned to the formation of a specific taxonomic group, such as animals, vertebrates, mammals, eutherian mammals, or primates (Jacques, Jeyakani, & Bourque, 2013). The connection between an ancient group of mobile DNA elements in multiple genomes and a certain kind of function, such as body plan development, forms the basis for the conclusion that those elements helped wire the control network at a particular evolutionary transition (Riegler, 2008). The evidence for positive selection operating on many recent mobile DNA element insertions indicates that the changes were truly of adaptive evolutionary benefit (Lowe, Bejerano, & Haussler, 2007; Rishishwar et al., 2018; Saber et al., 2016).

Mobile DNA elements contribute to genomic networks in other ways than just providing distributed TF binding sites. The protein coding sequences of mobile DNA elements have also been evolutionary precursors to a broad range of functionally different proteins

Table 15.1
Distributed genome network innovations attributed to mobile DNA elements

Organism	Transposable DNA-Modified System(s)	References
18 fungal genomes	Whole-genome architecture and transcriptional profiles	(Castanera et al., 2016)
Diatom *Phaeodactylum tricornutum*	Responses to nitrate starvation and exposure to diatom-derived reactive aldehyde-induced stress	(Maumus et al., 2009; Oliver, Schofield, & Bidle, 2010)
Leishmania	Post-transcriptional regulation	(Bringaud et al., 2007)
PLANTS		
Plants	Genome evolution	(Bennetzen & Wang, 2014; Qiu & Köhler, 2020)
Plants	Epigenetic controls	(Lisch & Bennetzen, 2011)
Plants	Stress responses	(Hou et al., 2019; Negi, Rai, & Suprasanna, 2016)
Peaches, almonds	Evolutionary genome differences	(Alioto et al., 2020)
Grasses	C4 photosynthesis	(Cao et al., 2016)
Maize	Abiotic stress response	(Lv et al., 2019; Makarevitch et al., 2015; Ramachandran et al. 2020)
Maize	Helitron transposons "reshuffle the transcriptome"	(Barbaglia et al., 2012)
Maize	25% of DNAseI-hypersensitive sites (actively transcribed loci) evolved from mobile DNA insertions	(Zhao et al., 2018)
Cotton	Fiber cell development	(Wang et al., 2016)
Tomato	Ripening	(Jouffroy et al., 2016)
Coffea	Drought stress response	(Lopes et al., 2013)
Arabidopsis	Abiotic stress (phosphate limitation, high salt, freezing temperatures, arsenic toxicity)	(Joly-Lopez et al., 2016; Joly-Lopez et al., 2017)
Arabidopsis	Flower development	(Baud et al., 2020; Muino et al., 2016)
Arabidopsis	Endosperm development (functionally comparable to mammalian placenta)	(Batista et al., 2019; Qiu and Köhler, 2020)
Pine	"Transposable element interconnected gene networks"	(Voronova et al., 2020)
ANIMALS		
Metazoa	Evolutionary innovation	(Nishihara, 2020; Piskurek & Jackson, 2012)
Drosophila	Malathion insecticide resistance	(Salces-Ortiz et al., 2020)

Drosophila	X chromosome dosage compensation	(Ellison & Bachtrog, 2019; Ellison & Bachtrog, 2013)
Drosophila	Early embryonic development	(Spirov, Zagriychuk, & Holloway, 2014)
Vertebrates	Evolutionary innovation	(Etchegaray et al., 2021; Warren et al., 2015)
Vertebrates	Body plan development	(McEwen et al., 2009; Woolfe & Elgar, 2008)
Zebrafish	p53 response network	(Micale et al., 2012)
Fish	Migratory behavior	(Carotti et al., 2021)
Mammals	TET-controlled passive DNA demethylation	(Mulholland et al., 2020)
Mammals	Estrogen receptor network	(Testori et al., 2012)
Mammals	Uterine development, pregnancy	(Lynch et al., 2011; Lynch et al., 2015)
Mammals	Placental development (species-specific elements in mouse and humans)	(Chuong, 2013; Chuong & Feschotte, 2013; Chuong et al., 2013; Chuong, 2018; Dunn-Fletcher et al., 2018; Frank & Feschotte, 2017; Sakurai et al., 2017; Sun et al., 2021; Zhang & Muglia, 2021)
Mammals	Convergent evolution of prolactin expression with different mobile DNA insertions	(Emera et al., 2012)
Mammals	Mammary gland development	(Nishihara, 2019)
Mammals	X-inactivation in females	(Elisaphenko et al., 2008; Kannan et al., 2015; Lyon, 2000; Lyon, 2003)
Mammals	Innate immunity	(Chen et al., 2019; Chuong, Elde, & Feschotte, 2016; Srinivasachar Badarinarayan & Sauter, 2021)
Mammals	Wnt5a expression in mammalian secondary palate controlled by complex enhancer evolved from "coordinately co-opted" transposable elements	(Nishihara et al., 2016)
Mammals	Basic mammalian morphology and body plan development	(Hirakawa et al., 2009; Okada et al., 2010)
Mammals	Brain and nervous system development	(Bejerano et al., 2006; Ferrari et al., 2021; Lapp & Hunter, 2016; McEwen et al., 2009; Nakanishi et al., 2012; Notwell et al., 2015; Policarpi et al., 2017; Santangelo et al., 2007; Sasaki et al., 2008; Tashiro et al., 2011)
Eutherian mammals	Eutherian-specific morphology and neural development	(Polychronopoulos et al., 2017)
Eutherian mammals	"at least 16% of eutherian-specific CNEs overlap currently recognized transposable elements in human"	(Mikkelsen et al., 2007)
Bat *Myotis velifer*	Cell migration, gastrulation (weaker signals for Wnt signaling, artery and heart valve morphogenesis, neural crest differentiation)	(Cosby et al., 2021)

(continued)

Table 15.1
(continued)

Organism	Transposable DNA-Modified System(s)	References
Mouse	RNA polymerase PolII- or PolIII-specific insulator-bounded subnuclear domains	(Roman, Gonzalez-Rico, & Fernandez-Salguero, 2011a; Roman et al., 2011b)
Mouse	Circadian rhythms	(Cosby et al., 2021)
Mouse	Mouse germline development and specificity	(Maezawa et al., 2020)
Mouse	Visual cortex development	(Lennartsson et al., 2015)
Primates	Primate-specific innovations arose chiefly by positively selected mobile DNA element insertions	(Jacques, Jeyakani, & Bourque, 2013; Rishishwar et al., 2018; Trizzino et al., 2017)
Anthropoids	Brain and eye development	(del Rosario, Rayan, & Prabhakar, 2014)
Human	Insulator-bound subnuclear domains	(Wang et al., 2015)
Human	c-Myc regulatory subnetwork	(Wang et al., 2009)
Human	P53 response network	(Cui, Sirotin, & Zhurkin, 2011)
Human	Germline development	(Liu & Eiden, 2011; Sakashita et al., 2020)
Human	Embryonic development	(Kunarso et al., 2010; Wang et al., 2014; Xiang & Liang, 2021)
Human	Stem cell pluripotency	(Izsvak et al., 2016; Lu et al., 2014; Römer et al., 2017; Santoni, Guerra, & Luban 2012; Sexton, Tillett, & Han, 2021; Torres-Padilla, 2020)
Human	Zygotic genome activation and preimplantation embryonic development	(Fu, Ma, & Liu, 2019; Fu et al., 2021; Grow et al., 2015)
Human	Cell type- and tissue-specific expression	(Huda et al., 2011a; Huda et al., 2011b; Jjingo et al., 2011)
Human	Tissue-specific expression of regulatory long non-coding lncRNAs	(Chishima, Iwakiri, & Hamada, 2018)

(Etchegaray et al., 2021; Naville et al., 2016; Volff, 2006). Of particular relevance to the elaboration of genomic networks are those cases involving mobile element proteins that bind to specific sequences at the ends of the DNA being inserted into a new target site. A number of these transposases and integrases have contributed to the evolution of regulatory proteins controlling transcription. One way this has happened is by forming a chimeric fusion protein of the transposase DNA-binding domain with the non-DNA-binding domains of an existing TF, such as the fusions in higher vertebrates to regulatory domains of abundant zinc finger proteins (Cosby et al., 2021; Ecco et al., 2016; Ecco, Imbeault, & Trono, 2017). Such a fusion automatically transforms the corresponding mobile DNA element termini throughout the genome into CRMs responding to the new chimeric TF. The most recent study on this topic reports the independent formation of such mobile DNA-host TF fusion proteins to have occurred at least 22 times in the coelacanth, 90 times in amphibians, 313 times in reptiles, 92 times in birds, 31 times in the platypus, 106 times in marsupials, and 928 times in eutherian mammals (Cosby et al., 2021).[3]

15.3.2   How the Formation and Function of Mobile Element-Based Networks Connect to a Changing Ecology in Evolution

Another outdated feature of the physicalist/reductionist neo-Darwinian account is the exclusion of any influence of life history events from any role in evolutionary change. Part of this position comes from making natural selection the sole outside influence on the course of evolution, and another part comes from a rejection of the Lamarckian notion (even though it was shared by Darwin!) that "use and disuse" has an influence on the acquisition, exaggeration, or disappearance of particular traits (Darwin, 1859; Lamarck, 1994). However, today we recognize that all organisms have sensory and signal transduction systems to monitor their internal processes and environmental interactions and adjust their physiologies, growth, reproduction, and self-repair functions accordingly (Bray, 2009; Regolin & Vallortigara, 2021; Shapiro, 2020). These sensory capabilities can also influence the transpositional activities of mobile DNA elements as well as the expression of any resulting genomic networks.

Table 15.2 lists some of the many conditions documented to stimulate mobile element spreading in a variety of organisms (see also Negi et al., 2016, and https://shapiro.bsd .uchicago.edu/Ecological_Factors_that_Induce_Mutagenic_DNA_Repair_or_Modulate_ NGE_Responses.html for additional references). There are two major factors stimulating mobile DNA activity evident in the examples from table 15.2: (1) biotic and abiotic stresses and (2) interspecific hybridization, which often leads to polyploidization (Vicient & Casacuberta, 2017). The significance of such inputs is to increase genomic innovation by mobile DNA when the conditions of life are most difficult. Note that interspecific hybridization is an indicator of such difficulty because it is most likely to occur when the within-species mating pool has declined. A more basic point to remember in thinking about evolutionary theory is that artificial generation of novel species has been practiced in agriculture for thousands of years by interspecific hybridization, never by selection alone (Anderson, 1954; Stebbins, 1951). By these observations alone, we are obliged to say that the selection-based neo-Darwinian paradigm has not been empirically verified.

Stress-induced genome reformatting is just the kind of feedback response we would expect if mobile DNA had the function of providing adaptive variability and genomic diversity when most necessary. In addition, the recently transposed elements provide a molecular genomic

Table 15.2
Stimuli reported to trigger increased mobile DNA activities

Organism	Stimulus	Reference(s)
BACTERIA		
Bacterium *Deinococcus geothermalis*	Oxidative stress	(Lee, Choo, & Lee, 2020; Lee et al., 2021)
Bacterium *Geobacillus kaustophilus*	Heat stress	(Suzuki et al., 2021)
Bacterium *Cupriavidus metallidurans*	Zinc exposure	(Vandecraen et al., 2016)
FUNGI		
Yeast *Saccharomyces cerevisae*	Adenine starvation	(Servant, Pennetier, & Lesage, 2008; Todeschini et al., 2005)
Yeast *Saccharomyces cerevisae*	Ionizing radiation	(Sacerdot et al., 2005)
Yeast *Saccharomyces cerevisae*	Interspecifc hybridization	(Smukowski Heil et al., 2021)
Yeast *Schizosaccharomyces pombe*	Environmental stress (heavy metals, caffeine, and the plasticizer phthalate)	(Esnault et al., 2019)
Yeast *Candida albicans*	Anti-fungal medication miconazole	(Zhu et al., 2014)
Fungal pathogen *Magnaporthe oryzae*	Heat shock, copper stress	(Chadha & Sharma, 2014)
Aspergillus oryzae	CuSO stress, heat shock for conidia (strong effect), acidic environment, oxidative stress, and UV irradiation (weak effect)	(Ogasawara et al., 2009)
Wheat fungal pathogen *Zymoseptoria tritici*	Nutrient starvation, host infection stress	(Fouché et al., 2020)
ALGAE		
Photosynthetic coral symbiont *Symbiodinium microadriaticum*	Heat stress	(Chen et al., 2018)
Diatom *Phaeodactylum tricornutum*	Nitrate limitation, exposure to diatom-derived reactive aldehydes that induce stress responses and cell death	(Maumus et al., 2009)
PLANTS		
Oats	Biotic and abiotic stresses, including UV light, wounding, salicylic acid, and fungal attack	(Chenais et al., 2012; Kimura et al., 2001; Pourrajab & Hekmatimoghaddam, 2021)
Wheat (*Triticum durum L.*)	Salt and light stress	(Woodrow ct al., 2011)
Solanaceae	Stress, hormones	(Grandbastien et al., 2005; Grandbastien, 2015)
Solanum chilense	"Multiple stress-related signalling molecules"	(Salazar et al., 2007)
Tomato	Drought stress and abscisic acid signaling	(Benoit et al., 2019)
Tobacco, tomatoes	Low temperature	(Pourrajab & Hekmatimoghaddam, 2021)
Tobacco	Fungal attack	(Pourrajab & Hekmatimoghaddam, 2021)
Tobacco	Tissue culture growth, wounding and methyl jasmonate	(Hirochika, 1993; Takeda et al., 1998)
Tobacco	The toxic fungal elicitor cryptogein and reactive oxygen species	(Anca et al., 2014)

Table 15.2
(continued)

Brassica	Heat stress	(Pietzenuk et al. 2016)
Arabidopsis	Heat stress	(Cavrak et al., 2014; Ito et al., 2016; Masuda et al., 2017; Matsunaga et al., 2015) (Gaubert et al., 2017)
Arabidopsis	Tissue culture growth	(Steimer et al., 2000)
Arabidopsis	Autopolyploidy	(Baduel et al., 2019)
Antirrhinum majus	Low temperature	(Pourrajab & Hekmatimoghaddam, 2021)
Sunflowers	Interspecific hybridization	(Michalak, 2010)
Andropogoneae (maize and sorghum)	Polyploidy	(Ramachandran et al., 2020)
Rice	Hybridization with *Zizania*	(Wang et al., 2010)
Rice	Early embryo development, tissue culture growth, stresses of gamma-ray irradiation, and high hydrostatic pressure	(Hirochika et al., 1996; Teramoto et al., 2014)
Rice	Etoposide DNA damage	(Yang et al., 2012)
Maize	Roundup herbicide stress	(Tyczewska et al., 2021)
Maize	Viral infection	(Johns, Mottinger, & Freeling, 1985; Paszkowski, 2015)
METAZOA		
Nematode *Caenorhabditis elegans*	Heat shock (males only)	(Kurhanewicz et al., 2020)
Drosophila	Heat shock	(Jardim et al., 2015; Pereira et al., 2018)
Drosophila	Geographic isolation on volcanic islands and stresses from vulcanism	(Craddock, 2016)
Drosophila	Interspecific hybridization	(Carnelossi et al., 2014; Gámez-Visairas et al., 2020; Romero-Soriano et al., 2016; Romero-Soriano & Garcia Guerreiro, 2016)
Vertebrates		(Pappalardo et al., 2021)
Antarctic teleost *Trematomus*	Cold shock	(Auvinet et al., 2018)
Human cancer cells	Arsenic, mercury, chemotherapy	(Clapes et al., 2021; Habibi et al., 2014; Karimi et al., 2014)

memory of the last stress experienced (Avramova, 2015; Kinoshita & Seki, 2014). In other words, living organisms have the ability to rewrite and rewire their genomes when necessary. Rather than being the passive beneficiaries of random mutations and natural selection, all organisms play an active role in their own hereditary variation and evolution by activating transposable elements in response to ecological challenges. Because the newly mobilized elements will respond to the same stimulus that activated their original transposition, they can also serve as a molecular record of past traumas (Shapiro, 2011; Shapiro, 2013). That is why so many adaptive stress responses are linked to mobile DNA-derived CRMs (Butelli et al., 2012; Grandbastien, 2015; Lv et al., 2019; Makarevitch et al., 2015; Mao et al., 2015; Negi et al., 2016; Wang et al., 2017). In *Arabidopsis*, for example, a recent study found that mobile

DNA plays a role in responses to phosphate limitation, tolerance to high salt concentration, freezing temperatures, and arsenic toxicity (Joly-Lopez et al., 2017).

15.4   Conclusions about a Genomic and Biological View of Evolution

Free from the a priori need for a purely physicalist account of evolutionary change, we are able to elaborate a more sophisticated and complex biological description of how evolution works. That description is fully goal-oriented in nature. At the end of the eighteenth century, Erasmus Darwin (Charles' grandfather) defined the "three great objects of desire" for every living organism as "lust, hunger, and security" (E. Darwin, 1794). Today, we would phrase this as "the three goals of all life are survival, growth, and reproduction." Under normal circumstances in a relatively stable ecology, these goals are met by each species' inherited physiology, sensory systems, regulatory networks, and behavior. Under the unusual circumstances following ecological disruption, however, those inherited capabilities may prove inadequate to ensure survival, growth, and reproduction, which may require hereditary variation and adaptive innovation. In meeting such challenges, the organisms will activate transposable DNA elements as well as many other natural cell and genetic engineering capabilities, such as symbiogenesis (Kozo-Polyansky, 1924), mutagenic DNA repair (Al Mamun et al., 2012), and/or rapid widespread genome restructuring (Heng, 2019; Shapiro, 2021; Umbreit et al., 2020). Trying to cope with the new ecology, these diverse inherent biological mechanisms of genome variation can produce evolutionary changes that range from microevolutionary adaptations altering a small number of phenotypes to macroevolutionary transformations that alter genome structure and generate new species or even higher taxa with multiple phenotypic novelties (Goldschmidt, 1940; Heng, 2019). Only a small fraction of the more complex macroevolutionary changes will succeed, but those that do can be the founders of novel taxa displaying major adaptive innovations. Taxonomic and adaptive radiations after mass extinctions indicate that there is no single outcome for successful ecologically-triggered macroevolutionary changes (Ezcurra & Butler, 2018; Grossnickle, Smith, & Wilson, 2019; Magallón, Sánchez-Reyes, & Gómez-Acevedo, 2019; Soltis & Soltis, 2014). The survivors that give rise to major lineages must provide a novel basis for ongoing adaptive changes. In other words, mobile DNA elements and other biological engines of genome change constitute essential survival tools in a dynamic ecology, and that is why they are ubiquitous in living organisms.

There are several conclusions to draw from this purposive view of how evolutionary change occurs. First of all, punctuated equilibrium is the predicted pattern for origination of new life forms at episodes of maximal ecological disruption, such as those marked by mass extinctions (Gould, 1983). We can therefore interpret the discontinuous nature of the fossil record and the genomic record documenting transpositional bursts distinguishing closely related lineages (Belyayev, 2014; El Baidouri & Panaud, 2013) as evidence in favor of episodic self-modification as a primary mode of evolutionary change. Secondly, the active participation of sensory-regulated cellular functions generating novel DNA configurations in evolutionary transformations (tables 15.1 and 15.2) tells us that evolution is inherently biological and cannot be reduced to purely physical explanations, as Coyne and his fellow neo-Darwinians hoped. In particular, this teleological perspective involves the operation of biological cognition (i.e., sensory-based control and adaptative responses) to regulate the processes of genome modification (Regolin & Vallortigara, 2021; Shapiro, 2020).

Recognition of the cognitive aspect of the evolutionary process opens a range of hitherto forbidden questions for scientific exploration. For example, we know that mobile DNA elements and other change operations occur nonrandomly and can be targeted to specific genome locations, as they are in the vertebrate adaptive immune system (https://shapiro.bsd .uchicago.edu/ExtraRefs.TargetingNaturalGeneticEngineeringInGenome.shtml) (Asif-Laidin et al., 2020; Birchler & Presting, 2012; Buerstedde et al., 2014; Craigie & Bushman, 2014; Hickey et al., 2015; Parks et al., 2009). Based on knowledge of this targeting potential, we are now able to ask whether there is any demonstrable connection between the ecological challenge that activates transposition of mobile DNA elements and the choice of target insertion sites that may facilitate successful adaptation to that particular challenge. In bacteria, there are distinct mutational spectra for different stress conditions (Maharjan & Ferenci, 2015; Maharjan & Ferenci, 2017), the cells target multiple antibiotic resistance determinants to specialized genomic structures (Cambray et al., 2011; Guerin et al., 2009; Parks et al., 2009), and some molecular geneticists even claim to have evidence for "directed mutation" in the activation of specific functions by mobile DNA element insertions (Saier, Kukita, & Zhang, 2017). While these directed-mutation claims require further scrutiny, pursuing this kind of previously taboo question has the potential to uncover another layer of biological control over the evolutionary process, just as McClintock's unexpected discovery of transposable controlling elements did. Perhaps a deeper theoretical question is whether the capacity for actively and purposefully generating hereditary variation is an essential feature of life. It may well be the case that living organisms that are incapable of actively modifying their genomes would be doomed to extinction. If that inference turns out to be correct, then genome self-modification, or rewriting, would be another essential goal-oriented function to incorporate into our definition of what it means to be alive.

Notes

1. Even important neo-Darwinists agreed with this view of purposive action throughout biology (see P. A. Corning, quoting Theodosius Dobzhansky, in ch. 2 of this volume).

2. Darwin himself gave up on pure phyletic gradualism by the time he published his sixth edition of *The Origin*, where he wrote of "sports" or discrete "variations which seem to us in our ignorance to arise spontaneously. It appears that I formerly underrated the frequency and value of these latter forms of variation, as leading to permanent modifications of structure independently of natural selection" (Darwin, *The Origin of Species by Means of Natural Selection, or the Preservation of Favoured Races in the Struggle for Life, 6th ed.*, ch. 15, p. 395).

3. Note that the differences in numbers reflect the number of genomes sequenced for each taxon rather than any clear difference in propensity for forming such fusion proteins.

References

Al Mamun, A. A., M. J. Lombardo, C. Shee, A. M. Lisewski, C. Gonzalez, D. Lin, R. B. Nehring, C. Saint-Ruf, J. L. Gibson, R. L. Frisch, O. Lichtarge, P. J. Hastings, and S. M. Rosenberg. 2012. "Identity and Function of a Large Gene Network Underlying Mutagenic Repair of DNA Breaks." *Science* 338(6112):1344–1348.

Alioto, T., K. G. Alexiou, A. Bardil, F. Barteri, R. Castanera, A. Cruz, A. Dhingra, H. Duval, I. Martí Á Fernández, L. Frias, B. Galán, J. L. García, W. Howad, J. Gómez-Garrido, M. Gut, I. Julca, J. Morata, P. Puigdomènech, P. Ribeca, M. J. Rubio Cabetas, A. Vlasova, M. Wirthensohn, J. Garcia-Mas, T. Gabaldón, J. M. Casacuberta, and P. Arús. 2020. "Transposons Played a Major Role in the Diversification between the Closely Related Almond and Peach Genomes: Results from the Almond Genome Sequence." *Plant J* 101(2):455–472.

Anca, I. A., J. Fromentin, Q. T. Bui, C. Mhiri, M. A. Grandbastien, and F. Simon-Plas. 2014. "Different Tobacco Retrotransposons Are Specifically Modulated by the Elicitor Cryptogein and Reactive Oxygen Species." *J Plant Physiol* 171(16):1533–1540.

Anderson, E. 1954. "Hybridization as an Evolutionary Stimulus." *Evolution* 8:378–388.

Asif-Laidin, A., C. Conesa, A. Bonnet, C. Grison, I. Adhya, R. Menouni, H. Fayol, N. Palmic, J. Acker, and P. Lesage. 2020. "A Small Targeting Domain in Ty1 Integrase Is Sufficient to Direct Retrotransposon Integration Upstream of Trna Genes." *Embo J* 39(17):e104337.

Auvinet, J., P. Graca, L. Belkadi, L. Petit, E. Bonnivard, A. Dettai, W. H. Detrich, 3rd, C. Ozouf-Costaz, and D. Higuet. 2018. "Mobilization of Retrotransposons as a Cause of Chromosomal Diversification and Rapid Speciation: The Case for the Antarctic Teleost Genus Trematomus." *BMC Genomics* 19(1):339.

Avramova, Z. 2015. "Transcriptional 'Memory' of a Stress: Transient Chromatin and Memory (Epigenetic) Marks at Stress-Response Genes." *Plant J* 83(1):149–159.

Baduel, P., L. Quadrana, B. Hunter, K. Bomblies, and V. Colot. 2019. "Relaxed Purifying Selection in Autopolyploids Drives Transposable Element Over-Accumulation Which Provides Variants for Local Adaptation." *Nat Commun* 10(1):5818.

Barbaglia, A. M., K. M. Klusman, J. Higgins, J. R. Shaw, L. C. Hannah, and S. K. Lal. 2012. "Gene Capture by Helitron Transposons Reshuffles the Transcriptome of Maize." *Genetics* 190(3):965–975.

Batista, R. A., J. Moreno-Romero, Y. Qiu, J. van Boven, J. Santos-González, D. D. Figueiredo, and C. Köhler. 2019. "The Mads-Box Transcription Factor Pheres1 Controls Imprinting in the Endosperm by Binding to Domesticated Transposons." *Elife* 8:e50541.

Baud, A., M. Wan, D. Nouaud, N. Francillonne, D. Anxolabéhère, and H. Quesneville. 2020. "Traces of Transposable Elements in Genome Dark Matter Coopted by Flowering Gene Regulation Networks." *bioRxiv*.

Beadle, G. W. 1948. "The Genes of Men and Molds." *Sci Am* 179(3):30–39.

Bejerano, G., C. B. Lowe, N. Ahituv, B. King, A. Siepel, S. R. Salama, E. M. Rubin, W. J. Kent, and D. Haussler. 2006. "A Distal Enhancer and an Ultraconserved Exon Are Derived from a Novel Retroposon." *Nature* 441(7089):87–90.

Belyayev, A. 2014. "Bursts of Transposable Elements as an Evolutionary Driving Force." *J Evol Biol* 27(12):2573–2584.

Bennetzen, J. L., and H. Wang. 2014. "The Contributions of Transposable Elements to the Structure, Function, and Evolution of Plant Genomes." *Annu Rev Plant Biol* 65:505–530.

Benoit, M., H. G. Drost, M. Catoni, Q. Gouil, S. Lopez-Gomollon, D. Baulcombe, and J. Paszkowski. 2019. "Environmental and Epigenetic Regulation of Rider Retrotransposons in Tomato." *PLoS Genet* 15(9):e1008370.

Birchler, J. A., and G. G. Presting. 2012. "Retrotransposon Insertion Targeting: A Mechanism for Homogenization of Centromere Sequences on Nonhomologous Chromosomes." *Genes Dev* 26(7):638–640.

Bolotin, E., K. Chellappa, W. Hwang-Verslues, J. M. Schnabl, C. Yang, and F. M. Sladek. 2011. "Nuclear Receptor Hnf4alpha Binding Sequences Are Widespread in Alu Repeats." *BMC Genomics* 12:560.

Bondos, S. E., and X. X. Tan. 2001. "Combinatorial Transcriptional Regulation: The Interaction of Transcription Factors and Cell Signaling Molecules with Homeodomain Proteins in Drosophila Development." *Crit Rev Eukaryot Gene Expr* 11(1–3):145–171.

Bornholdt, S. 2001. "Modeling Genetic Networks and Their Evolution: A Complex Dynamical Systems Perspective." *Biol Chem* 382(9):1289–1299.

Bourque, G., B. Leong, V. B. Vega, X. Chen, Y. L. Lee, K. G. Srinivasan, J. L. Chew, Y. Ruan, C. L. Wei, H. H. Ng, and E. T. Liu. 2008. "Evolution of the Mammalian Transcription Factor Binding Repertoire Via Transposable Elements." *Genome Res* 18(11):1752–1762.

Bourque, G., K. H. Burns, M. Gehring, V. Gorbunova, A. Seluanov, M. Hammell, M. Imbeault, Z. Izsvak, H. L. Levin, T. S. Macfarlan, D. L. Mager, and C. Feschotte. 2018. "Ten Things You Should Know About Transposable Elements." *Genome Biol* 19(1):199.

Bowman, J. L., L. N. Briginshaw, and S. N. Florent. 2019. "Evolution and Co-Option of Developmental Regulatory Networks in Early Land Plants." *Curr Top Dev Biol* 131:35–53.

Bray, D. 2003. "Molecular Networks: The Top-Down View." *Science* 301(5641):1864–1865.

Bray, D. 2009. *Wetware: A Computer in Every Living Cell*. New Haven, CT: Yale University Press.

Bray, D. 2012. "The Propagation of Allosteric States in Large Multiprotein Complexes." *J Mol Biol* 425(9):1410–1414.

Bray, D. 2015. "Limits of Computational Biology." *In Silico Biol* 12(1–2):1–7.

Bringaud, F., M. Muller, G. C. Cerqueira, M. Smith, A. Rochette, N. M. El-Sayed, B. Papadopoulou, and E. Ghedin. 2007. "Members of a Large Retroposon Family Are Determinants of Post-Transcriptional Gene Expression in Leishmania." *PLoS Pathog* 3(9):1291–1307.

Buerstedde, J. M., J. Alinikula, H. Arakawa, J. J. McDonald, and D. G. Schatz. 2014. "Targeting of Somatic Hypermutation by Immunoglobulin Enhancer and Enhancer-Like Sequences." *PLoS Biol* 12(4):e1001831.

Bukhari, A. I., J. A. Shapiro, and S. L. Adhya (Eds.). 1977. *DNA Insertion Elements, Plasmids and Episomes.* Cold Spring Harbor, NY: Cold Spring Harbor Press.

Butelli, E., C. Licciardello, Y. Zhang, J. Liu, S. Mackay, P. Bailey, G. Reforgiato-Recupero, and C. Martin. 2012. "Retrotransposons Control Fruit-Specific, Cold-Dependent Accumulation of Anthocyanins in Blood Oranges." *Plant Cell* 24(3):1242–1255.

Cambray, G., N. Sanchez-Alberola, S. Campoy, E. Guerin, S. Da Re, B. Gonzalez-Zorn, M. C. Ploy, J. Barbe, D. Mazel, and I. Erill. 2011. "Prevalence of Sos-Mediated Control of Integron Integrase Expression as an Adaptive Trait of Chromosomal and Mobile Integrons." *Mob DNA* 2(1):6.

Cao, C., J. Xu, G. Zheng, and X. G. Zhu. 2016. "Evidence for the Role of Transposons in the Recruitment of Cis-Regulatory Motifs During the Evolution of C4 Photosynthesis." *BMC Genomics* 17(1):201.

Carnelossi, E. A., E. Lerat, H. Henri, S. Martinez, C. M. Carareto, and C. Vieira. 2014. "Specific Activation of an I-Like Element in Drosophila Interspecific Hybrids." *Genome Biol Evol* 6(7):1806–1817.

Carotti, E., F. Carducci, A. Canapa, M. Barucca, S. Greco, M. Gerdol, and M. A. Biscotti. 2021. "Transposable Elements and Teleost Migratory Behaviour." *Int J Mol Sci* 22(2):602.

Castanera, R., L. Lopez-Varas, A. Borgognone, K. LaButti, A. Lapidus, J. Schmutz, J. Grimwood, G. Perez, A. G. Pisabarro, I. V. Grigoriev, J. E. Stajich, and L. Ramirez. 2016. "Transposable Elements Versus the Fungal Genome: Impact on Whole-Genome Architecture and Transcriptional Profiles." *PLoS Genet* 12(6):e1006108.

Cavrak, V. V., N. Lettner, S. Jamge, A. Kosarewicz, L. M. Bayer, and O. Mittelsten Scheid. 2014. "How a Retrotransposon Exploits the Plant's Heat Stress Response for Its Activation." *PLoS Genet* 10(1):e1004115.

Chadha, S. and M. Sharma. 2014. "Transposable Elements as Stress Adaptive Capacitors Induce Genomic Instability in Fungal Pathogen Magnaporthe Oryzae." *PLoS One* 9(4):e94415.

Chen, C., W. Wang, X. Wang, D. Shen, S. Wang, Y. Wang, B. Gao, K. Wimmers, J. Mao, K. Li, and C. Song. 2019. "Retrotransposons Evolution and Impact on Lncrna and Protein Coding Genes in Pigs." *Mob DNA* 10:19.

Chen, J. E., G. Cui, X. Wang, Y. J. Liew, and M. Aranda. 2018. "Recent Expansion of Heat-Activated Retrotransposons in the Coral Symbiont *Symbiodinium Microadriaticum.*" *ISME J* 12(2):639–643.

Chenais, B., A. Caruso, S. Hiard, and N. Casse. 2012. "The Impact of Transposable Elements on Eukaryotic Genomes: From Genome Size Increase to Genetic Adaptation to Stressful Environments." *Gene* 509(1):7–15.

Chishima, T., J. Iwakiri, and M. Hamada. 2018. "Identification of Transposable Elements Contributing to Tissue-Specific Expression of Long Non-Coding RNAs." *Genes (Basel)* 9(1):23.

Chuong, E. B. 2013. "Retroviruses Facilitate the Rapid Evolution of the Mammalian Placenta." *Bioessays* 35(10):853–861.

Chuong, E. B. 2018. "The Placenta Goes Viral: Retroviruses Control Gene Expression in Pregnancy." *PLoS Biol* 16(10):e3000028.

Chuong, E. B., N. C. Elde, and C. Feschotte. 2016. "Regulatory Evolution of Innate Immunity through Co-Option of Endogenous Retroviruses." *Science* 351(6277):1083–1087.

Chuong, E. B. and C. Feschotte. 2013. "Evolution. Transposons up the Dosage." *Science* 342(6160):812–813.

Chuong, E. B., M. A. Rumi, M. J. Soares, and J. C. Baker. 2013. "Endogenous Retroviruses Function as Species-Specific Enhancer Elements in the Placenta." *Nat Genet* 45(3):325–329.

Clapes, T., A. Polyzou, P. Prater, Sagar, A. Morales-Hernández, M. G. Ferrarini, N. Kehrer, S. Lefkopoulos, V. Bergo, B. Hummel, N. Obier, D. Maticzka, A. Bridgeman, J. S. Herman, I. Ilik, L. Klaeylé, J. Rehwinkel, S. McKinney-Freeman, R. Backofen, A. Akhtar, N. Cabezas-Wallscheid, R. Sawarkar, R. Rebollo, D. Grün, and E. Trompouki. 2021. "Chemotherapy-Induced Transposable Elements Activate Mda5 to Enhance Haematopoietic Regeneration." *Nat Cell Biol* 23(7):704–717.

Corning, P. A. 2007. "Control Information Theory: The 'Missing Link' in the Science of Cybernetics." *Syst Res Behav Sci* 24(3):297–311.

Cosby, R. L., J. Judd, R. Zhang, A. Zhong, N. Garry, E. J. Pritham, and C. Feschotte. 2021. "Recurrent Evolution of Vertebrate Transcription Factors by Transposase Capture." *Science* 371(6531).

Craddock, E. M. 2016. "Profuse Evolutionary Diversification and Speciation on Volcanic Islands: Transposon Instability and Amplification Bursts Explain the Genetic Paradox." *Biol Direct* 11:44.

Craigie, R., and F. D. Bushman. 2014. "Host Factors in Retroviral Integration and the Selection of Integration Target Sites." *Microbiol Spectr* 2(6).

Cui, F., M. V. Sirotin, and V. B. Zhurkin. 2011. "Impact of Alu Repeats on the Evolution of Human P53 Binding Sites." *Biol Direct* 6(1):2.

Darwin, C. 1859. *On the Origin of Species by Means of Natural Selection, or the Preservation of Favoured Races in the Struggle for Life.* London: John Russel.

Darwin, E. 1794. *Zoonomia; or, the Laws of Organic Life. Part I.* London: J. Johnson.

de Vries, H. 1905. *Species and Varieties, Their Origin by Mutation; Lectures Delivered at the University of California*. Chicago, IL: Open Court.

del Rosario, R. C., N. A. Rayan, and S. Prabhakar. 2014. "Noncoding Origins of Anthropoid Traits and a New Null Model of Transposon Functionalization." *Genome Res* 24(9):1469–1484.

Dunn-Fletcher, C. E., L. M. Muglia, M. Pavlicev, G. Wolf, M. A. Sun, Y. C. Hu, E. Huffman, S. Tumukuntala, K. Thiele, A. Mukherjee, S. Zoubovsky, X. Zhang, K. A. Swaggart, K. Y. B. Lamm, H. Jones, T. S. Macfarlan, and L. J. Muglia. 2018. "Anthropoid Primate-Specific Retroviral Element The1b Controls Expression of Crh in Placenta and Alters Gestation Length." *PLoS Biol* 16(9):e2006337.

Ecco, G., M. Cassano, A. Kauzlaric, J. Duc, A. Coluccio, S. Offner, M. Imbeault, H. M. Rowe, P. Turelli, and D. Trono. 2016. "Transposable Elements and Their Krab-Zfp Controllers Regulate Gene Expression in Adult Tissues." *Dev Cell* 36(6):611–623.

Ecco, G., M. Imbeault, and D. Trono. 2017. "Krab Zinc Finger Proteins." *Development* 144(15):2719–2729.

El Baidouri, M., and O. Panaud. 2013. "Comparative Genomic Paleontology across Plant Kingdom Reveals the Dynamics of Te-Driven Genome Evolution." *Genome Biol Evol* 5(5):954–965.

Elisaphenko, E. A., N. N. Kolesnikov, A. I. Shevchenko, I. B. Rogozin, T. B. Nesterova, N. Brockdorff, and S. M. Zakian. 2008. "A Dual Origin of the Xist Gene from a Protein-Coding Gene and a Set of Transposable Elements." *PLoS One* 3(6):e2521.

Ellison, C. and D. Bachtrog. 2019. "Recurrent Gene Co-Amplification on Drosophila X and Y Chromosomes." *PLoS Genet* 15(7):e1008251.

Ellison, C. E. and D. Bachtrog. 2013. "Dosage Compensation Via Transposable Element Mediated Rewiring of a Regulatory Network." *Science* 342(6160):846–850.

Emera, D., C. Casola, V. J. Lynch, D. E. Wildman, D. Agnew, and G. P. Wagner. 2012. "Convergent Evolution of Endometrial Prolactin Expression in Primates, Mice, and Elephants through the Independent Recruitment of Transposable Elements." *Mol Biol Evol* 29(1):239–247.

Esnault, C., M. Lee, C. Ham, and H. L. Levin. 2019. "Transposable Element Insertions in Fission Yeast Drive Adaptation to Environmental Stress." *Genome Res* 29(1):85–95.

Etchegaray, E., M. Naville, J. N. Volff, and Z. Haftek-Terreau. 2021. "Transposable Element-Derived Sequences in Vertebrate Development." *Mob DNA* 12(1):1.

Ezcurra, M. D., and R. J. Butler. 2018. "The Rise of the Ruling Reptiles and Ecosystem Recovery from the Permo-Triassic Mass Extinction." *Proc Biol Sci* 285(1880):20180361.

Ferrari, R., N. Grandi, E. Tramontano, and G. Dieci. 2021. "Retrotransposons as Drivers of Mammalian Brain Evolution." *Life (Basel)* 11(5).

Fouché, S., T. Badet, U. Oggenfuss, C. Plissonneau, C. S. Francisco, and D. Croll. 2020. "Stress-Driven Transposable Element De-Repression Dynamics and Virulence Evolution in a Fungal Pathogen." *Mol Biol Evol* 37(1):221–239.

Frank, J. A. and C. Feschotte. 2017. "Co-Option of Endogenous Viral Sequences for Host Cell Function." *Curr Opin Virol* 25:81–89.

Fu, B., H. Ma, and D. Liu. 2019. "Endogenous Retroviruses Function as Gene Expression Regulatory Elements During Mammalian Pre-Implantation Embryo Development." *Int J Mol Sci* 20(3).

Fu, H., W. Zhang, N. Li, J. Yang, X. Ye, C. Tian, X. Lu, and L. Liu. 2021. "Elevated Retrotransposon Activity and Genomic Instability in Primed Pluripotent Stem Cells." *Genome Biol* 22(1):201.

Gámez-Visairas, V., V. Romero-Soriano, J. Martí-Carreras, E. Segarra-Carrillo, and M. P. García Guerreiro. 2020. "Drosophila Interspecific Hybridization Causes a Deregulation of the Pirna Pathway Genes." *Genes (Basel)* 11(2).

Gaubert, H., D. H. Sanchez, H. G. Drost, and J. Paszkowski. 2017. "Developmental Restriction of Retrotransposition Activated in Arabidopsis by Environmental Stress." *Genetics* 207(2):813–821.

Goldschmidt, R. 1940/1982. *The Material Basis of Evolution, Reissued (the Silliman Memorial Lectures Series)*. New Haven, CT: Yale University Press.

Gould, S. J. 1983. "Punctuated Equilibrium and the Fossil Record." *Science* 219(4584):439–440.

Grandbastien, M. A. 2015. "Ltr Retrotransposons, Handy Hitchhikers of Plant Regulation and Stress Response." *Biochim Biophys Acta* 1849(4):403–416.

Grandbastien, M. A., C. Audeon, E. Bonnivard, J. M. Casacuberta, B. Chalhoub, A. P. Costa, Q. H. Le, D. Melayah, M. Petit, C. Poncet, S. M. Tam, M. A. Van Sluys, and C. Mhiri. 2005. "Stress Activation and Genomic Impact of Tnt1 Retrotransposons in Solanaceae." *Cytogenet Genome Res* 110(1–4):229–241.

Grossnickle, D. M., S. M. Smith, and G. P. Wilson. 2019. "Untangling the Multiple Ecological Radiations of Early Mammals." *Trends Ecol Evol* 34(10):936–949.

Grow, E. J., R. A. Flynn, S. L. Chavez, N. L. Bayless, M. Wossidlo, D. J. Wesche, L. Martin, C. B. Ware, C. A. Blish, H. Y. Chang, R. A. Pera, and J. Wysocka. 2015. "Intrinsic Retroviral Reactivation in Human Preimplantation Embryos and Pluripotent Cells." *Nature* 522(7555):221–225.

Guerin, E., G. Cambray, N. Sanchez-Alberola, S. Campoy, I. Erill, S. Da Re, B. Gonzalez-Zorn, J. Barbe, M. C. Ploy, and D. Mazel. 2009. "The Sos Response Controls Integron Recombination." *Science* 324(5930):1034.

Habibi, L., M. A. Shokrgozar, M. Tabrizi, M. H. Modarressi, and S. M. Akrami. 2014. "Mercury Specifically Induces Line-1 Activity in a Human Neuroblastoma Cell Line." *Mutat Res Genet Toxicol Environ Mutagen* 759:9–20.

Heng, H. H. 2008. "The Conflict between Complex Systems and Reductionism." *JAMA* 300(13):1580–1581.

Heng, H. H. 2019. *Genome Chaos: Rethinking Genetics, Evolution, and Molecular Medicine*: New York: Academic Press.

Hickey, A., C. Esnault, A. Majumdar, A. G. Chatterjee, J. R. Iben, P. G. McQueen, A. X. Yang, T. Mizuguchi, S. I. Grewal, and H. L. Levin. 2015. "Single-Nucleotide-Specific Targeting of the Tf1 Retrotransposon Promoted by the DNA-Binding Protein Sap1 of *Schizosaccharomyces Pombe*." *Genetics* 201(3):905–924.

Hirakawa, M., H. Nishihara, M. Kanehisa, and N. Okada. 2009. "Characterization and Evolutionary Landscape of Amnsine1 in Amniota Genomes." *Gene* 441(1–2):100–110.

Hirochika, H. 1993. "Activation of Tobacco Retrotransposons During Tissue Culture." *Embo J* 12(6):2521–2528.

Hirochika, H., K. Sugimoto, Y. Otsuki, H. Tsugawa, and M. Kanda. 1996. "Retrotransposons of Rice Involved in Mutations Induced by Tissue Culture." *Proc Natl Acad Sci U S A* 93(15):7783–7788.

Hou, J., D. Lu, A. S. Mason, B. Li, M. Xiao, S. An, and D. Fu. 2019. "Non-Coding RNAs and Transposable Elements in Plant Genomes: Emergence, Regulatory Mechanisms and Roles in Plant Development and Stress Responses." *Planta* 250(1):23–40.

Huda, A., N. J. Bowen, A. B. Conley, and I. K. Jordan. 2011a. "Epigenetic Regulation of Transposable Element Derived Human Gene Promoters." *Gene* 475(1):39–48.

Huda, A., E. Tyagi, L. Marino-Ramirez, N. J. Bowen, D. Jjingo, and I. K. Jordan. 2011b. "Prediction of Transposable Element Derived Enhancers Using Chromatin Modification Profiles." *PLoS One* 6(11):e27513.

Ito, H., J. M. Kim, W. Matsunaga, H. Saze, A. Matsui, T. A. Endo, Y. Harukawa, H. Takagi, H. Yaegashi, Y. Masuta, S. Masuda, J. Ishida, M. Tanaka, S. Takahashi, T. Morosawa, T. Toyoda, T. Kakutani, A. Kato, and M. Seki. 2016. "A Stress-Activated Transposon in Arabidopsis Induces Transgenerational Abscisic Acid Insensitivity." *Sci Rep* 6:23181.

Izsvak, Z., J. Wang, M. Singh, D. L. Mager, and L. D. Hurst. 2016. "Pluripotency and the Endogenous Retrovirus Hervh: Conflict or Serendipity?" *Bioessays* 38(1):109–117.

Jacob, F. and J. Monod. 1961. "Genetic Regulatory Mechanisms in the Synthesis of Proteins." *J Mol Biol* 3:318–356.

Jacques, P. E., J. Jeyakani, and G. Bourque. 2013. "The Majority of Primate-Specific Regulatory Sequences Are Derived from Transposable Elements." *PLoS Genet* 9(5):e1003504.

Jardim, S. S., A. P. Schuch, C. M. Pereira, and E. L. Loreto. 2015. "Effects of Heat and UV Radiation on the Mobilization of Transposon Mariner-Mos1." *Cell Stress Chaperones* 20(5):843–851.

Jjingo, D., A. Huda, M. Gundapuneni, L. Marino-Ramirez, and I. K. Jordan. 2011. "Effect of the Transposable Element Environment of Human Genes on Gene Length and Expression." *Genome Biol Evol* 3:259–271.

Johns, M. A., J. Mottinger, and M. Freeling. 1985. "A Low Copy Number, Copia-Like Transposon in Maize." *Embo J* 4(5):1093–1101.

Joly-Lopez, Z., E. Forczek, E. Vello, D. R. Hoen, A. Tomita, and T. E. Bureau. 2017. "Abiotic Stress Phenotypes Are Associated with Conserved Genes Derived from Transposable Elements." *Front Plant Sci* 8:2027.

Joly-Lopez, Z., D. R. Hoen, M. Blanchette, and T. E. Bureau. 2016. "Phylogenetic and Genomic Analyses Resolve the Origin of Important Plant Genes Derived from Transposable Elements." *Mol Biol Evol* 33(8):1937–1956.

Jordan, I. K., I. B. Rogozin, G. V. Glazko, and E. V. Koonin. 2003. "Origin of a Substantial Fraction of Human Regulatory Sequences from Transposable Elements." *Trends Genet* 19(2):68–72.

Jouffroy, O., S. Saha, L. Mueller, H. Quesneville, and F. Maumus. 2016. "Comprehensive Repeatome Annotation Reveals Strong Potential Impact of Repetitive Elements on Tomato Ripening." *BMC Genomics* 17(1):624.

Jurka, J., W. Bao, K. K. Kojima, O. Kohany, and M. G. Yurka. 2012. "Distinct Groups of Repetitive Families Preserved in Mammals Correspond to Different Periods of Regulatory Innovations in Vertebrates." *Biol Direct* 7(1):36.

Kannan, S., D. Chernikova, I. B. Rogozin, E. Poliakov, D. Managadze, E. V. Koonin, and L. Milanesi. 2015. "Transposable Element Insertions in Long Intergenic Non-Coding RNA Genes." *Front Bioeng Biotechnol* 3:71.

Karimi, A., Z. Madjd, L. Habibi, and S. M. Akrami. 2014. "Evaluating the Extent of Line-1 Mobility Following Exposure to Heavy Metals in Hepg2 Cells." *Biol Trace Elem Res* 160(1):143–151.

Kimura, Y., Y. Tosa, S. Shimada, R. Sogo, M. Kusaba, T. Sunaga, S. Betsuyaku, Y. Eto, H. Nakayashiki, and S. Mayama. 2001. "Oare-1, a Ty1-Copia Retrotransposon in Oat Activated by Abiotic and Biotic Stresses." *Plant Cell Physiol* 42(12):1345–1354.

Kinoshita, T., and M. Seki. 2014. "Epigenetic Memory for Stress Response and Adaptation in Plants." *Plant Cell Physiol* 55(11):1859–1863.

Kozo-Polyansky, B. M. 1924. *Symbiogenesis: A New Principle of Evolution*. (English translation by Victor Fet published 2010). Cambridge, MA: Harvard University Press.

Kunarso, G., N. Y. Chia, J. Jeyakani, C. Hwang, X. Lu, Y. S. Chan, H. H. Ng, and G. Bourque. 2010. "Transposable Elements Have Rewired the Core Regulatory Network of Human Embryonic Stem Cells." *Nat Genet* 42(7):631–634.

Kurhanewicz, N. A., D. Dinwiddie, Z. D. Bush, and D. E. Libuda. 2020. "Elevated Temperatures Cause Transposon-Associated DNA Damage in *C. elegans* Spermatocytes." *Curr Biol* 30(24):5007–5017 e5004.

Lamarck, J.-B. 1994. *Philosophie Zoologique, Original Edition of 1809 with Introduction by Andre Pichot*. Paris: Flammarion.

Lander, E. S., L. M. Linton, B. Birren, C. Nusbaum, M. C. Zody, J. Baldwin, K. Devon, K. Dewar, M. Doyle, W. FitzHugh, R. Funke, D. Gage, K. Harris, A. Heaford, J. Howland, L. Kann, J. Lehoczky, R. LeVine, P. McEwan, K. McKernan, J. Meldrim, J. P. Mesirov, C. Miranda, W. Morris, J. Naylor, C. Raymond, M. Rosetti, R. Santos, A. Sheridan, C. Sougnez, N. Stange-Thomann, N. Stojanovic, A. Subramanian, D. Wyman, J. Rogers, J. Sulston, R. Ainscough, S. Beck, D. Bentley, J. Burton, C. Clee, N. Carter, A. Coulson, R. Deadman, P. Deloukas, A. Dunham, I. Dunham, R. Durbin, L. French, D. Grafham, S. Gregory, T. Hubbard, S. Humphray, A. Hunt, M. Jones, C. Lloyd, A. McMurray, L. Matthews, S. Mercer, S. Milne, J. C. Mullikin, A. Mungall, R. Plumb, M. Ross, R. Shownkeen, S. Sims, R. H. Waterston, R. K. Wilson, L. W. Hillier, J. D. McPherson, M. A. Marra, E. R. Mardis, L. A. Fulton, A. T. Chinwalla, K. H. Pepin, W. R. Gish, S. L. Chissoe, M. C. Wendl, K. D. Delehaunty, T. L. Miner, A. Delehaunty, J. B. Kramer, L. L. Cook, R. S. Fulton, D. L. Johnson, P. J. Minx, S. W. Clifton, T. Hawkins, E. Branscomb, P. Predki, P. Richardson, S. Wenning, T. Slezak, N. Doggett, J. F. Cheng, A. Olsen, S. Lucas, C. Elkin, E. Uberbacher, M. Frazier, R. A. Gibbs, D. M. Muzny, S. E. Scherer, J. B. Bouck, E. J. Sodergren, K. C. Worley, C. M. Rives, J. H. Gorrell, M. L. Metzker, S. L. Naylor, R. S. Kucherlapati, D. L. Nelson, G. M. Weinstock, Y. Sakaki, A. Fujiyama, M. Hattori, T. Yada, A. Toyoda, T. Itoh, C. Kawagoe, H. Watanabe, Y. Totoki, T. Taylor, J. Weissenbach, R. Heilig, W. Saurin, F. Artiguenave, P. Brottier, T. Bruls, E. Pelletier, C. Robert, P. Wincker, D. R. Smith, L. Doucette-Stamm, M. Rubenfield, K. Weinstock, H. M. Lee, J. Dubois, A. Rosenthal, M. Platzer, G. Nyakatura, S. Taudien, A. Rump, H. Yang, J. Yu, J. Wang, G. Huang, J. Gu, L. Hood, L. Rowen, A. Madan, S. Qin, R. W. Davis, N. A. Federspiel, A. P. Abola, M. J. Proctor, R. M. Myers, J. Schmutz, M. Dickson, J. Grimwood, D. R. Cox, M. V. Olson, R. Kaul, C. Raymond, N. Shimizu, K. Kawasaki, S. Minoshima, G. A. Evans, M. Athanasiou, R. Schultz, B. A. Roe, F. Chen, H. Pan, J. Ramser, H. Lehrach, R. Reinhardt, W. R. McCombie, M. de la Bastide, N. Dedhia, H. Blocker, K. Hornischer, G. Nordsiek, R. Agarwala, L. Aravind, J. A. Bailey, A. Bateman, S. Batzoglou, E. Birney, P. Bork, D. G. Brown, C. B. Burge, L. Cerutti, H. C. Chen, D. Church, M. Clamp, R. R. Copley, T. Doerks, S. R. Eddy, E. E. Eichler, T. S. Furey, J. Galagan, J. G. Gilbert, C. Harmon, Y. Hayashizaki, D. Haussler, H. Hermjakob, K. Hokamp, W. Jang, L. S. Johnson, T. A. Jones, S. Kasif, A. Kaspryzk, S. Kennedy, W. J. Kent, P. Kitts, E. V. Koonin, I. Korf, D. Kulp, D. Lancet, T. M. Lowe, A. McLysaght, T. Mikkelsen, J. V. Moran, N. Mulder, V. J. Pollara, C. P. Ponting, G. Schuler, J. Schultz, G. Slater, A. F. Smit, E. Stupka, J. Szustakowski, D. Thierry-Mieg, J. Thierry-Mieg, L. Wagner, J. Wallis, R. Wheeler, A. Williams, Y. I. Wolf, K. H. Wolfe, S. P. Yang, R. F. Yeh, F. Collins, M. S. Guyer, J. Peterson, A. Felsenfeld, K. A. Wetterstrand, A. Patrinos, M. J. Morgan, P. de Jong, J. J. Catanese, K. Osoegawa, H. Shizuya, S. Choi, and Y. J. Chen. 2001. "Initial Sequencing and Analysis of the Human Genome." *Nature* 409(6822):860–921.

Lapp, H. E. and R. G. Hunter. 2016. "The Dynamic Genome: Transposons and Environmental Adaptation in the Nervous System." *Epigenomics* 8(2):237–249.

Lee, C., M. K. Bae, N. Choi, S. J. Lee, and S. J. Lee. 2021. "Genome Plasticity by Insertion Sequences Learned from a Case of Radiation-Resistant Bacterium *Deinococcus Geothermalis*." *Bioinform Biol Insights* 15:11779322211037437.

Lee, C., K. Choo, and S. J. Lee. 2020. "Active Transposition of Insertion Sequences by Oxidative Stress in *Deinococcus Geothermalis*." *Front Microbiol* 11:558747.

Lennartsson, A., E. Arner, M. Fagiolini, A. Saxena, R. Andersson, H. Takahashi, Y. Noro, J. Sng, A. Sandelin, T. K. Hensch, and P. Carninci. 2015. "Remodeling of Retrotransposon Elements During Epigenetic Induction of Adult Visual Cortical Plasticity by HDAC Inhibitors." *Epigenetics Chromatin* 8:55.

Lindblad-Toh, K., M. Garber, O. Zuk, M. F. Lin, B. J. Parker, S. Washietl, P. Kheradpour, J. Ernst, G. Jordan, E. Mauceli, L. D. Ward, C. B. Lowe, A. K. Holloway, M. Clamp, S. Gnerre, J. Alfoldi, K. Beal, J. Chang, H. Clawson, J. Cuff, F. Di Palma, S. Fitzgerald, P. Flicek, M. Guttman, M. J. Hubisz, D. B. Jaffe, I. Jungreis, W. J. Kent, D. Kostka, M. Lara, A. L. Martins, T. Massingham, I. Moltke, B. J. Raney, M. D. Rasmussen, J. Robinson, A. Stark, A. J. Vilella, J. Wen, X. Xie, M. C. Zody, J. Baldwin, T. Bloom, C. W. Chin, D. Heiman, R. Nicol, C. Nusbaum, S. Young, J. Wilkinson, K. C. Worley, C. L. Kovar, D. M. Muzny, R. A. Gibbs, A. Cree, H. H. Dihn, G. Fowler, S. Jhangiani, V. Joshi, S. Lee, L. R. Lewis, L. V. Nazareth, G. Okwuonu, J. Santibanez, W. C. Warren, E. R. Mardis,

G. M. Weinstock, R. K. Wilson, K. Delehaunty, D. Dooling, C. Fronik, L. Fulton, B. Fulton, T. Graves, P. Minx, E. Sodergren, E. Birney, E. H. Margulies, J. Herrero, E. D. Green, D. Haussler, A. Siepel, N. Goldman, K. S. Pollard, J. S. Pedersen, E. S. Lander, and M. Kellis. 2011. "A High-Resolution Map of Human Evolutionary Constraint Using 29 Mammals." *Nature* 478(7370):476–482.

Lisch, D. and J. L. Bennetzen. 2011. "Transposable Element Origins of Epigenetic Gene Regulation." *Curr Opin Plant Biol* 14(2):156–161.

Liu, M. and M. V. Eiden. 2011. "Role of Human Endogenous Retroviral Long Terminal Repeats (LTRs) in Maintaining the Integrity of the Human Germ Line." *Viruses* 3(6):901–905.

Lopes, F. R., D. Jjingo, C. R. da Silva, A. C. Andrade, P. Marraccini, J. B. Teixeira, M. F. Carazzolle, G. A. Pereira, L. F. Pereira, A. L. Vanzela, L. Wang, I. K. Jordan, and C. M. Carareto. 2013. "Transcriptional Activity, Chromosomal Distribution and Expression Effects of Transposable Elements in Coffea Genomes." *PLoS One* 8(11):e78931.

Lowe, C. B., G. Bejerano, and D. Haussler. 2007. "Thousands of Human Mobile Element Fragments Undergo Strong Purifying Selection Near Developmental Genes." *Proc Natl Acad Sci U S A* 104(19):8005–8010.

Lowe, C. B., M. Kellis, A. Siepel, B. J. Raney, M. Clamp, S. R. Salama, D. M. Kingsley, K. Lindblad-Toh, and D. Haussler. 2011. "Three Periods of Regulatory Innovation During Vertebrate Evolution." *Science* 333(6045):1019–1024.

Lu, X., F. Sachs, L. Ramsay, P. E. Jacques, J. Goke, G. Bourque, and H. H. Ng. 2014. "The Retrovirus HERVH Is a Long Noncoding RNA Required for Human Embryonic Stem Cell Identity." *Nat Struct Mol Biol* 21(4):423–425.

Lv, Y., F. Hu, Y. Zhou, F. Wu, and B. S. Gaut. 2019. "Maize Transposable Elements Contribute to Long Non-Coding RNAs That Are Regulatory Hubs for Abiotic Stress Response." *BMC Genomics* 20(1):864.

Lynch, V. J., R. D. Leclerc, G. May, and G. P. Wagner. 2011. "Transposon-Mediated Rewiring of Gene Regulatory Networks Contributed to the Evolution of Pregnancy in Mammals." *Nat Genet* 43(11):1154–1159.

Lynch, V. J., M. C. Nnamani, A. Kapusta, K. Brayer, S. L. Plaza, E. C. Mazur, D. Emera, S. Z. Sheikh, F. Grutzner, S. Bauersachs, A. Graf, S. L. Young, J. D. Lieb, F. J. DeMayo, C. Feschotte, and G. P. Wagner. 2015. "Ancient Transposable Elements Transformed the Uterine Regulatory Landscape and Transcriptome During the Evolution of Mammalian Pregnancy." *Cell Rep* 10(4):551–561.

Lyon, M. F. 2000. "Line-1 Elements and X Chromosome Inactivation: A Function for 'Junk' DNA?" *Proc Natl Acad Sci U S A* 97(12):6248–6249.

Lyon, M. F. 2003. "The Lyon and the Line Hypothesis." *Semin Cell Dev Biol* 14(6):313–318.

Maezawa, S., A. Sakashita, M. Yukawa, X. Chen, K. Takahashi, K. G. Alavattam, I. Nakata, M. T. Weirauch, A. Barski, and S. H. Namekawa. 2020. "Super-Enhancer Switching Drives a Burst in Gene Expression at the Mitosis-to-Meiosis Transition." *Nat Struct Mol Biol* 27(10):978–988.

Magallón, S., L. L. Sánchez-Reyes, and S. L. Gómez-Acevedo. 2019. "Thirty Clues to the Exceptional Diversification of Flowering Plants." *Ann Bot* 123(3):491–503.

Maharjan, R., and T. Ferenci. 2015. "Mutational Signatures Indicative of Environmental Stress in Bacteria." *Mol Biol Evol* 32(2):380–391.

Maharjan, R. P., and T. Ferenci. 2017. "A Shifting Mutational Landscape in 6 Nutritional States: Stress-Induced Mutagenesis as a Series of Distinct Stress Input-Mutation Output Relationships." *PLoS Biol* 15(6):e2001477.

Makarevitch, I., A. J. Waters, P. T. West, M. Stitzer, C. N. Hirsch, J. Ross-Ibarra, and N. M. Springer. 2015. "Transposable Elements Contribute to Activation of Maize Genes in Response to Abiotic Stress." *PLoS Genet* 11(1):e1004915.

Mao, H., H. Wang, S. Liu, Z. Li, X. Yang, J. Yan, J. Li, L. S. Tran, and F. Qin. 2015. "A Transposable Element in a NAC Gene Is Associated with Drought Tolerance in Maize Seedlings." *Nat Commun* 6:8326.

Marino-Ramirez, L., K. C. Lewis, D. Landsman, and I. K. Jordan. 2005. "Transposable Elements Donate Lineage-Specific Regulatory Sequences to Host Genomes." *Cytogenet Genome Res* 110(1–4):333–341.

Maslowska, K. H., K. Makiela-Dzbenska, and I. J. Fijalkowska. 2019. "The SOS System: A Complex and Tightly Regulated Response to DNA Damage." *Environ Mol Mutagen* 60(4):368–384.

Masuda, S., K. Nozawa, W. Matsunaga, Y. Masuta, A. Kawabe, A. Kato, and H. Ito. 2017. "Characterization of a Heat-Activated Retrotransposon in Natural Accessions of Arabidopsis Thaliana." *Genes Genet Syst* 91(6):293–299.

Matsunaga, W., N. Ohama, N. Tanabe, Y. Masuta, S. Masuda, N. Mitani, K. Yamaguchi-Shinozaki, J. F. Ma, A. Kato, and H. Ito. 2015. "A Small RNA Mediated Regulation of a Stress-Activated Retrotransposon and the Tissue Specific Transposition During the Reproductive Period in *Arabidopsis*." *Front Plant Sci* 6:48.

Maumus, F., A. E. Allen, C. Mhiri, H. Hu, K. Jabbari, A. Vardi, M. A. Grandbastien, and C. Bowler. 2009. "Potential Impact of Stress Activated Retrotransposons on Genome Evolution in a Marine Diatom." *BMC Genomics* 10:624.

McClintock, B. 1950. "The Origin and Behavior of Mutable Loci in Maize." *Proc Natl Acad Sci U S A* 36(6):344–355.

McClintock, B. 1952. "Controlling Elements and the Gene." *Cold Spring Harb Symp Quant Biol* 21:197–216.

McClintock, B. 1984. "The Significance of Responses of the Genome to Challenge." *Science* 226(4676):792–801.

McClintock, B. 1987. *Discovery and Characterization of Transposable Elements: The Collected Papers of Barbara McClintock*. New York: Garland.

McEwen, G. K., D. K. Goode, H. J. Parker, A. Woolfe, H. Callaway, and G. Elgar. 2009. "Early Evolution of Conserved Regulatory Sequences Associated with Development in Vertebrates." *PLoS Genet* 5(12):e1000762.

Micale, L., M. N. Loviglio, M. Manzoni, C. Fusco, B. Augello, E. Migliavacca, G. Cotugno, E. Monti, G. Borsani, A. Reymond, and G. Merla. 2012. "A Fish-Specific Transposable Element Shapes the Repertoire of P53 Target Genes in Zebrafish." *PLoS One* 7(10):e46642.

Michalak, P. 2010. "An Eruption of Mobile Elements in Genomes of Hybrid Sunflowers." *Heredity (Edinb)* 104(4):329–330.

Mikkelsen, T. S., M. J. Wakefield, B. Aken, C. T. Amemiya, J. L. Chang, S. Duke, M. Garber, A. J. Gentles, L. Goodstadt, A. Heger, J. Jurka, M. Kamal, E. Mauceli, S. M. Searle, T. Sharpe, M. L. Baker, M. A. Batzer, P. V. Benos, K. Belov, M. Clamp, A. Cook, J. Cuff, R. Das, L. Davidow, J. E. Deakin, M. J. Fazzari, J. L. Glass, M. Grabherr, J. M. Greally, W. Gu, T. A. Hore, G. A. Huttley, M. Kleber, R. L. Jirtle, E. Koina, J. T. Lee, S. Mahony, M. A. Marra, R. D. Miller, R. D. Nicholls, M. Oda, A. T. Papenfuss, Z. E. Parra, D. D. Pollock, D. A. Ray, J. E. Schein, T. P. Speed, K. Thompson, J. L. VandeBerg, C. M. Wade, J. A. Walker, P. D. Waters, C. Webber, J. R. Weidman, X. Xie, M. C. Zody, J. A. Graves, C. P. Ponting, M. Breen, P. B. Samollow, E. S. Lander, and K. Lindblad-Toh. 2007. "Genome of the Marsupial *Monodelphis Domestica* Reveals Innovation in Non-Coding Sequences." *Nature* 447(7141):167–177.

Muino, J. M., S. de Bruijn, A. Pajoro, K. Geuten, M. Vingron, G. C. Angenent, and K. Kaufmann. 2016. "Evolution of DNA-Binding Sites of a Floral Master Regulatory Transcription Factor." *Mol Biol Evol* 33(1):185–200.

Mulholland, C. B., A. Nishiyama, J. Ryan, R. Nakamura, M. Yiğit, I. M. Glück, C. Trummer, W. Qin, M. D. Bartoschek, F. R. Traube, E. Parsa, E. Ugur, M. Modic, A. Acharya, P. Stolz, C. Ziegenhain, M. Wierer, W. Enard, T. Carell, D. C. Lamb, H. Takeda, M. Nakanishi, S. Bultmann, and H. Leonhardt. 2020. "Recent Evolution of a TET-Controlled and DPPA3/STELLA-Driven Pathway of Passive DNA Demethylation in Mammals." *Nat Commun* 11(1):5972.

Nakanishi, A., N. Kobayashi, A. Suzuki-Hirano, H. Nishihara, T. Sasaki, M. Hirakawa, K. Sumiyama, T. Shimogori, and N. Okada. 2012. "A Sine-Derived Element Constitutes a Unique Modular Enhancer for Mammalian Diencephalic Fgf8." *PLoS One* 7(8):e43785.

Naville, M., I. A. Warren, Z. Haftek-Terreau, D. Chalopin, F. Brunet, P. Levin, D. Galiana, and J. N. Volff. 2016. "Not So Bad after All: Retroviruses and Long Terminal Repeat Retrotransposons as a Source of New Genes in Vertebrates." *Clin Microbiol Infect* 22(4):312–323.

Negi, P., A. N. Rai, and P. Suprasanna. 2016. "Moving through the Stressed Genome: Emerging Regulatory Roles for Transposons in Plant Stress Response." *Front Plant Sci* 7:1448.

Nishihara, H. 2019. "Retrotransposons Spread Potential Cis-Regulatory Elements During Mammary Gland Evolution." *Nucleic Acids Res* 47(22):11551–11562.

Nishihara, H. 2020. "Transposable Elements as Genetic Accelerators of Evolution: Contribution to Genome Size, Gene Regulatory Network Rewiring and Morphological Innovation." *Genes Genet Syst* 94(6):269–281.

Nishihara, H., N. Kobayashi, C. Kimura-Yoshida, K. Yan, O. Bormuth, Q. Ding, A. Nakanishi, T. Sasaki, M. Hirakawa, K. Sumiyama, Y. Furuta, V. Tarabykin, I. Matsuo, and N. Okada. 2016. "Coordinately Co-Opted Multiple Transposable Elements Constitute an Enhancer for *wnt5a* Expression in the Mammalian Secondary Palate." *PLoS Genet* 12(10):e1006380.

Notwell, J. H., T. Chung, W. Heavner, and G. Bejerano. 2015. "A Family of Transposable Elements Co-Opted into Developmental Enhancers in the Mouse Neocortex." *Nat Commun* 6:6644.

Ogasawara, H., H. Obata, Y. Hata, S. Takahashi, and K. Gomi. 2009. "Crawler, a Novel Tc1/Mariner-Type Transposable Element in *Aspergillus Oryzae* Transposes under Stress Conditions." *Fungal Genet Biol* 46(6–7):441–449.

Okada, N., T. Sasaki, T. Shimogori, and H. Nishihara. 2010. "Emergence of Mammals by Emergency: Exaptation." *Genes Cells* 15(8):801–812.

Oliver, M. J., O. Schofield, and K. Bidle. 2010. "Density Dependent Expression of a Diatom Retrotransposon." *Mar Genomics* 3(3–4):145–150.

Pal, C., K. Asiani, S. Arya, C. Rensing, D. J. Stekel, D. G. J. Larsson, and J. L. Hobman. 2017. "Metal Resistance and Its Association with Antibiotic Resistance." *Adv Microb Physiol* 70:261–313.

Pappalardo, A. M., V. Ferrito, M. A. Biscotti, A. Canapa, and T. Capriglione. 2021. "Transposable Elements and Stress in Vertebrates: An Overview." *Int J Mol Sci* 22(4).

Parks, A. R., Z. Li, Q. Shi, R. M. Owens, M. M. Jin, and J. E. Peters. 2009. "Transposition into Replicating DNA Occurs through Interaction with the Processivity Factor." *Cell* 138(4):685–695.

Paszkowski, J. 2015. "Controlled Activation of Retrotransposition for Plant Breeding." *Curr Opin Biotechnol* 32:200–206.

Pattee, H. H. (Ed.). 1973. *Hierarchy Theory: The Challenge of Complex Systems*. New York: Braziller.

Pauls, S., D. K. Goode, L. Petrone, P. Oliveri, and G. Elgar. 2015. "Evolution of Lineage-Specific Functions in Ancient Cis-Regulatory Modules." *Open Biol* 5(11):150079.

Pereira, C. M., T. J. R. Stoffel, S. M. Callegari-Jacques, A. Hua-Van, P. Capy, and E. L. S. Loreto. 2018. "The Somatic Mobilization of Transposable Element Mariner-Mos1 During the Drosophila Lifespan and Its Biological Consequences." *Gene* 679:65–72.

Pezzulo, G. and M. Levin. 2016. "Top-Down Models in Biology: Explanation and Control of Complex Living Systems above the Molecular Level." *J R Soc Interface* 13(124).

Pietzenuk, B., C. Markus, H. Gaubert, N. Bagwan, A. Merotto, E. Bucher, and A. Pecinka. 2016. "Recurrent Evolution of Heat-Responsiveness in *Brassicaceae Copia* Elements." *Genome Biol* 17(1):209.

Piskurek, O. and D. J. Jackson. 2012. "Transposable Elements: From DNA Parasites to Architects of Metazoan Evolution." *Genes (Basel)* 3(3):409–422.

Pittendrigh, C. S. 1958. *Adaptation, Natural Selection, and Behavior*, edited by A. Roe and George Gaylord Simpson. New Haven, CT: Yale University Press.

Polak, P., and E. Domany. 2006. "Alu Elements Contain Many Binding Sites for Transcription Factors and May Play a Role in Regulation of Developmental Processes." *BMC Genomics* 7:133.

Polavarapu, N., L. Mariño-Ramírez, D. Landsman, J. F. McDonald, and I. K. Jordan. 2008. "Evolutionary Rates and Patterns for Human Transcription Factor Binding Sites Derived from Repetitive DNA." *BMC Genomics* 9:226.

Policarpi, C., L. Crepaldi, E. Brookes, J. Nitarska, S. M. French, A. Coatti, and A. Riccio. 2017. "Enhancer SINEs Link Pol Iii to Pol Ii Transcription in Neurons." *Cell Rep* 21(10):2879–2894.

Polychronopoulos, D., J. W. D. King, A. J. Nash, G. Tan, and B. Lenhard. 2017. "Conserved Non-Coding Elements: Developmental Gene Regulation Meets Genome Organization." *Nucleic Acids Res* 45(22):12611–12624.

Pourrajab, F. and S. Hekmatimoghaddam. 2021. "Transposable Elements, Contributors in the Evolution of Organisms (from an Arms Race to a Source of Raw Materials)." *Heliyon* 7(1):e06029.

Qiu, Y., and C. Köhler. 2020. "Mobility Connects: Transposable Elements Wire New Transcriptional Networks by Transferring Transcription Factor Binding Motifs." *Biochem Soc Trans* 48(3):1005–1017.

Ramachandran, D., M. R. McKain, E. A. Kellogg, and J. S. Hawkins. 2020. "Evolutionary Dynamics of Transposable Elements Following a Shared Polyploidization Event in the Tribe Andropogoneae." *G3 (Bethesda)* 10(12):4387–4398.

Rebeiz, M., N. H. Patel, and V. F. Hinman. 2015. "Unraveling the Tangled Skein: The Evolution of Transcriptional Regulatory Networks in Development." *Annu Rev Genomics Hum Genet* 16:103–131.

Rebollo, R., M. T. Romanish, and D. L. Mager. 2012. "Transposable Elements: An Abundant and Natural Source of Regulatory Sequences for Host Genes." *Annu Rev Genet* 46:21–42.

Regolin, L., and G. Vallortigara. 2021. "Rethinking Cognition: From Animal to Minimal." *Biochem Biophys Res Commun* 564:1–3.

Reznikoff, W. S. 1992. "The Lactose Operon-Controlling Elements: A Complex Paradigm." *Mol Microbiol* 6(17):2419–2422.

Riegler, A. 2008. "Natural or Internal Selection? The Case of Canalization in Complex Evolutionary Systems." *Artif Life* 14(3):345–362.

Rishishwar, L., L. Wang, J. Wang, S. V. Yi, J. Lachance, and I. K. Jordan. 2018. "Evidence for Positive Selection on Recent Human Transposable Element Insertions." *Gene* 675:69–79.

Roman, A. C., F. J. Gonzalez-Rico, and P. M. Fernandez-Salguero. 2011a. "B1-SINE Retrotransposons: Establishing Genomic Insulatory Networks." *Mob Genet Elements* 1(1):66–70.

Roman, A. C., F. J. Gonzalez-Rico, E. Molto, H. Hernando, A. Neto, C. Vicente-Garcia, E. Ballestar, J. L. Gomez-Skarmeta, J. Vavrova-Anderson, R. J. White, L. Montoliu, and P. M. Fernandez-Salguero. 2011b. "Dioxin Receptor and Slug Transcription Factors Regulate the Insulator Activity of B1 SINE Retrotransposons Via an RNA Polymerase Switch." *Genome Res* 21(3):422–432.

Römer, C., M. Singh, L. D. Hurst, and Z. Izsvák. 2017. "How to Tame an Endogenous Retrovirus: HERVH and the Evolution of Human Pluripotency." *Curr Opin Virol* 25:49–58.

Romero-Soriano, V., N. Burlet, D. Vela, A. Fontdevila, C. Vieira, and M. P. Garcia Guerreiro. 2016. "Drosophila Females Undergo Genome Expansion after Interspecific Hybridization." *Genome Biol Evol* 8(3):556–561.

Romero-Soriano, V. and M. P. Garcia Guerreiro. 2016. "Expression of the Retrotransposon Helena Reveals a Complex Pattern of TE Deregulation in Drosophila Hybrids." *PLoS One* 11(1):e0147903.

Saber, M. M., I. Adeyemi Babarinde, N. Hettiarachchi, and N. Saitou. 2016. "Emergence and Evolution of Hominidae-Specific Coding and Noncoding Genomic Sequences." *Genome Biol Evol* 8(7):2076–2092.

Sacerdot, C., G. Mercier, A. L. Todeschini, M. Dutreix, M. Springer, and P. Lesage. 2005. "Impact of Ionizing Radiation on the Life Cycle of Saccharomyces Cerevisiae Ty1 Retrotransposon." *Yeast* 22(6):441–455.

Saier, M. H., Jr., C. Kukita, and Z. Zhang. 2017. "Transposon-Mediated Directed Mutation in Bacteria and Eukaryotes." *Front Biosci (Landmark Ed)* 22:1458–1468.

Sakashita, A., S. Maezawa, K. Takahashi, K. G. Alavattam, M. Yukawa, Y. C. Hu, S. Kojima, N. F. Parrish, A. Barski, M. Pavlicev, and S. H. Namekawa. 2020. "Endogenous Retroviruses Drive Species-Specific Germline Transcriptomes in Mammals." *Nat Struct Mol Biol* 27(10):967–977.

Sakurai, T., S. Nakagawa, H. Bai, R. Bai, K. Kusama, A. Ideta, Y. Aoyagi, K. Kaneko, K. Iga, J. Yasuda, T. Miyazawa, and K. Imakawa. 2017. "Novel Endogenous Retrovirus-Derived Transcript Expressed in the Bovine Placenta Is Regulated by WNT Signaling." *Biochem J* 474(20):3499–3512.

Salazar, M., E. González, J. A. Casaretto, J. M. Casacuberta, and S. Ruiz-Lara. 2007. "The Promoter of the TLC1.1 Retrotransposon from *Solanum chilense* Is Activated by Multiple Stress-Related Signaling Molecules." *Plant Cell Rep* 26(10):1861–1868.

Salces-Ortiz, J., C. Vargas-Chavez, L. Guio, G. E. Rech, and J. González. 2020. "Transposable Elements Contribute to the Genomic Response to Insecticides in *Drosophila Melanogaster*." *Philos Trans R Soc Lond B Biol Sci* 375(1795):20190341.

Santangelo, A. M., F. S. de Souza, L. F. Franchini, V. F. Bumaschny, M. J. Low, and M. Rubinstein. 2007. "Ancient Exaptation of a Core-Sine Retroposon into a Highly Conserved Mammalian Neuronal Enhancer of the Proopiomelanocortin Gene." *PLoS Genet* 3(10):1813–1826.

Santoni, F. A., J. Guerra, and J. Luban. 2012. "HERV-H RNA Is Abundant in Human Embryonic Stem Cells and a Precise Marker for Pluripotency." *Retrovirology* 9:111.

Sasaki, T., H. Nishihara, M. Hirakawa, K. Fujimura, M. Tanaka, N. Kokubo, C. Kimura-Yoshida, I. Matsuo, K. Sumiyama, N. Saitou, T. Shimogori, and N. Okada. 2008. "Possible Involvement of SINEs in Mammalian-Specific Brain Formation." *Proc Natl Acad Sci U S A* 105(11):4220–4225.

Schwarzer, W., and F. Spitz. 2014. "The Architecture of Gene Expression: Integrating Dispersed Cis-Regulatory Modules into Coherent Regulatory Domains." *Curr Opin Genet Dev* 27:74–82.

Servant, G., C. Pennetier, and P. Lesage. 2008. "Remodeling Yeast Gene Transcription by Activating the Ty1 Long Terminal Repeat Retrotransposon under Severe Adenine Deficiency." *Mol Cell Biol* 28(17):5543–5554.

Sexton, C. E., R. L. Tillett, and M. V. Han. 2021. "The Essential but Enigmatic Regulatory Role of HERVH in Pluripotency." *Trends Genet* 38(1):12–21.

Shapiro, J., and D. Noble. 2021. "What Prevents Mainstream Evolutionists Teaching the Whole Truth About How Genomes Evolve?" *Prog Biophys Mol Biol* 165:140–152.

Shapiro, J. A. 1983. *Mobile Genetic Elements*. New York: Academic Press.

Shapiro, J. A. 2011. *Evolution: A View from the 21st Century*. Upper Saddle River, NJ: FT Press Science.

Shapiro, J. A. 2013. "How Life Changes Itself: The Read-Write (RW) Genome." *Phys Life Rev* 10(3):287–323.

Shapiro, J. A. 2017. "Living Organisms Author Their Read-Write Genomes in Evolution." *Biology (Basel)* 6(4):42.

Shapiro, J. A. 2019. "No Genome Is an Island: Toward a 21st Century Agenda for Evolution." *Ann NY Acad Sci* 1447(1):21–52.

Shapiro, J. A. 2020. "All Living Cells Are Cognitive." *Biochem Biophys Res Commun* 564:134–149.

Shapiro, J. A. 2021. "What Can Evolutionary Biology Learn from Cancer Biology?" *Prog Biophys Mol Biol* 165:19–28.

Shapiro, J. A. 2022. *Evolution: A View from the 21st Century. Fortified*, 2nd ed. Chicago: Cognition Press.

Smukowski Heil, C., K. Patterson, A. S. Hickey, E. Alcantara, and M. J. Dunham. 2021. "Transposable Element Mobilization in Interspecific Yeast Hybrids." *Genome Biol Evol* 13(3):evab033.

Soltis, P. S., and D. E. Soltis. 2014. "Flower Diversity and Angiosperm Diversification." *Methods Mol Biol* 1110:85–102.

Spirov, A. V., E. A. Zagriychuk, and D. M. Holloway. 2014. "Evolutionary Design of Gene Networks: Forced Evolution by Genomic Parasites." *Parallel Process Lett* 24(2).

Sprague Jr., G. F. 1991. "Signal Transduction in Yeast Mating: Receptors, Transcription Factors, and the Kinase Connection." *Trends Genet* 7(11–12):393–398.

Srinivasachar Badarinarayan, S., and D. Sauter. 2021. "Switching Sides: How Endogenous Retroviruses Protect Us from Viral Infections." *J Virol* 95(12):e02299–20.

Stebbins, Jr., G. L. 1951. "Cataclysmic Evolution." *Scientific American* 184(4):54–59.

Steimer, A., P. Amedeo, K. Afsar, P. Fransz, O. Mittelsten Scheid, and J. Paszkowski. 2000. "Endogenous Targets of Transcriptional Gene Silencing in Arabidopsis." *Plant Cell* 12(7):1165–1178.

Sun, M. A., G. Wolf, Y. Wang, A. D. Senft, S. Ralls, J. Jin, C. E. Dunn-Fletcher, L. J. Muglia, and T. S. Macfarlan. 2021. "Endogenous Retroviruses Drive Lineage-Specific Regulatory Evolution across Primate and Rodent Placentae." *Mol Biol Evol* 38(11):4992–5004.

Sundaram, V., Y. Cheng, Z. Ma, D. Li, X. Xing, P. Edge, M. P. Snyder, and T. Wang. 2014. "Widespread Contribution of Transposable Elements to the Innovation of Gene Regulatory Networks." *Genome Res* 24(12):1963–1976.

Suryamohan, K., and M. S. Halfon. 2015. "Identifying Transcriptional Cis-Regulatory Modules in Animal Genomes." *Wiley Interdiscip Rev Dev Biol* 4(2):59–84.

Suzuki, H., T. Taketani, M. Tanabiki, M. Ohara, J. Kobayashi, and T. Ohshiro. 2021. "Frequent Transposition of Multiple Insertion Sequences in Geobacillus Kaustophilus Hta426." *Front Microbiol* 12:650461.

Takeda, S., K. Sugimoto, H. Otsuki, and H. Hirochika. 1998. "Transcriptional Activation of the Tobacco Retrotransposon Tto1 by Wounding and Methyl Jasmonate." *Plant Mol Biol* 36(3):365–376.

Tashiro, K., A. Teissier, N. Kobayashi, A. Nakanishi, T. Sasaki, K. Yan, V. Tarabykin, L. Vigier, K. Sumiyama, M. Hirakawa, H. Nishihara, A. Pierani, and N. Okada. 2011. "A Mammalian Conserved Element Derived from SINE Displays Enhancer Properties Recapitulating Satb2 Expression in Early-Born Callosal Projection Neurons." *PLoS One* 6(12):e28497.

Teramoto, S., T. Tsukiyama, Y. Okumoto, and T. Tanisaka. 2014. "Early Embryogenesis-Specific Expression of the Rice Transposon *Ping* Enhances Amplification of the MITE *mPing*." *PLoS Genet* 10(6):e1004396.

Testori, A., L. Caizzi, S. Cutrupi, O. Friard, M. De Bortoli, D. Cora, and M. Caselle. 2012. "The Role of Transposable Elements in Shaping the Combinatorial Interaction of Transcription Factors." *BMC Genomics* 13(1):400.

Todeschini, A. L., A. Morillon, M. Springer, and P. Lesage. 2005. "Severe Adenine Starvation Activates Ty1 Transcription and Retrotransposition in *Saccharomyces Cerevisiae*." *Mol Cell Biol* 25(17):7459–7472.

Torres-Padilla, M. E. 2020. "On Transposons and Totipotency." *Philos Trans R Soc Lond B Biol Sci* 375(1795):20190339.

Toussaint, M., and W. von Seelen. 2007. "Complex Adaptation and System Structure." *Biosystems* 90(3):769–782.

Trizzino, M., Y. Park, M. Holsbach-Beltrame, K. Aracena, K. Mika, M. Caliskan, G. H. Perry, V. J. Lynch, and C. D. Brown. 2017. "Transposable Elements Are the Primary Source of Novelty in Primate Gene Regulation." *Genome Res* 27(10):1623–1633.

Tyczewska, A., J. Gracz-Bernaciak, J. Szymkowiak, and T. Twardowski. 2021. "Herbicide Stress-Induced DNA Methylation Changes in Two Zea Mays Inbred Lines Differing in Roundup® Resistance." *J Appl Genet* 62(2):235–248.

Umbreit, N. T., C. Z. Zhang, L. D. Lynch, L. J. Blaine, A. M. Cheng, R. Tourdot, L. Sun, H. F. Almubarak, K. Judge, T. J. Mitchell, A. Spektor, and D. Pellman. 2020. "Mechanisms Generating Cancer Genome Complexity from a Single Cell Division Error." *Science* 368(6488):eaba0712.

van de Lagemaat, L. N., J. R. Landry, D. L. Mager, and P. Medstrand. 2003. "Transposable Elements in Mammals Promote Regulatory Variation and Diversification of Genes with Specialized Functions." *Trends Genet* 19(10): 530–536.

Vandecraen, J., P. Monsieurs, M. Mergeay, N. Leys, A. Aertsen, and R. Van Houdt. 2016. "Zinc-Induced Transposition of Insertion Sequence Elements Contributes to Increased Adaptability of *Cupriavidus metallidurans*." *Front Microbiol* 7:359.

Vicient, C. M., and J. M. Casacuberta. 2017. "Impact of Transposable Elements on Polyploid Plant Genomes." *Ann Bot* 120(2):195–207.

Volff, J. N. 2006. "Turning Junk into Gold: Domestication of Transposable Elements and the Creation of New Genes in Eukaryotes." *Bioessays* 28(9):913–922.

Voronova, A., M. Rendón-Anaya, P. Ingvarsson, R. Kalendar, and D. Ruņģis. 2020. "Comparative Study of Pine Reference Genomes Reveals Transposable Element Interconnected Gene Networks." *Genes (Basel)* 11(10):1216.

Wang, D., Z. Qu, L. Yang, Q. Zhang, Z. H. Liu, T. Do, D. L. Adelson, Z. Y. Wang, I. Searle, and J. K. Zhu. 2017. "Transposable Elements (TEs) Contribute to Stress-Related Long Intergenic Noncoding RNAs in Plants." *Plant J* 90(1):133–146.

Wang, J., N. J. Bowen, L. Marino-Ramirez, and I. K. Jordan. 2009. "A c-Myc Regulatory Subnetwork from Human Transposable Element Sequences." *Mol Biosyst* 5(12):1831–1839.

Wang, J., C. Vicente-Garcia, D. Seruggia, E. Molto, A. Fernandez-Minan, A. Neto, E. Lee, J. L. Gomez-Skarmeta, L. Montoliu, V. V. Lunyak, and I. K. Jordan. 2015. "MIR Retrotransposon Sequences Provide Insulators to the Human Genome." *Proc Natl Acad Sci U S A* 112(32):4428–4437.

Wang, J., G. Xie, M. Singh, A. T. Ghanbarian, T. Raskó, A. Szvetnik, H. Cai, D. Besser, A. Prigione, N. V. Fuchs, G. G. Schumann, W. Chen, M. C. Lorincz, Z. Ivics, L. D. Hurst, and Z. Izsvák. 2014. "Primate-Specific Endogenous Retrovirus-Driven Transcription Defines Naive-Like Stem Cells." *Nature* 516(7531):405–409.

Wang, N., H. Wang, H. Wang, D. Zhang, Y. Wu, X. Ou, S. Liu, Z. Dong, and B. Liu. 2010. "Transpositional Reactivation of the *Dart* Transposon Family in Rice Lines Derived from Introgressive Hybridization with *Zizania Latifolia*." *BMC Plant Biol* 10:190.

Wang, X., G. Ai, C. Zhang, L. Cui, J. Wang, H. Li, J. Zhang, and Z. Ye. 2016. "Expression and Diversification Analysis Reveals Transposable Elements Play Important Roles in the Origin of Lycopersicon-Specific lncRNAs in Tomato." *New Phytol* 209(4):1442–1455.

Warren, I. A., M. Naville, D. Chalopin, P. Levin, C. S. Berger, D. Galiana, and J. N. Volff. 2015. "Evolutionary Impact of Transposable Elements on Genomic Diversity and Lineage-Specific Innovation in Vertebrates." *Chromosome Res* 23(3):505–531.

Wiener, N. 1965. *Cybernetics or Control and Communication in the Animal and the Machine.* Cambridge, MA: MIT Press.

Woodrow, P., G. Pontecorvo, L. F. Ciarmiello, A. Fuggi, and P. Carillo. 2011. "Ttd1a Promoter Is Involved in DNA-Protein Binding by Salt and Light Stresses." *Mol Biol Rep* 38(6):3787–3794.

Woolfe, A., and G. Elgar. 2008. "Organization of Conserved Elements near Key Developmental Regulators in Vertebrate Genomes." *Adv Genet* 61:307–338.

Xiang, Y., and H. Liang. 2021. "The Regulation and Functions of Endogenous Retrovirus in Embryo Development and Stem Cell Differentiation." *Stem Cells Int* 2021:6660936.

Yang, X., Y. Yu, L. Jiang, X. Lin, C. Zhang, X. Ou, K. Osabe, and B. Liu. 2012. "Changes in DNA Methylation and Transgenerational Mobilization of a Transposable Element (*mPing*) by the Topoisomerase II Inhibitor, Etoposide, in Rice." *BMC Plant Biol* 12:48.

Zhang, X., and L. J. Muglia. 2021. "Baby's Best Foe-Riend: Endogenous Retroviruses and the Evolution of Eutherian Reproduction." *Placenta* 113:1–7.

Zhang, Y. D., Z. Chen, Y. Q. Song, C. Liu, and Y. P. Chen. 2005. "Making a Tooth: Growth Factors, Transcription Factors, and Stem Cells." *Cell Res* 15(5):301–316.

Zhao, H., W. Zhang, L. Chen, L. Wang, A. P. Marand, Y. Wu, and J. Jiang. 2018. "Proliferation of Regulatory DNA Elements Derived from Transposable Elements in the Maize Genome." *Plant Physiol* 176(4):2789–2803.

Zhu, C. X., L. Yan, X. J. Wang, Q. Miao, X. X. Li, F. Yang, Y. B. Cao, P. H. Gao, X. L. Bi, and Y. Y. Jiang. 2014. "Transposition of the Zorro2 Retrotransposon Is Activated by Miconazole in *Candida albicans*." *Biol Pharm Bull* 37(1):37–43.

16 Agency, Teleonomy, Purpose, and Evolutionary Change in Plant Systems

Anthony Trewavas

All those adaptations in the organism are purposeful which contribute to its maintenance and ensure its existence.
—von Sachs (1887 p. 601)

Julius von Sachs was a foremost plant physiologist of the nineteenth century and author of the first major work on plant physiology. He stated: "Concerning one point I should wish to anticipate: viz, the use of the word Purpose, a word which many fanatics of the theory of descent would if possible banish entirely from the language" (von Sachs, 1887, p.10). The strict interpretation of natural selection saw evolution (and life) as purposeless, and the modern synthesis has continued that view. As indicated in this chapter, it is not obvious that Darwin himself followed that strict interpretation.

If all organisms have a purposeful existence, can their interactions be any the less so? Would there not be some indication of convergent evolution? The driving force of all life—its purpose—is survival, its existence. If it wasn't, there would be no life. Pittendrigh (1958) coined the term *teleonomy* to indicate the purposefulness of adaptations in both structure and function as evidenced by goal-directed activity. Von Sachs had already recognized this to be the case some 77 years earlier. We have paraphrased Pittendrigh's comments about turtles to indicate relevance to plants. Compare these two statements: The seed germinates and makes a new plant vs. The seed germinates to make a new plant. The first is descriptive of what happens, but the second directly indicates purpose, that is part of the schedule which ensures that life continues because it has this goal of reproduction and survival to do so. The change is subtle but profound in understanding biology.

That drive to survive is easy to see with plants. Even in cities, plants often grow in very tiny niches of wind-blown soil. Single-celled algae often adorn fences, gates, and tree trunks and sometimes glass in wet habitats. The struggle for existence would not occur without this purposeful drive to survive.

16.1 Darwin's Failing: The Conditions of Life

In the preface of his major work, Darwin (1859) stated that selection was the main but not exclusive method of evolutionary change. The more minor alternative, which he later referred to as the "external conditions of life," is described in the compilations on animal and plant variation under domestication. These conditions for plants are to be found in Darwin (1868, reprint 2007, p. 198 and onwards). Changing the "conditions of life" elicits specific phenotypic changes in many plant species to what looks like the first step in evolutionary change. Variation during domestication is of course common. Once conditions that threaten survival are removed, many kinds survive, but when placed in real-world environments they quickly disappear. Horticulture is underpinned by plant domestication, inbreeding, or vegetative multiplication. Lamarck (1914) placed equivalent emphasis on variation by domestication, and Darwin is known to have read Lamarck's contributions. Domestication and animal breeding figure strongly in Darwin (1859) to indicate potential variation.

In letters, though, Darwin regretted his failure not to amplify the phrase "conditions of life" more (Darwin, 1876): "In my opinion, the greatest error which I have committed has been not allowing sufficient weight to the direct action of the environment, (i.e. food, climate) independently of natural selection." The modern synthesis of evolution has ignored Darwin's plaintive cry in this regard or maybe never read it.

Darwin's practical experience of working with plants and descriptions of plant behavior was extensive, as indicated by the numerous books he wrote about his experiments with them. He became familiar with the often-profound effects that the "conditions of life" have on plants. Lamarck too was originally a botanist and he commented (Larmarck, 1914, p. 215) on the remarkable change that resulted from domesticating plants by changing the "conditions of life" (see Box 16.1). It is possible that Lamarck, more than two centuries ago, may have thought that plant and animal evolution used the same mechanism. The easy modification of plant phenotypes by a changing environment then led to his belief in the inheritance of acquired characters. With a permanent change in the environment, this is probably correct for plant evolution as presently understood. But Lamarck was quite clear that this took place over very long periods of time. What Lamarck usefully said relevant to plant evolution is indicated in Box 16.1. He is often misquoted.

Lamarck's concerns included problems with the size of legs of giraffes: no doubt as they increased in length to run faster and avoid predators, but too long and the neck would have to lengthen coordinately to enable the consumption of water. Lamarck saw that a gradual increase in the leg length of the evolving giraffe, while no doubt improving escape from predators, would require an increasing neck length enabling it to drink. The 'need' is obvious and nothing to do with supposed internal attempts to stretch its neck (Cannon,1959).

16.2   The Extended Evolutionary Hypothesis

Dissatisfaction with the modern synthesis that seemingly created a completely purposeless world has now generated alternatives (Noble, 2015). The extended evolutionary hypothesis is one important proposal for a change in understanding (Noble, 2015). The following summary of evidence provided later in this paper, as well as a previous paper (Gilroy & Trewavas, 2022), indicates that this hypothesis with additional material more closely fits what is understood about plant evolution than just the modern synthesis on its own.

Box 16.1
What Did Lamarck Actually Say of Relevance to Plant Evolution? Direct Quotations from Lamarck (1914)

- Nature has successively produced the different living beings by proceedings from the simplest to the most compound . . . and becomes complicated in a most remarkable way (p. 241)

- I do not mean by this to say that the existing animals [plants?] form a very simple series but that they form a branched series which has no discontinuity in its parts. It is true that owing to extinction of some species there are some breaks (p. 243).

- It follows that the species which terminates each branch of the general series is connected at least on one side with other species which intergrade with it (p. 243).

- As the individuals of one of our species are subjected to changes in situation, in climate, mode of life and habits, receive influences which gradually change the consistence and proportion of their parts, their form, their faculties[,] even their structure, so that it follows that all of them participate in the changes to which they have been subjected (p. 243.).

- Very different situations and exposures cause simple variations in the individuals there which live and successively reproduce under the same circumstances, produce differences in form which become essential to their existence so that at the end of many successive generations these individuals which originally belonged to another species become finally transformed into a new species distinct from the other (p. 244).

- Species then have only relative stability and are invariable only temporarily (p. 247).

- The influence of circumstance is really continuously and everywhere active on living beings as its effects only become recognisable at the end of a long period (p. 252).

- In the plant where there are no movements and consequently no habits so-called, great changes in circumstance do not bring about less great differences in the development of their parts (p. 255).

- All botanists know that plants transplanted from their natal spot into gardens gradually undergo changes which in the end make them unrecognisable (p. 215).

- Is not wheat a plant brought by man to the state wherein we actually see it which otherwise I could not believe. Who can now say in what place its like lives in nature (p. 215).

The extended evolutionary hypothesis for plants contains the following ideas:

- The struggle for existence indicates that a fundamental driving force of life is survival to reproduction: without it, there would be no life. Survival is surely purposeful and a goal for any individual plant.

- Plants and bacteria are commonly omitted from much evolutionary discussion (Noble, 2015). Plants are easily the dominant form of life on Earth and their evolution is prominently constrained by unchanging physical laws.

- Plants are agents; they act to control their own response to what they perceive in the environment and do so by phenotypic and chemical plasticity, properties that describe behavior.

- Development is a process of construction; organism and environment are intimately interrelated; they are not separate entities (Waddington, 1957; Lewontin, 2000; Trewavas, 2009). Together they construct a holistic entity (organism ↔ environment) to form a niche. Lewontin (2000) described developmental construction as a Markov process involving reciprocal causation and recurrent learning.

• Many, if not all, plant behaviors are purposeful and goal-directed and thus teleonomic.

• Some plant species are biased in certain environments toward evolutionary change. Plasticity can influence which plant species can more rapidly adapt to a new set of circumstances because plasticity varies among species and individuals.

• Plants are intelligent organisms because they behave adaptively, a common definition of biological intelligence. Adaptive behavior acts as an evolutionary constraint on what can survive.

• Plant transgenerational inheritance has been frequently reported.

• There should be a change in emphasis away from genes toward complex traits; that is, toward complex interacting genetic networks that are constructed from hundreds or more different protein species.

• Plant convergent evolution has been frequently identified.

• Unchanging physical laws constrain plant evolutionary change. The process is not random but physically limited.

16.3   Plants Are the Pre-eminent Form of Life on Land

Most biological research and evolutionary studies tend to concentrate only on animals. With the publication of the distribution of biomass among the different biological groups, this attention can be seen as anomalous. Bar-On et al. (2018, with a very long supplement) extracted the representative quantities of biomass as presently understood. They report that worldwide, plants form 82% and marine bacteria 13% of biomass; the residue is distributed, relatively unevenly, among the remaining bacterial, fungal, and animal groups. However, the estimate of marine bacteria biomass is subject to considerable uncertainty and at best operates only at very low levels of metabolic activity.

Ninety-three percent of all biomass is land-based, and plant biomass now forms about 90–95% of this total. The approximate 90% was calculated by Bar-On et al. (2018) by assuming (incorrectly) that tree trunks have no metabolic activity. Interxylary phloem and ray parenchyma are metabolically-active cells that are located throughout the wood in most tree species. They store starch, water and nutrients and are essential for regulating xylem hydraulics and growth (Carlquist, 2007; 2013). The cambium, dividing cells that generate new phloem and xylem (wood); and the so-called cork cambium that generates bark are part of the trunk. The metabolic activity of tree trunks is not zero but so far as I am aware has not actually been measured.

The ratio of land-based plant biomass to land-based animal biomass is approximately 400–500. An alternative estimate can be made from the ratio of atmospheric oxygen (20%) from photosynthesis by plants to carbon dioxide (0.04%) from nearly all forms of respiration among animals, including night respiration of plants and carbon dioxide exchange from marine sources, fish, and bacteria. Adjusting for the different molecular weights of oxygen and carbon dioxide gives a ratio of 285. The biomass ratio of plant/humankind is more than 4,000.

Does this matter? It does when it distorts biological understanding. For example, "the modern synthesis of evolution has always had a strongly zoological basis, tending to ignore prokaryotes, unicellular organisms and plants even though these cover more than 80% of the whole duration of the evolutionary process long before 'zoology' could even have a meaning in evolutionary history" (Noble, 2015). The Weismann barrier figured strongly in the rejection of Lamarckian views by zoologists—plants never had such a barrier. Vegetative experience can be carried directly into the reproductive structures in the flower and trans-generational inheritances can be considered common in plants.

16.4   What Causes This Bias in Biological Understanding?

Bias results from a failure to recognize that we ourselves are animals and frame the biological world from an animal perspective only. To identify behavior, we require obvious movement, but within our own time frame. However, our time frame is limited physiologically. Images in the eye are replaced every tenth of a second to avoid adaptation; films, for example, have to run at 50 frames/second to make movement seem continuous.

Plants seem completely unresponsive compared to most animals. Animals and plants separated at the amoebic stage some two billion years ago or so (McFadden, 2014). The vast majority of animals have always had to move to find food, and that movement operates at about the same speed throughout the animal kingdom, no doubt for the same reason. The requirements for most plant life are satisfied by the physical resources—minerals, water, light, CO_2—which are reasonably abundant over much of the globe. Most plants respond to a large array of signals that in number probably dwarfs those perceived by most animals. Plants live in two very different environments (above and below ground), each of which has unique environmental issues. The local environment is more heavily perceived by a plant than by the average animal because an individual plant perceives it continuously and must assess the free space around itself for the needed resources, which can be heavily contested. Phenotypic modification follows from this assessment, whereas the animal in many local environments remains phenotypically unchanged.

The commonest response to a perceived shift in the local environment is a change in plant phenotype; this is an adaptive behavioral response. Intriguingly, most environmental signals start the processes of transduction by initiating cytosolic calcium transients within less than a tenth of a second, even though changes in structure and phenotype may take days or longer (Knight et al., 1991; Trewavas, 2014). Plants are the basis of most food chains, but the survival of the individual to reproduce is still the inherent driving force, as it is for all life. Accepting the inevitability of predation, the evolutionary solution of most plants was to become modular, with repetitions of leaves plus subtended buds and below-ground branch roots. Lose some and others replace them. Growth takes place via tip-based embryogenic meristems and the dormancy of buds when broken initiates new branches. Branching is a very efficient and competitive way to occupy local space.

The emphasis on evolutionary mechanisms and on understanding between plants and animals just has to be different. Much animal evolution has elaborated sensory systems and accuracy in perception and assessment, a necessary accompaniment to the changing needs of movement and the result of having to move to find food. Plant evolution is mostly directed toward more efficiently occupying local space and acquiring the resources contained in that space.

16.5   Physical Laws Constrain How Plants Grow and Develop

Most higher plant species have a stem and a root. Both are composed of cells that are surrounded by a relatively rigid cell wall and contain a cytoplasm operating at an atmospheric (turgor) pressure of about 7–8 atmospheres resulting from the accumulation of some 0.3M KCl. Growth is limited to small regions of the shoot and root. The wall is softer here, enabling division, cell expansion, and tissue specification. These meristematic areas, permanently embryogenic, are, of necessity, spatially limited and located in shoot and root tips. They enable the polarized growth so characteristic of many plants. Both stem and root have cells or structures (buds and pericycle cells) that basically are dormant meristems able to form branches when activated. The cambium, a meristematic ring inside the growing stem and root, is likewise permanently embryonic. It acts to generate vascular cells and expand the diameter of root and shoot. Its additional function is to direct resources between competing shoots and roots (Trewavas, 2014). Cambial activity generates xylem internal to the cambial ring. This tissue, by transporting water and minerals, continues to thicken its wall. The xylem helps keep the stem upright and strengthens the root system. Phloem composed eventually of anucleate cells is generated external to the cambial ring and transports organic chemicals throughout the whole plant. Shoot buds and root pericycle contain stem cell representatives that are left in mature tissues, thereby enabling the familiar regenerative capabilities involved in vegetative reproduction. Many species can regenerate roots from stem pieces. Fewer generate stems from root pieces, and some can be regenerated from leaves. Nevertheless, species are known that lack roots, others lack stems, and some lack leaves.

Physical constraints and unchanging physical laws, together with the recognition that plants only need the physical components of the environment, already hint at different emphases in evolutionary mechanisms.

16.6   Plants Exert Choice Over Where to Grow

Despite the inability of any individual plant to uproot itself and move elsewhere, plants do make choices as to where they grow (Bazzaz, 1991; 1996). For example, by the production of highly dispersible seeds, particularly among the r-strategy category of plants. Dispersal mechanisms include wind, or via moving water or being caught in the fur of a moving animal. Targeted dispersal is accomplished by offering loyal dispersers (e.g. bats, monkeys) rewards, like seed-enclosing fruits that lead to ultimate seed distribution through defecation. Other larger seeds can be collected and buried by some animals and birds at some distance from the parent. Explosive dehiscence can leave seeds on the surface either for collection or burial by falling leaves, or by burrowing animals, thus constructing a seed bank. A spoonful of soil can contain several hundred seeds (Darwin, 1859). Dormancy of such seeds is common and, while quickly broken in some good habitats, it can last for up to a century in moist soils providing a fail-safe or bet-hedging capability (Trewavas, 2014). Seeds of some species exhibit annual flushes in germinability lasting for up to a decade (Barton, 1961).

Another mechanism involves dispersal with beneficial resources enclosed in dung that accelerates early germination, growth and competitive ability. Dung movement by beetles helps distribution. Yet another method involves movement, by growth, of ground-based stems or soil-based rhizomes into less-contested areas. Timing of dormancy breakage to coincide with the availability of resources, or receipt of environmental signals indicating open space, can also be important. Because plants are found throughout much of the earth's surface, these dispersal mechanisms, while seemingly variable in targeting distribution, are very effective.

16.7 Visible Plant Behavior Is Phenotypic Plasticity

The interaction between a single plant and its known environment constructs a holistic structure best summarized as (phenotype ↔ environment) (Lewontin, 2000; 2001) and described in more detail in Gilroy and Trewavas (2022). The extent of phenotypic plasticity can be substantial and much more than most, if not all, animals. While virtually all parts of higher plants have been reported to be phenotypically plastic, individuals and individual species vary in the signals that induce plasticity and in the extent of tissue plasticity variation.

The following phenotypic changes in plants resulting from different environments have been recorded:

Plastic changes in leaf shape and branching

Frequency of patterns in shoot or root abscission numbers

Ratios of the shoot to root weight

Thinness or thickness changes in the root, shoot, petioles, hypocotyls

Leaf area ratios

Reproductive output and seed number

Anatomical variations such as cuticle thickness, cuticle chemical and reflective properties, palisade mesophyll depth, stomatal density, tissue cell numbers (Pham & McConnaughay, 2014).

These plastic changes can relate to observable environmental variation, but even individuals of a species vary phenotypically within this framework.

Figure 16.1 illustrates the phenotype changes in an inbred line of *Polygonum* when experiencing either drought or low light. These phenotypic changes are summarized in the legend. The most obvious plasticity in drought is the increase in root proliferation to expand the surface area for absorption of water. But other changes in the shoot are often initiated to reduce water loss and accelerate reproduction. In low light, primary changes in leaf area, numbers of leaves, and/or anatomical alterations inside the leaves improve capture and processing of the lower light now available.

Individuals that exhibit greater speed in changing plasticity can learn to adapt more quickly. These will increase rapidly in number if the environmental shift continues. Thus, in the long term, the mere prevalence of the shift should promote the selection of plants with the relevant characteristics to develop with greater rapidity, higher probability, or lower cost (Bateson, 1963).

16.8 The Ability to Behave Plastically Identifies Plants as Agents

Agents act autonomously to direct their own behavior to achieve both external and internal goals or norms (Barandiaran et al., 2009; Beer, 1995; Pfiefer & Scheier, 1999; Walsh, 2015)

Figure 16.1
Plant plasticity activated by environmental conditions.
Genetic replicates of the same inbred line of *Polygonum cespitosum* construct dramatically different juvenile phenotypes in contrasting naturalistic treatments. The plant on the left was grown in dry soil and full sunlight. There is multiple branching and numerous reproductive axes, narrow leaves with thick mesophyll and cuticle and high biomass allocation to root tissue, a typical response to drying conditions. The plant on the right was grown in moist soil and with 79% reduction in photosynthetically active radiation (shade), with the R/FR ratio reduced to 0.70. This plant expresses a less branched upright habit with elongated internodes, large broad leaves with thin mesophyll and cuticle, and high biomass allocation to the leaves maximizing photosynthetic area and more limited reproductive axes (Sultan, 2010). Photo credit: Tim Horgan-Kobeliski, Courtesy Sultan lab, Wesleyan University.

while in continuous long-term interaction with the real-world environment (Beer, 1995). Plants are described as agents because they plastically control and modify their phenotype to improve survival (Sultan, 2015). Behavior is adjusted as the environment changes but within the context of achieving longer-term goals, as was illustrated in figure 16.1, which shows how juveniles adjust their phenotype in response to either the overriding stimulus of water deficit or low light. The phenotypic responses detailed in the figure 16.1 legend are typical of many plants. Environmental information is assembled by the individual that then self-organizes the phenotypic change.

Although phenotypic plasticity is obvious, changes in cellular chloroplast movements, leaf blade rotation, or natural pesticide synthesis to kill or damage predating insects are described as *physiological plasticity*, because they are reversible. Despite the phenotypic response being slow, plants can be superlatively sensitive to some environmental parameters like touch or slight movement (Bunning et al., 1948; Trewavas & Knight, 1994) or slight variations in light regime (Massonnet et al., 2010). Brief stroking with sheets of paper can substantially reduce the height of a growing hypocotyl (Bunning et al., 1948).

Real-world circumstances result in continued variation and precise composition of all of the abiotic and biotic constituents of the environment. Individual plants experience environmental abiotic constituents that can vary in longevity, intensity, speed of change, and unpredictability from seconds to hours, days, and months (Caldwell & Pearcy, 1994). An important requirement for any plant is to be able to distinguish between ephemeral environmental changes (a passing animal, a sunfleck, or a brief gust of wind) compared to a more persistent environmental change such as trampling or drought. Initiating major changes to the phenotype in response to an ephemeral signal would be energetically wasteful, and of course phenotypic change itself may take days or weeks. Ephemeral signals do have effects but are quickly reversible.

Recognition of the persistence of a stimulus becomes important to plant survival and competitive ability. Oscillations of cellular signals such as $[Ca^{2+}]_i$ could help distinguish ephemeral signals from the persistence of change, as suggested later. Successful environmental navigation results in individuals with variable numbers of organs and differing degrees of historical plastic variation (Herrera, 2017).

16.9   Behavioral Plasticities Are Complex Traits Constructed From Large Numbers of Interacting Proteins But Vary Among Individuals

Only about an estimated 1% of plant genes mendelize; the remaining 99% contribute to traits that govern behavior. It is likely that any trait is constructed from the complex interactions of tens, hundreds, or even thousands of proteins interacting in a complex fashion to form an interactome (Braun et al., 2013; Zhao et al., 2019). Hubs are connected to many proteins, edges to many fewer. Hubs are commonly essential proteins; mutation can lead to radical change or even death. The average number of interactions of any one protein with others is reckoned to be about seven to ten and obeys a power law (Trewavas, 2017).

All that genes do is specify a norm of reaction: the ability of an individual plant to change behavior (phenotype) when subjected to numerous different environmental situations (Lewontin, 2000; 2001). The first detailed investigation was reported in the classic experiments of Clausen et al. (1940). These authors vegetatively cloned individuals of *Achillea* and other plants and grew them in different environments, usually different altitudes. There was no uniform or linear response to environmental change between the individuals; the norms of reaction crossed each other (Lewontin, 2000). Many other investigations have used different plant species and different environments, but the conclusions are the same. Each individual plant constructs traits that differ from those of its neighbors, thus indicating the complexity of plants.

16.10   Purpose and Teleonomy: Goals Require Negative Feedback for Their Operation

The modern synthesis used two metaphors for explanation (Lewontin, 2000): genes, an explanation that originated with their discovery by Mendel; and environment, Darwin's primary contribution. Mutations were the final contribution to the modern synthesis and were assumed to appear at random, thus eliminating notions of direction in evolutionary terms. The simple assumption was that the genes proposed, while the environment disposed (Lewontin, 2000). This construct gave rise to notions of selfish genes as the dominating, determinist influence in evolutionary change, and evolutionary change itself as stochastic. Evolution at all levels resulted simply from shifts in populations described by population genetics. Evolutionary alternatives such as Lamarck's early views emphasizing adaptation to the environment (the "conditions of life") became the butt of contempt, usually by those who never read Lamarck. The delightful text by Cannon (1959) indicated their ignorance.

What was omitted was any contribution by the organism itself to its own evolution. The word *purpose*, along with *teleonomy* (teleology), were expunged from the literature, although Rosenblueth et al. (1943) mounted an early defense to retain the term and showed that teleonomy [as teleology] involved negative feedback (a cybernetic principle) and goal-orientation.

Homeostasis is a prominent goal-directed behavior using negative feedback as control (Cannon, 1932; Gilroy & Trewavas, 2022). Evidence that the leaves on trees can exert a degree of control over their internal temperature, improving photosynthetic yield in crowded environments, has been reported (Helliker & Richter, 2008). Competition for light can be fierce. The internal temperature variations are wider than those found in mammals, but the authors identified numerous sources of negative feedback that will exert control. Some 70,000 species of eudicots are trees. Similarly, flowers of the sacred lotus (*Nelumbo lucifera*) thermoregulate to promote controlled floral development and rates of pollen tube growth, and fertilization (e.g., Dieringer et al., 2014). Feedback processes linking thermal control to metabolic activities, such as through alternative oxidases, were identified.

The well-known tropic curvatures of plant stems to directional light and gravity stimuli use negative feedback. The goals are either to improve exposure to light or to recover an upright position enabling better leaf position for photosynthesis and enabling subsequent flowers to be seen by pollinators. But less-well-known are tropic responses to water, touch (thigmotropic bending or mechano-tropic bending), electrotropism, and chemotropism; the last is used to direct root growth toward patches rich in nitrogen (Cahill & McNickle, 2011). Each tropic response operates with purpose; teleonomy, when seen against a background of competition for resources (Russell, 1946). "In tropistic movements plants appear to exhibit a sort of intelligence: their movement is of subsequent advantage to them" (Went & Thimann, 1937). The "advantage" refers to survival advantage. Plants growing in arid climates or desert climates often have roots extending purposely in the soil to 10–50 meters using hydrotropic perception. Such plants are thereby enabled to search out potential pockets of water at great depth and thus survive (Russell, 1946).

16.11   Goals and Teleonomy in Regenerative Ability

Behavior is most easily described as purposive, and thus teleonomic, when it seems directed toward a goal or end state; "directiveness" was used by Russell (1946) in an early text on this subject. More recent research has established that a flowering individual can be generated from a single leaf protoplast; from an isolated single vegetative cell; from callus; from an isolated seed embryo or from cultured, isolated, or regenerated embryos; from cuttings from stems, roots, or isolated leaves; and from individuals damaged to varying degrees through disease or pests (Trewavas, 2004). The purposeful routes toward the goal are multiple, but the goal remains constant: survival, thence to reproduction.

16.12   Teleonomy of Behavior in the Vegetative State

The need for a growing plant is to establish a vegetative structure that accumulates sufficient reserves to provision seeds, and the correlative goal is to ensure that the stem reaches a suitable height so that subsequent flowers can be seen by pollinators. Provision of light, CO_2, minerals, and water at reasonable temperatures and in approximate balance are sufficient for good vegetative growth of root and shoot (Bloom et al., 1985; White, 1937). But achieving growth can be countered by mismatch: excess or inadequate provision of the physical requirements; extremes of temperature or water such as flooding, drought, and salt stress; strong gusty wind; heavy rain; trampling; and finally the biotic problems of herbivory, disease, and competition. Plasticity is then used to potentially improve survival probability.

A gravity stimulus sensed through statoliths directs most root growth downward into the soil, but the shoot can also experience wind, heavy rain, or trampling that interferes with the teleonomic goal of flowers and reproduction. Gravitropic responses use differential growth to recover the vertical position. But if statoliths were the only mechanism involved in controlling stem reorientation, it would simply lead to permanent oscillations around a mean, which is not observed. Proprioception, an agency property (Moulia et al., 2021), uses gravicline sensing and feedback of local mechanical perturbations to ensure a new stable vertical position. All tropic phenomena are goal-directed, purposive, and involve control by negative feedback.

16.13   Necessary Teleonomy Involving the Flower

The goals of a flower are directed toward efficient cross-fertilization and seed dispersal. Self-fertilization is a fail-safe mechanism that occurs in about 50% of all angiosperm species, but is a last resort if productive pollen from a compatible individual is not received (Lin et al., 2015). Most flowers are adapted to promote transfer of pollen to another individual flower of the same species and to receive pollen back from a similar source. Cross-fertilization is improved by attracting visiting pollinator insects such as bees. Flowers of some species specifically detect the sound frequency of a flying bee and vibrate with sympathetic resonance, increasing the supply of nectar within one to two minutes (Nepi et al., 2018; Veits et al., 2019). Flowers construct an electrical field and bees learn to recognize its specific form (Clarke et al., 2013). Bumblebees carry a very substantial electrical charge, and entry to the flower modifies the field and causes an increase in scent production (Montgomery et al., 2021). These mechanisms increase visits by other bees/insects and thus have an identical control structure, as indicated in figure 16.2. All are clearly purposive and goal-directed: both partners benefit.

Action potentials are generated when the transfer of sufficient compatible pollen is achieved, modifying the metabolic status of the ovary in preparation for fertilization (Fromm et al., 1995). Large floral displays in which individual flowers change color when adequate pollen has been received are known to occur; bees are directed to those still needing visits. This color change can revert when sufficient pollen has been received (Willmer et al., 2009). To ensure that a visitor carries pollen away, many species of flowers have evolved stamen touch responses. Rapid movement deposits pollen directly on the visitor, sometimes with some force (Braam, 2005). Flowers with poricidal (enclosed) anthers resonate to the bee buzz which tribologically charges the pollen to increase adherence, after which the pollen is released in a shower covering the visitor, ensuring goal fulfillment (Corbet & Huang, 2014; Mesquita-Neto et al., 2018).

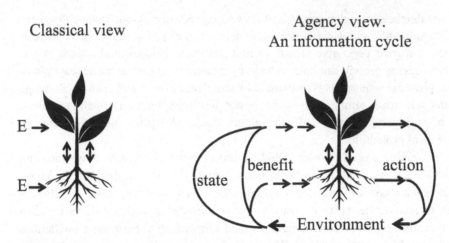

Figure 16.2
Recurrent learning and decision responses of plant agents between the individual and its environment. The process of construction: a comparison of the common experimental view of the environment–plant interaction, compared to a Markov decision process, the agency view. In classical terms, the plant simply reacts to a laboratory-controlled change in the environment. The Markov decision process sees the [environment↔plant] interaction as a recurrent learning process with reciprocal causation. The cyclical flow of information between plant and environment represents the understanding of the numerous interactions between a developing plant and its continually changing environment. The state of the plant indicates environment interaction at any one time point.

16.14   Development as Construction: How Plants Make Their Root and Shoot Niches via the Markov Mechanism—"Laying Down a Path in Walking" (Varela, 1987)

In real-world environments, plant agents purposefully modify and construct their environments in a continual two-way exchange of information (Lewontin, 2001; Waddington, 1975; Trewavas, 2009). The Markov chain process enables agents to take actions that construct their local environment and receive environmental benefits back (see figure 16.2). Because this process involves progressive and multiple negative feedback, it can be described as purposeful, and progressively it constructs a niche, the eventual biological goal. It is thus teleonomic. Environment and phenotype form an interrelated and continuous dialectic, which is summarized as [environment ↔ phenotype]. The construction of the root and shoot niche illustrates this process in action. The intricacies of root and shoot construction have been greatly expanded upon elsewhere (Gilroy & Trewavas, 2022). Only brief summaries are presented here.

16.15   Construction of the Root Niche

Natural soil is a highly heterogeneous and complex mixture of silt, stones of variable sizes, clay and sand, decaying organic matter (usually patchily distributed), pockets of free or bound minerals, bacteria, fungi, and fauna. Any growing root has to navigate these circumstances and exploit them for its own benefit, thereby improving the chances of survival. The construction of the root niche involves a changing composition of root secretions containing some hundreds of chemicals and enzymes that modify the soil and attract more than a thousand bacterial species to form a microbiome. Information from these bacterial members leads to adjustment of secretion composition to repel those which may be harmful and attract those that benefit (Hartman & Tringe, 2019; Ahmad et al., 2006; Pascale et al., 2020). Secretion of acid and enzymes helps mobilize phosphate from insoluble sources. The microbiome includes bacteria that penetrate the root, and some end up inside cells fixing nitrogen for carbohydrates. Changes in soil circumstances (e.g., drought) lead to changes in secretion, often of carbohydrates to entice bacteria that aid in such environmental changes to the benefit of both.

Lewontin described the construction of development as a Markov process (figure 16.2). The Markov process (figure 16.2) describes the recurrent learning and reciprocal causation of the changing program that oversees the developing microbiome and soil-penetrating root. The most remarkable example of these symbiotic connections, involving some 90% of all eudicot species and many gymnosperms, is the construct of a mycorrhizal network whose hyphae provide phosphate and iron for carbohydrate and lipid (Smith & Read, 2008; Genre et al., 2020). Through this network, which also connects numerous species together, carbohydrate can flow to kin preferentially, and information on predation and disease to enable partners to take pre-emptive action (Song et al., 2010, 2015; Bibikova et al., 2013; Tederson et al., 2020). Fifty thousand mycorrhizal species are involved. Some free riders occur, but when such are recognized, action is taken by the plant to rigidly control the exchange or eliminate the aberrant partner from its root system (Keirs et al., 2011). Root sensing of aliens, soil objects, and reactive behavior has been well documented (Novoplansky, 2019).

16.16   Construction of the Shoot Niche

Shoot niches form in crowded circumstances with individuals perceiving light competition and adjusting their phenotype in an attempt to outgrow competitors, thereby enabling eventual flowers to be seen by pollinators. Leaves touching each other induce a similar response. Interaction also occurs via the synthesis of volatile chemicals (ethylene, methylsalicylate, methyljasmonate) by an attacked individual, which more quickly induces defense responses in the more remote parts of the plant than communication through the circulatory system. But these are sensed by close neighbors who also can take pre-emptive defense actions. Kin benefit more than alien species (Karban et al., 2013).

Defense mechanisms are also initiated when an individual plant either experiences direct damage and constructs long-range electrical currents (elevating $[Ca^{2+}]_i$) and/or senses the chewing frequency of caterpillars (Appel & Cocroft, 2014; Choi et al., 2016; 2017). The attacked plant can distinguish the kind of pest caterpillar from an analysis of its salivary juice chemistry (Acevedo et al., 2018). Other species, upon experiencing predation, increase sugar production from glands to attract parasitoids, such as ants, wasps, and ladybugs; this response is known to increase fitness. Other detections include the mating pheromones of a predator initiating defense reactions: commonly natural pesticides (Helms et al., 2017), ovicidal chemicals against egg layers, and leaf abscission are all used (Mescher & de Moraes, 2015). Figure 16.3 summarizes the information flow and teleonomic nature of predation defense.

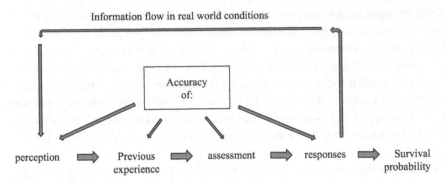

Information flow in real world conditions

Figure 16.3
Direction of information flow during herbivory in real-world conditions (another Markov process). The figure is a simplified version of the reaction of a plant to herbivory. Attack by a predator is *perceived* initially by changes in $[Ca^{2+}]_i$ and associated electrical changes, leading to the generation of volatiles such as methyl jasmonate or methyl salicylate that initiate some defense responses (benefit) that modifies the state. This process is likely a continual cycle of assessment of damage and adjustment of volatile synthesis and production of defense responses, and continues cyclically until herbivory diminishes or stops, as indicated in figure 16.2; this is another Markov chain. Much if not all plant development in wild conditions can be characterized as a Markov chain.

16.17   Summary of Consequences of Construction of Root and Shoot Niche

The information in this chapter identifies plant agency involving goal-directed behavior. Individual plants act on other individuals and organisms purposefully to achieve the initial goal of improved survival. Plant agents do not meander aimlessly like driftwood on water; the interactions described in this chapter bear strongly on mechanisms of selection and, in turn, a direction in the evolutionary process (Pigliucci & Müller, 2010; Sultan, 2010). The intimate relationship between the plant root, the environment, and the niche the plant constructs is described as an *affordance* (Gibson, 1979) and is what philosophers refer to as *enactivism* (Thompson, 2007).

16.18   Biological Intelligence Is Adaptive Behavior and Its Goal Is to Improve the Probability of Survival

Anna Anastasi (1986, p. 19), a well-known and prominent psychologist, stated the nature of biological intelligence: "Intelligence is not an entity within the organism but a quality of behaviour. Intelligent behaviour is essentially adaptive insofar as it represents effective ways of meeting the demands of a changing environment. Such behaviour varies with the species and with the context in which the individual lives." Identification of biological intelligence with adaptive behavior is common (Gilroy & Trewavas, 2022). Adaptive behavior, as Dobzhansky (1956) clearly indicated, is behavior that improves the probability of survival. The ability to adapt to requisite circumstances constrains what can evolve. Those individuals with weaker adaptive behavior, or none at all, simply die out.

It is easy for us as animals to see intelligent behavior in other animals: the rabbit running from the fox, the octopus changing color to meld with its background, the cow distinguishing between plants for better food, among many other possible examples. The expression of intelligence is a specific behavior resulting from specific environmental problems (organism ↔ environment) and thus a holistic structure. All organisms live in an environment and depend on that environment—no environment, no organism.

Recognition of plant intelligence suffers the same complication as recognition of behavior: failure to recognize that we as animals pass judgment on organisms that are not animals and whose behavior operates on a totally different time scale. As indicated earlier, changes in phenotype are the common plant response to environmental problems (e.g., figure 16.1) and represent obviously intelligent behavior in a specific environment. In real-world circumstances, competition is fierce and resources strongly contested. Resource limitation is the common consequence, and it is in these circumstances that specific changes in plasticity actually evolve; they do so because the probability of survival is increased thereby (Chapin, 1980). The growing plant can be significantly modeled as an economic system, and discussion of resource budgets, cost-benefit analyses, and marginal rates of return figure strongly in understanding behavior when specific resources run short (Bloom et al., 1985). Intelligent behavior is required at all levels in any economic system, model or organism, if the entity is to survive.

Chemical changes are used against pests, but the synthesis of natural pesticides requires heavy use of carbon and nitrogen. Consequently, seed yield decreases substantially. The compensation is increased survival probabilities compared to those plants that fail to synthesize these chemicals, are slow to do so, or do so to an inadequate level. Experimental situations don't help plant biologists recognize plant intelligence either, as it is very difficult to establish real adaptive situations in the laboratory that see the majority of plants die off. While experimental plant biologists may be familiar with adaptive behavior, the realization that it is about survival and that it is intelligent, is rarely reached. When experimental environments tend to be uniform, the contribution of the environment to the experimental phenotype tends to disappear from recognition. Experimental environments are merely one out of thousands. No environment = no plant.

Human intelligence as a psychological topic is distinct from biological intelligence; its concern is with scholastic ability. Human societies are socially protective organizations and together with modern medicine and modern agriculture to a large extent obviate biological adaptive behavior with its emphasis on survival probabilities.

16.19   Transgenerational Inheritance Contradicts the Modern Synthesis

Transgenerational inheritance makes real sense for many, but not all, plant species because their ability to escape changing environments is quite limited. Thus, pre-arming of progeny with the relevant adapted traits should improve survival probabilities when environments change over the long term. Environmental experience in plants can be passed to their siblings (Herman & Sultan, 2011). One early example was published more than 50 years ago (Durrant, 1962; 1971). Flax was grown on soil with an ideal NPK (nitrogen, phosphate, potassium) application which, with light and water, enabled good plant growth; the flax generated bushy phenotypes. The alternative was a number of unbalanced mineral applications, such as NK, which are detrimental for growth. A sparse phenotype was the response. Siblings of the two phenotypes remained bushy or sparse for up to six generations or more regardless of what mineral mixture they were subsequently grown on. The common supposition is that the transgenerational inheritance in subsequent generations is epigenetic.

Somatic cells generate the flower's germ cells. Consequently, transgenerational inheritance is frequently reported and is of course Lamarckian in character (Jablonka & Raz, 2009). In total, some 60 examples have been published (Jablonka & Raz, 2009; Kellenberger et al. 2018; Herman & Sultan, 2011; Singh & Roberts, 2015). However, this inheritance continues only for one to nine generations before disappearing when the environment reverts to normality. The following parental environmental experiences are expressed by siblings: shade, drought, nutrient status, viral and bacterial disease, temperatures (both hot and cold), seasonal influences, serpentine soils and high metal concentrations, herbivory, and salt stress. DNA methylation or histone marks are the suggested epigenetic changes that enable reversibility if the environmental change is reversed. The induction of systemic defenses after pathogen challenge is passed across multiple generations via a salicylic acid-dependent, RNA-directed DNA methylation (Luna & Ton, 2012). Some transgenerational events last only one generation; in addition to epigenesis, transfer of different acquired sRNAs, mRNAs, or proteins into the developing seed may help explain. Transgenerational inheritance becomes critical when the environmental shift is permanent. Currently, nearly 20 transgenerational cases have been shown to increase sibling fitness.

16.20   Oscillations in $[Ca^{2+}]_i$ May Be the Interpreter of Persistent Environmental Change and Prefigure Transgenerational Inheritance

Environmental changes inducing plant phenotypic and chemical plasticity in plants are initiated and interpreted through intracellular calcium $[Ca^{2+}]_i$ (Sanders et al., 1999; Vadassery & Oelmuller, 2009; Zeng et al., 2015; Zhu, 2016; Tian et al., 2020). Calcium spikes, waves, and oscillations all occur, and when inhibited (using either molecular or chemical inhibition) the response itself is likewise inhibited. The initial environmental signal is interpreted by specific receptors, which induce $[Ca^{2+}]_i$ transients, or spikes, that usually start within less than a tenth of a second and can finish within 30 seconds. For certain phenotypic responses, this single $[Ca^{2+}]_i$ spike may be sufficient, but for others, repetition in the form of oscillations may be necessary. Approximately 1% of the genome of the plant *Arabidopsis thaliana* is devoted to proteins containing just one of the many Ca^{2+}-responsive motifs, the EF-hand (Day et al., 2006). Many downstream interpreting proteins have been identified (reviewed in Dodd et al., 2010), and this network can be embedded in an integrated hormone response network involving several thousand proteins (Bender & Snedden, 2013; La Verde et al., 2018; Altmann et al., 2020). The information passing through this network continues into gene regulation networks, ion pumps, metabolic enzymes, and developmental regulators (Monshausen et al., 2007; 2011; Simeunovic et al., 2016; Jose et al., 2020; Tian et al., 2020). Transmission over long distances can involve oxidative signals coupled to calcium waves (e.g., Campbell et al., 1996; Choi et al., 2016; Nguyen et al., 2018; Toyota et al., 2018; Johns et al., 2021).

Ephemeral environmental signals are likely to induce a single $[Ca^{2+}]_i$ but it would be costly to initiate long-term and complex development on this basis. $[Ca^{2+}]_i$ oscillations are associated with definite and longer-term changes in the environment. The known environmental conditions that seemingly initiate $[Ca^{2+}]_i$ oscillations are new stable environments resulting from drought, cold, salinity, light, circadian change, and altered hormone levels (Evans et al., 2001): Darwin's "conditions of life." Although the expression of some genes can be altered by a single $[Ca^{2+}]_i$ transient, increasingly more genes were expressed as oscillation numbers increased (Lenzoni et al., 2018; Whalley et al., 2011), revealing a mechanism by which to link the extent of the response to the duration of the environmental change.

Determinate numbers of $[Ca^{2+}]_i$ oscillations are required for substantive changes in response by two carnivorous plants, the Venus flytrap and the sundew. In both cases, five to six oscillations are needed to definitely indicate a struggling insect and to initiate digestion (Williams & Pickard, 1972; Bohm et al., 2016). Some 30 nuclear oscillations are induced by rhizobial secretion of chemical signals to indicate that finally, sufficient rhizobia are outside the root to initiate root hair modification and initiate eventual nitrogen fixation in a nodule (Miwa et al., 2006). $[Ca^{2+}]_i$ oscillations are also essential in developing cells and tissues (Felle, 1988; Allen et al., 2001; Watahiki et al., 2004; Watahiki & Trewavas, 2019; Monshausen et al., 2007) that interpret the environmental circumstances as good for growth and only stop when development ceases. Perhaps one of the most dramatic changes occurs when a growing pollen tube closely approaches the ovary; the two synergid cells synchronize their $[Ca^{2+}]_i$ oscillations to those of the growing pollen tube (Iwano et al., 2012; Ngo et al., 2014), indicating recognition and then fertilization.

16.21 Convergent Evolution: Is There an Evolutionary Direction?

Neo-Darwinism advocates a trial-and-error scenario in which species blindly and randomly reconfigure, while the external environment accepts or rejects the product. The proposed means of change are mutation, recombination, genetic drift, and natural selection. The first three of these are considered to be stochastic, unpredictable, and possibly random in character; thus, any direction in evolution on this basis can be rejected out of hand. If the tape of life could be replayed, would it lead to the apparent reappearance of familiar species, including humankind? Stephen Jay Gould (1989) concluded not: "replay the tape a million times and I doubt that anything like *Homo sapiens* would evolve again."

Evolution does, however, often seem to repeat itself or *converge,* concluded Conway-Morris (2003, p. xii): there is a "recurrent tendency of biological organisation to arrive at the same 'solution' to a particular 'need'." Evolutionary convergence is more common than many biologists think (Blount et al., 2018). But these authors identify other instances where they consider that repetition has apparently *not* occurred, quoting as one plant example the desert plant *Welwitschia*. This unusual gymnosperm may be the only extant member of its genus, but extinct members are known; fossils might reveal that its genus was extensive and only part of a much wider gymnosperm group living in desert conditions.

Fifteen examples of convergent evolution in plants have been collated (Trewavas, 2014). The old-world (fleshy-stemmed) euphorbias and stapelias simulate members of the new-world cactaceae. Similarly, the thick, layered leaves of the aloes of Africa resemble those of the agaves of America; the alpine veronicas of New Zealand mimic the thujas of North America and even *Lycopodium* (Henslow, 1895). The phyllids of mosses, the thalli of liverworts, the microphylls and megaphylls of lycopods, horsetails, ferns, gymnosperms, and the stems and roots of angiosperms are often morphologically similar and thus seemingly convergent (Niklas, 1992). Physical principles are directly applicable to the fundamental nature of the plant structure and may constrain the potential variety here.

Plant biology is replete with examples of convergent evolution arising from polyphyletic sources and genetically remote groups (Trewavas, 2014). For example, C4 photosynthesis has supposedly originated at least 40–45 times, reflecting either very low CO_2 or water deficit (Sage, 2004). Mycorrhizal symbiosis has arisen separately many times (Genre et al., 2020). Symbiotic nitrogen fixation has evolved at least twice with different symbionts. There are six different parasitic plant lifestyles; there are carnivorous plants in both monocots and eudicots (Trewavas, 2014). These two major angiosperm groups are thought to have separated 170 million years ago. The "Conway-Morris need" was either for water, photosynthetic carbon, and/or nitrogenous materials of one kind or another.

16.22   Adaptive Teleonomic Requirements Constrain Evolutionary Change in Plant Systems and Contradict the Modern Synthesis That Relies on Random Mutations

Nearly all plants acquire their energy, minerals, water, and CO_2 from the physical, abiotic environment. Throughout evolutionary time, what is obvious is the improvement of the competitive facility for controlling surrounding space and acquiring physical resources (Niklas, 1992). An upright stem, as indicated earlier, is dependent on strong cell walls and balancing turgor pressure. Physical principles dictate the requirements of wall stiffness.

Competition for light has been prominent in plant evolution. Every vascular plant lineage has seen the emergence of the arborescent lifestyle to satisfy a requirement to erect leaves above the ground in eudicots, some monocots, gymnosperms, and the earlier Lycopodiales and Equisitales. Simple engineering principles apply to height, trunk strength or width, taper, and branch weight and thus govern the form any tree and bush can assume (Niklas, 1995, 2000). Wet environments can weaken wood strength, requiring plants to dynamically compensate their structure during growth (Henslow, 1908). Swamp cypress, for example, has a highly ridged, spreading base to support a relatively thin trunk. Root spread and soil physical structure also determine the stable trunk height of the individual. Leaf-like structures have evolved in all phyla, algae, mosses, liverworts, ferns, horsetails, lycopods, and seed plants (Niklas, 1992). Large leaves require strong or equally large petioles to support them. Stiffness allometries in angiosperms and ferns substantiate this biomechanical convergence (Niklas, 1991). Propagation by seed started first in the lycopods and secondary thickening via cambium among phyletically distant plant groupings. "There seem to be a limited number of ways in which photoautotrophs can be constructed" (Niklas, 1992), so this constrains what can appear and survive. Stress and strain in stem and petiole structures, gas diffusion, water flow, heat loss, and the effects of gravity are governed by unchanging physical principles and act to constrain what kind of plant can evolve. When new species evolve in distinct and physically unconnected parts of the world, similar environmental situations generate plants that morphologically look similar to each other because physical constraints dictate what can evolve, giving rise to a kind of convergent evolution. Periodic geological or asteroidal disturbance can lead to large-scale species extinction. Nevertheless, recovery involves a recurrence of similar evolutionary themes; anagenesis within new or surviving lineages rather than cladogenesis (Niklas, 1992).

Although a random element in natural selection could account for some plant evolution, the overriding dominance of basic physical laws has constrained others into an identifiable path that indicates the intimate relationship of plants with their physical environment. Plants that flouted these basic physical principles, and arose by chance variation in their genome, simply became extinct. Many comparative morphological, anatomical, and eco-physiological studies support the view that organs dissimilar in their developmental origins, but fulfilling the same function, should exhibit similar form and structure. Thorns on different species are a good example: plants with thorns often have different developmental origins and disappear in humid environments (Henslow, 1895).

16.23   Does Phenotypic Plasticity Direct Evolutionary Change?

George Henslow (1895, 1908), a plant ecologist, disagreed with Darwin's natural selection as applicable to plant evolution and argued strongly for Lamarckian views and the "conditions of life." The environment, he considered, generates definite natural variations, which are usually in the direction of environmental adaptation and to the appearance of subsequent forms that are well adapted in the primary ecosystems: deserts, arctic-alpine regions, saline and freshwater habitats, tundra, and prairies (Henslow, 1895: preface). "When many individuals of the same variety and in natural circumstances are exposed during several generations to any change in their physical conditions of life, all or nearly all the individuals are modified in the same manner" (Henslow, 1895). Transgenerational inheritance mechanisms are then likely to have evolved in individuals that experienced these variations on a less-than-permanent basis. Nevertheless, the specific changes in plasticity as a result of environmental modification represent the start of any evolutionary change, because these changes are now found in those plants that live permanently in those environments. Environmental reversions or variabilities may account for the retention of plasticity responses in many temperate plants: a kind of evolutionary memory but only recently discussed. Learning and memory have both been identified as potential mechanisms of evolutionary change and thus identify the evolutionary process as having a degree of intelligent assessment (Parter et al., 2008; Watson & Szathmáry, 2016).

Henslow (1895) criticized the Darwinian notion of indefinite variation, used by Darwin to support natural selection, because the claim was based on domesticated organisms. Many domesticated individuals and genotypes that survive in controlled environments fail to do so in real-world habitats. Domestication is a man-made manipulation used for agriculture, horticulture, and pets. Many garden plants, wheat, and maize do not survive in fallow fields (Calvo et al., 2020).

The conditions of life induce specific plastic changes in the plant phenotype (Gilroy & Trewavas, 2022). Spiny *Ononis spinosa* loses its spines in high humidity; *Ononis repens* is a wild, spineless version. *Ampelopsis vietchii* produces tiny part-developed adhesive pads that only mature when mechanically stimulated (usually by a wall); *Ampelopsis hederacea* forms no such pads until mechanically stimulated. The glycophyte *Plantago major* can respond to limited salty conditions by increasing succulence, whereas the halophyte *Plantago maritima* can tolerate high salt conditions and is succulent but tends to lose succulence on ordinary soil. Low light is thought responsible for the evolution of the climbing habit. Periwinkle, normally a ground plant, can climb a stick in very low light, possibly reflecting the greatly increased circumnutatory activity in low light (Stolarz, 2009). Experimental windy conditions severely reduce plant height in the laboratory (Knight et al., 1992; Braam, 2005). The dwarf *Salix herbacea*, which hardly grows to 6 cm in height, is found on high arctic/alpine and windy mountainsides. The lowland *Salix babylonica* grows to 7 meters or more. Many other examples are summarized by Henslow (1895). What was an inducible trait in one species is now permanently expressed in an adjacent one. These are all indicative of how the evolutionary process may have started—but thousands of years are required to confirm any theories.

16.24   Concluding Remarks

There are apparent constraints on the evolution of plants that contradict the tenets of the modern synthesis, which relies on random mutations, genetic drift, and population genetics. Plant evolution is not a random process. It is constrained by an overriding requirement to obey unchanging physical laws. The physical structure of 95% of all plant species has to obey fundamental physical and engineering principles to create a simple stable structure. This need not be surprising given the use of wood by mankind to create stable structures. But plants are living entities and change their structure as they develop.

There are obvious adaptive capabilities to be seen in phenotypic plasticity. *Adaptive behavior* is defined as intelligent behavior; in competitive circumstances, an inability to adapt to those circumstances and thus to act intelligently will likely lead to premature mortality. An additional constraint thus limits what can evolve. Development is a process of construction, as illustrated by the Markov process which with recurrent learning and reciprocal causation indicates its purposive nature. Root systems construct a purposeful niche in soils of varying kinds that improves growth yield, enhances resistance to disease, and gains information about predation. Plants do make choices as to what habitat they will occupy, and transgenerational inheritance will likely improve that choice. If purposive teleonomic plants interact competitively with each other, an emergent ecological structure is a possible consequence. Ecological memory indicates that such emergent structures do form and that they constrain what species can be constituents of any particular structure (Ogle et al., 2015). Because of the need for competition to be part of this emergent structure, the extent of phenotypic and competitive change by any species will be restricted; convergent evolution may be one consequence of this.

Along with the required adherence to physical laws and now ecological constraints, some form of extended evolutionary hypothesis more accurately describes plant evolution (Noble, 2015; Corning, 2020).

References

Acevedo, F. E. E., Peiffer, M., Ray, S., Meagher, R., Luthe, D. S., & Felton, G. W. (2018). Intra-specific differences in plant defence induction by fall army worm strains. *New Phytologist* 218, 310–321.

Ahmad, F., Ahmad, I., & Khan, M. S. (2006). Screening of free-living rhizospheric bacteria for their multiple plant growth-promoting activities. *Microbiological Research* 18, 173–181.

Allen, G. J., Chu, S. P., Harrington, C. L., Schumaker, K., Hoffmann, T., Tang, Y. T., et al. (2001). A defined range of guard cell calcium oscillation parameters encodes stomatal movements. *Nature* 411, 1053–1057.

Altmann, M., Altmann, S., Rofriguez, P. A., Weller, B., Vergara, L. E., Palme, J., et al. (2020). Extensive signal integration by the phytohormone protein network. *Nature* 583, 271–276.

Anastasi, A. (1986). Intelligence as a quality of behaviour. In R. J. Sternberg & D. K. Detterman (Eds.), *What is intelligence? Contemporary viewpoints on its nature and definition*, 19–23. Norwood, NJ. Ablex Publishing.

Appel, H. M., & Cocroft, R. B. (2014). Plants respond to leaf vibrations caused by insect chewing. *Oecologia* 175, 1257–1266.

Bar-On, Y. M., Phillips, R., & Milo, R. (2018). The biomass distribution on Earth. *Proceedings of the National Academy of Sciences U S A* 115, 6506–6511.

Barandiaran, X., Di Paolo, E., & Rohde, M. (2009). Defining agency, individuality, normativity, asymmetry and spatiotemporality in action. *Adaptive Behaviour* 17(5), 367–386. https://doi.org/10.1177/1059712309343819

Barton, L. V. (1961). *Seed preservation and longevity.* London: Hill.

Bateson, G. (1963). The role of somatic change in evolution. *Evolution* 17, 529–539.

Bazzaz, F. A. (1991). Habitat selection in plants. *American Naturalist* 137, s117–s130.

Bazzaz, F. A. (1996). *Plants in changing environments.* Cambridge, UK: Cambridge University Press.

Beer, R. D. (1995). A dynamical systems perspective on agent/environment interaction. *Artificial Intelligence* 72, 173–215.

Bender, K. W., & Snedden, W. A. (2013). Calmodulin-related proteins step out of the shadow of their namesake. *Plant Physiology* 163, 486–495.

Bibikova, Z., Gilbert, L., Toby, J. A., Bruce, J. A. G., Birkett, M., et al. (2013). Underground signals carried through common mycelial networks warn neighbouring plants of aphid attack. *Ecology Letters* 16, 835–843.

Bloom, A. J., Chapin, S., & Mooney, H. A. (1985). Resource limitation in plants—an economic analogy. *Annual Review Ecology and Systematics* 16, 363–392.

Blount, Z. D., Lenski, R. E., & Losos, J. B. (2018). Contingency and determinism in evolution: replaying life's tape. *Science* 362, 655–665.

Bohm, J., Scherer, S., Krol, E., Kreuzer, I., von Meyer, K., Lorey, C., et al. (2016). The Venus flytrap *Dionaea muscipula* counts prey-induced action potentials to induce sodium uptake. *Current Biology* 26(3), 286–295.

Braam, J. (2005). In touch: plant responses to mechanical stimuli. *New Phytologist* 165, 373–389.

Braun, P., Auburg, S., van Leene, J., de Jaeger, G., & Lurin, C. (2013). Plant protein interactomes. *Annual Review of Plant Biology* 64, 161–187.

Bunning, E., Haag, L., & Timmermann, G. (1948). Weitere untersuchungen uber die formative wirkung des lichtes und mechanischer reize auf Pflanzen. *Planta* 36, 178–187.

Cahill, J. F., & McNickle, G. G. (2011). The behavioral ecology of nutrient foraging in plants. *Annual Review of Ecology Evolution and Systematics* 42, 281–311.

Caldwell, M. M., & Pearcy, R. W. (1994). *Exploitation of environmental heterogeneity by plants.* London: Academic Press.

Calvo, P., Gagliano, M., Souza, G. M., & Trewavas, A. J. (2020). Plants are intelligent, here's how. *Annals of Botany* 125, 11–28.

Campbell, A. K., Trewavas, A. J., & Knight, M. R. (1996). Calcium imaging shows differential sensitivity to cooling and communication in luminous transgenic plants. *Cell Calcium* 19, 211–218.

Cannon, H. G. (1959). *Lamarck and modern genetics.* Manchester, UK: Manchester University Press.

Cannon, W. B. (1932). *The wisdom of the body.* New York: Norton.

Carlquist, S. (2007). Bordered pits in ray cells and axial parenchyma: the histology of conduction, storage, and strength in living wood cells. *Botanical Journal of the Linnean Society* 153, 157–168.

Carlquist, S. (2013). Interxylary phloem: Diversity and functions. *Brittonia* 65, 477–495.

Chapin, F. C. (1980). The mineral nutrition of wild plants. *Annual Review of Ecology and Systematics* 11, 233–260.

Choi, W-G., Hilleary, R., Swanson, S. J., Kim, S-H., & Gilroy, S. (2016). Rapid long distance electrical and calcium signalling in plants. *Annual Review of Plant Biology* 67, 287–307.

Choi, W-G., Miller, G., Wallace, I., Harper, J., Mittler, R., & Gilroy, S. (2017). Orchestrating rapid long-distance signaling in plants with Ca^{2+}, ROS and electrical signals. *Plant Journal* 90, 698–707.

Clarke, D., Whitney, H., Sutton, G., & Robert, D. (2013). Detection and learning of floral electric fields by bumblebees. *Science* 340, 66–69.

Clausen, J., Keck, D. D., & Hiesey, W. H. (1940). *Experimental studies on the nature of species. 1 Effect of varied environments on Western North American plants.* Carnegie Institute of Washington Publications No 520. Washington, DC: Carnegie Institute of Washington.

Conway-Morris, S. (2003). *Life's solution: Inevitable humans in a lonely universe*. Cambridge: Cambridge University Press.

Corbet, S. A., & Huang, S-Q. (2014). Buzz pollination in eight bumblebee-pollinated *Pedicularis* species: does it involve vibration-induced triboelectric charging of pollen grains? *Annals of Botany* 114, 1665–1674.

Corning, P. A. (2020). Beyond the modern synthesis: a framework for a more inclusive biological synthesis. *Progress in Biophysics and Molecular Biology* 53, 5–12.

Darwin, C. (1859). *The origin of species by means of natural selection*. London: John Murray.

Darwin, C. (1868). *Variation of animals and plants under domestication, Volume 2*. Reprinted 2007. Middlesex, UK: The Echo Library.

Darwin, C. (1876). Letter to Moritz Wagner, 13 October 1876. Darwin Correspondence Project. https://www.darwinproject.ac.uk/letter/DCP-LETT-10643.xml

Day, I. S., Reddy, V. S., Ahad Ali, G., & Reddy, A. S. N. (2002). Analysis of EF-hand-containing proteins in *Arabidopsis*. *Genome Biology* 3, R0056.

Dieringer, G., Leticia-Cabrera, R., & Mottaleb, M. (2014). Ecological relationship between floral thermogenesis and pollination in *Nelumbo lutea*. *American Journal of Botany* 101, 357–364.

Dobzhansky, D. (1956). What is an adaptive trait? *American Naturalist* 90, 337–347.

Dodd, A. N., Kudla, J., & Sanders, D. (2010). The language of calcium signalling. *Annual Review of Plant Biology* 61, 593–620.

Durrant, A. (1962). The environmental induction of heritable change in *Linum*. *Heredity* 17, 27–61.

Durrant, A. (1971). Induction and growth of flax genotrophs. *Heredity* 27, 277–298.

Evans, N. H., McAinsh, M. R., & Hetherington, A. M. (2001). Calcium oscillations in higher plants. *Current Opinion in Plant Biology* 4, 415–420.

Felle, H. (1988). Auxin causes oscillations of cytosolic free calcium and pH in *Zea mays* coleoptiles. *Planta* 174, 495–499.

Fromm, J., Hajirezaei, M., & Wilke, I. (1995). The biochemical response of electrical signalling in the reproductive system of *Hibiscus* plants. *Plant Physiology* 109, 375–384.

Genre, A., Lanfranco, L., Perotto, S., & Bonfante, F. (2020). Unique and common traits in mycorrhizal symbioses. *Nature Reviews Microbiology* 18, 649–660.

Gibson, J. J. (1979). *The ecological approach to visual perception*. Boston: Houghton Mifflin.

Gilroy, S., & Trewavas, A. J. (2022). Agency, teleonomy and signal transduction in plant systems. *Biological Journal of the Linnean Society*, blac021. https://doi.org/10.1093/biolinnean/blac021.

Gould, S. J. (1989). *Wonderful life. The Burgess Shale and the nature of history*. New York: W. W. Norton.

Hartman, K., & Tringe, S. G. (2019). Interactions between plants and soil: shaping the root microbiome under abiotic stress. *Biochemical Journal* 476, 2705–2724.

Helliker, B. R., & Richter, S. I. (2008). Subtropical to boreal convergence of leaf temperatures. *Nature* 454, 511–514.

Helms, A. M., de Moraes, C. M., Troger, A., Alborn, H. T., Francke, W., Tooker, J. F., & Mescher, M. C. (2017). Identification of an insect olfactory cue that primes plant defenses. *Nature Communications* 8, 337.

Henslow, G. (1895). *The origin of plant structures by self-adaptation to the environment*. London: Kegan Paul, Trench, Trubner.

Henslow, G. (1908). *The heredity of acquired characters in plants*. London: John Murray.

Herman, J. J., & Sultan, S. E. (2011). Adaptive transgenerational plasticity in plants: case studies, mechanisms and implications for natural populations. *Frontiers in Plant Science* 2, 102.

Herrera, C. M. (2017). The ecology of sub-individual variability in plants: patterns, processes and prospects. *Web Ecology* 17, 51–64.

Iwano, M., Ngo, Q. A., Entani, T., Shiba, H., Nagal, T., Miyawaki, A., et al. (2012). Cytoplasmic Ca^{2+} changes dynamically during the interaction of the pollen tube with synergid cells. *Development* 139, 4202–4209.

Jablonka, E., & Raz, G. (2009). Transgenerational and epigenetic inheritance. Prevalence, mechanisms and implications for the study of heredity and evolution. *Quarterly Review of Biology* 84, 131–176.

Johns, S., Hagihara, T., Toyota, M., & Gilroy, S. (2021). The fast and furious: rapid long-range signalling in plants. *Plant Physiology* 185, 694–706.

Jose, J., Ghantasala, S., & Choudhury, S. R. 2020. *Arabidopsis* transmembrane receptor-like kinases (RLKs): a bridge between extracellular signals and intracellular regulatory machinery. *MDPI: International Molecular Science* 21, 4000.

Karban, R., Shiojiri, K., Ishizaki, S., Wetzel, W. C., & Evans, R. Y. (2013). Kin recognition affects plant communication and defence. *Proceedings of the Royal Society Series B* 280, 20123062.

Keirs, E. T., Duhamel, M., Beesetty, Y., & Mensah, J. A. (2011). Reciprocal rewards stabilise cooperation in the mycorrhizal symbiosis. *Science* 333, 881–882.

Kellenberger, R. T., Desurmont, G. A., Schluchter, P. M., & Schiestl, F. P. (2018). Transgenerational inheritance of herbivory-induced phenotypic changes in *Brassica rapa*. *Scientific Reports* 8, 3536.

Knight, M. R., Campbell, A. K., Smith, S. M., & Trewavas, A. J. (1991). Transgenic aequorin reports the effect of touch, cold shock and elicitors on cytoplasmic calcium. *Nature* 352, 524–526.

Knight, M. R., Smith, S. M., & Trewavas, A. J. (1992). Wind-induced plant motion immediately increases cytosolic calcium. *Proceedings of the National Academy of Sciences USA* 89, 4967–4971.

Lamarck, J. B. (1914). *Zoological philosophy*. Translated by Hugh Elliott. London: MacMillan.

Laverde, V., Dominici, P., & Astegno, A. (2018). Toward understanding plant calcium signalling through calmodulin-like proteins: a biochemical and structural perspective. *MDPI International Journal of Molecular Science* 19, 1331.

Lenzoni, G., Liu, J., & Knight, M. R. (2018). Predicting plant immunity gene expression by identifying the decoding mechanism of calcium signatures. *New Phytologist* 217, 1598–1609.

Lewontin, R. C. (2000). *The triple helix*. Boston: Harvard University Press.

Lewontin, R. C. (2001). Gene, organism and environment. In: S. Oyama, P. E. Griffiths, R. D. Gray (Eds.), *Cycles of Contingency*, 59–67. Cambridge, MA: MIT Press.

Lin, Z., Eaves, D. J., Sanchez-Moran, F., Franklin, C. H., & Franklin-Tong, V. E. (2015). The *Papaver rhoeas S* determinants confer self-incompatibility to *Arabidopsis thaliana* in planta. *Science* 350, 684–687. https://www.science.org/doi/10.1126/science.aad2983

Luna, E., & Ton, J. (2012). The epigenetic machinery controlling transgenerational systemic acquired resistance. *Plant Signaling and Behavior* 7, 615–618.

Massonet, C., Vile, D., Fabre, J., Hannah, M. A., Caldana, C., Lisee, J., et al. 2010. Probing the reproducibility of leaf growth and molecular phenotypes: a comparison of three *Arabidopsis* accessions cultivated in ten different laboratories. *Plant Physiology* 152, 4142–4157.

McFadden, G. I. (2014). Origin and evolution of plastids and photosynthesis in eukaryotes. *Cold Spring Harbor Perspectives in Biology* 6, a016105.

Mescher, M. C., & de Moraes, C. M. (2015). Role of plant sensory perception in plant/animal interactions. *Journal of Experimental Botany* 66, 425–433.

Mesquita-Neto, J. N., Bluthgen, N., & Schlinwein, C. (2018). Flowers with poricidal anthers and their complex interaction networks; disentangling legitimate pollination and illegitimate visitors. *Functional Ecology* 32, 2321–2322.

Miwa, H., Sun, J., Oldroyd, G. E, & Downie, J. A. (2006). Analysis of calcium spiking using a cameleon calcium sensor reveals that nodulation gene expression is regulated by spike number and the developmental status of the cell. *Plant Journal* 48, 883–894.

Monshausen, G. B., Bibikova, T. N., Messerli, M. A., Shi, C., & Gilroy, S. (2007). Oscillations in extracellular pH and reactive oxygen species modulate tip growth of *Arabidopsis* root hairs. *Proceedings of the National Academy of Sciences U S A* 104, 20996–21001.

Monshausen, G. B., Miller, N. D., Murphy, A. S., & Gilroy, S. (2011). Dynamics of auxin dependent Ca^{2+} and pH signaling in root growth revealed by integrating high-resolution imaging with automated computer vision-based analysis. *Plant Journal* 65, 309–318.

Montgomery, C., Vuts, J., Woodcock, C. M., Withall, D. M., Birkett, M. A., Pickett, J. A., & Robert, D. 2021. Bumblebee electric charge stimulates floral volatile emissions in *Petunia integrifolia* but not in *Antirrhinum majus*. *The Science of Nature* 108, 44.

Moulia, B., Douady, S., & Hamant, O. (2021). Fluctuations shape plants through proprioception. *Science* 372, 359–369.

Nepi, M., Grasso, D. A, & Mancuso, S. (2018). Nectar in plant-insect mutualistic relationships: from food reward to partner manipulation. *Frontiers in Plant Science* 2018, 01063.

Ngo, Q. A., Vogler, H., Lituiev, D. S., Nestorova, A., & Grossniklaus, U. (2014). A calcium dialog mediated by the Feronia signal transduction pathway controls plant sperm delivery. *Developmental Cell* 29, 491–500.

Nguyen, C. T., Kurenda, A., Stolz, S., Chételat, A., & Farmer, E. E. (2018). Identification of cell populations necessary for leaf-to-leaf electrical signaling in a wounded plant. *Proceedings of the National Academy of Sciences* 115(40), 10178–10183.Niklas, K. J. (1991). Flexural stiffness allometries of angiosperms and fern petioles and rachises: evidence for biomechanical convergence. *Evolution* 45, 734–750.

Niklas, K. J. (1992). *Plant biomechanics: an engineering approach to plant form and function*. Chicago, IL: Chicago University Press.

Niklas, K. J. (1995). Size-dependent allometry of tree height, diameter and trunk taper. *Annals of Botany* 75, 217–227.

Niklas, K. J. (2000). The evolution of plant body plans—a biomechanical perspective. *Annals of Botany* 85, 411–438.

Noble, D. (2015). Evolution beyond neo-Darwinism: a new conceptual framework. *Journal of Experimental Biology* 218, 7–13.

Novoplansky, A. (2019). What plant roots know. *Seminars in Cell and Developmental Biology* 92, 126–133.

Ogle, K., Barber, J. J., Barron-Gafford, G. A., Bentley, L. P., Young, J., Husman, T., et al. (2015). Quantifying ecological memory in plant and ecosystems processes. *Ecology Letters* 18, 221–235.

Parter, M., Kashtan, N., & Alon, U. (2008). Facilitated variation: how evolution learns from past environments to generalise to new environments. *PLoS Computational Biology* 4, e1000206.

Pascale, A., Proletti, S., Pantelides, I. S., & Stringlis, I. A. (2020). Modulation of the root microbiome by plant molecules: the basis for targeted disease suppression and plant growth promotion. *Frontiers in Plant Science* 10, 1741.

Pfeifer, R., & Scheier, C. (1999). *Understanding intelligence*. Cambridge, MA: MIT Press.

Pham, B., & McConnaughay, K. (2014). Plant phenotypic expression in variable environments. In R. Monson (Ed.), *Ecology and the environment: the plant sciences* 8, 119–141. Berlin: Springer-Verlag,

Pigliucci, M., & Müller, G. (2010). Elements of an extended evolutionary synthesis. In M. Pigliucci & G. B. Müller (Eds.), *Evolution—the extended synthesis*, 3–17. Cambridge, MA: MIT Press.

Pittendrigh, C. S. (1958). Adaptation, natural selection and behavior. In A. Roe & G. C. Simpson (Eds.), *Behavior and evolution*, 390–416. New Haven, CT: Yale University Press.

Rosenblueth, A., Weiner, N., & Bigelow, J. (1943). Behavior, purpose and teleology. *Philosophy of Science* 10, 18–24.

Russell, E. S. (1946). *The directiveness of organic activities*. Cambridge: Cambridge University Press.

Sage, R. (2004). The evolution of C4 photosynthesis. *New Phytologist* 161, 341–370.

Sanders, D., Brownlee, C., & Harper, J. F. (1999). Communicating with calcium. *Plant Cell* 11, 691–706.

Simeunovic, A., Mair, A., Wurzinger, B., & Teige, M. (2016). Know where your clients are: subcellular localisation and targets of calcium-dependent protein kinases. *Journal of Experimental Botany* 67, 3856–3872.

Singh, P., & Roberts, M. R. (2015). Keeping it in the family: transgenerational memories of plant defence. *CAB Reviews: Perspectives in Agriculture Veterinary Science Nutrition & Natural Resources* 10, 028.

Smith, S. M., & Read, D. J. (2008). *Mycorrhizal symbiosis* (3rd ed.). London: Academic Press.

Song, Y. Y., Simard, S. W., Carroll, A., Mohn, W. W., & Zeng, R. S. (2015). Defoliation of interior Douglas fir elicits carbon transfer and stress signalling to ponderosa pine neighbours through ectomycorrhizal networks. *Scientific Reports* 5, 8495.

Song, Y. Y., Zeng, R. S., Xu, J. F., Li, J., Shen, X., & Yihdego, W. C. (2010). Interplant communication of tomato plants through underground common mycorrhizal networks. *PLoS One* 5, e13324.

Stolarz, M. (2009). Circumnutation as a visible plant action and reaction: physiological, cellular and molecular basis for circumnutations. *Plant Signaling and Behaviour* 4, 380–387.

Sultan, S. E. (2010). Plant developmental responses to the environment: eco-devo insights. *Current Opinion in Plant Biology* 13, 96–101.

Sultan, S. E. (2015). *Organism & environment*. Oxford: Oxford University Press.

Tederson, L., Bahram, M., & Zobel, M. (2020). How mycorrhizal associations drive plant populations and community biology. *Science* 367, eaba 1223.

Thompson, E. (2007). *Mind in life*. Cambridge, MA: Belknap Press.

Tian, W., Wang, C., Gao, Q., Li, L., & Luan, S. (2020). Calcium spikes, waves, oscillations in plant development and biotic interactions. *Nature Plants* 6, 750–759.

Toyota, M., Spencer, D., Sawai-Toyota, S., Jiaqui, W., Zhang, T., Koo, A. J., Howe, G. A., & Gilroy, S. (2018). Glutamate triggers long-distance calcium-based plant defence signalling. *Science* 361, 1112–1115.

Trewavas, A. J. (2004). Aspects of plant intelligence: an answer to Firn. *Annals of Botany* 93, 353–357.

Trewavas, A. J. (2009). What is plant behaviour? *Plant Cell and Environment* 32 (6), 606–616.

Trewavas, A. J. (2014). *Plant behaviour and intelligence*. Oxford: Oxford University Press.

Trewavas, A. J. (2017). The foundations of plant intelligence. *Interface Focus* 7, 20160098.

Trewavas, A. J., & Knight, M. R. (1994). Mechanical signalling, calcium and plant form. *Plant Molecular Biology* 26, 1329–1341.

Vadassery, J., & Oelmuller, R. (2009). Calcium signalling in pathogenic and beneficial plant microbe interactions. *Plant Signaling & Behavior* 4, 1024–027.

Varela, F. J. (1987). Laying down a path in walking. In W. I. Thompson (ed.), *Gaia: a way of knowing. Political implications of the new biology*, 48–64. New York: Lindisfarne Press.

Veits, M., Khait, I., Obolski, U., Zinger, E., Boonman, A., Goldshtein, A., et al. (2019). Flowers respond to pollinator sound within minutes by increasing nectar sugar concentration. *Ecology Letters* 22, 1483–1492.

von Sachs, J. (1887). *Lectures on the physiology of plants*. Oxford: Clarendon Press.

Waddington, C. H. (1957). *The strategy of the genes*. London: George Allen & Unwin.

Waddington, C. H. (1975). *The evolution of an evolutionist*. Edinburgh: Edinburgh University Press.

Walsh, D. M. (2015). *Organisms, agency and evolution*. Cambridge: Cambridge University Press.

Watahiki, M., & Trewavas, A. J. (2019). Systems variation, individuality and hormones. *Progress in Biophysics & Molecular Biology* 146, 3–22.

Watahiki, M. K., Trewavas, A. J., & Parton, R. M. (2004). Fluctuations in the pollen tube tip-focused calcium gradient are not reflected in nuclear calcium level: a comparative analysis using recombinant yellow cameleon calcium reporter. *Sexual Plant Reproduction* 17, 125–130.

Watson, R. A., & Szathmáry, E. (2016). How can evolution learn? *Trends in Ecology and Evolution* 31, 147–157.

Went, F., & Thimann, K. V. (1937). *Phytohormones*. New York: Macmillan.

Whalley, H. J., Sargent, A. W., Steele, J. F. C., Lacoere, T., Lamb, R., Saunders, N. J., Knight, H., & Knight, M. R. (2011). Transcriptomic analysis reveals calcium regulation of specific promoter motifs in *Arabidopsis*. *Plant Cell* 23, 4079–4095.

White, H. L. (1937). The interaction of factors in the growth of *Lemna*: XI. The interaction of nitrogen and light intensity in relation to growth and assimilation. *Annals of Botany* 1, 623–647.

Williams, S. E., & Pickard, B. G. (1972). Receptor potentials and action potentials in *Drosera* tentacles. *Planta* 103, 193–221.

Willmer, P., Stanley, D. A., Steijven, K., Matthews, I. M., & Nuttman, C. V. (2009). Bidirectional flower colour and shape changes allow a second opportunity for pollination. *Current Biology* 19, 919–923.

Zeng, H., Xu, L., Singh, A., Wang, H., Du, L., & Poovaiah, B. W. (2015). Involvement of calmodulin and calmodulin-like proteins in plant responses to abiotic stresses. *Frontiers in Plant Science* 6, 600.

Zhu, J-K. (2016). Abiotic stress signalling and responses in plants. *Cell* 167, 313–324.

Zhao, J., Lei, Y., Hong, J., Zheng, C., & Zhang, L. (2019). AraPPINet: An updated interactome for the analysis of hormone signaling crosstalk in *Arabidopsis thaliana*. *Frontiers in Plant Science* 2019, 00870.

17 Agency, Goal Orientation, and Evolutionary Explanations

Tobias Uller

Overview

The skepticism toward teleology in the natural sciences may give the impression that all reference to agency and goals in evolutionary explanations is just convenient shorthand. In this chapter I suggest that agential concepts may in fact serve several distinct epistemic functions. Firstly, the gene's-eye view demonstrates that agential concepts can promote the intelligibility of evolutionary theories, thereby facilitating the application of those theories to explain natural phenomena. Secondly, agential concepts can structure evolutionary investigation according to particular criteria of explanatory adequacy. These explanatory agendas admit nonselective causal influence on adaptive evolution, which begs the question of how developmental and selective explanations should be integrated. Thirdly, a more radical proposal is that organismal goals themselves can be explanatory for evolutionary change. Such naturalistic teleological explanation is motivated by an explanatory gap left by causal explanations and encourages development of theories and models that allow the principles of evolution to depend on organismal activities that originate as a result of the organisms' internal organization.

17.1 Introduction

Anyone who has ever visited a rain forest or taken a virtual tour through a cell would surely agree that living systems are extraordinarily complex—far too complex to grasp in their entirety. Scientific explanation must therefore rely on simplified representations that leave out detail and even distort reality. A generic representation provides principles from which can be developed more specific representational theories and models that are applied to explain natural phenomena (figure 17.1). Such explanations commonly refer to variables, events, entities, or states of affairs that make a difference as to whether or not a phenomenon obtains. Explanations that concern humans and our societies may also refer to goals and purposes; perhaps your opening this book is explained by your goal to learn something new, or intention to see if your work was cited; the human striving for greater equality and justice can be invoked to explain the rise of modern democracy; and so on.

In contrast to the social sciences, the natural sciences consider goals and purposes unacceptable. Matter is not imbued with agency, and planets do not move around the sun because it is their function or purpose to do so. Such phenomena are supposed to be explained by mechanisms, causes, and forces. Moreover, goals lie in the future, whereas causes must precede their effects. This is perhaps not so problematic for human affairs, because we can think about the future and decide what to do on the basis of our mental representations, thereby keeping the cause–effect relation in the right order. But in the absence of the human capacity for deliberation, rational choice, and cumulative culture, any reference to goals may simply appear unscientific.

At first sight, evolutionary biology appears to adhere to the mechanistic ideal: mutations occur without regard to their effects, and selection favors current, not future, utility. Yet, the goal-oriented nature of development, physiology, and behavior is hard to deny, and talk of agency, goals, and purpose is rife within biology (Box 17.1). One of the most famous perspectives on the evolutionary process—the gene's-eye view—even seems to grant purposive agency to DNA; genes are said to have goals and interests, play strategies, and be in conflict with each other. Perhaps such use of agential concepts is just convenient and innocuous shorthand. This chapter explores the alternative explanation that agential concepts really do serve important epistemic functions in evolutionary biology.[1] Three broad possibilities will be considered (figure 17.1).

Firstly, agential concepts may further epistemic goals by making evolutionary theories intelligible. Thinking of genes, biological processes, or organisms as goal-oriented can help scientists reason intuitively and make qualitative predictions, something that often is necessary to develop formal models. Secondly, agential concepts can set explanatory standards. By drawing attention to the adaptive biases imposed on evolution by development and behavior, agential concepts structure scientific investigation around particular sets of problems (including agency itself) and associated criteria of explanatory adequacy. Thirdly, if biological evolution is a consequence of the goal-oriented activities of individual organisms, naturalistic teleological explanation could perhaps be expanded beyond human cultural evolution, and make it scientifically legitimate to account for biological evolution by referring to organismal goals.

17.2 Agential Concepts Can Make Evolutionary Theories Intelligible

The gene's-eye view is a useful starting place to explore the epistemic functions of agential concepts; if nothing else, then simply because a molecule seems a rather unlikely candidate for exhibiting goals and purposes. Proponents of the gene's-eye-view have indeed been quick to point out that ascribing agency to genes is merely a convenient shorthand, which they could translate "into respectable terms if we wanted to" (Dawkins, 1976: 88).[2] While there is some controversy over whether or not genes can ever be considered agents (Okasha, 2018), it is widely accepted that thinking of genes as having goals, interests, intentions, or strategic repertoires can help scientists think clearly about evolution (Burt & Trivers, 2006; Ågren, 2021).

There is undoubtedly something right about this suggestion. But what exactly is it about the agential metaphors that make them helpful? If purposive agency does not correspond to an actual property of genes, the epistemic function of agential concepts seems unlikely to mediate the relationship between theory and phenomenon; it is not really the goal-oriented

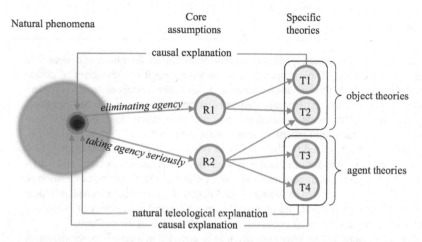

Figure 17.1
Scientific understanding requires application of an appropriate explanatory model or theory to natural phenomena. The complexity of biological evolution means that scientists must rely on simplified generic representations (R1, R2), with particular core assumptions, from which more specific scientific theories (T1–T4) or models can be developed. The standard theories in evolutionary biology are object theories: they explain by externally imposing a set of rules or principles on the objects or entities of the theory (e.g., individuals or alleles). An alternative set of theories are agent theories, which explain by accounting for the interplay between what entities do and the principles used to explain their behavior (this is explained more fully in this chapter). The three broad categories of epistemic functions of agential concepts discussed in this chapter are illustrated in this figure. As discussed in section 17.2, applying agential concepts to living systems or their parts (e.g., genes) can facilitate evolutionary explanation by making evolutionary theory (e.g., T1) intelligible to the scientists who wish to use it to explain natural phenomena. Section 17.3 demonstrates how agential concepts (e.g., phenotypic accommodation) can structure investigation of biological evolution by influencing how scientists choose to represent the evolutionary process. These representations can admit, for example, development, extragenetic inheritance, and niche construction to explain (alone or jointly with natural selection) adaptive evolution. Such theories may be developed from existing object theories (e.g., T2), which means that the interpretative understanding of a single representative model can differ depending on which core assumptions particular scientists hold onto. Section 17.4 discusses how representations of evolution that are structured around organismal agency also may motivate the development of agent theories, which are incompatible with the core assumptions of traditional representations of the evolutionary process (arrow connecting R2 and T3). As agent theories are fundamentally different from object theories, they may admit both causal explanation and naturalistic teleological explanations, where the latter involves the demonstration that the attainment of an evolutionary event was conducive to organismal goals ("naturalistic teleological explanation" arrow from agent theories to natural phenomena).

activities of genes that explain why ants cooperate, for example. A more plausible alternative is that applying agential concepts to genes makes evolutionary theory itself intelligible.

According to philosopher of science Henk De Regt, scientific understanding requires application of an appropriate explanatory model or theory (of which there are many) to natural phenomena (De Regt, 2018; Fig. 1).[3] Scientists cannot apply a theory, or cannot apply it appropriately, unless the theory is intelligible to them. This makes the intelligibility of scientific theories central to understanding how scientists can deliver adequate explanations. De Regt defines *scientific intelligibility* as "the value that scientists attribute to the cluster of qualities of a theory (in one or more of its representations) that facilitate the use of the theory" (De Regt, 2018: 40). Drawing on physics, he further considers a theory to be (at least to a first approximation) intelligible to a scientist if that scientist can "recognize qualitatively characteristic consequences of the theory without performing exact calculations" (De Regt, 2018: 102).

Box 17.1
What Is Biological Agency?

In daily speech, reference to agency, goals, and purpose often implies intentions or desires. The biological agency concept(s) of this chapter is much broader than that, although it of course includes the sophisticated cognitive abilities of humans. An inclusive concept of agency considers it a dynamical property of a system (e.g., an organism) that makes the system able to "transduce, configure, and respond to the conditions it encounters," and to maintain "functional stability in response to conditions that would otherwise compromise their viability" (Sultan et al. 2022). Agential systems are characterized by their ability to initiate activity from within their own boundaries, to sustain and transform themselves through novel structures, functions, and activities, often in ways that ensure their continued existence (Walsh 2015). In contrast to a storm or a biogeochemical cycle, an organism can change what it does to navigate obstacles and overcome challenges that threaten its survival. We observe this as goal-oriented activities or behaviors, which are characteristic features of all organisms, the *E. coli* as well as the elephant.

These behavioral repertoires are often considered to be encoded in a genetic program, which makes organisms appear goal-oriented (or *teleonomic;* Mayr 1974) without them really being agents by their own making. That is, agency is not a property of organisms as much as a property of their genomes, much like how a piece of hardware can respond to commands and carry out its functions only because of its software. But in contrast to the programs written by human software developers, the presence of organismal goal-orientation and the organisms' pursuit of particular goals are not designed, but externally provided by natural selection on random genetic variation (more on that in the main part of the chapter).

An alternative way to think about organismal agency is to attribute it, not to a program, but to a particular kind of closed organisation where "the processes and constraints . . . logically and materially entail each other" (Jaeger 2021, p.8; Mossio et al. 2009; Montevil & Mossio 2015). A system in which processes that are essential for the organism's continued existence regulate and sustain each other can be considered to demonstrate an intrinsic orientation toward goals (or an *internal teleology*) because the organism must act upon the world to stay alive (Mossio & Bich 2017). Organisms are agents because of what they do, and they must continue to be agents to stay alive. Because the closed organization of living systems is not reducible to genes, or even gene regulatory networks, it seems problematic to explain organismal agency by natural selection of random genetic variation alone, and thus this perspective tends to direct attention to the transformation of biological processes during evolution.

From these preliminaries, it follows that intelligibility cannot be intrinsic to a theory, but must depend on the cognitive ability of and conceptual tools available to scientists. This, in turn, makes scientific intelligibility highly context-dependent. Scientists choose theories in part because of their ability to put those theories to work, making intelligibility an important determinant of the success and propagation of scientific theories. The context-dependence of intelligibility explains why a theory that required great intellectual effort in the 1960s can be textbook material today.[4] At the same time, a theory that is easy to grasp may continue to be applied even if a harder or unfamiliar theory would produce a deeper understanding of the phenomenon to be explained.

De Regt illustrates the relationship between intelligibility of theories and understanding of phenomena with the kinetic theory of gases (De Regt, 2018: 138). The kinetic theory represents real gases as an aggregation of particles that obey Newtonian mechanics. On its own, this generic representation merely provides the principles of the theory: even elementary gas laws (e.g., Boyle's law) can only be explained by developing more specific representational models that rely on further idealization, such as assuming that the particles are smooth and hard elastic spheres. To use the kinetic theory (i.e., to apply it to explain phenomena), scientists must make decisions that are tailored to their explanatory aims. These decisions would be difficult to make unless the theory were intelligible to the scientist in the sense described earlier; only by being able to "recognize qualitatively characteristic consequences of the theory" can the scientist make informed decisions about the idealizations and approximations that go into building a model.

Is this model of scientific understanding applicable to evolutionary biology? More specifically, does assigning agency and goals to genes facilitate the intelligibility of evolutionary theory? That is, does it help scientists grasp the consequences of the theory without writing down the mathematics? There seem to be good reasons to believe that it does.

One reason is that the traditional workhorse of evolutionary explanation—theoretical population genetics—is something that many biologists find hard going. Dobzhansky, for example, apparently found it hard enough to hum through the equations as he went through Sewall Wright's papers (Provine, 2003). Yet, population genetic models require nothing more than the algebra skills many biologists would have acquired before entering college. This suggests that it is not (only) the mathematical operations that make population genetics hard, but that the theory is difficult to understand. A skilled population geneticist intuitively "knows" how to use the basic principles of population genetics in order to construct specific models, tailored to a particular explanation of a particular phenomenon. In contrast, biologists who lack this intuition[5] may not even know where to begin, even if they are perfectly apt at algebra.

Scientific theories that are difficult to grasp motivate the use of visual or conceptual tools to further their intelligibility (De Regt, 2018). The "adaptive landscape" introduced by Sewall Wright is a good example of a visual tool: it makes it easier to infer consequences of evolutionary theory without performing calculations (Provine, 1986). Thinking of genes as agents that pursue goals arguably serves a similar function: it makes it easier to intuitively grasp the possible consequences of an intervention on relevant variables, such as fitness or the rate of dispersal. Here is theoretical evolutionary biologist John Maynard Smith speaking:

I am prepared to think as loosely as necessary to give me an idea when I'm confronted with a new biological problem. If it helps me think to say, "If I was a gene, I would do so-and-so" then I think that is OK. But when I've got an idea, I want to be able to write down the equation and show that the idea works. . . . I'm all for loose thinking. We all need ideas (Maynard Smith, 1998, quoted in Ågren, 2021).

Maynard Smith here interprets his "loose thinking" of what a gene should do as a way to generate possible solutions to a problem that can be formalized mathematically. This suggests that applying agential concepts to genes enabled Maynard-Smith to argue qualitatively and to recognize possible consequences of evolutionary theory without formal calculation.

If this interpretation is correct, agential concepts are not just a convenient shorthand, but rather conceptual tools that make evolutionary theory intelligible (at least to some scientists).[6] This conclusion is compatible with standard justification for agential thinking, which emphasizes its metaphorical nature (e.g., Haig, 1997). However, it identifies a distinct epistemic function of agency or goal-directedness. This function makes no assumption about the metaphysical status of genes as agents; indeed, biologists could deny any biological entity agency and yet apply agential concepts to make biological theories intelligible. In practice, such a situation may be hard to sustain, as metaphors that make scientific theories intelligible have a tendency to shape metaphysical views and vice versa (Godfrey-Smith, 2009; De Regt, 2018).[7]

17.3   Agential Concepts Can Set Explanatory Standards

While genetic agency is widely acknowledged to be a metaphor, biologists really do consider individual organisms to be active, purposive agents (e.g., Dobzhansky, 1968; Waddington, 1968; Mayr, 1974). Even a very demanding concept of agency will apply to humans, and less demanding concepts will grant agency to the smallest autonomous living system—the cell—and perhaps other organismal systems (Box 17.1). The dilemma is that, to be a respectable natural science, evolutionary biologists seemingly must deny the organism's goal-oriented activities any explanatory relevance for adaptation and diversification. That is, they must demonstrate that their evolutionary explanations not only are compatible with the causal-mechanistic ideal, but also that the structure of evolutionary theory itself ensures "an impermeable barrier between individual agency and evolutionary transformation" (Riskin, 2020: 273).

Ernst Mayr's exemplar of how organismal agency can be at once accepted as factual but denied any explanatory relevance for evolution is an influential attempt to erect such a barrier (Laland et al., 2011; Corning, 2019). Acknowledging that organisms appear to be purposive agents, Mayr insisted that their goal-oriented activities imply the presence of a program, encoded by the inherited genome (Mayr 1961, 1974, 1988). Organisms are literally programmed to ensure that their development, physiology, and behavior are directed toward outcomes that serve their survival and reproductive interests (organismal activities and behaviors are *teleonomic* rather than *teleological*[8]; Mayr, 1988). In Mayr's view, developmental biologists and physiologists are concerned with explanations of how the genetic program is decoded, or how the program works. Such explanations rely on "proximate" causes that readily fit within the mechanistic ideal. For example, a mechanistic explanation for a lizard escaping a predator may refer to the visual input stimulating the sensory system and the brain, which in turn triggers the muscles and tendons to cause movement of the limbs. Evolutionary biologists, in contrast, are concerned with *why* particular programs exist and thus they rely on "ultimate causes." Lizards that do not run from predators fail to pass on their genes, making the survival difference between genes that code for skittish and docile lizards an explanation for why lizards run from predators, as well as an explanation for why the behavior appears goal-oriented.

This neo-Darwinian representation of evolution by natural selection solves the evolutionary biologist's dilemma. Firstly, it mechanizes historical explanations by enabling more specific representational theories that explain adaptive (and nonadaptive) evolution by relying solely on causes and forces. This is well exemplified by the explanations delivered by theoretical population genetic models.[9] Secondly, the structure of evolutionary theory seemingly rules out *a priori* goal-oriented processes—organismal development, physiology, or behavior—from evolutionary explanation (Mayr, 1961).[10] Those processes do not directly change allele frequencies, and thus may seem unable to account for directional, adaptive change. Only one difference-maker for adaptive bias remains: fitness differences between genotypes. Thus, according to this perspective, evolutionary biologists can consider organismal agency a real phenomenon, yet shrug it off as an intermediate, proximate, expression of a genetic program; an expression of past natural selection rather than a cause of future adaptation.[11] Under the neo-Darwinian representation of the evolutionary process, evolutionary explanation is genes and natural selection all the way down.

The elimination of organismal agency and goal-oriented processes from evolutionary theory has successfully structured evolutionary inquiry around a set of problems for which the theory is appropriate. Yet, two limitations are difficult to avoid.

Firstly, some evolutionary problems inevitably fall outside the theory's domain. The origin of novelty is one familiar example. An explanation for a novel morphological feature, such as the flower, requires attention to the sequence of morphological transformation over evolutionary time and the genetic and developmental changes that were responsible for this change (Calcott, 2009). This is not something that a population genetic model can explain: someone who explained the evolution of flowers in terms of fitness advantages and shifting allele frequencies would simply be off the mark. Selective explanations are valid explanations in their own right, but they are not adequate explanations for the flower as an evolutionary novelty (Love, 2008).

Secondly, to account for adaptation in terms of fitness differences, the neo-Darwinian representation must make assumptions about the evolutionary process that are at odds with biological reality (Walsh, 2015; Uller & Helanterä, 2019; see Potochnik, 2017, for a general discussion). The insistence that natural selection of genetic variation is the only legitimate difference-maker for adaptive evolution (e.g., Charlesworth et al., 2017) is a consequence of these idealizations,[12] not a fundamental feature of the causal fabric of the world. Failure to appreciate that criteria of explanatory adequacy depend on how the evolutionary process is represented by theories and models can make it appear as if the privileged role of genes and natural selection is indispensable to *any* evolutionary explanation. This, in turn, may result in an overreliance on fitness-based explanations and neglect of alternatives (e.g., Gould & Lewontin, 1979; Lloyd, 2005), slow acceptance of phenomena that do not fit assumptions (e.g., extragenetic inheritance; Jablonka & Lamb, 2014), and limited explanatory power as a result of failure to account for adaptive biases imposed by goal-oriented processes (e.g., development; West-Eberhard, 2003; Gerhart & Kirschner, 2007). The genetic representation of development and evolution also carries several conceptual difficulties and inconsistencies (Keller, 2000; Oyama, 2000; Griffiths & Stotz, 2013).

As these inconsistencies, explanatory gaps, and deficits are built into the structure of evolutionary theory itself, they cannot easily be amended without alternative representations that pick out different causal patterns. Such attempts have often been motivated by understanding the evolutionary consequences of organismal agency and goal-oriented processes. For example, to Mary-Jane West-Eberhard, adaptive evolution begins with phenotypic accommodation—adaptive mutual adjustment among variable parts during development—in response to genetic or environmental perturbation (West-Eberhard, 2003: 51, 140–141). To West-Eberhard and similar-minded biologists, development is not just a conservative force that constrains evolution, as the neo-Darwinian representations depict it, but also what makes the generation of adaptive variation possible (e.g., Salazar-Ciudad 2007; Uller et al., 2018). Such "facilitated variation" (Gerhart & Kirschner, 2007) contributes to evolution by providing natural selection with a source of putative adaptive phenotypes, which makes development a co-determinant of the rate and direction of adaptive change. Others (e.g., Lewontin,

1983; Odling-Smee, 1988; Edelaar & Bolnick, 2019) have emphasized that organisms can influence their own individual fitness by modifying selective environments, which makes such "niche construction" a codeterminant of adaptation (Odling-Smee et al., 2003).

Concepts such as phenotypic accommodation, facilitated variation, and niche construction capture aspects of the responsive, self-organizing nature of development that allows organisms to maintain functional stability even when exposed to conditions that threaten their persistence. These concepts thus direct attention to "internal" or "agential" sources of consistent bias in evolution that may account for the evolution of particular adaptations, diversification, or evolvability (Sultan et al., 2022).[13] In so doing, those concepts set an alternative explanatory agenda; they structure scientific investigation of evolution according to criteria of explanatory adequacy that are different from those of the neo-Darwinian representation of evolution by natural selection (Love, 2008; Brigandt, 2010). These alternative criteria in turn determine what biological fields may be deemed relevant for a scientific understanding of evolution. A scientist who considers development a source of adaptive bias in evolution will draw on different sets of knowledge, concepts, and methods from other disciplines than will a scientist who considers natural selection alone responsible for the diversity and adaptive fit of organisms.

A characteristic feature of such evolutionary research is that it admits developmental and behavioral difference-makers (e.g., developmental plasticity, habitat choice) alongside fitness differences as explanans for adaptation and diversification. A major challenge, therefore, is to distribute causal responsibility between transformational and selective processes, and among the different types of difference-makers involved (for a general discussion, see Love, 2017).[14] The developmental causes of adaptive bias that refer to, for example, exploratory processes of bone and tissue growth are not easily compared to the causes of fitness differences, which begs the question of how different causes should be integrated in an evolutionary explanation.[15] Moreover, the processes that generate phenotypic variation and differential fitness can be intertwined or modify each other on the relevant time scales (Watson & Thies, 2019), which makes it difficult to attribute adaptive change to either one or the other (Uller & Helanterä, 2019).

Another challenge is to decide when the addition of particular difference-makers results in better or worse explanations (Ylikoski & Kuorikoski, 2010). For example, the striking morphological diversity and convergence of cichlid fish in African lakes (Kocher et al., 1993) may be explained in part by the phylogenetically shared developmental biases of fish feeding on different diets, and in part by consistent fitness differences between fish with alternative morphologies (e.g., Muschick et al., 2011; Conith & Albertson, 2021; review in Schneider & Meyer, 2017). An explanation that refers jointly to developmental bias and natural selection can arguably be preferable over an explanation that refers to either natural selection or developmental bias. But exactly when is the first explanation better than the second or vice versa? Choosing between alternative explanations may be more difficult than it seems, as it requires an evaluation of the explanatory power of different representations of the evolutionary process (Baedke et al., 2020; Uller et al., 2020).

17.4   Agential Concepts Can Explain Evolutionary Change

The extensions to evolutionary theory discussed in the previous section have been accompanied by a lively debate about the reinterpretation of development and other "intrinsic" or "proximate" causes in evolution (Laland et al., 2014). Yet, these alternative explanatory agendas still explain in terms of causes and mechanisms: the goal-oriented developmental processes and feeding behaviors of cichlids may be sources of adaptively relevant causes, but it is not the goals themselves that account for the convergence in the cichlids' morphology. To some, this stance does not go far enough in how the agential perspective should influence evolutionary explanation. One critic is philosopher of science Denis Walsh. Walsh argues that if we accept that adaptive evolution is a consequence of organisms' goal-oriented activities, there is a counterfactual relation between goals and organismal activities that can exploited to explain evolutionary events, and therefore to reinstall teleology as a mode of explanation in evolutionary biology (Walsh, 2015, 2018; see also Jaeger, 2021).

To explain just how seriously he wishes us to take agency, Walsh distinguishes what he calls object theories from agent theories (Walsh, 2015, ch. 10). *Object theories* are those that aim to explain what is happening to a set of objects within a system by externally imposing a set of rules or principles. Population genetics is a good example: the alleles are the objects of interest, and those alleles mutate and change in frequency according to principles that are external to the alleles and exist independently of them. The result is an explanatory asymmetry:

The principles—e.g. laws of nature, initial conditions, and the space of possible configurations—explain the changes to the objects in the domain, but the objects do not explain the principles (Walsh, 2015: 212).

Agent theories are different from object theories, because

in an agent theory the entities in the domain include both agents and the principles we use to explain their dynamics. The agents' activities are generated endogenously; agents cause their own changes in state in response to the conditions they encounter. These conditions, in turn, are largely of the agent's making (Walsh 2015: 212).

This explanatory symmetry between entities and principles means that the explanatory objective of an agent theory is different from that of an object theory: an agent theory explains the dynamics of a set of entities by accounting for the interplay between what these entities do and the principles used to explain their behavior. Contemporary theory on the evolutionary causes and consequences of goal-oriented processes (e.g., developmental plasticity) do not meet these criteria; a tell-tale is that such models and explanations often rely on traditional tools and theories, such as population or quantitative genetics (e.g., Lande, 2009; Chevin et al., 2010; Levis & Pfennig, 2016).

Why would an understanding of biological evolution demand agent theories rather than object theories? One reason is that biological evolution appears to be an open-ended process. As organisms evolved from single-celled organisms into multicellular organisms, and eventually into organisms with symbolic means of transmitting information, they changed the evolutionary process itself. The evolution of multicellularity or symbolic communication is not merely the evolution of another unit of selection or an adaptation to transmit information; by changing the principles of the evolutionary dynamics, these innovations opened up other opportunities for evolution that were previously impossible.

Another reason for agent theories is even more fundamental, because it addresses why biological evolution can be open-ended (Jaeger, 2021). Not all entities that can evolve by natural selection exhibit this open-endedness; algorithms in computer programs do not, for example. This suggests that biological evolution is possible because organisms are living beings, not because they fulfill abstract principles of evolution by natural selection (e.g., as summarized by Lewontin, 1970).[16] On the agential view, evolution happens because organisms engage in goal-oriented activities. The organizational account of biological agency further tells us that these responses are initiated not by a program but from within the organism itself, and hence cannot be predicted even with full knowledge of the population's selective history. For example, understanding why African cichlids evolved a similar suite of morphologies in different lakes would require both an understanding of the cichlids' developmental and behavioral repertoire (what they are capable of) and what their surroundings—rocks, plants, sand, snails, other fishes—offer or furnish to the cichlids as they pursue their goals.[17] In practice, a decent knowledge of cichlid biology may suffice. However, the philosophical point is that it is these repertoires and affordances that explain why the cichlids evolved similar adaptations, not "natural selection" (Walsh, 2015).

If organisms' goal-oriented engagement with their affordances is what enables adaptive evolution, would not citing those stable endpoints—or goals—also explain why evolution proceeded in one direction rather than in another (Walsh, 2015; Jaeger, 2021; Sultan et al., 2022)?[18] This is, perhaps, not so different from legitimate teleological explanations for adaptive cultural change in humans, a kind of explanation that can be preferred over cultural selection explanations (Chellappoo, 2022). Jaeger goes as far as to conclude that "naturalistic teleological explanation is a necessary part of any agential theory of evolution, because of the immanence of rules which are generated by the agents themselves" (Jaeger, 2021: 31). It is important to stress that none of this implies that the evolutionary process itself is goal-oriented; the goals and purposes in naturalistic teleological (or teleonomic) explanations would be those of organisms (and perhaps their parts or collectives of organisms that exhibit organizational closure; see Box 17.1).[19] Neither does naturalistic teleology imply that goals somehow cause their means, because teleological explanations are not causal (this is not as fatal as it may seem, since natural sciences already do admit "becauses" without "causes"; Lange, 2016). Granting all this, what can an explanation that refers to organismal goals contribute to the scientific understanding of evolution?[20]

One possibility is that naturalistic teleological explanations fill an explanatory gap that mechanistic explanations simply cannot fill. Consider again the evolutionary convergence of cichlid jaw morphologies. As mentioned above, an explanation for this convergence could cite a number of different sources of adaptive (and perhaps nonadaptive) bias. These include diet choices and preferences, the developmental genetics and plasticity of craniofacial development, and fitness differences between individuals with different morphologies. But the consistent biases imposed by the fishes' search for food or the development of their mouth parts may seem fortuitous—and hardly capable of promoting consistent adaptive bias in evolution—unless they are understood as means conducive to the goals that fish pursue. That is, a purely causal account of adaptive convergence seems to leave an explanatory gap; it refers to the adaptive biases on phenotypic evolution caused by development or differential fitness, but it struggles to make sense of why those biases (and not others) exist. Indeed, it is this explanatory gap that makes it tempting to explain any adaptive bias caused by developmental plasticity in terms of past natural selection on random genetic variation (e.g., Wray et al., 2014). However, if Walsh and others are right about agent theories, the regular attainment of particular phenotypes in cichlid evolution can be fully accounted for neither by chance and natural selection, nor by the addition of mechanistic developmental causation, but must credit the goals, affordances, and repertoires of individual fish. Goals can be legitimate difference-makers (again, not in a *causal* explanation) because the presence of goals is a natural consequence of how organisms are organized, and those goals impose a certain order on the world that, in the case of cichlids, resulted in the repeated evolution of a similar set of morphologies.

Whatever one makes of this case for naturalistic teleological explanation, biologists do recognize that it can matter to evolutionary dynamics whether or not organismal activities are oriented toward goals. For example, the niche construction literature emphasizes that organismal activities that are goal-oriented result in more consistent selective pressures than do other sources of selection (Clark et al., 2020), and that goal-oriented modification of environments can result in highly regular sequences of adaptive change (Laland et al., 2017). Similarly, it has been suggested that the evolutionary consequences of developmental plasticity will depend on whether or not individual responses are directed toward goals, or what goals those responses actually serve (Feiner et al., 2020).[21] Thus, insofar as biological systems exhibit the organizational closure that makes them goal-oriented, reference to those goals can perhaps help biologists understand patterns of evolution that may be difficult to grasp by explaining them solely in terms of mechanisms and causes.

17.5 Conclusions

The goal-oriented processes that we observe as development and behavior have proven difficult to fit within the explanatory standards of the dominant evolutionary theories. This has not prevented biologists from making liberal use of agential concepts, and those concepts can indeed fulfill legitimate epistemic functions even when the objects, such as genes, do not fulfill the criteria for biological agency. Biological agency is hard to deny for organisms, however, and some contemporary research on biological evolution is in fact organized around concepts that refer to organisms' ability to initiate activities from within their own boundaries, to sense and respond to the conditions they encounter, and to maintain their functional stability when perturbed. This suggests that biological agency can exercise a substantial influence on evolutionary biology by influencing the kinds of problems biologists address; what knowledge, concepts, and methods they need to import from other disciplines; and what they consider a satisfactory explanation.

Acknowledgments

I am grateful to the editors for the invitation to contribute to this volume, and to Nathalie Feiner, Juan Gefaell, Denis Walsh, and the editors for feedback on the manuscript. I gratefully acknowledge the financial support of the John Templeton Foundation (#62220). The opinions expressed in this paper are those of the author and not those of the John Templeton Foundation.

Notes

1. This chapter is concerned with the explanatory roles played by the goal-oriented activities of organisms, not with the potential utility of ascribing agency to natural selection or attributing goal-oriented properties to the evolutionary process itself.

2. This does not mean it is just loose talk; to Dawkins (1976), the agency metaphor appears to communicate a deep metaphysical commitment about the world.

3. Scientific understanding also requires that an explanation based on the application of a theory fulfill the fundamental scientific values of empirical adequacy and internal consistency (De Regt, 2018: 93).

4. Conversely, a scientific theory that was widely used in the past, often to good effect, can seem entirely unintelligible to the scientists of today.

5. The use of "intuition" here follows De Regt (2018: 109–113).

6. Not all scientists will find the same conceptual tools useful; thus, we may expect differences in opinion regarding the scientific value of agential thinking to reflect the background, cognitive ability, and skill sets of individual scientists.

7. It has been pointed out there are significant risks associated with applying agential concepts to genes or other entities that do not have those properties (see, e.g., Godfrey-Smith, 2009: ch.7). Perhaps the success of agential concepts in making evolutionary theory intelligible has contributed to illegitimate metaphysical views of genes among both scientists and the general public. These views may in turn have influenced what biologists consider fundamental or indispensable to evolutionary explanation.

8. On the teleonomy concept, see ch. 1 (the introduction of this book) by Corning et al.

9. Not unlike statistical mechanics in physics, population geneticists represent evolutionary change in terms of forces affecting the spread and maintenance of alleles coding for alternative versions of a trait. This representation of evolution can answer a range of what-if-things-were-different questions about adaptive and nonadaptive change; it is quantitative, predictive, and empirically testable—all hallmarks of good natural science. For example, by assigning fitness values to genotypes, population genetic models demonstrate under what conditions natural selection will maintain more than one genotype. Models like these bring understanding because they help us grasp why different genotypes (and hence phenotypes) can coexist by demonstrating how interventions on variables like fitness or population size influence the composition of genotypes within a population.

10. Explanations that cite organismal development, physiology, or behavior are considered to violate the distinction between proximate and ultimate causation (Mayr, 1961). For more recent examples of how this distinction has been used to identify "inadequate" evolutionary explanations, see Scott-Phillips et al., 2011; Dickins & Rahman, 2012; For counterpoints, see Mesoudi et al., 2013; Laland et al., 2015. Different interpretations of the status of the proximate-ultimate distinction in evolutionary biology are discussed in Laland et al., 2011; Laland et al., 2013 and the commentaries and author response to this paper in the same issue of *Biology & Philosophy*; and by Pigliucci & Scholl, 2015.

11. Biases imposed by mutation, development, or inheritance that are not fitness-enhancing are also easily neglected under this representation, as they will appear inconsequential for the "interesting bits" of evolution (e.g., adaptation; see Stoltzfus, 2021).

12. Key assumptions that underlie the explanatory standards of the neo-Darwinian representation of the evolutionary process concern the nature of genes (Oyama, 2000) and the autonomy of variation, fitness, and inheritance (Walsh, 2015; Uller & Helanterä, 2019).

13. To those who emphasize the role of the organism in evolution, metaphors such as genetic "programs" and "blueprints" discourage such work because they explain *away* agency rather than single out agency as a fundamental property of living systems.

14. Love et al. (2017) discuss these challenges from the perspective of integrating genetic and physical explanations for the origins of novelties.

15. There is some debate over whether or not explanations that refer to natural selection are causal (Otsuka, 2016; Walsh, 2018), but this does not deny that there are challenges of explanatory integration (on the contrary, it would arguably make matters worse).

16. Although abstract criteria for evolution by natural selection are helpful in understanding how evolution works, it makes a difference to evolution how those principles are instantiated by the evolving entities.

17. *Affordance* is a concept used to describe this complementarity of organism and environment that are salient to an agent's pursuit of its goal.

18. "Because an agent is capable of attaining and maintaining stable endpoints that reliably secure its stability, one can cite the stable endpoint to which the system tends in explaining its activities" (Sultan et al., 2022). For Walsh's defense of naturalistic teleological explanations, see Walsh, 2012, 2015, 2018. See also Mossio and Bich (2017) and Jaeger (2021).

19. Organisms, parts of organisms (e.g., metabolic processes within cells), and collectives (e.g., social insect colonies) are candidates for exhibiting closure (Mossio et al., 2009; Montévil & Mossio, 2015). Whether or not they do is an empirical issue, and an important one for understanding the role of agency in evolution. On this organizational account of biological agency, organismal agency can impose regularities in evolution and may therefore be responsible for macroevolutionary trends, but for the evolutionary process itself to be goal-oriented, it too should exhibit closure.

20. To even grant the possibility of naturalistic teleological explanation in science may seem heretical. Some evolutionary biologists are understandably concerned that it blurs the distinction between scientific and unscientific explanation, and thus can be exploited by creationists, for example. Others may welcome naturalistic teleological explanation exactly because it *scientifically* addresses features of the living world that have been left unexplained by past scientific theories, and hence left vulnerable to exploitation by those hostile to science and scientific knowledge in general, and evolution in particular.

21. Of course, there is nothing to prevent biologists from rejecting all these arguments for taking agency seriously, but still agreeing that goal-oriented activities have special evolutionary consequences. In the neo-Darwinian explanatory framework, the evolutionary consequences of goal-oriented activities would be fully accounted for by random genetic variation and natural selection, while the proximate development and behavior of organisms are optional causal detail (Wray et al., 2014).

References

Ågren, A. 2021. *The gene's-eye view of evolution.* Oxford University Press.

Baedke, J., A. Fábregas-Tejeda, and F. Vergara-Silva. 2020. Does the extended evolutionary synthesis entail extended explanatory power? *Biology & Philosophy* 35: 20.

Brigandt, I. 2010. Beyond reduction and pluralism: toward an epistemology of explanatory integration in biology. *Erkenntnis* 73: 295–311.

Burt, A., and Trivers, R. (2006). *Genes in conflict.* Harvard University Press.

Calcott, B. 2009. Lineage explanations: explaining how biological mechanisms change. *British Journal for the Philosophy of Science* 60: 51–78.

Charlesworth, D., N. H. Barton, and B. Charlesworth. 2017. The sources of adaptive variation. *Proceedings of the Royal Society B-Biological Sciences* 284: 20162864.

Chellappoo, A. 2022. When can cultural selection explain adaptation? *Biology & Philosophy* 37: 2.

Chevin, L.-M., R. Lande, and G. M. Mace. 2010. Adaptation, plasticity, and extinction in a changing environment: Towards a predictive theory. *PLoS Biology* 8: e1000357.

Clark, A. D., D. Deffner, K. Laland, J. Odling-Smee, and J. Endler. 2020. Niche construction affects the variability and strength of natural selection. *American Naturalist* 195: 16–30.

Corning, P. A. 2019. Teleonomy and the proximate–ultimate distinction revisited. *Biological Journal of the Linnean Society* 127: 912–916.

Conith, A. J. and R. C. Albertson. 2021. The cichlid oral and pharyngeal jaws are evolutionarily and genetically coupled. *Nature Communications* 12: 5477.

Dawkins, R. 1976. *The selfish gene.* Oxford University Press.

De Regt, H. W. 2018. *Understanding scientific understanding.* Oxford University Press.

Dickins, T. E., and Q. Rahman. 2012. The extended evolutionary synthesis and the role of soft inheritance in evolution. *Proceedings of the Royal Society B-Biological Sciences* 279: 2913–2921.

Dobzhansky, T. 1968. On some fundamental concepts of Darwinian biology. In T. Dobzhansky, M. K. Hecht, and W. C. Steere (eds.), *Evolutionary biology* (vol. 2), pp. 1–34. Springer.

Edelaar, P., and D. I. Bolnick. 2019. Appreciating the multiple processes increasing individual or population fitness. *Trends in Ecology & Evolution* 34: 435–446.

Feiner, N., I. S. C. Jackson, K. L. Munch, R. Radersma, and T. Uller. 2020. Plasticity and evolutionary convergence in the locomotor skeleton of Greater Antillean Anolis lizards. *Elife* 9: e57468.

Gerhart, J., and M. Kirschner. 2007. The theory of facilitated variation. *Proceedings of the National Academy of Sciences of the United States of America* 104: 8582–8589.

Godfrey-Smith, P. 2009. *Darwinian populations and natural selection.* Oxford University Press.

Gould, S. J., and R. C. Lewontin. 1979. Spandrels of San-Marco and the Panglossian paradigm—a critique of the adaptationist program. *Proceedings of the Royal Society of London Series B-Biological Sciences* 205: 581–598.

Griffiths. P., and K. Stotz. 2013. *Genetics and philosophy: an introduction.* Cambridge University Press.

Haig, D. 1997. The social gene. In J. R. Krebs and N. B. Davies (eds.), *Behavioral ecology* (4th ed.), pp. 284–304. Blackwell Scientific.

Jablonka, E., and M. J. Lamb. 2014. *Evolution in four dimensions: genetic, epigenetic, behavioral and symbolic variation in the history of life* (2d ed.). MIT Press.

Jaeger, J. 2021. *The fourth perspective: evolution and organismal agency.* OSF Preprints. February 26. https://doi .org/10.31219/osf.io/2g7fh

Keller, E. F. 2000. *The century of the gene.* Harvard University Press.

Kocher, T. D., J. A. Conroy, K. R. McKaye, and J. R. Stauffer. 1993. Similar morphologies of cichlid fish in Lakes Tanganyika and Malawi are due to convergence. *Molecular Phylogenetics and Evolution* 2: 158–165.

Laland, K. N., J. Odling-Smee, and J. Endler. 2017. Niche construction, sources of selection and trait coevolution. *Interface Focus* 7: 20160147.

Laland, K. N., J. Odling-Smee, W. Hoppitt, and T. Uller. 2013. More on how and why: cause and effect in biology revisited. *Biology & Philosophy* 28: 719–745.

Laland, K. N., K. Sterelny, J. Odling-Smee, W. Hoppitt, and T. Uller. 2011. Cause and effect in biology revisited: Is Mayr's proximate-ultimate dichotomy still useful? *Science* 334: 1512–1516.

Laland, K. N., T. Uller, M. W. Feldman, K. Sterelny, G. B. Muller, A. Moczek, E. Jablonka, and J. Odling-Smee. 2014. Does evolutionary theory need a rethink? *Nature* 514: 161–164.

Laland, K. N., T. Uller, M. W. Feldman, K. Sterelny, G. B. Muller, A. Moczek, E. Jablonka, and J. Odling-Smee. 2015. The extended evolutionary synthesis: its structure, assumptions and predictions. *Proceedings of the Royal Society B-Biological Sciences* 282.

Lande, R. 2009. Adaptation to an extraordinary environment by evolution of phenotypic plasticity and genetic assimilation. *Journal of Evolutionary Biology* 22: 1435–1446.

Lange, M. 2016. *Because without cause: non-causal explanations in science and mathematics.* Oxford University Press.

Levis, N. A., and D. W. Pfennig. 2016. Evaluating 'Plasticity-First' evolution in nature: key criteria and empirical approaches. *Trends in Ecology & Evolution* 31: 563–574.

Lewontin, R. C. 1970. The units of selection. *Annual Review of Ecology and Systematics* 1: 1–18.

Lewontin, R. C. 1983. Gene, organism, and environment. In D. S. Bendell (ed.), *Evolution from molecules to men*, pp. 273–285. Cambridge University Press.

Lloyd, E. A. 2005. *The case of the female orgasm: bias in the science of evolution.* Harvard University Press.

Love, A. C. 2008. Explaining evolutionary innovations and novelties: criteria of explanatory adequacy and epistemological prerequisites. *Philosophy of Science* 75: 874–886.

Love, A. C. 2017. Building integrated explanatory models of complex biological phenomena: From Mill's methods to a causal mosaic. In M. Massimi and J.-W. Romeijn (eds.), *EPSA Philosophy of Science: Düsseldorf 2015. The European Philosophy of Science Association Proceedings*, Vol. 5, pp. 221–232. Springer.

Love, A. C., T. A. Stewart, G. P. Wagner, and S. A. Newman. 2017. Perspectives on integrating genetic and physical explanations of evolution and development. *Integrative and Comparative Biology* 57: 1258–1268.

Maynard Smith, J. 1998. The units of selection. In G. R. Bock and J. A. Goode (eds.), *The limits of reductionism in biology*, pp. 203–210. Novartis Foundation, Wiley.

Mayr, E. 1961. Cause and effect in biology—kinds of causes, predictability, and teleology are viewed by a practicing biologist. *Science* 134: 1501–1506.

Mayr, E. 1974. Teleological and teleonomic: a new analysis. *Boston Studies in the Philosophy of Science* 14:91–117.

Mayr, E. 1988. *Toward a new philosophy of biology: observations of an evolutionist.* Harvard University Press.

Mcsoudi, A., S. Blanchet, A. Charmantier, E. Danchin, L. Fogarty, E. Jablonka, K. N. Laland, T. J. H. Morgan, G. B. Mueller, F. J. Odling-Smee, and B. Pujol. 2013. Is non-genetic inheritance just a proximate mechanism? A corroboration of the extended evolutionary synthesis. *Biological Theory* 7: 189–195.

Montévil, M., and M. Mossio. 2015. Biological organisation as closure of constraints. *Journal of Theoretical Biology* 372: 179–91.

Mossio, M., and L. Bich. 2017. What makes biological organisation teleological? *Synthese* 194: 1089–1114.

Mossio, M., Saborido, C., and A. Moreno. 2009. An organizational account of biological functions. *British Journal for the Philosophy of Science* 60: 813–41

Muschick, M., M. Barluenga, W. Salzburger, and A. Meyer. 2011. Adaptive phenotypic plasticity in the Midas cichlid fish pharyngeal jaw and its relevance in adaptive radiation. *BMC Evolutionary Biology* 11: 116.

Odling-Smee, F. J. 1988. Niche-constructing phenotypes. In H. C. Plotkin (ed.), *The role of behaviour in evolution*, pp. 73–132. MIT Press.

Odling-Smee, F. J., K. N. Laland, and M. Feldman. 2003. *Niche construction: the neglected process in evolution.* Princeton University Press.

Okasha, S. 2018. *Agents and goals in evolution.* Oxford University Press.

Otsuka, J. 2016. A critical review of the statisticalist debate. *Biology & Philosophy* 31: 459–482.

Oyama, S. 2000. *The ontogeny of information: Developmental systems and evolution.* Duke University Press.

Pigliucci, M. and R. Scholl. 2015. The proximate–ultimate distinction and evolutionary developmental biology: causal irrelevance versus explanatory abstraction. *Biology and Philosophy* 30: 653–670.

Potochnik, A. 2017. *Idealization and the aims of science.* University of Chicago Press.

Provine, W. B. 1986. *Sewall Wright and evolutionary biology.* Chicago University Press.

Provine, W. B. 2003. Origins of the genetics of natural populations series. In R. C. Lewontin, J. Moore, W. B. Provine, and B. Wallace (eds.), *Dobzhansky's genetics of natural populations I–XLIII*, pp. 1–92. Columbia University Press.

Riskin, J. 2020. Biology's mistress, a brief history. *Interdisciplinary Science Reviews* 45: 268–298.

Salazar-Ciudad, I. 2007. On the origins of morphological variation, canalization, robustness, and evolvability. *Integrative and Comparative Biology* 47: 390–400.

Schneider, R. F., and A. Meyer. 2017. How plasticity, genetic assimilation and cryptic genetic variation may contribute to adaptive radiations. *Molecular Ecology* 26: 330–350.

Scott-Phillips, T. C., T. E. Dickins, and S. A. West. 2011. Evolutionary theory and the ultimate-proximate distinction in the human behavioral sciences. *Perspectives on Psychological Science* 6: 38–47.

Stoltzfus, A. 2021. *Mutation, randomness, and evolution.* Oxford University Press.

Sultan, S. E., A. P. Moczek, and D. Walsh. 2022. Bridging the explanatory gaps: what can we learn from a biological agency perspective? *BioEssays* 44: 2100185.

Uller, T., N. Feiner, R. Radersma, I. S. C. Jackson, and A. Rago. 2020. Developmental plasticity and evolutionary explanations. *Evolution & Development* 22: 47–55.

Uller, T., and H. Helanterä. 2019. Niche construction and conceptual change in evolutionary biology. *The British Journal for the Philosophy of Science* 70: 351–375.

Uller, T., A. P. Moczek, R. A. Watson, P. M. Brakefield, and K. N. Laland. 2018. Developmental bias and evolution: a regulatory network perspective. *Genetics* 209: 949–966.

Waddington, C. H. 1968. *Towards a theoretical biology, Vol. I.* Edinburgh University Press.

Walsh, D. 2012. Mechanism and purpose. A case for natural teleology. *Studies in the History and Philosophy of Biology and the Biomedical Sciences* 43: 173–181.

Walsh, D. 2015. *Organisms, agency and evolution.* Cambridge University Press.

Walsh, D. 2018. Objectcy and agency: Towards a methodological vitalism. In D. J. Nicholson and J. Dupré (eds.), *Everything flows. Towards a processual philosophy of biology*, pp. 167–185. Oxford University Press.

Watson, R. A., and C. Thies. 2019. Are developmental plasticity, niche construction and extended inheritance necessary for evolution by natural selection? The role of active phenotypes in the minimal criteria for Darwinian individuality. In T. Uller and K. N. Laland (eds.), *Evolutionary causation.*, pp. 197–226. MIT Press.

West-Eberhard, M. J. 2003. *Developmental plasticity and evolution.* Oxford University Press.

Wray, G. A., H. E. Hoekstra, D. J. Futuyma, R. E. Lenski, T. F. C. Mackay, D. Schluter, and J. E. Strassmann. 2014. Does evolutionary theory need a rethink?—No, all is well. *Nature* 514: 161–164.

Ylikoski, P., and J. Kuorikoski. 2010. Dissecting explanatory power. *Philosophical Studies* 148: 201–219.

18 Evolutionary Foundationalism: The Myth of the Chemical Given

Denis M. Walsh

"A picture held us captive." (Wittgenstein, 1953: 115)

Overview

The modern synthesis theory of evolution has been subject to unprecedented critical scrutiny in recent years. Detractors routinely charge that it is incapable of accounting fully for the component processes of evolution: inheritance, development, innovation, and adaptively biased change. I offer a diagnosis of these deficiencies, and briefly outline an alternative approach. I argue that modern synthesis evolution fails because it is a foundationalist theory. The component processes of evolution are contact phenomena: they take place at the interface between the purposive organism and its conditions of existence. Foundationalist theories are incapable of representing contact phenomena *as* contact phenomena. I illustrate the structure and shortcomings of foundationalist theories by discussing their role in orthodox accounts of "mental" phenomena—perception, cognition, intentionality. These deficiencies carry over directly to the deficiencies of the modern synthesis in accounting for inheritance, development, innovation, and adaptively biased change. Thus, the problems of the modern synthesis are structural and systemic. I argue that an adequate account of these processes requires an agent theory, which is significantly different in its structure from foundationalist theories.

18.1 Introduction

The principal objective of a theory of evolution is to explain the fit and diversity of organic form. I shall concentrate here on *fit*, by which I mean adaptedness, or what Darwin meant by "those endless forms most beautiful and most wonderful" (Darwin, 1859/1968: 450). Adaptedness is special kind of phenomenon; it consists in an organism's ability to cope successfully with its conditions of existence. No other phenomenon in the natural sciences is like that. It is hardly surprising that adaptation should require a special kind of explanation.

The modern synthesis theory of evolution has adopted a distinctive approach to explaining the adaptedness of organisms. It accords special theoretical significance to entities internal to organisms (i.e., genes) and to features external to organisms (e.g., environments). Genes build organisms according to a program that they encode, and then the environment selects from among the variant organisms that the genes build those individuals that are best suited to surviving and reproducing in the environment. The result is an increased relative frequency of those individuals whose genes encode programs for those traits that better suit the exigencies of that environment. In this way, form becomes adapted.[1] This is a twice-told tale, that stakes its place as well-worn biological lore. But it has come in for an unprecedented amount of critical scrutiny in recent years (Pigliucci, 2009; Laland et al., 2014). It is increasingly being suggested that the unalloyed modern synthesis approach of genes and environments is incapable of accounting for all we need to know about adaptive evolution (Laland et al., 2015; Uller et al., 2020, and ch. 17 in this volume; Sultan et al., 2022).[2] Be that as it may, most commentators who call for a revision of the modern synthesis accept the gene as the canonical unit of evolutionary theorizing. The dynamics of genes alone may not offer a wholly comprehensive account of adaptive evolution, but it will likely be the core of any eventual theory that does (Pigliucci, 2009; Laland et al., 2014).

That alone attests to the tenacity of the gene's-eye view of evolution. Despite the challenges, is still widely promulgated and extolled (Charlesworth et al., 2017; Ågren, 2021).[3] It is, in Wittgenstein's vivid phrase, "a picture [that] holds us captive." Part of its longevity is undoubtedly due to its past successes, and part perhaps to the lack of a viable alternative. I want to argue here that the draw of the modern synthesis derives in large part from its structure: it is a foundationalist theory. Foundationalist theories constitute a widespread approach to understanding the structure and dynamics of complex entities.[4] Foundationalist theories have been remarkably successful but, I shall argue, they suffer a systematic deficiency when applied to a particular kind of system—an agential system. Identifying modern synthesis evolutionary biology as a foundationalist theory will help us understand its structure, but more significantly to identify the sources of its widely recognized faults. That in turn might serve to loosen the grip in which it holds us. I'm trying "to show the fly the way out of the fly-bottle," in Wittgenstein's (1953: 310) resonant phrase.

I go about this in a fairly unorthodox way. I think there is an object lesson for biology to be drawn from certain parallels with longstanding issues in philosophy. In particular, there are analogies between the structure of gene-centered modern synthesis evolutionary biology and a battery of standard philosophical approaches to human thought, perception, knowledge, reference, meaning, and the like. (For want of a generally accepted umbrella term, I'll call these *mental phenomena*). These analogies can be salutary. They allow us to approach the structure of gene-centered evolutionary theory at a helpful level of abstraction. In particular, these approaches to mental phenomena have a systematic deficiency that is quite reminiscent of those infecting the modern synthesis.

Like the modern synthesis approach to evolution, they are foundationalist theories. Foundationalist theories of mental phenomena are elegant, compelling, tractable, and flawed. They fail for a very specific reason: they leave out of the account, or misconstrue, the contribution of agents to the phenomena they seek to explain. I suggest that the widely recognized shortcomings of the modern synthesis gene-centered approach arise from the same source. Not surprisingly, the alternative approach I recommend in each case is an agent theory. An agent theory starts by recognizing the system in question—the human thinker, knower, perceiver, or the organism—as an agent. Agents are natural purposive systems, actively and adaptively engaged in their circumstances. Agent theories are very different in their structure from foundationalist theories. For now, I want to raise awareness of the distinction, and offer some considerations in support of an agent theory of adaptive evolution.[5]

I start with a cursory survey of foundationalist thinking in philosophical approaches to mental phenomena—it won't take too long—and then proceed to argue for the analogy with modern synthesis thinking. I then argue that the deficiencies of each arise from their common structure. I then gesture to an alternative picture, the agential view. It does not hold us captive the way the foundational picture does, but it might help to show the fruit fly the way out of the bottle.

18.2   The Foundationalist Stratagem

As humans, we undertake a variety of relations with our world of the sort that involve the mind. We experience it, we think about it, refer to it, describe it, understand it, desire things from it, theorize about it, act on it (and much else). These phenomena—thought, perception, reference, desire, meaning, action, and others—connect the agent to the world. Understanding these phenomena is among the traditional and enduring concerns of philosophy.

One prominent strategy for understanding them is to start from a kind of decomposition between the agent and the world. There is the agent endowed with an inner mental life, and there is an external environment, wholly separate and independent from the inner realm of the agent's thought. Having sundered the mental from the external world, we concentrate on the inner realm: on those mental processes occurring within the agent more or less independently of the environment. We think of reason, thought, perception, language, and the like as being generated within an internal arena, a mental space. The activities of the mental realm are governed by a set of rules, or procedures internal to the workings of the mind.[6] The outputs of these internal mental machinations are thoughts, beliefs, perceptions, actions, locutions, knowledge claims, and so on. Once they have been constructed by the mind, they are directed toward the external world. Success in these mental endeavors consists in effecting an appropriate match between the deliverances of the inner mental processes and the external environment. A belief is true, a perception veridical, an action appropriate, a locution meaningful, if it puts the agent in the right kind of correspondence with the external world.

According to this approach, our contact with the world is mediated through inner entities or episodes constructed by the mind. "We grasp the world through something, what is outside by something inner" (Taylor, 2006: 25). Charles Taylor describes the general strategy: "Knowledge of things outside the mind/agent/organism only comes about through . . . mental images, or conceptual schemes within the mind/agent/organism. The input is combined, computed over, or structured by the mind to construct a view of what lies outside" (Taylor, 2006: 27). The channels of communication between the mind and the world are narrow. We take in the world passively, through perception; its role is merely to report or deliver information. We act on the world through motor outputs that merely implement those commands encoded in thought.

This is the stratagem that, following Taylor (2006) and Dreyfus and Taylor (2015), I shall call *foundationalism*. It is the picture of the mind that Wittgenstein claimed holds us captive. It has five basic features: (i) the separation of the internal workings of the agent from the external influences of the world, and (ii) a set of foundational inner primitive entities (*givens*); (iii) the construction of other complex inner entities, according to inner rules; and (iv) the consequences of these inner processes stand proxy for (are about) the external world; and

(v) the inner realm and the outer communicate through narrow and transparent channels. The givens play a pivotal role in foundational theories. These are the basic (foundational) structures of the inner realm. They are primitively units of knowledge, meaning, sensory experience, and thought, over which mental procedures operate, and upon which more complex mental states are built. Their status as these basic units needs no special justification. Whatever warrant the more complex structures might require are inherited from the basic warrant of the givens, and the rules for manipulating them. That is a little abstract, but a crude survey of some traditional foundationalisms might fill in some of the detail.

18.2.1 Epistemology

Descartes' account of knowledge is the *locus classicus* of epistemic foundationalism. Descartes' objective was to understand how an agent might know anything about the world. His strategy was to find a foundational epistemic unit, whose status as a unit of knowledge is unquestionably guaranteed, upon which reason can build a more substantive edifice of knowledge. Descartes convinces himself that he cannot be mistaken about whether he is thinking, and that if he thinks he must exist. In this way the *cogito* delivers a foundational epistemic simple—a given: the indubitable knowledge that he is a thinking thing. From there, further knowledge claims can be built up through the application of a procedure that guarantees its products as known propositions. Reason provides the guarantee, Descartes surmised, because a beneficent God would not endow him with an unreliable faculty for deriving beliefs from beliefs: "And since God does not wish to deceive me, he surely did not give me the kind of faculty which would ever enable me to go wrong while using it correctly" (Descartes, 1641/1998, Med. 4, AT 7:53f, CSM 2:37).

Descartes' internalist epistemology displays the hallmarks of foundationalism: (i) the separation of the workings of the inner realm from the external world, (ii) the positing of an inner given, (iii) a set of inner rules for constructing more complex inner entities from simpler ones, (iv) the contact with the external world at the periphery, and (v) perception and motor implementation are mere conduits of thought, but they do not participate in it. Inner thoughts stand for outer things: "I can be certain that I can have no knowledge of what is outside me, except by means of the ideas I have within me" (Descartes, 1641/1998, *Correspondence*).

18.2.2 Perception

How can experience deliver perceptions about the world? Empiricist-inspired accounts of perception suggest that our perceptual apparatus passively delivers information to us in the form of inner sensations or experiences. The foundational entities of sensory experiences are *qualia* (qualitative sensations), or sense data. They are the smallest units of phenomenal experience. These units are used by the mind to construct a complex representation of the external world. These representations constitute veridical perceptions to the extent that they conform to the external world.

In sense datum theory, we clearly see foundationalism at work. The sensory simples—sense data—are inner givens.[7] "The core notion of a sense datum is an immediately given, minimal perceptual object, consisting in the case of vision of a shaped patch of color" (Hadfield, 2021).[8] Crucially, sense data come with a guarantee. They are self-intimating; they are known to us by experience: *esse ist percipe*.[9] Sense data form the basis of our mind's construction of its perceptual representation of the world. Wilfred Sellars describes empiricist theory of how we come to know about the world through perception: "These 'takings' are, so to speak, the unmoved movers of empirical knowledge, the 'knowings in presence' which are presupposed by all other knowledge. . . . Such is the framework in which traditional empiricism makes its characteristic claim that the perceptually given is the foundation of all empirical knowledge" (Sellars, 1956: 299).

18.2.3 Cognition

How is it possible to think about the world? The computational/representational theory of mind tells us that thinking is a kind of inner operation—computation—that proceeds according to fixed rules (logic, mathematics, probability, Bayesian reasoning).[10] The operands are mental entities, pieces of syntax that are manipulated by the mind. These states have their syntactic properties in virtue of their realization in the brain. It is their semantic properties that connect them, and thoughts constructed out of them, to the external world (Chirumuuta, 2020). In virtue of their semantic properties, they stand for features of the external world. These semantic properties are given to them primitively.[11]

This is but a small sample of foundationalisms; we could extend them at will. There are foundationalist theories of mental content (Rowlands et al., 2020), linguistic reference and meaning (Farkas, 2008); action, rational justification, moral justification (Finlay & Schroeder, 2017); and science (Putnam, 1982), to name but a few. The point is that they share a structure, an objective, and a strategy. Foundationalisms involve:

1. *separation*: the separation of the inner realm from the outer realm.
2. *givenism*: the positing of inner fundamental units.
3. *construction*: the construction and manipulations of inner entities by inner procedures.
4. *proxyism*: the inner states stand for outer features of the world.
5. *mediation*: the inner communicates with the outer realm through narrow channels.

On the foundationalist view, the inner space has theoretical primacy. The phenomena that the theories explain are constructed from primitive inner foundational units. The deliverances of the mind are internal; they connect with the world only incidentally, after the mental work has been done. The objective of a foundationalist theory of mental phenomena is to understand how the mind builds complex states out of simple inner givens. "The aim of foundationalism is to peel back all the layers of inference and interpretation and get back to something genuinely prior to them all—a brute Given" (Taylor, 2006: 43).

18.3 Underdetermination

Foundationalist theories of human mental phenomena have a systemic, structural shortcoming. They routinely fail to characterize the phenomena they seek to explain. In general, they face a problem of underdetermination. The inner processes by which the phenomenon in question—a belief, a perception, a thought about the world, a rational justification—is generated fail to guarantee that the phenomenon actually obtains.[12] A few examples might help to bring this general shortfall into view.

18.3.1   Skepticism

Foundationalist epistemology is famously beset by the problem of skepticism. The internal operations of the mind that generate beliefs—candidates for knowledge—operate in a way that is insulated from the outside world. But knowing consists precisely in the right kind of sensitivity to the world.[13] The problem is that the inner operations of the mind fail to guarantee that the appropriate relation between the thinker and the world obtains. This is the lesson of Putnam's (1975) famous brain-in-a-vat argument. If I were a brain in a vat in Alpha Centauri being manipulated by an evil demon to believe I was sitting in the sunshine in Stadpark in Bochum, I would believe that I am, just as I do now. In one case I know that I am in Stadpark; in the other I don't. Real and counterfeit knowledge states are generated by the same internal mental processes, and nothing that is available to the knower can distinguish one from the other. Skepticism of this sort is a form of underdetermination, in that foundationalist accounts of how knowledge is generated fail to articulate the conditions sufficient for genuine knowledge.

18.3.2   Illusion

Foundationalist theories of perception suffer an analogous problem. The representation of the world drawn from sensory experience is constructed largely in isolation of the external world itself.[14] The same process may produce a grossly false perceptual state, and nothing about the perceiver's own condition could determine the veridical from the illusory. Perceptual processes fail to guarantee that their success conditions are met. This too is a form of underdetermination.

18.3.3   Solipsism

In foundationalist theories of mind, each thinker has immediate, noninferential introspective access to her own mind's inner workings. The access is parochial, to say the least. Others' minds are not available to the thinker in the same way that her own is. In order to attribute mental life to others we must do something else: make an inference, construct a theory of mind, project from our first personal experience. But all these ways of accessing others' minds are in no way as immediate, nor do they come with the same guarantees. So, for all our account of the way we know about minds goes, one's own mind might be the only one in the universe. Foundationalism seems to lead to an absurd form of solipsism. The reason is, again, that the account of what it is to have access to a mind underdetermines the conditions for knowing minds in general.

Underdetermination problems for foundationalist theories can be manufactured at will. Foundationalist (i.e., internalist) theories of meaning underdetermine the conditions for our words and utterances to be about things in the world (Farkas, 2008). Foundationalist reasons for action underdetermine that our actions in the world are justified. Foundationalist accounts of motivating reason underdetermine what is reasonable, and on and on. The point of the foregoing is to demonstrate that foundationalist theories suffer from a systematic kind of shortcoming. We want our theories to explain our contact—perceptual, doxastic, epistemic— with the world. And in each case the foundationalist account falls short of capturing what this kind of contact consists of.[15] "Once the foundationalist arguments . . . are seen to fail, we are left with the image of the self-enclosed subject, out of contact with the transcendent world" (Taylor, 2006: 40). That's enough sophomore philosophy for one sitting—but what has it got to do with evolution?

18.4   The Modern Synthesis Stratagem

The modern synthesis approach to explaining adaptation adheres quite closely to the foundationalist recipe we have just surveyed. It begins by separating the organism from its environment (Lewontin, 2000). It then posits a foundational, primitive unit of evolution: a given. Thereafter it proceeds to explain the features of organisms by appeal to internal processes operating over these givens. There could be no more definitive encapsulation of the modern synthesis stratagem than that recently offered in its defense by Ågren:

> By this reasoning, organisms are nothing but temporary occurrences—present in one generation, gone in the next. And, as a consequence, organisms cannot be the ultimate beneficiary in evolutionary explanations.
>
> Instead, this role is filled by the gene. Genes are considered immortal and they pass on their intact structure from generation to generation. . . . [N]atural selection is conceptualized as a struggle between genes, usually through the effects they have on organisms, for replication and transmission to the next generation (Ågren, 2021: 2).

It will be helpful for what is to follow if we trace out the foundationalist features of the gene's-eye, modern synthesis view of evolution in a little more detail.

18.4.1   Two Spaces

The gene-centered modern synthesis starts with the construction of two spaces: genotype space and phenotype space (Lewontin, 1974). In the case of adaptive evolution, the phenomena to be explained occupy phenotype space. We want our evolutionary theory to explain why organisms manifest their particular structures, processes, and behaviors, and why they are so well suited to their conditions of existence. The evolutionary processes that explain these phenomena occur for the most part in genotype space. We account for adaptive evolution by citing the way that genetic processes of replication, mutation, translation, recombination, and selective changes in relative gene frequencies result in adaptive changes to organisms over time. There are narrow channels of communication between genotype space and phenotype space. Organismal development maps the information from genotypes onto phenotypes, but it introduces no evolutionary changes of its own. Uniquely, natural selection translates phenomena of phenotype space into genotype space. As organisms differentially survive and reproduce under natural selection, their genes differentially persist and replicate. Change in relative gene frequencies ensues.

There are two points worth stressing about the two-space nature of the gene's-eye modern synthesis. The first is the primacy of genotype space. Evolution *is* change in genotype space. Change in phenotype space is neither necessary (Kimura, 1974) nor sufficient for evolution.[16] The second is that the channels that mediate between genotype space and phenotype space are narrow.[17] The Weismann barrier and the central dogma of molecular biology purportedly entail that those processes that occur in phenotype space that alter, influence, or produce an organism's structure, form, or behavior are not generally registered as changes in genotype space. These are merely acquired traits; they do not evolve. The two points are related. The barriers between genotype space and phenotype space secure the conviction that while the processes that occur within genotype space—replication, translation, mutation—are genuinely evolutionary processes, those that occur within phenotype space—adaptive innovations, developmental plasticity, ecological transmission, niche construction, social learning—generally are not. Consequently, the study of evolution, properly construed, is the study of the dynamics of genotype space. Phenotype space matters only insofar as (and to the limited degree that) it has implications for changes in genotype space.

18.4.2   The Given

Genes—aka "replicators"—are the givens of evolution: "Replicators exist. That's fundamental" (Dawkins, 1999: 262). They are the primitive, foundational entities of evolution. They encode information or a program for building biological form, and for the intergenerational constancy of form. To that extent they exert a particular control over phenotypes. These capacities are vouchsafed by their chemical constitution; genes are chemical givens. Furthermore, evolutionary change is defined and measured in the currency of genes. "Evolution is the external and visible manifestation of the survival of alternative replicators . . . Genes are replicators; organisms . . . are best not regarded as replicators; they are vehicles in which replicators travel about" (Dawkins, 1982: 82). Genes are ideally suited for this role as the markers of evolution because they are so persistent; they are "practically immortal," as Ågren is at pains to emphasize. Genes are, to paraphrase Sellars' paraphrase of Aristotle, the "unmoved movers" of evolution.[18]

18.4.3   The Component Processes of Evolution

The component processes of evolution are defined over genes and, with one notable exception, occur exclusively within genotype space. Adaptive evolution requires that organisms develop into entities that engage with their conditions of existence, pass on their capacities to do so to their offspring, produce evolutionary novelties, and interact with their conditions in ways that bias evolutionary change. In the modern synthesis view, genes are intimately involved in each of these processes. Development is the translation of genetic information into genotypes. Inheritance is the replication and transmission of information encoded in genotypes. Evolutionary novelty comes about through the mutation or recombination of genes. Natural selection differentially promotes those genes that code for characters that suit their bearers to their environments.

In this respect, the gene-centered modern synthesis approach to adaptive evolution conforms to the structure of a foundationalist theory. It bears the foundationalist hallmarks:

1. *Separation*: the separation of the inner realm from the outer realm, in this case genotype space and phenotype space.

2. *Givenism*: the positing of inner fundamental units, in this case genes.

3. *Construction*: the construction and manipulation of inner entities by inner procedures; in this case, the genetic processes of replication, translations, mutation, and differential retention.

4. *Proxyism*: the inner states (genetic information) stand for outer features of the world (phenotypic traits); in this case, the dynamics of phenotype space are explained by the dynamics of genotype space.

5. *Mediation*: the inner realm of genes communicates with the outer realm through narrow channels (development and selection).

If the modern synthesis picture of evolution is satisfactory, then the entities and processes defined over genotype space must be sufficient to describe and explain the phenomena we observe in phenotype space. The challenge is articulated powerfully by Richard Lewontin:

A description and explanation of genetic change in a population is a description and explanation of evolutionary change only insofar as we can link those genetic changes to the manifest diversity of living organisms. . . . To concentrate only on genetic change . . . is to forget entirely what it is we are trying to explain in the first place (Lewontin, 1974: 23).

What we are trying to explain in the first place—adaptive evolution—is a phenomenon of phenotype space. The gene-centered modern synthesis attempts to explain them by appeal to the processes of genotype space.

18.5   Evolutionary Underdetermination

In its broadest form, we can see the battery of recent challenges to the modern synthesis theory of evolution as originating from a concern that the processes of genotype space do not adequately explain the phenomena of phenotype space. In fact, modern synthesis genotype space theory comes up short in systematic ways that should be familiar to us from our brief survey of foundationalist theories of mental phenomena. Gene-centered modern synthesis processes underdetermine the target phenomena to be explained.

The case for underdetermination starts with a simple question: "What has to happen for evolution to happen?" From the perspective of phenotype space, adaptive evolution requires:

1. persistence of form (inheritance) across generations,
2. development of organisms,[19]
3. novel traits (capable of persisting across generations), and
4. adaptive bias to form.

The battery of recent challenges to the modern synthesis (e.g., Jablonka, 2017; Uller et al., 2020; Weiner & Katz, 2021) can be interpreted as claims that the genotype space processes that stand proxy for these conditions underdetermine them. Here is a quick overview.

18.5.1   Inheritance

Those who argue for extragenetic forms of inheritance routinely claim that genetic inheritance underdetermines the kinds of transgenerational transmission of form that is necessary for evolution (Jablonka & Avital, 2006; Jablonka & Lamb, 2010; Jablonka, 2017). From the phenotype-space perspective, transgenerational stability of form is just what inheritance *is*. There are many more processes that secure it than just the transmission of genes. Epigenetic inheritance provides a vivid case in point. Epigenetic marks are applied to DNA and to chromatin in response to an organism's (or a cell's) environmental, metabolic, endocrine, immune state.[20] These marks regulate the expression of genes. Crucially, the effects that they exert over form are transgenerationally stable (Lacal & Ventura, 2018; Weiner & Katz, 2021). Epigenetic phenomena have been shown to secure the persistence of characters sometimes (for example in *Caenorhabditis elegans*) for up to 40 generations *without* genetic change (Heard & Martienssen, 2014).[21] These stable characters in turn can affect the capacity of an organism to cope (or not) with its conditions. That is to say, epigenetic phenomena underwrite inheritable differences in fitness. Simply put, there is much more to transgenerational stability of form than is countenanced in genetic transmission.[22] The gene-centered modern synthesis view radically underdetermines inheritance. This is widely accepted even in mainstream evolutionary biology, of course:

The weight of theory and empirical evidence indicates that nongenetic inheritance is a potent factor in evolution that can engender outcomes unanticipated under the Mendelian-genetic model (Bonduriansky & Day, 2009: 103).[23]

One way around the welter of evidence that genetic inheritance is not all there is to inheritance is simply to deny that extragenetic inheritance is, properly speaking, inheritance at all. After all, inheritance has come to be redefined in the modern synthesis as the transmission of genes. It is difficult to see what might motivate such a dogmatic stance. One suggestion might be that only genes are directly transmitted, and only genes last long enough to be genuine units of inheritance; other modes of resemblance are indirect and transitory (Ågren, 2021). Genes are certainly passed on from parent to offspring directly, and they are long-lasting, but there is nothing in the role that inheritance plays in evolution that suggests anything about how intergenerational resemblance is secured or for how many generations resemblances must persist. In fact, one might argue that inheritance should in part be short-lived. The optimal inheritance system must effect a balance between reversibility and fidelity. An inheritance system that it is too labile (i.e., reversible) may fail to lock in those traits whose effects are beneficial over long stretches of time. But such a system may be capable of effecting changes in form whose significance may be relevant only over shorter time scales. Conversely, a highly rigid (i.e., high fidelity), permanent inheritance system may reliably lock in traits whose impacts persist across multiple generations but may be too inflexible to deal with changes that pertain only to shorter time spans.

Danchin et al. (2019) argue that an optimal inheritance system for evolution would be responsive to the temporal scale of change and persistence of conditions in the environment. The idea is that an inheritance system should combine robustness with sensitivity. Genetic inheritance alone is extremely robust, but highly insensitive.

While the strength of genetic inheritance undoubtedly is in its stability, its weakness lies in its lack of reversibility, making it inappropriate to allow adaptation to relatively fast changing environmental characteristics. These conditions suggest that genetic and epigenetic inheritance systems constitute complementary mechanisms of adaptation to an environment whose many changes occur along different time scales (Danchin et al., 2019: 5).

Taken together, the entire suite of inheritance mechanisms—genetic, epigenetic, parental effects, niche construction, ecological and cultural transmissions, learning—offers the phenomenon of intergenerational transmission the requisite balance between fidelity and responsiveness. Multiple modes of inheritance are much better for adaptive evolution than just one.

The take-home message is that the phenomenon of inheritance that is required for phenotypic evolution—as seen from phenotype space—is greatly different from the phenomenon of genetic inheritance as it appears in models of genotype-space dynamics. In particular, transmission of replicators vastly under-represents the kinds and modes of transgenerational persistence of form that contribute to adaptive evolution. Gene transmission underdetermines inheritance.

18.5.2 Development

According to traditional two-space modern synthesis evolutionary biology, development is the expression of phenotypic information encoded in genes. Perhaps more than any other feature of gene-centered evolutionary biology, this conception has been shown to be the most woefully deficient (West-Eberhard, 2003; Gilbert & Epel, 2009). Genotype radically underdetermines phenotype. This has been repeatedly demonstrated by evolutionary developmental biology, by ecological developmental biology, by developmental systems theory (Oyama et al., 2000), by niche construction theory (Odling-Smee et al., 2003), and by the plasticity of development.

Organismal development is highly adaptively plastic (Gilbert & Epel, 2009; Sultan, 2015).[24] Plasticity ushers in two crucial evolutionary phenomena. The first is robustness. Organisms of the same sort reliably develop the same phenotypes not because of genetic and environmental similarity, but despite the vagaries of each (Wagner, 2011). Developmental plasticity buffers organisms against genetic and environmental perturbations. It confers on organisms a wide phenotypic repertoire, the capacity to marshal their developmental resources in the reliable production of their normal phenotypes. The flip side of this development repertoire is that it confers on organismal development the capacity to innovate, to produce new stable phenotypes in new conditions (Ledón-Rettig et al., 2008).

An elegant example of the latter is furnished by Standen et al. (2014). These authors demonstrated that *Polypterus* fish raised in terrestrial environments have the capacity to develop pectoral girdles and limbs suited to locomotion on land.[25] The various ways in which developmental plasticity might contribute to adaptive evolution are well rehearsed.[26]

Developmentally induced or constrained variation can be a major contributor to phenotypic evolution (Uller et al., 2020; Hu et al., 2020; Jablonka, 2017). Selectable, transgenerationally stable variants can arise and be sustained by development without corresponding changes in genotype space. The important point for our purposes is that developmental plasticity introduces a radical form of genetic underdetermination; genes underdetermine form.

18.5.3 Innovation

The only process in genotype space that introduces genuine evolutionary novelties is random genetic mutation.[27] Yet, genetic mutation radically underdetermines the source of evolutionary novelties.

Much novelty in evolution thus appears to be possible without the need to evolve novel genes, pathways, or cell types. Exactly why, how, and when evolutionary innovations occur and unfold the way they do has thus mostly eluded conventional molecular, population, and quantitative genetic approaches toward understanding the evolutionary process (Sultan et al., 2022: 4).

Here too, the plasticity of development plays an ineliminable role. Plasticity serves to integrate and accommodate the adaptive response of organismal subsystems (Pfennig et al., 2007; Pfennig et al., 2010).

Moczek and colleagues stress that developmental systems bias the production of novel forms (Laland et al., 2015; Hu et al., 2020; Uller et al., 2020; Salazar-Cuidad, 2021).

Just as developmental systems can produce phenotypic adaptive novelties, so too can organisms' ecological relations, and the manipulation of their environments (Gilbert & Epel, 2009; Sultan, 2015). Moreover, these novelties are frequently adaptively biased (Hu et al., 2020; Uller et al., 2020). Furthermore, they can be intergenerationally stable. Once again, we encounter a radical form of underdetermination. Genetic mutation does not adequately account for the source of evolutionary novelty.

18.5.4 Adaptive Bias

Modern synthesis orthodoxy holds that differential retention of genes is the only process that makes evolution adaptive: "allele frequency traits caused by natural selection is the credible process underlying evolution of adaptive organismal traits" (Charlesworth et al., 2017). In this view, selection, and only selection, causes biased changes in genotype space. This, in turn, is the exclusive source of adaptive evolutionary changes of form. The thrust of the foregoing discussion, however, suggests that genuinely adaptive evolutionary change may be implemented by any number of biological processes. Epigenetic phenomena (Sultan et al., 2022), developmental plasticity (Newman, 2022), and niche construction (Odling-Smee et al., 2003) are all sources of adaptive bias of form.[28] The upshot, again, is that the dynamics of gene space radically underdetermine the adaptive dynamics of phenotypic evolution.

There is a common theme here: underdetermination. In each of these cases, the evolutionary phenomena of phenotype space are underdetermined by the processes of genotype space that were traditionally thought to explain them. These claims are not particularly unusual or new; they are well-established, legion, and growing by the day. Taken as a whole, they recommend more than a piecemeal mend-and-make-do. They suggest a fundamental flaw in the general picture of evolution.[29] The foundationalist gene-centered modern synthesis approach to evolution fails systematically to account for the phenomena of adaptive phenotypic evolution. It is tempting to suppose that it fails for the same reason that foundationalist theories of mental phenomena do. In each case it is their foundationalist structure—the picture that holds us captive—that leads us astray.

18.6 Contact Phenomena

Where do foundationalist theories of mental and biological phenomena go wrong? Charles Taylor (2006) and Taylor and Dreyfus (2015) locate the fault of foundational theories of human mental phenomena in a mischaracterization of what is to be explained. Knowing, thinking, perceiving, and the like, are not in the first instance merely the external manifestations of internal mental processes. They are the result of the agent's intimate engagement with her world. Human agents are active, adaptive entities, in constant commerce with their conditions of existence. The transaction consists in the way that agents engage with their settings in pursuit of their goals. Agents respond to their conditions by exploiting them, ameliorating them, accentuating them, avoiding them, adjusting them, constructing them, and changing themselves in accordance with their exigencies. As a consequence, the world that humans inhabit is, on account of their own activities, one that is imbued with significance for the pursuit of their goals. Humans operate in "a world organized in terms of their needs, interests, and bodily capacities without their minds needing to impose a meaning on a meaningless Given" (Dreyfus, 2005: 49). We do not think about the world by representing it; we think about the world by living in it. "My first understanding of reality is not a picture I am forming of it, but the sense given to a continuing transaction with it" (Taylor, 2006: 47). What is to be explained is the agent's experience of, and response to, the world not as it is in itself, but as it is transduced by her purposive transaction with it.

Gareth Matthews provides an illustrative example in Merleau-Ponty's understanding of what it is to perceive the world.

The 'world' for us is more than simply the spatial container of our existence. It is the sphere of our lives as active, purposive beings. . . . The world is the place we 'inhabit', rather than simply a set of objects that we represent to ourselves in a purely detached way (Matthews, 2002: 49).

This relation of "inhabitation" between human agents and the world is crucial to understanding thought. To perceive the world is to respond to the ways its features strike us given our goals and capacities (Nöe, 2004; Thompson, 1995). To know something about the world is to have a capacity to exploit it. To have a belief about the world is to be poised to act in ways that take into account its implications. Perception, thought, knowledge, and so on are modes of transactional contact that we as agents have with the world (Varela, Thompson, & Rosch, 2016). In a word, they are "contact" phenomena (Taylor & Dreyfus, 2015).

Foundationalisms, as we have seen, seek to account for these phenomena by sundering the internal mental realm from the external world, and then by adverting to "internal" mental processes only. In the process, they lose the contact phenomena. The problem for foundationalist theories is that contact phenomena are not internal, nor are they external. They are wholly determined neither by the inner workings of agents, nor by their external environments. They are located at the interface of the agent and the conditions the agent encounters. They are products of the agent's purposive transactions with its environment. In the course of these transactions, agents transduce the features of their conditions of existences into *saliences*, opportunities for or impediments to the pursuit of their goals. These contact phenomena are not represented or countenanced in any way in the foundationalist approach to understanding human minds. It is for this reason that foundationalist theories of human mental phenomena underdetermine the features they seek to explain.

18.7 The Active Role of Organisms

Similar considerations apply to foundationalist approaches to evolution. The foregoing examples of the processes that drive evolution in phenotype space—extragenetic modes of inheritance, developmental plasticity, evolutionary novelty, and adaptive bias—properly construed, are not the consequence of internal genetic processes. They too are contact phenomena. The gene-centered modern synthesis view of evolution fails to give an adequate account of these processes simply because as a foundationalist theory it does not recognize evolutionary contact phenomena.

Organisms are purposive systems. They respond to their conditions as impediments to, or as propitious for, the pursuit of their goals. In the process, they transduce features of their environments into saliences; a bacterium may experience O_2 as a toxin or as a metabolic requirement depending on its goals and capacities.[30] In this way, organisms experience their environments as relevant to their purposes. For organisms, the world is "more than the spatial container" of their existence. It is a system of opportunities, challenges, benefits, and threats to the pursuit of their goals: in a word, "affordances."[31] This engagement of organisms with the demands of their conditions of existence—their affordances—underwrites their successful development in the face of uncertainty, the stability of form across generations, evolutionary novelty, and adaptively biased change. We fail to understand the component processes of evolution unless we understand them as contact phenomena. Inheritance, development, innovation, and adaptive bias arise from organisms' making of, and response to, their system of affordances. These processes happen in the way they do because of the active role of organisms in constituting them.

18.7.1 Inheritance

As we have seen, there is more to inheritance than genetic inheritance. Modes of extra-genetic inheritance are a consequence of organisms' interacting with—marshaling—the resources of their genomes, their developmental systems, and their environments in ways that secure the transgenerational persistence of form. These effects are well documented. The role may involve epigenetic marks that respond to environmental conditions.

Consistent components of complex traits, such as those linked to human stature/height, fertility, and food metabolism or to hereditary defects, have been shown to respond to environmental or nutritional condition and to be epigenetically inherited (Trerotola et al., 2015: 1)

This is not merely a phenomenon of animals; it is also common in plants.

Stressful parental . . . environments can dramatically influence expression of traits in offspring, in some cases resulting in phenotypes that are adaptive to the inducing stress (Herman et al., 2016).

It is widely evident in eukaryotic microbes (Weiner & Katz, 2021). It may involve the transmission of immune responses:

the transfer of parental immunological experience to enhance the offspring immune defence is present in both vertebrates and invertebrates, and can be inherited for multiple generations (Linn & Spagopoulou, 2018: 205).

It may further involve the transmission of learned behaviors, microbiota, and environmental structures.

In short, inheritance, construed as transgenerational constancy of phenotype, arises from the active interaction of organisms with their experienced conditions—including genetic, epigenetic, environmental—in securing the stability of form across generations. It is a contact phenomenon.

18.7.2 Development

Similar considerations apply to development. The plasticity of organismal development is ubiquitous. Organisms respond to their conditions by directing their development toward the production of stable, advantageous phenotypes (Newman, 2022). This may come about by co-opting old processes for new, by regulating the activities of gene regulatory networks, by regulating the transcription of genes into proteins. In fact, plasticity-led evolution is increasingly drawing the attention of mainstream evolutionary researchers (Levis & Pfennig, 2020). Plasticity is manifest at practically every level of biological organization. It is evident not just in development, as traditionally conceived, but also in protein synthesis (Steward et al., 2020), metabolism, immunity, and endocrine function, among others. Alternative splicing appears to be a mechanism by which organisms enlist their protein transcription machinery in response to their conditions.

Cells respond to signals from the environment through changes in gene expression and protein activity, both of which play important roles in the regulation of alternative splicing (Edwalds-Gilbert, 2010).

Thus, organisms have the capacity to regulate the processes by which they build themselves, and by which they interact with their conditions.

Similar considerations apply to innovation. As plasticity is the principal source of phenotypic stability *and novelty*, new evolutionary phenotypes tend to appear first from the plastic responses of organisms to their conditions of existence (West-Eberhard, 2003). They may only later be routinized as genetic changes.

18.7.3   Adaptive Bias

Similarly, the adaptive bias of form is not the sole province of some external force—selection—that pushes genes around genotype space (Walsh, 2019). Rather, it is a consequence of the organismal processes that regulate inheritance, development, and the origin of novelties. The source of the adaptive bias of evolution is to be found in the contribution of organisms as adaptive, responsive systems. In each of these cases the gap between genes and what they are called upon to explain is filled by the active role of organisms in responding to the salience of their conditions—genetic, epigenetic, environmental—to the pursuit of their goals (Sultan et al., 2022).

To say that organisms are active participants in inheritance, development, innovation, and adaptation is to say that they are *natural agents*. This notion of a natural agent requires some care. Organisms are not cognitive agents.[32] They do not think about their worlds or themselves; they do not make conscious decisions, or entertain beliefs and desires, or act under the guise of the good. These expressly mental capacities are not necessary for agency. Rather, organisms are agents in the sense that they pursue purposes, respond adaptively to their circumstances, maintain stability through their activities, alter their relations to their conditions of existence, alter their conditions, and innovate. These are all manifest properties of living things, and none of them requires cognition, conation, or volition.[33] Nevertheless, the agential properties of organisms are implicated in producing the component processes of evolution.

The lacunae in modern synthesis foundationalist approaches to the component processes of evolution—inheritance, development, innovation and adaptive bias—are filled by the active, purposive participation of organisms. Organisms make themselves; in the process they make *for* themselves, and respond to, a system of threats and opportunities for the pursuit of their purposes. Inheritance, development, innovation, and adaptive bias ensue. The consequence of all this purposive activity, seen from the perspective of populations (or lineages) is what we know as adaptive evolution. This is simply to say that organisms are agents, and their agency is what makes adaptive evolution happen.

18.8   What Kind of Process Is Evolution?

Two pictures of evolution emerge, the foundational and the agential. In the foundational picture, evolution takes place in two discrete realms: genotype space and phenotype space. The phenomena to be explained (in adaptive evolution at least) are phenomena of phenotype space. Yet, they are explained by appeal to the processes that occur within genotype space. Genotype space has theoretical priority because the canonical units (genes) and processes that create phenotype space (replication, recombination, translation, mutation) reside wholly in genotype space. Genes are primitive units of evolution. They are molecular entities. Their status as fundamental units of evolution is grounded in their molecular properties; they are chemical givens. Because of their molecular properties, they exert a distinctive control over the production of phenotype, over the inheritance of form, and over the origin of novelties. Evolution happens because of what genes do.

We have reasons to believe that evolutionary foundationalism of this sort fails to meet the challenge of explaining adaptive evolution. The diagnostic mark of its inadequacy is the presence of explanatory gaps, the inability of gene-centered approaches to account fully for the component processes of evolution: inheritance, development, innovation, or adaptive bias (Sultan et al., 2022). One explanation of these deficiencies is that evolutionary foundationalism has mischaracterized its explanatory target. Adaptive evolution is not the dynamics of genotype space. It is not primarily a process in which a certain kind of molecule is copied, translated into form, mutated, and differentially preserved. Evolution is not a molecular phenomenon; there is no chemical given.

The deficiencies of evolutionary foundationalism at least provide some impetus for entertaining an alternative. In the agential view, evolution is a fundamentally organismal—rather than genetic—phenomenon. Organisms are agents; they are naturally purposive systems. They are self-building, self-synthesizing, self-regulating systems. They are robust adaptive systems that respond to their conditions in ways that secure their own persistence and well-functioning. They do this by altering their structures, their activities, and their circumstances in ways that promote the attainment of their goals. In doing so, they experience their worlds as salient to the attainment of those goals. Organisms respond to their conditions by regulating their development, by innovating, by routinizing novel developmental processes, and by securing the conditions for the intergenerational stability of form. In this way, adaptive evolution is a consequence of organisms' purposive engagement with their conditions of existence. Organisms make contact with the world, and in doing so they enact evolution: "the organism cannot be regarded as simply the passive object of autonomous internal and external forces; it is also the subject of its own evolution" (Lewontin, 1985: 89). Adaptive evolution is an agential phenomenon. As such it calls for an agential theory, as opposed to a foundationalist theory.

The agential view has the virtue of explaining why modern synthesis evolutionary foundationalism falls so drastically short of its objectives. According to evolutionary foundationalism, organisms aren't active or purposive. At least insofar as they are, it isn't particularly significant for evolution. More importantly, there are no agents and no experienced environments in the ontology of evolutionary foundationalism. There are genes, internal processes that operate over genes (replication, translation, mutation), the phenotypes they build, and there are external environments. Set against the austere ontology of the gene's-eye modern synthesis view, the explanation of evolution would appear to require a richer ontology of *contact phenomena*. These are neither internal states of the organism, nor external states of the environment. They are transactional, constituted by the purposive engagement of organisms with their conditions. The component processes of evolution—inheritance, development, innovation, biased change—are all consequences of organisms' purposive transactions with their conditions of existence. If modern synthesis evolutionary foundationalism cannot countenance these sorts of contact phenomena, then it lacks the resources to explain evolution.[34] No wonder, then, that gaps appear in its attempts to explain the phenomena. This, I think, is the cardinal lesson to be drawn from the welter of recent challenges to the modern synthesis: It's the wrong kind of theory.

18.9   Conclusion

The agency view is not intended as a polemic. It is offered here as a positive, albeit inchoate, alternative to the gene's-eye view of evolution. It points to an understanding of the ineliminable role of organisms in adaptive evolution. In the agential view, organisms enact evolution (Thompson, 2017). That is to say, adaptive evolution happens because of what organisms do.[35] What they do is engage with their conditions of existence, exploit them, respond to them, ameliorate them, and change them, and they change themselves in response to their conditions. They enlist and regulate their genetic, epigenetic, developmental, immunological, and behavioral resources in pursuit of their basic biological goals. We lack a general term for this complex, disparate suite of organismal purposive activities; perhaps the "struggle for life" will do. As a result of this struggle for life, lineages of organisms change over time and this change manifests as adaptive evolution. This is not a radical view, or a new one. It is a venerable one, and as old as the theory of evolution itself: "all these things follow inevitably from the *struggle for life*" (Darwin, 1859/1968: 115, emphasis added).[36]

Acknowledgments

I thank Peter Corning for very helpful comments on the penultimate draft and Fermín Fulda for a helpful discussion on an earlier version. I'm grateful to Tobias Uller and Stuart Newman for sharing as yet unpublished works with me.

Notes

1. There is, of course, a lively debate about what should and should not be included in the modern synthesis. Many authors (Corning, 2022; Noble & Noble, ch. 12 in this volume) point out that at its inception the modern synthesis was a much more inclusive body of thought than it later became, especially in the hands of Dobzhansky (1937) and Huxley (1942) who coined the term. It should be remembered, however, that even these authors defined evolution in genetic terms. I intend my characterization of the core of the modern synthesis to apply to all its versions.

2. Various essays collected in Huneman and Walsh (2017) offer a range of opinions on the breadth and severity of these challenges.

3. Notable detractors from the gene's-eye view include Noble (2006, 2016) and Corning (2020).

4. They are an instance of what Winther (2011) calls "part-whole science" and what Cartwright (1999) calls "the analytic method."

5. Agent theories of evolution are discussed by Walsh (2015, 2018). See Jäger (2021) for a compelling and comprehensive defense.

6. Rules of reason, for example, or rules for abstracting sensory content from perceptual experiences (e.g., Locke, 1975, or James, 1890). See Winther (2014) for a discussion of James on abstraction and Ayers (1991) for Locke on perception.

7. Indeed, the philosophical term *given* was coined by Sellars (1956) for these very things.

8. As Merleau-Ponty (1962/1945:3) describes the position: "sensation should be sought on the hither side of any qualified content, since red and blue, in order to be distinguishable as two colours, must already form some picture before me."

9. Just as Descartes' *cogito* gave us a fundamental unit of knowledge, sense data give us a fundamental unit of sensory experience, about which we cannot be wrong.

10. Chirimuuta's (2020) discussion of the "abstraction argument" offers a similar account of computational theories of cognition. See also Cantwell Smith's (2019) discussion of reasoning as reckoning.

11. This is to say that the computational processes of the mind do not confer semantic properties on them. Perhaps, as the teleosemantics tradition has it, they derive their contents from their evolutionary functions (Millikan, 1984; Shea, 2018), or by their histories of recruitment (Dretske, 1995), or by their causal relations to the world (Fodor, 1987).

12. I am merely pointing out a common shortfall here, or a perduring philosophical challenge. It is a further philosophical question whether we should, or could, do anything about it.

13. Quite how to characterize that relation is itself a pressing philosophical problem (see Williamson, 1995).

14. Perception takes sensory input from the outside world and then "throws the world away," in Andy Clark's (1997) colorful phrase.

15. The recognition of this has spawned all manner of philosophical externalisms—reliabilism in epistemology, direct reference and direct perception theories, externalism in intentional and semantic content, externalism in motivating reasons—in each of which features of the external environment figure in the realization of the mental state being explained. I shall be concentrating on another strategy later in this discussion.

16. Witness phenomena like the Flynn effect in which IQ scores rise by an average of 13.8 points each generation (Trahan et al., 2014), or the increase in body mass index (BMI) over generations in recent years in Western societies. These are clearly biased, transgenerational changes in phenotype space, but they are not generally considered *evolutionary* changes.

17. In just the same way that the channels that mediate between the world and the mind are in foundationalist approaches to human agential phenomena.

18. They are occasionally "moved" by random mutation, of course.

19. I propose to read *development* here very broadly. In this sense all organisms develop as they move through their life histories from inception to reproduction to death.

20. For a discussion of some transgenerational consequences of maternally induced immune activation, see Pollok and Weber-Stadlbauer (2020).

21. For in-depth discussion of the evolutionary significance (or otherwise) of epigenetic inheritance, see Gapp (2020) and Sarkies (2020). See especially Jablonka (2017).

22. Uller and Helanterä (2017) argue that inheritance should be considered a developmental phenomenon.

23. Admittedly, Bonduriansky and Day (2018) seek to secure special theoretical privilege for genetic inheritance in explaining evolutionary dynamics.

24. Levis and Pfennig (2020) survey the evidence for plasticity-led evolution.

25. I thank Armin Moczek for bringing this to my attention.

26. See, for example, Kirschner and Gerhard (2010), Wagner (2012), West-Eberhard (2003), and Pfennig et al. (2010).

27. New phenotypes may also arise from new combinations of genes.

28. See Weiner and Katz (2021) for a discussion of the way in which epigenetic mechanisms act as drivers of adaptation and diversity in the evolution of microbial eukaryotes.

29. The urgency of the need for an alternative comes across clearly, for example, in Jablonka (2017).

30. In the sense that aerobes and anaerobes have different goals and capacities.

31. I take the term from Gibson (1979). For an extended discussion of the organisms' world as affordances, see Walsh (2013).

32. Except for the cognitive ones (like us).

33. See Fulda (2017) for an extensive discussion of noncognitive agency.

34. See Uller (chapter 17 in this volume) for a parallel argument for adopting the agential perspective in evolution.

35. This viewpoint lends special relevance to West-Eberhard's dictum that "genes are followers in evolution, not leaders" (2003).

36. I sketched early parts of this paper in Jardin de Luxembourg, Paris, and more or less finished it in Stadpark, Bochum. I thank the custodians of these public spaces for providing these serene places conducive to thought. Most of this work was done during a fellowship at the Institute for Advanced Study, Paris. I thank the IEA for the opportunity. I was greatly helped by audiences at the Sciences in Context seminar series, CRI, Paris, and at KLI, Vienna. Fermín Fulda offered helpful criticisms of an early draft. I am grateful for the invitation from Peter Corning and Dick Vane-Wright, and for their patience. I would also like to thank Tobias Uller and Stuart Newman for sharing their contributions to this volume with me.

References

Ågren, A. (2021). *The gene's-eye view of evolution*. Oxford: Oxford University Press.

Ayers, M. (1991). *Locke: Epistemology and ontology* (2 vols.). London: Routledge.

Bonduriansky, R., & Day, T. (2009). Nongenetic inheritance and its evolutionary implications. *Annual Review of Ecology, Evolution, and Systematics*, 40, 103–125.

Bonduriansky, R., & Day, T. (2018). *Extended heredity*. Princeton, NJ: Princeton University Press.

Cantwell Smith, B. (2019). *The promise of artificial intelligence: Reckoning and judgement*. Cambridge, MA: MIT Press.

Cartwright, N. (1999). *The dappled world. A study of the boundaries of science*. Cambridge: Cambridge University Press.

Chirimuuta, M. (2020). The reflex machine. *Perspectives on Science*, 28 (3), 421–457.

Charlesworth, D., Charlesworth, B., & Barton, N. (2017). The sources of adaptive variation. *Proceedings of the Royal Society B, Biological Sciences*, 284, 20162864. https://doi.org/10.1098/rspb.2016.2864

Clark, A. (1997). *Being there: Putting brain, body, and world together again*. Cambridge, MA: Bradford Books.

Corning, P. (2020). Beyond the modern synthesis: A framework for a more inclusive biological synthesis. *Progress in Biophysics and Molecular Biology*, 153, 5–12.

Corning, P. (2023). Teleonomy in evolution: "The ghost in the machine" (ch. 2 in this volume).

Danchin, E., Pocheville, A., & Huneman, P. (2019). Early-in-life effects and heredity: Reconciling neo-Darwinism with neo-Lamarckism under the banner of the inclusive evolutionary synthesis. *Philosophical Transactions of the Royal Society B*, 374 (1770), 20180113.

Darwin, C. (1859/1968). *The origin of species*. London: Penguin.

Dawkins, R. (1982). *The extended phenotype*. Oxford: Oxford University Press.

Dawkins, R. (1999). *The extended phenotype: The long reach of the gene* (2d ed.). Oxford: Oxford University Press.

Descartes, R. (1641/1998). Meditations on first philosophy. In D. Clark (Ed.), *Meditations and other metaphysical writings* (Desmond M. Clarke, trans.) (pp. 113–145). London: Penguin.

Dobzhansky, T. (1937). *Genetics and the origin of species*. New York: Columbia University Press.

Dretske, F. (1995). *Naturalizing the mind*. Cambridge, MA: MIT Press.

Dreyfus, H. (2005). Overcoming the myth of the mental. *Proceedings and Addresses of the American Philosophical Association*, 79, 47–65.

Dreyfus, H., & Taylor, C. (2015). *Retrieving realism*. Cambridge, MA: Harvard University Press.

Edwalds-Gilbert, G. (2010). Regulation of mRNA splicing by signal transduction. *Nature Education*, 3 (9), 43.

Farkas, K. (2008). Semantic internalism and externalism. In E. Lepres & B. Smith (Eds.), *The Oxford handbook of philosophy of language* (pp. 323–340). Oxford: Oxford University Press.

Finlay, S., & Schroeder, N. (2017). Reasons for action: Internal vs. external. In E. N. Zalta (Ed.), *The Stanford encyclopedia of philosophy*. https://plato.stanford.edu/archives/fall2017/entries/reasons-internal-external

Fodor, J. (1987). *Psychosemantics*. Cambridge, MA: MIT Press.

Fulda, F. (2017). Natural agency: The case of bacterial cognition. *Journal of the American Philosophical Association*, 3, 69–90.

Gapp, K. (2020). Unconventional forms of inheritance. *Seminars in Cell and Developmental Biology*, 97, 84–85.

Gibson, J. J. (1979). *The ecological approach to visual perception*. London: Routledge.

Gilbert, S., & Epel, D. (2009). *Ecological developmental biology*. Sunderland, MA: Sinauer.

Hadfield, G. (2021). Sense data. In E. N. Zalta (Ed.), *Stanford encyclopedia of philosophy*. https://plato.stanford .edu/archives/fall2021/entries/sense-data

Heard, E., & Martienssen, R. A. (2014). Transgenerational epigenetic inheritance: Myths and mechanisms. *Cell*, 157, 95–109.

Herman, J. J., Sultan, S., Horgan-Kybelski, T., & Riggs, C. (2016). Adaptive transgenerational plasticity in an annual plant: Grandparental and parental drought stress enhance performance of seedlings in dry soil. *Integrative and Comparative Biology*, 52, 77–88.

Hu, T., Linz, D., Parker, E. S., & Moczek, A. (2020). Developmental bias in horned dung beetles and its contributions to innovation, adaptation, and resilience. *Evolution & Development*, 22 (1–2), 165–180.

Huneman, P., & Walsh, D. M. (2017). *Challenging the modern synthesis: Adaptation, development, inheritance*. Oxford, UK: Oxford University Press.

Huxley, T. (1942). *Evolution: The modern synthesis*. New York: Harper & Row.

Jablonka, E. (2017). The evolutionary implications of epigenetic inheritance. *Interface Focus*, 7, 20160135.

Jablonka, E., & Avital, E. (2006). Animal innovation: The origins and effects of new learned behaviours. *Biology and Philosophy*, 21, 135–141.

Jablonka, E., & Lamb, M. (2010). Transgenerational epigenetic inheritance. In M. Pigliucci & G. Müller (Eds.), *Evolution: The extended synthesis* (pp. 137–174). Cambridge, MA: MIT Press.

Jäger, J. (2021). The fourth perspective: Evolution and organismal agency. OSF Preprints. February 26. https:// doi.org/10.31219/osf.io/2g7fh

James, W. (1890/2007). *Principles of psychology*. Vol. 1. 1890. New York: Cosimo.

Kimura, M. (1974). *The neutral theory of evolution*. Cambridge: Cambridge University Press.

Kirschner, M., & Gerhard, J. C. (2010). Facilitated variation. In M. Pigliucci & G. B. Muller (Eds.), *Evolution: The extended synthesis* (pp. 253–280). Cambridge, MA: MIT Press.

Lacal, I., & Ventura, R. (2018). Epigenetic inheritance: Concepts, mechanisms and perspectives. *Frontiers in Molecular Neuroscience*. https://doi.org/10.3389/fnmol.2018.00292

Laland, K. N., Uller, T., Feldman, M., Sterelny, K., Müller, G. B., Moczek, A. P., Jablonka, E., & Odling-Smee, J. (2014). Does evolutionary theory need a rethink? (Yes, urgently.). *Nature*, 514 (7521), 161–164. https://doi.org /10.1038/514161a

Laland, K. N., Uller, T., Feldman, M., Sterelny, K., Müller, G. B., Moczek, A. P., Jablonka, E., & Odling-Smee, J. (2015). The extended evolutionary synthesis: Its structure, assumptions and predictions. *Proceedings of the Royal Society B, Biological Sciences*, 282 (1813), 20151019.

Ledón-Rettig, C. C., Pfennig, D. W., & Nascone-Yoder, N. (2008). Ancestral variation and the potential for genetic accommodation in larval amphibians: Implications for the evolution of novel feeding strategies. *Evolution & Development*, 10, 316–325.

Levis, N., & Pfennig, D. (2020). Plasticity-led evolution: A survey of developmental mechanisms and empirical tests. *Evolution & Development*, 22, 71–87.

Lewontin, R. C. (1974). *The genetic basis of evolutionary change*. New York: Columbia University Press.

Lewontin, R. C. (1985). The organism as subject and object of evolution. In R. Levins & R. C. Lewontin (Eds.), *The dialectical biologist* (pp. 85–106). Cambridge, MA: Harvard University Press.

Lewontin, R. C. (2000). *The triple helix: Gene, organism, and environment*. Cambridge, MA: Harvard University Press.

Linn, M., & Spagopoulou, F. (2018). Evolutionary consequences of epigenetic inheritance. *Heredity*, 121, 205–209.

Locke, J. (1689/1975). *An essay concerning human understanding* (P. H. Nidditch, Ed.). London: Clarendon.

Matthews, G. (2002). *The philosophy of Merleau-Ponty*. London: Routledge.

Merleau-Ponty, M. (1962/1945). *Phenomenology of perceptions* (C. Smith, Trans.). London: Routledge.

Millikan, R. (1984). *Language, thought, and other biological categories*. Cambridge, MA: MIT Press.

Newman, S. (2023). "Form, function, agency: Sources of natural purpose in animal evolution" (ch. 11 in this volume).

Noble, D. (2006). *The music of life: Biology beyond the genes*. Oxford, UK: Oxford University Press.

Noble, D. (2016). *Dance to the tune of life: Biological relativity*. Cambridge: Cambridge University Press.

Noble, D., & Noble, R. (2023). "How purposive agency became banned from evolutionary biology" (ch. 12 in this volume).

Nöe, A. (2004). *Action in perception*. Cambridge, MA: MIT Press.

Odling-Smee, F. J., Laland, K., & Feldman, M. (2003). *Niche construction: The neglected process in evolution*. Princeton, NJ: Princeton University Press.

Oyama, S., P. Griffiths, & R. Gray. (2000). *Cycles of contingency*. Cambridge, MA: MIT Press.

Pfennig, D. W., Wund, M. A., Schlichting, C., Snell-Rood, E. C., Cruikshank, T. C., Schlichting, C., & Moczek, A. (2010). Phenotypic plasticity's impacts on diversification and speciation. *Trends in Ecology and Evolution*, 25, 459–467.

Pfennig, D. W., Wund, M. A., Snell-Rood, E. C., Cruickshank, T., Ciliberti, S., Martin, O. C., & Wagner, A. (2007). Innovation and robustness in complex regulatory gene networks. *Proceedings of the National Academy of Sciences*, 104 (34), 13591–13596.

Pigliucci, M. (2009). An extended synthesis for evolutionary biology. *Annals of the New York Academy of Sciences*, 1168, 218–228.

Pollok, D., & Weber-Stadlbauer, U. (2020). Transgenerational consequences of maternal immune activation. *Seminars in Cell and Developmental Biology*, 97, 181–188.

Putnam, H. (1975). The meaning of "meaning." *Mind, language and reality: Philosophical papers*, Vol. 1. Cambridge: Cambridge University Press.

Putnam. H. (1982). Why there isn't a ready-made world. *Synthese*, 51, 141–167.

Rowlands, M., Lau, J., & Deutsch, M. (2020). Externalism about the mind. In E. N. Zalta (Ed.), *The Stanford encyclopedia of philosophy*. https://plato.stanford.edu/archives/win2020/entries/content-externalism/

Salazar-Ciudad, I. (2021). Why call it developmental bias when it is just development? *Biology Direct*, 16. https://biologydirect.biomedcentral.com/articles/10.1186/s13062-020-00289-w

Sarkies, P. (2020). Molecular mechanisms of epigenetic inheritance: Possible evolutionary implications. *Seminars in Cell and Developmental Biology*, 97, 106–115.

Sellars, W. (1956). Empiricism and philosophy of mind. *Minnesota Studies in the Philosophy of Science*, 1, 253–329.

Shea, N. (2018). *Representation in cognitive science*. Oxford: Oxford University Press.

Standen, E. M., Du, T. Y., & Larsson, H. C. E. (2014). Developmental plasticity and the origin of tetrapods. *Nature*, 513, 54–58.

Steward, R. A., de Jong, M. A., Ostra, V., & Wheat, C. W. (2020). Alternative splicing in seasonal plasticity and the potential for adaptation to environmental change. *Nature Communications*, 13, 55.

Sultan, S. E., Moczek, A., & Walsh, D. (2022). Bridging the explanatory gaps: What can we learn from a biological agency perspective? *BioEssays*, 44, 2100185.

Sultan, S. E. (2015). *Organism and environment: Ecological development, niche construction and adaptation*. Oxford: Oxford University Press.

Taylor, C. (2006). Merleau-Ponty and the epistemological picture. In T. Carman (Ed.), *Cambridge companion to Merleau-Ponty* (pp. 1–25). Cambridge: Cambridge University Press.

Taylor, C., & Dreyfus, H. (2015). *Retrieving realism*. Cambridge, MA: Harvard University Press.

Thompson, E. (1995). *Colour vision: A study in cognitive science and philosophy of science*. London: Routledge.

Thompson, E. (2017). *Mind and world*. Oxford: Oxford University Press.

Trahan, L., Stuebing, K., Hiscock, M. K., & Fletcher, J. M. (2014). The Flynn effect: A meta-analysis. *Psychological Bulletin*, 140 (5), 1332–1360.

Trerotola, M., Relli, V., Simeone, P., & Alberti, S. (2015). Epigenetic inheritance and the missing heritability. *Human Genomics*, 9 (1), 17. https://www.ncbi.nlm.nih.gov/pmc/articles/PMC4517414

Uller, T. (2022). "Agency, goal orientation, and evolutionary explanations" (ch. 17 in this volume).

Uller, T., & Helanterä, H. (2017). Heredity and evolutionary theory. In P. Huneman & D. M. Walsh (Eds.), *Challenging the modern synthesis* (pp. 280–316). Oxford: Oxford University Press.

Uller, T., Moczeck, A, Watson R. A., Brakefield, P., & Laland, K. (2018). Developmental bias and evolution: A regulatory network perspective. *Genetics*, 209 (4), 949–966.

Uller, T., Feiner, N., Radersma, R., Jackson, I. S. C., & Rago, A. (2020). Developmental plasticity and evolutionary explanations. *Evolution & Development*, 22, 47–55.

Varela, F., Thompson, E., & Rosch, E. (2016). *The embodied mind: Cognitive science and human experience*. Cambridge, MA: MIT Press.

Wagner, A. (2011). *The origin of evolutionary innovations: A theory of transformative change in living systems*. Oxford: Oxford University Press.

Wagner, A. (2012). The role of robustness in phenotypic adaptation and innovation. *Proceedings of the Royal Society B, Biological Sciences, 279,* 1249–1258.

Walsh, D. M. (2013). Adaptation and the affordance landscape. In G. Barker, E. Desjardins, & T. Pearce (Eds.), *Entangled life* (pp. 213–236). Dordrecht: Springer.

Walsh, D. M. (2015). *Organisms, agency, and evolution.* Cambridge: Cambridge University Press.

Walsh, D. M. (2018). Objectcy and agency. In D. Nicholson & J. Dupré (Eds.), *Everything flows: Toward a process ontology for biology* (pp. 167–185). Oxford: Oxford University Press.

Walsh, D. M. (2019). The paradox of population thinking. In T. Uller & K. Laland (Eds.), *Evolutionary causation: Biological and philosophical reflections* (pp. 227–246). Cambridge, MA: MIT Press.

West-Eberhard, M. J. (2003). *Developmental plasticity and evolution.* Oxford: Oxford University Press.

Weiner, A. K. M., & Katz, L. A. (2021). Epigenetics as driver of adaptation and diversification in microbial eukaryotes. *Frontiers in Genetics, 12,* 642220. https://doi.org/10.3389/fgene.2021.642220

Williamson, T. (1995). Is knowing a state of mind? *Mind, 104,* 533–565.

Winther, R. (2011). Part-whole science. *Synthese, 398* (178), 397–427.

Winther, R. (2014). James and Dewey on abstraction. *The Pluralist, 9,* 1–28.

Wittgenstein, L. (1953). *Philosophical investigations.* (E. Anscombe, Trans.). Oxford: Oxford University Press.

Contributors

František Baluška, University of Bonn
Peter A. Corning, Institute for the Study of Complex Systems
Dominik Deffner, Max Planck Institute for Human Development
Simona Ginsburg, The Open University of Israel
Francis Heylighen, Vrije Universiteit Brussel
Abir U. Igamberdiev, Memorial University of Newfoundland
Eva Jablonka, Tel Aviv University
Stuart A. Kauffman, Institute for Systems Biology
Kalevi Kull, University of Tartu
Michael Levin, Tufts University
William B. Miller Jr., Paradise Valley
Stuart Newman, New York Medical College
Denis Noble, University of Oxford
Raymond Noble, University College London
Samir Okasha, University of Bristol
Robert Pascal, Aix-Marseille Université
Addy Pross, Ben-Gurion University of the Negev
Arthur S. Reber, University of British Columbia
Andrea Roli, Campus of Cesena
James A. Shapiro, University of Chicago
Anthony Trewavas, University of Edinburgh
Tobias Uller, Lund University
Richard I. Vane-Wright, Natural History Museum London
Denis M. Walsh, University of Toronto

Index

Note: Figures, Tables, and Boxes are indicated by italicized page numbers. Lower case n followed by a number indicates an endnote.

Aboutness (intentionality), in consciousness, 123
Abram, David, 84–85
Abstract Kantian wholes, 153–154
Actions, ontology of, 86–87, 90–91
Actor-network theory, 86
Adaptation
 biological intelligence and behavior of, 313–314, 318
 Darwinian preadaptation, 147
 definition of, 16, 267
 features of, 240
Adaptedness of organisms, 341–342
Adaptive bias
 development as source of, 332
 from organism interactions, 355
 underdetermination and, 352
Adaptive choice, animals and, 128
Affordances, 336n17
 complexity and, 156
 in evolution of biosphere, 146–147
 function and, 164
 plants and, 312
Agency, 3. *See also* Agents; Purposive agency; Relational agency
 actions and, 90–91
 biological, 275, 326, *328*
 in biology, 18, 25n5
 challenges and, 89–90
 cognition and, 25n5
 in consciousness, 123–124
 continuum of, 177–179
 environmental conditions and, 89–90
 in evolution, 6, 156
 free-will and, 135, 276
 GDB and, 134–136, 176–177
 metazoan animals and cell, 207–211, 215
 morality and, 135
 neo-Darwinism and, 81, 275–276
 niche construction and, 213–214
 objects compared to, 83

reactions, condition-action rules and, 88–89
 spontaneous activity and, 134–135
 super-, 97–98
 teleonomy and, 88–89, 156
Agent-based modeling, 86
Agential concepts
 epistemic functions of, 325
 evolutionary change explained with, 332–335
 for evolutionary theories, 326–330, 336n7
 explanatory standards set by, 330–332
 foundationalism compared to, 342–343, 356–357
 gene's-eye view in, 326–327
 metaphorical nature of, 329–330
 object theories compared to, 333–334
Agents. *See also* Agency; Organisms
 animism and, 83
 challenges and, 89–90
 definition of, 83, 305–306
 goal-directed, 176–177, 180
 natural selection from interactions of, 11–12
 novel, 177
 organisms as, 355, 356
 plants as, 305–307
 selective, 61
 synergy and, 91
 types of, 86
Agents and Goals in Evolution (Okasha), 12
Agent theories. *See* Agential concepts
Ågren, J. Arvid, 60, 326, 342, 347–348, 350
Alphabetic writing, 84–85
AM (arbuscular mycorrhiza) fungal networks, 46
Amoebae, 38–41
Amoeboflagellates, 38
Ampelopsis hederacea, 317
Ampelopsis vietchii, 317
Anastasi, Anna, 312
Anatomical homeostasis, 182, 212
Andropogoneae, *285*
Anesthetics, protozoa and protist sensitivity to, 47

Animals. *See also* Metazoan animals
  adaptive choice and, 128
  cellular cognition in, 41
  developmental toolkit genes in, 202–203, 212–213
  electricity in, 222–223
  GDB in humans compared to, 246–249
  mobile DNA-derived networks and, *280–282*
  movement of, 221–222
  plant evolution compared to, 303
  soul of, 119–120
  total plant biomass compared to, 302–303
Animal Traditions (Avital and Jablonka), 20
Animism, relational agency roots in, 83–84
Anthropomorphism, 177
Anticipation
  biological meaning from, 168–169
  retrocausality and, 105–107
Anti-reductionism, 258
Antirrhinum majus, 285
Aperiodic crystal, 108–109
Arabidopsis, 280, 285, 285–286
  thaliana, 314
Arbitrariness, code and, 167
Arbuscular mycorrhiza (AM) fungal networks, 46
Archaea, 33–36, 42, 46–47, 145–146, 205, 208, 277
Arias Del Angel, Juan Antonio, 208, 210
Aristotle, 81, 84, 110, 348
  on heredity, 106
  on soul levels, 119–120
  on teleonomy, 105–106
Artificial selection, 15, 59, 74
  Darwin, C., on, 223
  niche construction and, 65
Ascendency, 108
Asgard archaea superphylum, 35
Ashby, W. Ross, 92, 96, 176
Aspergillus oryzae, 284
Attractors, 88–89, 92–94, 97–99, 105, 107–108,
  110–114, 142, 155, 177, 187–188
Australopithecines, 23
Autocatalytic
  cycle, 110
  feedback, 108–109, 112
  growth, 24
  sets, 105, 144–146, *154*
Autonomy, 107, 210, 258, 336
Autopoiesis, 3, 18–19, 99–100, 106, 170
Autoregression, 70–72
Avital, Eytan, 20, 349
Axiom of choice, 148–149
Axiom of extensionality, 148

Bacteria, 5, 19–20, 25, 33–34, 36, 38, 42, 46–47,
  89–92, 94–95, 97, 145–146, 177, 208–210, 212,
  246, 259, 265–267, 278, 284, 287, 301–302,
  310–311, 314
Balleine, Bernard W., 121, 124, 127–129
Baluška, František, 33
Barbieri, Marcello, 106–107, 109–111, 167
Bar-On, Yinon M., 302
Barrett, Louise, 244, 248
Basin of attraction, 88
Bateson, Patrick, 4, 17
Bauer, Ervin S., 108–109, 112

Beer, Randall D., 266, 305–306
Behavior. *See also* Goal-directed behavior
  bioelectricity for control of, 185
  biological intelligence and adaptive, 313–314, 318
  in evolution, 3–4, 18, 73
  evolutionary change and, 3, 18
  goal-directed, 12
  morphogenesis and prediction of, 184
  natural selection and, 18–19
  plants and complex, 307
  teleonomic, 19, 119, 134
  UAL predictions on, 126–127
  of woodpecker finch, 19
Behavioral ecology, 165
Behaviorism, 36
Berg, Leo, 111
Bernard, Claude, 120
Bigelow, Julian, 161, 176–178
Bioelectricity, for teleonomic control of growth and
  form, 185–190
Bioexceptionalism, 162
Biological agency, 275, 326, *328*
Biological intelligence, adaptive behavior and,
  313–314, 318
Biological Journal of the Linnean Society, 232
Biological meaning, 161
  from anticipation and need, 168–169
  approaches to, 163–170
  from code, 166–167
  from coexistence, 166
  cognition and, 164
  from communication, 168
  evolution and, 163–164
  from evolutionary past, 166
  from function, 163, 164
  mediatedness and, 169–170
  from purpose, 164–165
  from signaling, 167–168
  of species, 166
  from survival, 165–166, 171
  from umwelt, 169, 171
Biology. *See also* Evolutionary biology
  agency in, 18, 25n5
  chemistry, physics and, 258–259
  complexity in chemistry, biology and, 259–260, 270
  complex systems, 278
  in eighteenth century, 222–223
  GDB in, 175–176
  mechanistic, 36, 326
  multiscale competency in, 175
  in nineteenth century, 223–224
  purposive agency's restoration in, 232
  reductionist approach to, 258, 271
  in seventeenth century, 221–222
  theoretical, 162–163
  in twentieth century, 224–225
  in twenty-first century, 156
  "upside-down," 34
Biosemiotic studies, 162
Biosphere
  affordances in evolution of, 146–147
  evolution of, 141, 143–147, 153, 157
  Gödel's incompleteness theorem and evolving, 149–150
  Kantian wholes in, 144–145

as nested set of Kantian wholes, 155–156
small-molecule autocatalytic sets and origin of life in, 145–146
Biotechnology, 156
Birth defects, 180
Biston betularia (peppered moth), 15–16
Body mass index (BMI) increases, 358n16
Bogert, Charles M., 73
Bogert effect, 73
Bohr, Niels, 107
Bonner, John Tyler, 205
Boolean networks, 98
Bose-Einstein condensates, 109
Brain, bioelectric signaling in, 185–187
Brain from Inside Out, The (Buzsáki), 269
Brain-in-a-vat argument, of Putnam, 346
Brassica, 285
Bray, Dennis, 247, 278, 283
Buzsáki, György, 269
Byrne, Richard W., 21

$[Ca^{2+}]_i$ intracellular calcium oscillations, 314–315
Cactospiza pallidus (woodpecker finch), 19
Caenorhabditis elegans, 285, 350
Calcium, $[Ca^{2+}]_i$, 307, 311–312, 314–315
Camerarius, Rudolf Jakob, 44
Campbell, John H., 15
Cancer, tumorigenesis and, 189–190
Candida albicans, 284
Cannon, H. Graham, 300, 307
Cashmore, Anthony, 276
Categorizing sensory states (CSSs), 125
Causation, 3, 13, 15–16, 64, 75, 87, 107, 144, 153, 244, 276, 301, 311, 318, 334, 336
CBC (cellular basis of consciousness), 33, 35
Cell-cell transport, TNTs and, 45–46
Cells
dynamic kinetic chemistry in cytoskeleton of, 262–263
metabolism of, 95
metazoan animals and agency of, 207–211, 215
in relational agency, 95–96
self-construction of, 152
Cellular basis of consciousness (CBC), 33, 35
Cellular sentience
in animals and humans, 41
cognition and, 34
complexity of, 35
in protozoa and other protists, 37–41
Central dogma, 13, 222, 347
CGE (competitive general equilibrium), 142–143
Challenges, agency and, 89–90
Chalmers, David J., 265
Chance and Necessity (Monod), 12, 156
Charpentier, Emmanuelle, 265
Chemical DKS system, 5, 257, 260–265, 267–271
Chemical organization theory (COT), 92, 93, 99
Chemistry. *See also* Dynamic kinetic chemistry
biology and, 258–259
complexity in physics, biology and, 259–260, 270
life as, 258
stability types in, 260–261
Chemotaxis, 45, 89, 208, 267
Chemotropism, 308
Cheney, Dorothy L., 168

Chimeric transposase-transcription factor hybrids, 283
Choanoflagellate protozoa, 41, 200
Choice
axiom of, 148–149
plants and, 304–305
Chromatin, in metazoan animals, 206
Chromosome rearrangements, 277
Churchland, Paul M., 248
Ciliates, 38
Cis-regulatory modules (CRMs), 278–279, 283
Clark, Andrew D., 59, 66–71, 73–74, 335
Classical conditioning, 20
Code, biological meaning from, 166–167
Codepoiesis, 105–107, 110, 111
Coevolution, 80, 97
Coexistence, biological meaning from, 166
Cognition
agency and, 25n5
animal and human cellular, 41
biological meaning and, 164
cellular, 34–35
consciousness and, 45
Darwin, C., on, 266–267, 271
defining, 266–267
DKS and, 267–270
eukaryotic, 35–36
in evolution, 4–6, 19–21, 33–34
evolutionary change and, 286–287
foundationalism and, 345
fungal networks for supra-organismal, 46
"inside"-"outside" approach to, 268–269
physical basis for, 266–270
in plants, 20
in protozoa and protists, 37–41
in sexual reproduction of plants, 44–45
Shettleworth on, 266
social learning and, 20–21
stages of cellular, 47
tool-use and, 21
Cognitive ethology, goal attributions in, 246–249
"Collective brain," 24
Collective intelligence, teleonomy and, 176–179
Communication, biological meaning from, 168
Competency, in biological systems, 177–178
Competition, in plants, 316–317
Competitive general equilibrium (CGE), 142–143
Complex adaptive systems, 86
Complexity
affordances and, 156
in biological systems, 177–178
of cellular sentience, 35
in evolution, 21–22
niche construction and, 74–75
in object-based worldview, 82
of organisms, 61–62
in physics, chemistry and biology, 259–260, 270
in plant behavior, 307
scientific explanatory models of evolutionary biology and, 325, *327*
self-organization and, 259
in universe, 144
Complexity theory, 259
Complex systems biology, 278
Component processes of evolution, 348–349

Condition-action rules, 87–89
Consciousness
  cellular basis of, 33, 35
  characteristics of, 123–124
  cognition and, 45
  ETM and, 123–124
  evolution of, 227
  function of, 121–122
  fungal networks for supra-organismal, 46
  HIT model of, 119, 128–129, 134
  neural architecture of, 122
  Schrödinger on, 258
  supracellular, 45–46
  UAL and evolution of, 121–126, 130, *131*, 134
Construction, foundationalism and, 345, 348
Contact phenomena, foundationalism and, 352–353, 357
"Control information," 25n6
Controlling elements. *See* Transposable elements
Convergent evolution, 210, *281*, 301, 318, 332–334
  in plants, 302, 315–316
Conway-Morris, Simon, 315–316
Cooperation
  among amoebae, 40
  in complex adaptive systems, 86
  in evolution, 4, 22–23
  GDB and, 177
  innovation and, 23
Corning, Peter A., 1, 11, 80, 91, 107, 112, 114, 136n1, 156, 165, 171, 176, 237, 259, 276, 318, 330, 336n8, 357n1, 357n3
Corporations, as Kantian wholes, 154
Cosmides, Leda, 165
COT (chemical organization theory), 92, 93, 99
Counteractive niche construction, 62
Coyne, Jerry, 276
Creativity, emergent, 152–153
Crick, Francis, 13, 258, 263
CRISPR-Cas9, 265
CRMs (*cis*-regulatory modules), 278–279, 283
Croone, William, 222–223
Cryptodifflugia operculata, 38
CSSs (categorizing sensory states), 125
Cultural diversity, niche construction and, 74–75
Cultural evolution, 4, 16, 21, 63
Cupriavidus metallidurans, 284
Cyanobacteria, 34
Cybernetics, teleonomy and, 177, 192
Cytoskeleton, dynamic kinetic chemistry in cell, 262–263

Damasio, Antonio, 125, 129
Danchin, Étienne, 1, 62, 350
Danio rerio (zebrafish), 203
Darwin, Charles, 3, 214, 215
  on artificial selection, 223
  on biological meaning, 163–164
  on cognition, 266–267, 271
  on evolution of mind, 271
  on "external conditions of life," 299–300
  gemmule theory of, 223, 230–232
  on natural selection, 15, 59–60, 164, 223
  on phyletic gradualism, 277, 278, 287n2
  on physics, chemistry and biology, 270

  on plant domestication, 300
  on purposive agency, 223
  on sexual selection, 226–227
  on speed of evolution, 228
Darwin, Erasmus, 286
Darwinian preadaptation, 147
Darwinism (Wallace), 223
Data, hypotheses and, 233
Davidson, Donald, 246–247
Dawkins, Richard, 1, 21, 67, 165, 250
Deacon, Terrence W., 4, 162, 165
De Anima (Aristotle), 105
Death, DKS and, 270
Deffner, Dominik, 59
Deinococcus geothermalis, 284
Delafield-Butt, Jonathan T., 125
Dennett, Daniel C., 132, 134, 176, 248–250, 264
Depew, David, 4
De ratione motus musculorum (Croone), 222
De Regt, Henk, 327–328
Descartes, René, 11, 81–83, 221–224, 232, 269, 344, 358n9
Descent of Man, and Selection in Relation to Sex, The (Darwin, C.), 223, 226
Desire
  in GDB, 119, 128, 134
  in UAL, 129
Determinism, 142
Development
  as adaptive bias source, 332
  as continuous regenerative repair, 182
  defects, 180
  ontogenetic, 98
  from organism interactions, 354–355
  of plants, 304
  plasticity of, 351
  regulative, 181
  underdetermination and, 351
"Developmental analogy-homology paradox," 213
Developmental selection and selective attention, in consciousness, 123
Developmental systems of evolution, 61
Developmental toolkit genes, in animals, 202–203, 212–213
Dickinson, Anthony, 121, 124, 127–129, 246–248
Dicrocoelium dendriticum, 251
Dictyostelium (Mycetozoa), fruiting bodies in, 209–210
Difflugia tuberspinifera, 38
Digraphs, 153, *154*
Dinoflagellates, warnowiid, 40
Dittrich, Peter, 92, 99
DKS. *See* Dynamic kinetic stability
DNA. *See also* Mobile DNA elements
  editing, 265
  methylation, 314
  RNA in genome rearrangement and, 111–112
  structure of, 108–109
Dobzhansky, Theodosius, 12, 18–19, 165, 175, 287n1, 312, 329–330, 357n1
Domestication
  niche construction and, 65
  plant, 300, 317
"Domestication syndrome," 74
Doolittle, Ford, 263
Dor, Daniel, 133

Doudna, Jennifer, 265
Dreyfus, Hubert, 343–344, 352–353
Driesch, Hans, 165
Drosophila pseudoobscura, 251, *285*
Dualism, 83, 276
Duplication, 14, 105, 187, 206
Dynamical patterning modules, in metazoan animals, 204
Dynamical systems, 88, 92
Dynamic kinetic chemistry, 5, 257, 262–263
Dynamic kinetic stability (DKS), 5
  chemical systems in, 261–262, *262*
  cognition and, 267–270
  death and, 270
  definition of, 257
  as energized state, 260–263
  environment linked to, 268
  Gaia hypothesis and, 263
  homeostasis and, 265
  molecular replication and, 263–264, 269
  nested systems in, 263, *263*
  persistence principle and, 264–265
  thermodynamics and, 260–261
  water-fountain analogy for, 261–262, 268

ECM (extracellular matrix), 201
Ecological causes of evolution, 13
Ecological inheritance, 62–63
Ecology, 142–143
Eelkema, Rienk, 261–262
EES. *See* Extended evolutionary synthesis
Efficiency, 108–110, 112
Ehrlich, Paul, 43
Eigen, Manfred, 167
Eldredge, Niles, 3, 17, 211, 213, 228, 230
Electricity, animal, 222–223
Electrotropism, 308
"Eliminative materialism," 248
Embodiment and agency, in consciousness, 123–124
Embryogenesis, 180
Embryology, 111
Emergence
  levels of organization, 22, 80, 110, *181*, 188
  life and cognition, 257–259, 263, 265, 269
  statistical mechanics of, 150–152, 155–156
Emergent creativity, 152–153
Emmeche, Claus, 162, 164, 170
Enactivism, 312
Encyclopedia of Theory in Science, Technology, Engineering, and Mathematics (SAGE), 276
Endosomal sorting complex required for a transport (ESCRT) complex, 35
Entelechy, 105–106
Environment
  agency and conditions in, 89–90
  DKS linked to, 268
  GDB modification of, 335
  neo-Darwinism on, 307
  organisms and, 59–61
  of plants, 303
Environmental regulation, niche construction and, 67
Epigenetic inheritance, 14, 111–112
Epigenetic landscape, 60, 98
Epistemic cut, 106, 108
Epistemic foundationalism, 344

Equilibrium, in microeconomics, 142–143
Equilibrium state, 262
van Esch, Jan, 261–262
ESCRT (endosomal sorting complex required for a transport) complex, 35
Ethology, goal attributions in cognitive, 246–249
ETMs (evolutionary transition markers), 123–125
Eugenics, 227
Eukaryotes, 4, 24, 33, 42, 206, 208, 358
Eukaryotic cells, 33
  cognition of, 35–36
  emergence of first, 34–35
  sexual reproduction of, 41–42
Evo-devo. *See* Evolutionary developmental biology
Evolution
  agency in, 6, 156
  agential concepts for theories of, 326–330, 336n7
  behavior in, 3–4, 18, 73
  biological meaning and, 163–164
  of biosphere, 141, 143–147, 153, 157
  cognition in, 4–6, 19–21, 33–34
  complexity in, 21–22
  component processes of, 348–349
  of consciousness, 127, 227
  of consciousness and UAL, 121–126, 130, *131*, 134
  convergent, plants and, 315–316
  cooperation in, 22–23
  cultural, 4, 16, 21
  cumulative cultural, 63
  definition of, 15, 276
  developmental systems of, 61
  ecological causes of, 13
  exploitative systems of, 60
  final state in, 112–114
  GDB origin in, 179–180
  genes as "givens" of, 348
  genetic causes of, 13
  goal-directedness in, 110–111
  of humankind, 23–24
  of integrated functionality, 152–153
  Lamarck on plant, 300, *301*
  of language, 133–134
  limits of, 147–149
  of mind, 271
  natural selection and, 16
  niche construction cause-effect relationships with, 64–65
  organism and environment interactions in, 60–61
  organism's active role in, 353–355
  phenotypic plasticity in, 107
  of plants compared to animals, 303
  of social learning, 74
  speed of, 228
  synergy in, 22–23
  teleonomy in, 11–25, 110–112
  underdetermination and, 349–352
Evolutionary biology. *See also* Neo-Darwinism
  complexity in, 21–22
  goal attributions in, 249–253
  historical developments in, 1–2
  mechanistic, 36, 326
  problems and misunderstandings in, 33–34
  scientific explanatory models of complexity in, 325, *327*
  scientific intelligibility and, 329
  survival in, 15

Evolutionary change, 275
  agential concepts explaining, 332–335
  behavior and, 3, 18
  cognition and, 286–287
  culture in, 21
  cumulative, 24
  life history inputs in, 283
  neo-Darwinism on, 276–277
  niche construction and, 64
  phenotypic plasticity in plants and, 317–318
  punctuated equilibrium and, 286
  speed of, 228
  traditional view of, 60
Evolutionary developmental biology (evo-devo)
  concept of, 80
  neo-Darwinism challenged by, 2, 14
Evolutionary Genetics (Maynard Smith), 234n1
Evolutionary past, biological meaning from, 166
Evolutionary psychology, 165
Evolutionary stasis, 211–213
"Evolutionary Systems" (Waddington), 60
Evolutionary transition, 80, 98
Evolutionary transition markers (ETMs), 123–125
Evolution of the Sensitive Soul (Ginsburg and
  Jablonka), 227
Evolution: The Extended Synthesis (Pigliucci and
  Müller), 224
Evolution: The Modern Synthesis (Huxley), 1, 5,
  224–230, 234n2, 357n1
"Exaptations," 19
Exosomes, 14, 232
Exploitative systems of evolution, 60
Extended evolutionary hypothesis, for plants, 300–302
Extended evolutionary synthesis (EES), 1, 96, 99, 213
Extensionality, axiom of, 148
Extracellular matrix (ECM), 201

Facilitated variation, 214, 332
FACs (function-amplifying centers), 206–207
Famintsyn, A. S., 25n2
Field effects, in metazoan animals, 211–212
Final state, in evolution, 112–114
First order of cellular cognition (FiOCC), 47
Fish, swim bladder in, 147
Fisher, Ronald A., 165
Fitness
  adaptedness of organisms and, 341–342
  concept of, 165
  multilevel selection theory (MLS) and, 209
  relational agency and, 93–94
  unity of purpose and, 250–252
Flexible value attribution, desires, and goals, in
  consciousness, 123
Flowers, teleonomy and, 309, *310*
Flynn effect, 358n16
FoOCC (fourth order of cellular cognition), 47
Foundationalism
  agential concepts compared to, 342–343, 356–357
  cognition and, 345
  component processes of evolution and, 348–349
  construction and, 345, 348
  contact phenomena and, 352–353, 357
  deficiencies of, 356
  epistemic, 344
  on evolution and active role of organisms, 353–355

features of, 343–344
  givenism and, 345, 348
  illusion and, 346
  mediation and, 345, 349
  mental phenomena and, 342
  neo-Darwinism and, 6, 341–342, 347–349
  perception and, 344–345
  proxyism and, 345, 348
  sense datum theory, 344–345
  separation and, 345, 348
  skepticism and, 346
  solipsism and, 346–347
  underdetermination and, 345–347, 349–352
Fourth order of cellular cognition (FoOCC), 47
Freedom, 123, 133, 146, 149
Free-will, agency and, 134–135, 276
Frogs
  anatomical homeostasis in, 212
  leg regeneration in, 182–183
  metamorphosis in, 184
Fröhlich, Herbert, 109
Function
  affordances and, 164
  biological meaning from, 163, 164
  goal-directedness compared to, 238, 240–241, 243
  Kantian wholes and concept of, 152, 156
  metazoan animals and cell differentiation by
    appropriation of unicellular, 204–207
  precision and, 211
  proper, 204
Functional integration, Kantian wholes and, 152–153
Function-amplifying centers (FACs), 206–207
Fungal mind, 46
Fungal networks, for supra-organismal cognition and
  consciousness, 46
Funktionkreis, 107–108

Gaia hypothesis, 94, 263
Galápagos Islands, 17, 19
Galileo, 110, 221
Gamow, George, 166
Gánti, Tibor, 121–123
Garson, Justin, 238
Gastrulation, in metazoan animals, 200
GDB. *See* Goal-directed behavior
Gemmule theory, of Darwin, C., 14, 223, 230–232
Gene coexpression networks, in metazoan animals,
  205–207
Gene-culture coevolution, 63
Gene-editing, 191, 265
Gene regulatory networks (GRNs), 205
Genes. *See also* Mobile DNA elements
  as "givens" of evolution, 348
  neo-Darwinism on, 307
  organization of, 108–109
  in relational agency, 95–96
Gene's-eye view, 326–327, 347–348
Genetic assimilation, 60, 107, 133, 225–226
Genetic causes of evolution, 13
Genome
  alteration and reconstruction of, 2, 14
  interspecific hybridization activating change in, 283, *285*
  McClintock on, 229, 277–278
  mobile DNA elements contributing to, 278–279,
    280–282, 283

RNA in rearrangement of, 111–112
self-modification, 107
stress-induced reformatting of, 283, 285
structure/composition of, 277
as "two-way read-write system," 14
Genome sequencing, 3, 233, 234n3, 276, 279, 287n3
Genome-wide association studies, 229, 232–233
Genomic networks, 278–279, *280–282,* 283
Genomics/phylogenomics, 1, 14, 40, 207, 212, 214, 233
Genotype space, 347–349, 351, 356
Geobacillus kaustophilus, 284
Geoffroy Saint-Hilaire, Étienne, 203
Ghost in the Machine, The (Koestler), 11
Gilroy, Simon, 20, 300, 305, 308, 310, 312, 317
Ginkgo biloba, 42, *43*
Ginsburg, Simona, 119, 227
Givenism, foundationalism and, 345, 348
Glisson, Francis, 222
Global accessibility and broadcast, in consciousness, 123
Global neural network (GNW) model, 125–126
Goal attributions, 237
  in cognitive ethology, 246–249
  contexts of, 238, 253
  in evolutionary biology, 249–253
  traditional debate on, 239–246
Goal-directed behavior (GDB)
  agency and, 134–136, 176–177
  in biology, 5, 175–176
  continuum of, 177–179
  cooperation and, 177
  definition of, 119, 127
  desire-driven, 119, 128, 134
  Dickinson and Balleine on, 127–129
  environmental modification by, 335
  evolutionary origin of, 179–180
  IIIT model and, 128–129
  in humans compared to animals, 246–249
  imaginative, 132–134
  intentions and desires driving, 121
  Mayr on, 12, 238
  morphogenesis and, 188
  S-R system and, 128
Goal-directedness, 88–89. *See also* Teleonomy
  codepoiesis and, 111
  Dickinson's criteria for, 247–248
  in evolution, 110–111
  function compared to, 238, 240–241, 243
  inner representations and, 245–246
  intentional view of, 241–242, 244
  Mayr on teleonomy and, 239–240, 245
  Nagel on, 241–244
  as natural, 243–244
  negative feedback and, 240
  objectiveness of, 243, 246, 253
  program view of, 242, 253
  of self-maintaining systems, 105
  systems view of, 242–243, 244
  traditional debate over, 239–246
  unity of purpose and, 250–253
Gödel numbering, 106
Gödel's incompleteness theorem, 149–150
Goethe, Johann W. von, 203
Goldenfeld, Nigel D., 259
Goldschmidt, Richard, 277, 286
Golgi, Camillo, 44

Goodwin, Brian, 162, 167, 203, 211
Gordon, Deborah M., 22
Gould, Stephen Jay, 3, 5, 19, 111, 147, 165, 203, 211, 224, 228, 230–231, 286, 315, 331
Grafen, Alan, 249
Grand, Steve, 264
Grant, B. Rosemary, 17
Grant, Peter R., 17
Gravitation, law of, 141
GRNs (gene regulatory networks), 181, 205
Growth of Biological Thought, The (Mayr), 230

Habit, definition of, 171
Habitat choice, niche construction and, 67
Habitat modification, niche construction and, 68
Haeckel, Ernst, 36–38, 42, 111
Haig, David, 252, 254n8, 329
Haldane, John Burdon Sanderson, 165
Hare, Brian, 74
Harvey, William, 232
Heap's law, 150
Hedonic interface theory (HIT) model, 119, 128–129, 134
Heinrich, Bernd, 20
van Helmont, Jan, 258, 271
Henslow, George, 315–318
Heredity
  Aristotle on, 106
  Shapiro on, 2, 14
  unlimited, 122
Heylighen, Francis, 79
Hierarchy, 130, 263
Histone marks, 206, 314
HIT (hedonic interface theory) model, 119, 128–129, 134
Homeostasis, 41, 47, 118–121, 171, 179, 188, 190, 215
  anatomical, 182, 212
  DKS and, 265
  negative feedback and, 308
Homo (genus), 132
Homo erectus, 23, 135
Homo sapiens, 315
Hooker, Joseph Dalton, 270
Human Genome Project, 233
Humans
  cellular cognition in, 41
  complexity in universe and, 144
  "domestication syndrome" and, 74
  evolution of, 23–24
  GDB in animals compared to, 246–249
  as Kantian wholes, 152
  mobile DNA-derived networks and, *282*
  soul of, 119–120
  unity of purpose and, 250–251
"Human Self-Domestication Hypothesis," 74
Huxley, Julian, 1, 5, 221, 234n2, 357n1
  on Darwin, C.'s, interpretation of sexual selection, 226–227
  eugenics and, 227
  on evolution of consciousness, 227
  on hypermutation, 228–229
  neo-Darwinism and, 224–230
  on purposive agency, 227
  on speed of evolution, 228
  on Waddington, 225–226
  on Weismann's ideas, 225, 227, 230–231

Hypermutation, 228–229
Hypotheses, data and, 233

Icthyosporeans, 200
Igamberdiev, Abir U., 105
IGDB (imaginative goal-directed behavior), 132–134
Illusion, foundationalism and, 346
Imaginative goal-directed behavior (IGDB), 132–134
Immune cells, protozoan nature of, 43
Inceptive niche construction, 62
Incompleteness theorem, Gödel's, 149–150
Individuality, 36
Individuation, 97
Information, teleonomy as, 25n6
Inherent/inherency, 15, 42, 47, 109–110, 129, 175, 199–200, 202–208, 210–211, 213–215, 268–269, 275, 286, 303
Inheritance
 ecological, 62–63
 epigenetic, 14, 111–112
 Lamarckian modes of, 14, 226
 from organism interactions, 354
 transgenerational, in plants, 313–315
 underdetermination and, 349–351
Inner representations, goal-directedness and, 245–246
Innovation(s), 2, 4, 6, 13, 18–19, 24, 201, 203, 211–213, 286, 333, 341, 348, 354–357
 cooperation and, 23
 mobile DNA elements and genomic network, 278–279, 280–282, 283
 underdetermination and, 351–352
"Inside," approach to cognition, 268–269
Integrated functionality, evolution of, 152–153
Intelligence, 175. See also Cognition
 artificial, 25, 176, 179, 182
 definition of, 179
 plants and biological, 313–314, 318
 teleonomy and collective, 176–179
Intentionality (aboutness), in consciousness, 123
Intentional view of goal-directedness, 241–242, 244
Internal quantum state (IQS), 108
Internal selection, 97
Interspecific hybridization, genome change activated by, 283, 285
Interval scale, 148
Intracellular calcium $[Ca^{2+}]_i$ oscillations, 314–315
IQS (internal quantum state), 108
Ironus sp., 38, 39

Jablonka, Eva, 4, 14, 20, 119, 227, 314, 331, 349, 351, 358n21, 358n29
Jaeger, Johannes, 328, 333–334, 336n18
James, William, 121–123, 136, 161, 179, 184, 358n6
Johnson, Samuel, 268
"Jumping genes." See Mobile DNA elements
Just, Ernest Everett, 215

Kagan, G. Ya., 234n2
Kant, Immanuel, 120, 202, 238
Kantian wholes
 abstract, 153–154
 in biosphere, 144–145
 concept of function and, 152, 156
 corporations as, 154

functional integration and, 152–153
 nested sets of, 155–156
 TAP process and, 155
Kauffman, Stuart A., 1, 3, 5, 17–18, 80, 96, 98, 109–110, 112, 141, 162, 164, 258–259
Keller, Evelyn Fox, 245, 258
Kennedy, John S., 247
Khrennikov, Andrei, 109
Kidney tubules, in newts, 180
Kinetic perfection of biosystems, 112–113
Kinetic selection, 265, 269
Kinetic stability, 260–261, 261. See also Dynamic kinetic stability
Kinetic theory of gases, 328–329
Kingdon, Jonathan, 4, 11, 16, 24
Knowledge, entelechy and, 105–106
Koch, Robert, 43
Koestler, Arthur, 11
Koonin, Eugene V., 2, 13–14, 122
Krohs, Ulrich, 164, 167
Kull, Kalevi, 161

Lacrymaria olor, 38, 39
Laland, Kevin, 1, 4, 13–14, 16, 61–65, 69, 73–74, 80, 120, 213, 330, 332, 335, 336n10, 342, 352
Lamarck, Jean Baptiste de, 3, 4, 226, 231, 276, 307
 on biological meaning, 163–164
 natural selection and, 18
 on plant evolution, 300, 301
 use-disuse idea of, 223, 283
Lamarckism
 exclusion of, 225–226
 gene-editing and, 191
 modes of inheritance, 14, 226
 natural selection and, 18
 Weismann on, 231
Landscape
 adaptive, 329
 attractor, 98
 epigenetic, 69, 98
 fitness, 112, 155
Language, evolution of, 133–134
Leeuwenhoeck, Antony, 36–38
"Letter on the Protozoa" (Leeuwenhoeck), 36
Levin, Michael, 5, 36–37, 175, 202, 205, 211–212, 278
Lewontin, Richard, 60 61, 203, 301, 305, 307, 310–311, 331, 334, 347, 349, 356
Liberman, Efim, 112
Life
 artificial, 176
 nature of, 257–258
Life history inputs, in evolutionary change, 283
Literacy, 133
lncRNAs, 3
Long terminal repeats (LTRs), 277–278
Loreto, Vittorio, 150, 152
Loreto-Strogatz urn model, 150, 152
Lotka, Alfred James, 108, 112, 260
Lovelock, James E., 263
LTRs (long terminal repeats), 277–278
Lycopodium, 315
Lying, 133–134
Lyon, Pamela, 177, 259, 266–267, 271

MacColl, A. D., 73
Magnaporthe oryzae, 284
Maize, 229, 277, *280, 285*
Margulis, Lynn, 2, 4, 13, 24–25, 263
Markov process, 310, 311
Materialism, 270
Matsuno, Koichiro, 106, 108–110
Matter, 82
Matthews, Gareth, 353
Maturana, Humberto, 3, 18, 176, 192, 266, 270
Maximum useful energy flow transformation
    principle, 112–113
Maxwell, James Clerk, 62
Maynard Smith, John, 22, 80, 98, 122, 133, 166–167,
    234n1, 254n5, 329
Mayr, Ernst, 2, 14, 17, 79–80, 106, 166, 230,
    237–246, 253, 254n2, 258, *328,* 336n10
  on adapted features, 240
  on agential concepts and explanatory standards, 330
  on behavior in evolution, 3–4, 18
  on GDB, 12, 238
  on genetic and ecological causes of evolution, 13
  on goal-directedness and teleonomy, 239–240, 245
  on negative feedback, 240
  on purposive behavior, 240, 246, 330
  on teleonomy, 12–13
McClintock, Barbara, 2, 6, 14, 229, 275, 277–278, 287
McFarland, David J., 248
McShea, Daniel, 21, 176, 254n1
Meaning, 161–162. *See also* Biological meaning
Meaning of Meaning (Ogden and Richards), 162
Meaning-relation, 162
Mechanism, in seventeenth century, 221–222
Mechanistic biology, 36, 326
Mechano-tropic bending, 308
Medaka fish *(Oryzias latipes),* 203
Mediatedness, biological meaning and, 169–170
Mediation, foundationalism and, 345, 349
Memetics, 165
Memory, 175, 188
Mental phenomena, foundationalism and, 342
Mereschkovsky, Konstantin C., 25n2
Merleau-Ponty, Maurice, 353, 358n8
Messenger RNA (mRNA) synthesis, 278
Metabolism, cell, 95
Metazoan animals, 199
  cell agency in, 207–211, 215
  cell differentiation by appropriation of unicellular
    functions in, 204–207
  chromatin in, 206
  developmental toolkit genes from, 202–203,
    212–213
  dynamical patterning modules in, 204
  ECM and, 201
  emergence of, 200
  FACs in, 206–207
  field effects in, 211–212
  gastrulation in, 200
  gene coexpression networks in, 205–207
  liquid-tissue effects and, 200–201, 210, 213
  MLS theory and, 209
  morphological elaboration in, 202–203
  pattern formation and, 202
  Wnt protein and, 200–201

Metchnikoff, Ilya, 37, 43
Michod, Richard, 23
Microeconomics, CGE and, 142–143
Microgenetics (Rapoport), 109
Miller, William B., Jr., 33
Mind, 82–83, 265, 267, 271. *See also* Cognition
Mind of the Raven, The (Heinrich), 20
MIUs (motor integrating units), 132
MLS (multilevel selection) theory, 80, 209
Mobile DNA elements ("jumping genes"). *See also*
    Transposable elements
  animal characteristics attributed to, *280–282*
  forms of, 277–278
  future research on, 287
  genomic network innovations attributed to,
    278–279, *280–282,* 283
  human characteristics attributed to, *282*
  plant characteristics attributed to, *280*
  protein coding sequences of, 279, 283
  stimuli triggering, 283, *284–285*
Moczek, Armin, 352, 358n25
Modern synthesis. *See* Neo-Darwinism
Molecular replication, DKS and, 263–264,
    269
Monod, Jacques, 1–2, 4, 12–13, 156, 176–177,
    237–238, 258, 278
Morality, agency and, 135
Morphogenesis, 175
  behavior prediction in, 184
  bioelectricity for control of, 185
  future research on teleonomy and, 190–192
  GDB and, 188
  novelty in, 177
  plasticity in, *181*
  regeneration and, 181–185, *183*
  robustness in, 177, *181*
  scaling of goals in, 187–190
Morphology, 1, 18, 41, 181–184, 187–188, 191, 200,
    208, 212–213, 275, *281,* 333
Morphospace, multicellular navigation through,
    179–180, 182–185
Moss, Lenny, 202
Mossio, Matteo, 97, 144–145, 152, 208, *328,*
    336nn18–19
Motion, laws of, 141, 142
Motor integrating units (MIUs), 132
Movement, of animals, 221–222
mRNA (messenger RNA) synthesis, 278
Müller, Gerd, 1, 2, 14, 80, 201, 203, 213, 224,
    312
Multiagent simulations, 86
Multicellular organisms, protozoa as basis of,
    36–37, 47
Multilevel autoregressive time-series model, for
    temporal dataset, 70–72
Multilevel selection (MLS) theory, 80, 209
Multiscale competency, in biology, 175
Muscles, movement of, 222
Mutation
  neo-Darwinism on, 307, 351
  random, 275–277
  rates, 228–229
Mutual interdependence, 60
Myxobacteria, fruiting bodies in, 209–210

Nagel, Ernest, 238–246
Nagel, Thomas, 258
"Natural genetic engineering," 2, 5, 14, 229
Natural kind, 201, 243–246, 253
Natural selection, 275
  agent interactions and, 11–12
  behavior and, 18–19
  complexity in universe and, 144
  Darwin, C., on, 59–60, 164, 223
  definition of, 15
  evolution and, 16
  Lamarck and, 18
  neo-Darwinism emphasis on, 276–277, 330
  niche construction and, 64
  peppered moth and, 15–16
  phenotypic plasticity in, 107
  purpose and, 164
  relational agency and, 93–94
  residual variation and, 71
  survival and, 17
  teleonomy and, 15–17, 114
  temporal selection gradient dataset on niche
    construction and, 66–72
  temporal stability and, 71
  Weismann on, 224
Nature, 223
Nature of life, 257–258
Nebela vas, 38, 39
Need, biological meaning from, 168–169
Needham, Dorothy M., 222
Negative feedback, 240, 308
Neo-Darwinism (modern synthesis), 1. See also
    Relational agency
  on adaptedness of organisms, 341–342
  agency and, 81, 275–276
  biological meaning from survival in, 165
  component processes of evolution in, 348–349
  concept of, 79, 276
  criticism against, 24, 80
  on environment, 307
  evo-devo challenging, 2, 14
  on evolutionary change, 276–277
  foundationalism and, 6, 341–342, 347–349
  on genes, 307
  gene's-eye view in, 347–348
  hardening of, 230–232
  historical development of, 224–225
  Huxley and, 224–230
  hypermutation rates and, 228–229
  on mutations, 307, 351
  natural selection emphasis of, 276–277, 330
  object-based worldview and, 81–85
  population in, 82
  prokaryote findings challenging, 13–14
  relational approaches to extending, 80–81, 98–99
  static entities in, 98
  transgenerational inheritance contradicting, 313–314
  two-space nature of, 347–349, 351, 356
  underdetermination and, 349–352
  Waddington distancing from, 225
Nernst, Walther, 109
Nested sets of Kantian wholes, 155–156
Neuron doctrine concept, 44
Neurons, 43–47, 62, 177, 186–188, 192, 205, 207,
    212, 276

Neuroscience, 187–188, 191, 266
Newman, Stuart A., 199
Newton, Isaac, 81, 98, 141–142, 221
Newtonian paradigm, 5, 141, 143, 146, 149–150, 153
Newts, kidney tubules in, 180
Niche construction, 4, 16, 59, 269, 332
  agency and, 213–214
  artificial selection and, 65
  categorization of, 66–67
  counteractive, 62
  cultural diversity, complexity and, 74–75
  definition of, 80
  domestication and, 65
  ecological inheritance and, 62–63
  environmental regulation and, 67
  evolutionary change and, 64
  evolution cause-effect relationships with, 64–65
  future research on, 74
  habitat choice and, 67
  habitat modification and, 68
  inceptive, 62
  natural selection and, 64
  overview of, 61–64
  perturbational, 62
  relocating, 62
  residual variation and, 71
  selective feedback due to, 63
  semantic information governing, 62
  temporal selection gradient dataset on natural
    selection and, 66–72
  temporal stability and, 71
  universality of, 61
Noble, Denis, 1–4, 14, 17, 21, 25, 34, 175, 177, 221,
    276–277, 300–301, 303, 318, 357n1
Noble, Raymond, 5, 21, 24, 34, 221, 357n1
Nominal scale, 148
Nonequilibrium thermodynamics, 108–109
Non-LTR retrotransposons, 278
Novelty, 5, 15, 25n5, 33, 106, 183, 211, 214, 249,
    348, 351–355
  explanations for, 331
  in morphogenesis, 177
Nowak, Martin A., 16, 23
Nucleoside triphosphates, 113–114

Object-based worldview, 81–85
Object theories, agential concepts compared to, 333–334
Odling-Smee, John, 4, 13–14, 16, 61–64, 67–68,
    213, 260, 332, 351–352
Odum, Howard T., 108, 112
Ogden, Charles K., 162
Okasha, Samir, 5, 12, 16–18, 25n5, 80, 237, 326
Olynthus, 42, 42
Ononis repens, 317
Ononis spinosa, 317
Ontogenetic development, 98
Ontology of actions, 86–87, 90–91
Ontology of objects, 81–85
Oocytes, 41–42
Operant conditioning, 20
Organic selection theory, 3, 18
Organisms
  adaptedness of, 341–342
  adaptive bias from active interaction of, 355
  as agents, 355, 356

complexity of, 61–62
development from active interaction of, 354–355
environment and, 59–61
evolution and active role of, 353–355
inheritance from active interaction of, 354
multicellular, protozoa as basis of, 36–37, 47
protozoan basis of multicellular, 36–37, 47
purposive behavior of, 353–354
Origin of life, as small-molecule autocatalytic sets, 145–146
Origin of Species, The (Darwin, C.), 15, 223, 228, 266, 277
Orthogonality, 243
Oryzias latipes (medaka fish), 203
Otto, Sijbren, 264
Outcome devaluation, 124
"Outside," approach to cognition, 268–269
Oyama, Susan, 61–62, 331, 336n12, 351

Packard, Andrew, 125
Paleontology, 111
Panksepp, Jaak, 125, 129, 135–136
Parmenides, 81
Partial order scale, 148
Particles, 82
Pascal, Robert, 257
Pattee, Howard H., 106, 278
Pattern formation, metazoan animals and, 202
Pauling, Linus, 258
Peppered moth *(Biston betularia)*, 15–16
Perception, foundationalism and, 344–345
Percept unification and differentiation, in consciousness, 123
Persistence principle, DKS and, 264–265
Persuadability, in biological systems, 178–179
Perturbational niche construction, 62
Peterson, Erik L., 162, 225
Phagocytella theory, 37, *37*
Phenotype space, 347–349, 351, 356
Phenotypic accommodation, 331–332
Phenotypic plasticity, 2, 14, *65*, 107
  plants and, 305, *306*, 317–318
Photosynthesis, 34, 91, 265, *280*, 302, 308, 316
Phyletic gradualism, 275, 277, 278, 287n2
Physical reductionism, 276
Physico-Chemical Factors of Biological Evolution (Shnoll), 112–113
Physics
  biology and, 258–259
  complexity in chemistry, biology and, 259–260, 270
  stability types in, 260–261
Physiological plasticity, 306
Physiology, 3, 179–180, 187, 213, 223–224, 233, 286, 299, 326, 330, 336
Physiome Project, 233
Piaget, Jean, 4, 169
Pigliucci, Massimo, 1, 80, 224, 312, 336n10, 342
Pittendrigh, Colin, 1, 12, 106, 120, 161, 237, 275, 299
Placozoa, 200–203, 207, 214
Planaria
  bioelectric circuits in, 186–187
  regeneration in, 183–184
Plantago major, 317

Plantago maritima, 317
Plants
  affordances and, 312
  as agents, 305–307
  animal evolution compared to, 303
  biological intelligence and adaptive behavior in, 313–314, 318
  choice and, 304–305
  cognition in, 20
  competition in, 316
  complexity in behavior of, 307
  "conditions of life" for, 300
  convergent evolution and, 315–316
  domestication of, 300, 317
  environments of, 303
  extended evolutionary hypothesis for, 300–302
  flowers and teleonomy in, 309, *310*
  in food chain, 303
  Lamarck on evolution of, 300, *301*
  Markov process and, 310, 311
  mobile DNA-derived networks and, *280*
  negative feedback and, 308
  phenotypic plasticity and, 305, *306,* 317–318
  physical laws on growth and development of, 304, 316–318
  physiological plasticity and, 306
  predation defense in, 311, *312*
  regeneration in, 308
  root and shoot niche construction in, 310–312
  seed dispersal mechanisms of, 304
  sexual reproduction of, 44–45
  soul of, 119–120
  sperm cells of, 42, *43*
  total animal biomass compared to, 302–303
  transgenerational inheritance in, 313–315
  vegetative growth in, 308–309
Plant Sexual Reproduction (journal), 44
Plasticity
  of development, 351
  in morphogenesis, *181*
  phenotypic, 2, 14, *65,* 107, 305, *306,* 317–318
  physiological, 306
Plato, 81, 84
Plotkin, Henry, 4
Popper, Karl, 132
Population, in neo-Darwinism, 82
Population genetics, 329, 336n9
Portia spiders, 248–249
Posthumanism, 86
Preadaptation, Darwinian, 147
Precision, field effects in, 211–212
Predation defense, in plants, 311, *312*
Preferences, 175
Prigogine, Ilya, 87, 260
Program view of goal-directedness, 242, 253
Prokaryotic cells, 33
  eukaryotic emergence from, 34–35
  neo-Darwinism challenged by findings on, 13–14
Proper function, 204
Proprioception, 309
Pross, Addy, 1, 5, 257
Protein coding sequences, of mobile DNA elements, 279, 283
Protein synthesis regulation, 278

Protists, 34
　anesthetic sensitivity of, 47
　cognition in, 37–41
　Haeckel introducing term of, 36
Protozoa, 33
　anesthetic sensitivity of, 47
　basic morphotypes of, 38
　as basis of multicellular organisms, 36–37, *37*, 47
　choanoflagellate, 41, 200
　cognition in, 37–41
　immune cells and, 43
　neurons and, 44
　sex cells and, 46–47
　unicellular, 36
Proxyism, foundationalism and, 345, 348
Punctuated equilibrium, 213, 230, 277, 286
Purpose, 161. *See also* Biological meaning
　biological meaning from, 164–165
　as fundamental to living things, 237–238
　natural selection and, 164
　unity of, 250–253
Purposefulness, 171, 275
Purposive agency, 5, 221
　biology and restoration of, 232
　category mistake of eliminating, 232–233
　Darwin, C., on, 223
　evolution of consciousness and, 227
　Huxley on, 227
Purposive behavior, 24, 223, 240, 245–246, 330,
　　353–354
Purposiveness. *See* Teleonomy
Putnam, Hilary, 345–346

Qualia, 106, 109, 129, 344
Quantum field theory, 86–87, 142
Quantum mechanics, 81, 86–87, 108, 142, 156

Ramón y Cajal, Santiago, 44
Random mutation, 275–277
Rapoport, Iosif A., 109–110
Ratio scale, 148
Ravens (bird), 20
Reactions, condition-action rules as, 88–89
Reactive oxygen species (ROS), 47
Reber, Arthur S., 33
Recombination, in sexual reproduction, 227
Reductionism, 258, 271, 276
Regeneration
　field effects in, 211–212
　morphogenesis and, 181–185, *183*
　in plants, 308
Regulative development, 181
Relational agency, 79
　animism roots of, 83–84
　cells and genes in, 95–96
　challenges and, 89–90
　coevolution and, 97
　concept of, 81
　COT and, 93
　evolutionary transition and, 98
　fitness and, 93–94
　natural selection and, 93–94
　ontology of actions and, 86–87, 90–91
　reactions, condition-action rules and, 88–89

recent approaches to, 85–86
self-maintaining systems and, 91–94, 96
self-organization and, 96–97
symbiogenesis and, 98
symbiosis and, 97
synergy and, 97
systems biology and, 97
teleonomy and, 97
variation and, 94
Relocating niche construction, 62
Repetitive DNA, 2–3
Replication, DKS and molecular, 263–264, 269
Reproductive success, 165
Residual variation, 71
Retrocausality, anticipation and, 105–107
Rice, *285*
Richards, Ivor Armstrong, 162
Richerson, Peter, 4, 16, 19, 21, 23–24, 63, 74
Ristau, Carolyn A., 246–247
RNA
　in genome rearrangement, 111–112
　replication experiment, 264
　template-replicating, 146
RNA-world hypothesis, 263–264
Robustness, in morphogenesis, 177, *181*
Roff, Derek A., 249
Roli, Andrea, 141
Romanes, George, 36, 38, 40, 223–224
Root niche construction, in plants, 310–312
ROS (reactive oxygen species), 47
Rosen, Robert, 17, 106–107, 110, 113–114, 157n2,
　　169, 192
Rosenblueth, Arturo, 161, 176–178, 308
Russell, Edward Stuart, 4, 308
Ryle, Gilbert, 11

Saccharomyces cerevisae, 284
Sachs, Julius von, 299
Sagan, Dorion, 2, 13, 24–25
SAGE *Encyclopedia of Theory in Science, Technol-
　ogy, Engineering, and Mathematics*, 276
Salix babylonica, 318
Salix herbacea, 318
Sapp, Jan, 2, 4, 13–14
Scaling of goals, 177, 187–190, *189*, 192
Schizosaccharomyces pombe, 284
Schrödinger, Erwin, 61, 108–109, 142, 166, 258
Scientific intelligibility, 327–329
Scrub jays, food-catching behavior of, 247
Sebeok, Thomas A., 162, 163, 168
Sebeok (Umiker-Sebeok), Jean, 163
Second order of cellular cognition (SeOCC), 47
Selective agents, 61
Selective feedback, due to niche construction, 63
Self, sense of, 269
Self-generated stimuli, world-generated stimuli
　compared to, 135
Self-growing logos, 107
Selfish Gene, The (Dawkins), 1, 21
Self-Made Man (Kingdon), 4, 24
Self-maintaining systems, 91–94, 96, 99
　goal-directedness of, 105
Self-organization, 80, 92, 96–97
　complexity and, 259

Self–other distinction from a point of view, in consciousness, 124
Sellars, Wilfred, 345, 348, 358n7
Semantic information, niche construction governed by, 62
Semiology, 162–163
Semiotics, 161–162
Sense datum theory, 344–345
Sensory integrating units (SIUs), 132
SeOCC (second order of cellular cognition), 47
Separation, foundationalism and, 345, 348
Set theory, limits of, 147–149
Sex cells, protozoan nature of, 46–47
Sexual reproduction
  of eukaryotic cells, 41–42
  of plants, 44–45
  recombination in, 227
Sexual selection, 15, 223
  Darwin, C., on, 226–227
Seyfarth, Robert M., 168
Shapiro, James A., 1–5, 14, 21, 24, 36–37, 107, 224, 266–267, 275
Shettleworth, Sara J., 266
Shnoll, Simon E., 112–113
Shoot niche construction, in plants, 310–312
Signaling, biological meaning from, 167–168
Signal transduction, 20, 265, 283
Simard, Suzanne W., 22, 46
Simpson, George Gaylord, 1, 3, 12, 18
Sinnott, Edmund W., 165
SIUs (sensory integrating units), 132
Skepticism, foundationalism and, 346
Skinner, Burrhus Frederic, 15
Small-molecule autocatalytic sets, 145–146
Smith, John Maynard, 22, 80, 98, 122, 133, 166–167, 234n1, 254n5, 329
Social learning
  cognition and, 20–21
  ecological inheritance and, 63
  evolution of, 74
Society for the Protection of Underground Networks (SPUN), 46
Sociobiology, 165
Sociobiology (Wilson), 21
Solanum chilense, 284
Solipsism, foundationalism and, 346–347
Sommerhoff, Gerd Walter Christian, 238, 242–243
Soul, Aristotle on levels of, 119–120
Spatial nonlocality, 142
Spatial variation, temporal selection gradient dataset and, 69
Species, biological meaning of, 166
Speed of evolution, 228
Spencer, Herbert, 161, 231
Sperm cells, 41–42, *43*
Spiegelman, Sol, 264
Spirit, 84
Spongilla lacustris, 207
Spontaneous activity, 134–135
SPUN (Society for the Protection of Underground Networks), 46
S-R (stimulus-response) system, 128
Stability, physio-chemical types of, 260–261
Standen, Emily M., 351

Stasis, evolutionary, 211–213
Statistical mechanics of emergence, 150–152, 155–156
Stentor, 38
Stigmergy, 210
Stimulus-response (S-R) system, 128
Stress, 175, 283–285
Structure of Evolutionary Theory, The (Gould), 230
Sultan, Sonia E., 60–61, 64, *306,* 312–314, 332, 334, 336n18, 342, 351–352, 355–356
Sunflowers, *285*
Super-agency, 97–98
Supracellular consciousness, 45–46
Survival
  biological meaning from, 165–166, 171
  in evolutionary biology, 15
  interest, 161
  natural selection and, 17
  teleonomy and, 161
Swammerdam, Jan, 222
Swim bladder, in fish, 147
Symbiogenesis, 2, 13, 24, 25n2, 79–80, 98, 107, 286
Symbiosis, 2, 13, 97
Synergism hypothesis, 22–23
Synergism Hypothesis, The (Corning), 22
Synergistic selection, 22, 259
Synergistic Selection (Corning), 23
Synergy
  agents and, 91
  definition of, 80
  in evolution, 22–23
  functional, 4, 11
  relational agency and, 97
Systems biology, 79–80, 97, 156, 278
Systems view of goal-directedness, 242–243, 244
Szathmáry, Eörs, 4, 22, 80, 98, 122, 133, 259, 264, 317

TADs (topologically associating domains), 206
TAP (theory of the adjacent possible) process, 150–152, 155
Taylor, Charles, 343–345, 347, 352–353
Teleodynamics, 165
Teleology
  biological meaning form purpose and, 164–165
  definition of, 170
  internal, 12, 165, 171, *328*
  reconceptualization of, 161
  teleonomy compared to, 237
Teleonomic character, 258
"Teleonomic selection," 19, 156
Teleonomy, 6
  agency and, 88–89, 156
  anticipation, retrocausality and, 105–107
  Aristotle on, 105–106
  behavior and, 19, 119, 134
  bioelectricity for control of growth and form in, 185–190
  birth defects and, 180
  codepoiesis and, 105–107
  collective intelligence and, 176–179
  cybernetics and, 177, 192
  definitions of, 1–2, 12–13, 80, 171
  epigenetic inheritance and, 111–112
  in evolution, 11–25, 110–112
  final state and, 112–114

Teleonomy (cont.)
  flowers and, 309, *310*
  frameworks of, 120–121
  future research on morphogenesis and, 190–192
  importance of, 24
  as information, 25n6
  Mayr on goal-directedness and, 239–240, 245
  memory and, 188
  natural selection and, 15–17, 114
  negative feedback and, 240, 308
  origin of, 1, 12, 120, 237, 299
  phenotypic plasticity and, 107
  purposive behavior and, 246
  regeneration and, 181–185, *183*
  relational agency and, 97
  scaling of goals and, 187–190, 192
  survival and, 161
  teleology compared to, 237
  unlimited heredity and, 122
Templating, field effects in, 211–212
Temporal depth, in consciousness, 123
Temporal selection gradient dataset
  main findings with, 68–69
  multilevel autoregressive time-series model for
    reanalysis of, 70–72
  on niche construction and natural selection in the
    wild, 66–72
  selection gradients and categorization protocol in,
    66–67
  spatial variation and, 69
  statistical analyses with, 68
Temporal stability, 71
TFs (transcription factors), 205–207, 278–279, 283
Theoretical biology, 162–163
Theory of the adjacent possible (TAP) process,
  150–152, 155
Thermodynamics
  Nernst's third law of, 109
  nonequilibrium, 108–109
  second law of, 260, 264, 267
  stability and, 260–261
  TAP process and fourth law of, 151–152
Thigmotropic bending, 308
Third order of cellular cognition (ThOCC), 47
Thompson, D'Arcy W., 203
Thompson, Evan, 176, 269, 312, 353, 357
"Thoughtful," IGDB and, 132–133
Timakov, Vladimir Dmitrievich, 229, 234n2
Time series, 70–72
Tinbergen, Nikolaas, 120–121
TNTs (tunneling nanotubes), 45–46
Tobacco, *284*
Tomasello, Michael, 132
Tomatoes, *284*
Tooby, John, 165
Tool-use, cognition and, 21
Topologically associating domains (TADs), 206
Trace conditioning, 124
Transcription factors (TFs), 205–207, 278–279, 283
Transgenerational inheritance, in plants, 313–315
Transposable elements (transposons), 6, 275,
  277–278, 283, *284–285*, 285. *See also* Mobile
  DNA elements
Trewavas, Anthony, 6, 20, 247, 299, 305

Trichoplax adhaerens, 207
Trivers, Robert, 251–252, 326
Troland, Leonard Thompson, 263
Tropical Nature and Other Essays (Wallace), 223
Truth, 133–134
Tumorigenesis, cancer and, 189–190
Tunneling nanotubes (TNTs), 45–46
Turing instabilities, 202
Turing machine, 156
Turner, J. Scott, 176, 191, 265
Two-space modern synthesis, 347–349, 351, 356

UAL. *See* Unlimited associative learning
Uexküll, Jakob von, 5, 106–107, 164, 169, 171
Ulanowicz, Robert E., 108, 112
Uller, Tobias, 325
Umiker-Sebeok, Jean, 163
Umwelt, biological meaning from, 169, 171
Underdetermination
  adaptive bias and, 352
  development and, 351
  foundationalism and, 345–347, 349–352
  inheritance and, 349–351
  innovation and, 351–352
Unicellular protozoa, 36
Unity of purpose, 250–253
Unlimited associative learning (UAL), 5, 119
  architecture of, 125, *126*
  behavioral predictions in, 126–127
  CSSs and, 125
  desire in, 129
  as ETM, 124–125
  evolution of consciousness and, 121–126, 130, *131,*
    134
  GNW model compared to, 125–126
  HIT model and, 128–129
  qualities of, 124
  unitary scheme of, 129–130
Unlimited heredity, 122
"Upside-down" biology, 34

Vachellia drepanolobium (whistling thorn acacia), 20
Vane-Wright, Richard I., 1
Varela, Francisco, 3, 18, 106, 176, 192, 266,
  269–270, 310, 353
Variation, relational agency and, 94
Variation of Animals and Plants Under Domestica-
  tion, The (Darwin, C.), 223
Ventricular hypertrophy, 204
Volvox algae, 37
Vrba, Elisabeth, 19, 111, 147, 203, 213
de Vries, Hugo, 277

Waddington, Conrad, 4, 25, 61, 162–163, 238, 301,
  310, 330
  genetic assimilation and, 60, 107, 225–226
  Lamarckian modes of inheritance and, 226
  neo-Darwinism and, 225
Wallace, Alfred Russel, 221, 223–224, 226, 231
Walsh, Denis M., 3, 4, 6, 12, 14, 17–18, 25n5, 81,
  98, 146, 148, 156, 305, *328,* 331, 333–335,
  336n12, 336n15, 336n18, 341
Warnowiaceae (dinoflagellates), 40
Water-fountain analogy, DKS and, 261–262, 268

Watson, James Dewey, 258, 263
Weber, Bruce, 4
Weismann, August, 221, 223–224, 225, 227, 230–231
Weismann barrier, 14, 222–223, 225, 227, 231–232, 303, 347–348
Welwitschia, 314
West-Eberhard, Mary-Jane, 2, 14, 25, 135, 203, 331, 351, 355, 358n26, 359n35
What Is Life? (Margulis and Sagan), 24–25
What Is Life? (Schrödinger), 61, 108–109
Wheat, *284*
Whistling thorn acacia *(Vachellia drepanolobium)*, 20
Whiten, Andrew, 4, 14, 19–21, 24, 63
Wiener, Norbert, 161, 176–178, 276
Williams, George C., 60, 237
Wilson, David Sloan, 16, 80
Wilson, Edward O., 21, 165

Wiring transcriptional networks, 279
Wittgenstein, L., 341–343
Wnt protein, metazoan animals and, 200–201
Woese, Carl, 259, 263
Woodpecker finch *(Cactospiza pallidus)*, 19
World-generated stimuli, self-generated stimuli compared to, 135
Wright, Larry, 238, 240–241
Wright, Sewall, 329
Written word, 84–85

Xavier, Joana C., 145–146

Zahavi, Amotz, 165
Zebrafish *(Danio rerio)*, 203, *281*
Zeno (student of Parmenides), 81
Zipf's law, 150
Zymoseptoria tritici, 284